Applied Multiple Regression/Correlation Analysis for the Behavioral Sciences

Applied Multiple Regression/Correlation Analysis for the Behavioral Sciences

Jacob Cohen
New York University

Patricia Cohen
New York State Department of Mental Hygiene
and
Columbia University School of Public Health

 LAWRENCE ERLBAUM ASSOCIATES, PUBLISHERS
1975   Hillsdale, New Jersey

DISTRIBUTED BY THE HALSTED PRESS DIVISION OF

JOHN WILEY & SONS
New York   Toronto   London   Sydney

Lawrence Erlbaum Associates, Inc., Publishers
62 Maria Drive
Hillsdale, New Jersey 07642

Distributed solely by Halsted Press Division
John Wiley & Sons, Inc., New York

Library of Congress Cataloging in Publication Data

Cohen, Jacob, 1923–
    Applied multiple regression/correlation analysis
for the behavioral sciences.

    Bibliography: p.
    Includes index.
    1. Regression analysis.   2. Correlation (Statistics)
I. Cohen, Patricia, joint author.   II. Title.
HA33.C63   519.5'36   75-12848
ISBN 0-470-16360-7

Printed in the United States of America

to Gideon Moses Cohen

(another collaborative product)

Contents

Preface

This book had its origin about a dozen years ago, when it began to become apparent to the senior author that there were relationships between regression and correlation on the one hand and the analysis of variance on the other which were undreamed of (or at least did not appear) in the standard textbooks with which he was familiar. On the contrary, the texts, then as now, treated these as wholly distinct systems of data analysis intended for types of research which differed fundamentally in design, goals, and types of variables. Some research, both statistical and bibliographic, confirmed the relationships noted, and revealed yet others. These relationships served to enrich both systems in many ways, but it also became clear that multiple regression/correlation was potentially a very general system for analyzing data in the behavioral sciences, one that could incorporate the analysis of variance and covariance as special cases. An article outlining these possibilities was published in the *Psychological Bulletin* (Cohen, 1968), and the volume and sources of reprint requests and several reprintings suggested that a responsive chord had been struck among behavioral scientists in diverse areas. It was also obvious that for adequacy of both systematic coverage and expository detail, this book-length treatment was needed.

In 1969, the authors were married and began a happy collaboration, one of whose chief products is this book. (Another is saluted on the dedication page.) During the preparation of the book, the ideas of the 1968 paper were expanded, further systematized, tried out on data, and hardened in the crucible of our teaching and consulting. We find the system which has evolved surprisingly easy to teach and learn, and this book is an effort to so embody it. We omit from this preface, except incidentally, a consideration of this book's scope, orientation, and organization, since Chapter 1 is largely devoted to these issues.

To describe the primary audience for whom this book is intended requires two dimensions. Substantively, this book is addressed to behavioral and social scientists. These terms have no sharply defined reference, but we intend them in the most inclusive sense to include the academic sciences of psychology, sociology, economics, branches of biology, political science, and anthropology, and also various applied research fields: education, clinical psychology and psychiatry, epidemiology, industrial psychology, business administration, social work, and political/social survey, market and consumer research. Although the methods described in this book are applicable in other fields (for example, industrial engineering, agronomy), our examples and atmospherics come from behavioral–social science.

The other dimension of our intended audience, amount of background in statistics and research, covers an equally broad span. This book is intended to be both a textbook for students and a manual for research workers, and thus requires a somewhat different approach by these two readerships. However, one feature of this book will be appreciated by a large majority of both groups of readers: its orientation is nonmathematical, applied, and "data-analytic." This orientation is discussed and justified in the introductory chapter (Section 1.2.1) and will not be belabored here. Our experience has been that with few exceptions, both students and research practitioners in the behavioral and social sciences approach statistics with considerable wariness (to say the least), and require a verbal–intuitive exposition, rich in redundancy and concrete examples. This we have sought to supply.

As a textbook, whether used in a course at the graduate or advanced undergraduate level, it is assumed that the students have already had a semester's introductory statistics course. Although Chapter 2 begins "from scratch" with bivariate correlation and regression, and reviews elementary statistical concepts and terminology, it is not really intended to be a thorough, basic exposition, but rather to refresh the reader's memory. Students without a nodding acquaintance with the analysis of variance will find some portions of Chapter 1 difficult; returning to this material later in the course should clear matters up. Increasingly, statistical offerings in graduate programs are so organized as to include a course in correlation/regression methods. This book is intended to serve as a text for such courses. It may also be used in courses in multivariate methods—although largely devoted to multiple regression/correlation analysis, the final chapter links it to and reviews the other multivariate methods.

As a manual, this book provides an integrated conceptual system and practical working methods for the research worker. The last five years has seen a rapidly growing interest in multiple regression/correlation methods, reflected in journal articles and books addressed to psychologists and sociologists. Much of this material is valuable, while some of it is misguided or simply incorrect. Some of the more valuable contributions are presented mathematically, thus limiting their access. Taken as a whole, the recent literature is lacking in the combination of an integrated conceptual system with easily understood practical working methods which is necessary for the method to realize its potential as a general data-analytic system. We have tried to provide this. Chapter 1 begins with an outline of this system, and was written primarily with the experienced research worker or advanced graduate student in mind. He or she will find much of Chapters 2 and 3 elementary, but they are worth skimming, since some of the topics are treated from a fresh perspective which may be found insight provoking. Chapter 4 considers sets of independent variables as units of analysis, and is basic for much of what follows. Beyond that, he may follow his specific interests in the chapters and appendices by reference to the table of contents and a carefully prepared index. To the stat buff or teacher, we recommend reading the chapters in order, and the appendices at the point in the text where they are referenced.

We acknowledge, first of all, the many students, colleagues and researchers seeking counsel whose stimulation so importantly shaped this book. A small subset of these are Elmer L. Struening, Mendl Hoffman, Joan Welkowitz, Claudia Riche, and Harry Reiss, but many more could be named. Special thanks are due to the members of the Society of Multivariate Experimental Psychology for the useful feedback they supplied when portions of the book were presented at their annual meetings during the last few years. We are very grateful to Joseph L. Fleiss for a painstaking technical critique from which the book greatly profited. Since we remained in disagreement on some points, whatever faults remain are our sole responsibility. Gerhard Raabe provided valuable advice with regard to the material on computer programs in Appendix 3. Patra Lindstrom did a most competent job in typing the manuscript.

JACOB COHEN
PATRICIA COHEN

PART I

BASICS

1
Introduction

1.1 MULTIPLE REGRESSION/CORRELATION AS A GENERAL DATA-ANALYTIC SYSTEM

1.1.1 Overview

Multiple regression/correlation analysis (MRC) is a highly general and therefore very flexible data-analytic system that may be used whenever a quantitative variable (the dependent variable) is to be studied as a function of, or in relationship to, any factors of interest (expressed as independent variables). The sweep of this statement is quite intentional:

1. The form of the relationship is not constrained; it may be simple or complex, for example, straight line or curvilinear, general or conditional, or combinations of these possibilities.

2. The nature of the research factors expressed as independent variables is also not constrained; they may be quantitative or qualitative, main effects or interactions in the analysis of variance (AV) sense, or covariates as in the analysis of covariance (ACV). They may be characterized by missing data. They may be correlated with each other, or uncorrelated (as in balanced factorial design AV). They may be naturally occurring ("organismic") properties like sex or diagnosis or IQ, or they may be consequences of planned experimental manipulation ("treatments"). They may be single variables or groups of variables. In short, virtually any information whose bearing on the dependent variable is of interest may be expressed as research factors.

The MRC system presented in this book has other properties that make of it a powerful analytic tool: it yields measures of the magnitude of the "whole" relationship of a factor to the dependent variable, as well as of its partial (unique, net) relationship, i.e., its relationship over and above that of other research factors (proportions of variance and coefficients of correlation and

3

regression). It also comes fully equipped with the necessary apparatus for statistical hypothesis testing, estimation, and power analysis.

In short, and at the risk of sounding like a television commercial, it is a versatile, all-purpose system for analyzing the data of the behavioral, social, and biological sciences and technologies.

We risk this hyperbole to heighten the contrast of the MRC system as developed in this book with the limited perspective of the typical treatments of MRC in the standard applied statistics textbooks in these fields, which both determine and reflect methodological practice. In psychology, which is perhaps the most quantitatively conscious of the behavioral sciences, textbook examples and research applications of MRC are dominated by the psychotechnological perspective of forecasting outcomes in educational or personnel selection and vocational guidance. The very terminology which has come to be popularly used betrays this preoccupation: "criterion" (dependent) variables are "predicted" by "predictor" (independent) variables, and the accuracy is assessed by means of the "standard error of prediction." Even when put to other uses, stereotypy is induced either implicitly or explicitly by limiting MRC to straight-line relationships among equal-interval scales on which the observations are assumed to be normally distributed. In this narrow view, MRC takes its place as one of a group of specialized statistical tools, its use limited to those occasional circumstances when its unique function is required and its working conditions are met.

Viewed from this traditional perspective, it is hard to see why anyone would want a whole textbook devoted to MRC, a monograph for specialists perhaps, but why a textbook? No such question arises with regard to textbooks entirely devoted to the analysis of variance and covariance, because of its presumed generality. This is ironic, since, as we will show, AV/ACV is in fact a special case of MRC!

Technically, AV/ACV and conventional multiple regression analysis are special cases of the "general linear model" in mathematical statistics.[1] The MRC system of this book generalizes conventional multiple regression analysis to the point where it is essentially equivalent to the general linear model. It thus follows that any data analyzable by AV/ACV may be analyzed by MRC, while the reverse is not the case. This is illustrated, for example, by the fact that when one seeks in an AV/ACV framework to analyze a factorial design with unequal cell frequencies, the nonindependence of the factors necessitates moving up to the more general multiple regression analysis to achieve an exact solution.

Historically, MRC arose in the biological and behavioral sciences around the turn of the century in the study of the natural covariation of observed characteristics of samples of subjects (Galton, Pearson, Yule). Somewhat later, AV/ACV grew out of the analysis of agronomic data produced by controlled variation of treatment conditions in manipulative experiments (Fisher). The

---

[1] For the technically minded, we point out that it is the "fixed" version of these models to which we address ourselves, which is the way they are most often used.

systems developed in parallel, and from the perspective of the research workers who used them, largely independently of each other. Indeed, MRC, because of its association with nonexperimental, observational, survey-type research, came to be looked upon as less respectable than AV/ACV, which was associated with experiments.

Close examination suggests that this guilt (or virtue) by association is unwarranted—the result of the confusion of data-analytic method with the logical considerations which govern the inference of causality. Experiments in which different treatments are applied to randomly assigned groups of subjects permit direct inference of causality, while the observation of associations among variables in a group of randomly selected subjects does not. Thus, the finding of significantly more cell pathology in the lungs of rats reared in cigarette smoke-filled environments than for normally reared control animals is in a logically superior position to draw the causal inference than is the finding that, for a random sample of postmortem cases, the lung cell pathology is significantly higher for divorced men than for married men. But each of these researches may be analyzed by either AV (a simple pathology mean difference and its t test) *or* MRC (a simple correlation between group membership and pathology and its identical t test). The logical status of causal inference is a function of how the data were produced, not how they are analyzed. Once a data-structure exists, it may be analyzed either by AV/ACV or by MRC.

Authors who make sharp theoretical distinctions between correlation and fixed-model AV (or fixed-model regression) are prone to claim that correlation and proportion of variance (squared correlation) measures lack meaning for the latter because these measures depend on the specific levels of the research factor chosen (fixed) by the investigator and the (fixed) number of cases at each level. Concretely, consider an experiment where random samples of subjects are exposed to several levels of magnitude of a sensory stimulus, each sample to a different level, and their responses recorded. Assume that, over the purists' objections, we compute a correlation between stimulus condition and response, and find it to be .70, that is, about half ($.70^2$ = .49) of the response variance is accounted for by the stimulus conditions. They would argue, quite correctly, that the selection of a different set of stimulus values (more or less varying, occurring elsewhere in the range), or a different distribution of relative sample sizes, would result in a larger or smaller proportion of the response variance being accounted for. Therefore, they would argue, the .49 (or .70) figure can not be taken as an estimate of "the relationship between stimulus and response" for this sensory modality and form of response. Again, we must agree. Therefore, they would finally argue, these measures are meaningless. Here, we beg to disagree. We find such measures to be quite useful, provided that their dependence on the levels and relative sample sizes of the research factor is understood. When necessary, one simply attaches to them, as a condition or qualification, the distribution of the research factor. We find such qualifications no more objectionable, in principle, than the potentially many others (apparatus, tests, time of day, subjects, experimenters) on which research results may depend. Such measures, qualified as necessary, may mean more or less, depending on substantive considerations, but they are hardly meaningless. (For an example where the research factor is religion, and further discussion of this issue, see Section 5.3.1.)

On the contrary, one of the most attractive features of MRC is its automatic provision of proportion of variance and correlation measures of various kinds. These are measures of "effect size," of the magnitude of the phenomena being studied. We venture the assertion that, despite the preoccupation (some critics would substitute "obsession") of the behavioral and social sciences with quantitative methods, the level of consciousness in many areas of just how big things are is at a surprisingly low level. This is because concern about the statistical significance of effects (whether they exist at all) has tended to preempt attention to their magnitude. That significant effects may be small, and nonsignificant ones large, is a truism. Although not unrelated, the size and statistical significance of effects are logically independent features of data from samples. Yet many research reports, at least implicitly, confuse the issues of size and statistical significance, using the latter as if it meant the former. At least part of the reason for this is that traditional AV/ACV yields readily interpretable F and t ratios for significance testing, but offers differences between group means as measures of effect size.[2] Now, a difference between means is a reasonably informative measure of effect size when the dependent variable is bushels of wheat per acre, or dollars of annual income, or age at time of marriage. It is, however, less informative when the dependent variable is a psychological test score, a sociological index, or the number of trials to learn a maze. Many of the variables in the behavioral and social sciences are expressed in units which are arbitrary and/or ad hoc, or otherwise unfamiliar. To report, for example, that law students show a 9.2-point higher mean than medical students on a scale measuring attitude toward the United Nations is to convey very little about whether this constitutes a large or trivial difference. However, to report that the law student–medical student distinction accounts for 4% of the attitude score variance conveys much more. Further, to report that the law students' mean on another scale, attitude toward public service, is 6.4 points higher than the medical students' mean not only fails to convey a useful sense of the size of this difference (as before), but is not even informative as to whether this is a smaller or larger difference than the other, since the units of the two scales are not directly comparable. But reporting that this distinction accounts for 10% of the public service attitude score variance not only expresses the effect size usefully, but *is* comparable to the 4% found for the other variable. Since various types of proportion of variance, that is, the squares of simple, multiple, partial and semipartial correlation coefficients, may be routinely determined in the MRC

---

[2] This is not to say that the more useful measures of proportion of variance have not been proposed in AV contexts; see, for example, Hays (1973), Cohen (1965, 1969), and Section 5.3.4. But they are neither integral to the AV tradition nor routinely presented in research reports where the data are analyzed by AV/ACV.

system, the latter has "built-in" effect size measures which are unit-free and easily understood and communicated. Each of these comes with its significance test value for the null hypothesis (F or t), and no confusion between the two issues of "whether" and "how much" need arise.

1.1.2 Multiple Regression/Correlation and the Complexity of Behavioral Science

The greatest virtue of the MRC system is its capacity to mirror, with high fidelity, the complexity of the relationships that characterize the behavioral sciences. The word "complexity" is itself used here in a complex sense to cover several issues.

Multiplicity of Influences

The behavioral sciences inherited from older branches of empirical inquiry the simple experimental paradigm: vary a single presumed causal factor (C) and its effects on the dependent variable (Y), while holding constant other potential factors. Thus, $Y = f(C)$; variation in Y is a function of controlled variation in C. This model has been, and continues to be, an effective tool of inquiry in the physical sciences and engineering, and in some areas of the behavioral sciences. Probably because of their much higher degree of evolution, functional areas within the physical sciences and engineering typically deal with a few distinct causal factors, each measured in a clear-cut way, and each in principle independent of others.

However, as one moves from the physical sciences through biology and across the broad spectrum of the behavioral sciences ranging from physiological psychology to cultural anthropology, the number of potential causal factors increases, their representation in measures becomes increasingly uncertain, and weak theories abound and compete. Consider a representative set of dependent variables: epinephrine secreted, distant word associations, verbal learning, school achievement, psychosis, anxiety, aggression, attitude toward busing, income, social mobility, birth rate, kinship system. A few moments' reflection about the causal nexus in which each of these is embedded suggests a multiplicity of factors, and possibly further multiplicity in how any given factor is represented. Given several reserach factors C, D, E, etc. to be studied, one might use the single-factor paradigm repeatedly in multiple researches, that is, $Y = f(C)$, then $Y = f(D)$, then $Y = f(E)$, etc. But MRC makes possible the use of paradigms of the form $Y = f(C, D, E$, etc.$)$, which are far more efficient than the strategy of studying multiple factors one at a time.

Correlation among Research Factors and Partialling

A far more important type of complexity than the sheer multiplicity of research factors lies in the effect of relationships among them.

The simpler condition is that in which the factors $C, D, E, \ldots$ are statistically unrelated (orthogonal) to each other, as is the case in true experiments where they are under the experimenter's manipulative control. The overall importance of each factor (for example, the proportion of Y variance it accounts for) can be unambiguously determined, since its orthogonality with the other factors assures that its effects on Y can not overlap with the effects of the others. Thus, concretely, consider a simple experimental inquiry into the proposition "don't trust anyone over 30" in which the persuasibility (Y) of male college students is studied as a function of the apparent age $(C: C_1$ = under 30, C_2 = over 30) and sex $(D: D_1$ = male, D_2 = female) of the communicator of a persuasive message. The orthogonality of the research factors C and D is assured by having equal numbers of subjects in the four "cells" $(C_1 D_1, C_1 D_2, C_2 D_1, C_2 D_2)$; no part of the difference in overall Y means for the two communicator ages can be attributed to their sexes, and conversely, since the effect of each factor is balanced out in the determination of the other. If it is found that C accounts for 10% of the Y variance, and D for 5%, no portion of either of these amounts can be due to the other factor. It thus follows that these amounts are additive: C and D together account for 15% of the Y variance.[3]

Complexity arises when one departs from manipulative experiments and the orthogonality of factors which they make possible. Many issues in behavioral sciences are simply inaccessible to true experiments, and can only be addressed by the systematic observation of phenomena as they occur in their natural flux. In nature, factors which impinge on Y are generally correlated with each other as well. Thus, if persuasibility (Y) is studied as a function of authoritarianism (C), intelligence (D), and socioeconomic status (E) by surveying a sample with regard to these characteristics, it will likely be found that C, D, and E are to some degree correlated with each other. If, taken singly, C accounts for 8%, D 12%, and E 6% of the Y variance, because of the correlations among these factors, it will not be the case that together, in an MRC analysis, they account for 8 + 12 + 6 = 26% of the Y variance. It will almost certainly be less (in this case, but may, in general, be more—see Section 3.4). This is the familiar phenomenon of redundancy among correlated explanatory variables with regard to what they explain. The Y variance accounted for by a factor is overlapped to some degree with others. This in turn implies the concept of the variance accounted for by a factor *uniquely,* relative to what is accounted for by the other factors. In the above example, these unique proportions of Y variance may turn out to be: C 4%, D 10%, and E 0%. This is a rather different picture than that provided by looking at each factor singly. For example, it might be argued that E's apparent influence on Y when appraised by itself is "spurious," being entirely attributable to its relationship to C and/or D.

---

[3] The reader familiar with AV will recognize this as a balanced 2 × 2 factorial design. To avoid possible confusion, it must be pointed out that the orthogonality of C and D is a fact which is wholly independent of the possibility of a $C \times D$ interaction. Interactions are research factors in their own right and in balanced designs are also orthogonal to all other research factors. If the $C \times D$ interaction were to be included as a third factor and found to account for 2% of the variance, this amount is wholly its own, and all three factors combined would account for 17% of the Y variance. See the section "General and Conditional Relationships," and Chapter 8, which is devoted entirely to interactions.

MRC's capability for assessing unique variance, and the closely related measures of *partial* correlation and regression coefficients it provides, is perhaps its most important feature, particularly for observational (nonexperimental) studies. Even a small number of research factors define many alternative possible causal systems or theories. Selection among them is greatly facilitated by the ability, using MRC, of partialling from the effects of any research factor those of any desired set of other factors. It is a copybook maxim that no correlational method can establish causal relations, but a given causal theory may be invalidated by the skillful use of this feature of MRC. It can show whether a set of observational data for Y and the correlated research factors C and D are consistent with any of the following possibilities:

1. C and D each bears causally on Y.
2. C is a surrogate for D in relationship with Y, that is, when D is partialled from C, C retains no variance in Y.
3. D suppresses the effect of C on Y, that is, when D is partialled from C, the unique variance of C in Y is greater than the proportion it accounts for when D is ignored (see the discussion of "suppression" in Section 3.4).

The possibility for complexity in causal structures is further increased when the number of research factors increases beyond two, yet the partialling inherent in MRC is a powerful adjunct to good theory for disentangling them. Further, partialling is at the base of a series of data-analytic procedures of increasing generality, which are realizable through MRC: general ACV, the Analysis of Partial Variance, and the Sequential Analysis of Partial Variance (see Chapter 9). Most generally, it is the partialling mechanism more than any other feature which makes it possible for the MRC system to mirror the complexity of causal relationships encountered in the behavioral sciences.

Form of Information

The behavioral and social sciences utilize information in various forms. One form which research factors may take is quantitative, and of any of the following levels of measurement (Stevens, 1951):

1. *Ratio scales.* These are equal interval scales with a true zero point, making such ratio statements as "J has twice as much X as K" sensible. X may here be, for example, inches, pounds, seconds, foot-candles, voltage, size of group, dollars, distance from hospital, years in prison, or literacy rate.

2. *Interval scales.* These have equal intervals but are measured from an arbitrary zero point, that is, the value of X that denotes absence of the property is not defined. Most psychological measures and sociological indices are at this level, for example, the scores of tests of intelligence, special abilities, achievement, personality, temperament, vocational interest, and social attitude. A physical example is temperature measured in Fahrenheit or Centigrade units.

3. *Ordinal scales.* Only the relative position within a collection are signified by the values of ordinal scales, neither conditions of equal intervals nor a true

zero obtaining. Whether simple rank order values are used, or they are expressed as percentiles, deciles, or quartiles, these properties of ordinal scales are the same.

The above scheme is not exhaustive of quantitative scales, and others have been proposed. For example, psychological test scores are unlikely to measure with exactly equal intervals and it may be argued that they fall along a continuum between interval and ordinal scales. Also, some rating scales frequently used in applied psychological research are not covered by the Stevens scheme since they have a defined zero point but intervals of dubious equality, for example, 0-never, 1-seldom, 2-sometimes, 3-often, 4-always.

Nominal scales. Conventional MRC analysis has generally been restricted to quantitative scales with (more or less) equal intervals. But much information in the behavioral sciences is not quantitative at all, but qualitative or categorical, or, using Steven's (1951) formulation, measured on "nominal" scales. Whether they are to be considered a form of measurement at all is subject to debate, but they undoubtedly constitute information. Some examples are: ethnic group, place of birth, religion, experimental group, marital status, psychiatric diagnosis, type of family structure, choice of political candidate, sex. Each of these represents a set of mutually exclusive categories which accounts for all the cases. The categories of true nominal scales represent distinguishable qualities, without natural order or other quantitative properties. Thus, nominal scales are sets of groups which differ on some qualitative attribute.

When research factors expressed as nominal scales are to be related to a dependent variable Y, past practice has been to put them through the mill of AV, whose grist is Y values organized into groups. But the qualitative information which constitutes nominal scales may be expressed quantitatively, and used as independent variables in MRC. (Chapter 5 is devoted to this topic, and answers such questions as, "How do you score religion?")

The above does not exhaust the forms in which information is expressed, since mixtures of scale types and other irregularities occur in practice. For example, interviews and questionnaires often require for some items the provision of categories for "does not apply" and/or "no response"; some questions are asked only if a prior question has had some specified response. As uninformative as such categories may seem at first glance, they nevertheless contain information and are capable of expression in research factors (see Chapter 7).

The above has been presented as evidence for that aspect of the complexity of the behavioral sciences which resides in the great variety of forms in which their information comes. Beginning with Chapter 5, we shall show how information in any of these forms may be used as research factors in the MRC system. The traditional restriction of MRC to equal interval scales will be shown to be quite unnecessary. The capacity of MRC to use information in almost any form, and to mix forms as necessary, is an important part of its adaptive flexibility. Were it finicky about the type of input information it could use, it could hardly function as a *general* data-analytic system.

Shape of Relationship

When we come to scrutinize a given relationship expressed by $Y = f(C)$, it may be well described by a straight line on the usual graph, for example, if Y and C are psychological tests of abilities. Or, adequate description may require that the line be curved, for example, if C is age, or number of children, such may be the case. Or, the shape may not be definable, as when C is a nominal scale, for example, diagnosis, or college major. When there are multiple research factors being studied simultaneously, each may relate to Y (and each other) in any of these ways. Thus, when we write $Y = f(C, D, E, \ldots)$, f (as a function of) potentially covers very complex functions, indeed. Yet such complex functions are readily brought under the sway of MRC.

How so? Most readers will know that MRC is often (and properly) referred to as *linear* MRC and may well be under the impression that correlation and regression are restricted to the study of straight-line relationships. This mistaken impression is abetted by the common usage of "linear" to mean "rectilinear," and "nonlinear" to mean "curvilinear." We are thus confounded by what is virtually a pun. What is literally meant by "linear" is any relationship of the form

(1.1.1) $$Y = a + bU + cV + dW + eX + \cdots ,$$

where the lower-case letters are constants (either positive or negative) and the capital letters are variables. Y is said to be "linear in the variables U, V, etc." because it is confected by taking certain constant amounts (b, c, etc.) of each variable, and the constant a, and simply adding them together. Were we to proceed in any other way, the resulting function would not be linear in the variables, by definition. But in the fixed-model framework in which we operate, there is no constraint on the nature of the variables. That being the case, they may be chosen so as to define relationships of *any* shape, rectilinear or curvilinear, or of no shape at all (as for unordered nominal scales), and all the complex combinations of these which multiple factors can produce.

Multiple regression equations are, indeed, linear; they are exactly of the form of Eq. (1.1.1). Yet they can be used to describe such complex relationships as the length of psychiatric hospital stay as a function of symptom ratings on admission, diagnosis, age, sex, and average length of prior hospitalizations (if any). This relationship is patently not rectilinear, yet readily described by a linear multiple regression equation.

To be sure, not all or even more relationships studied in the behavioral sciences are of this order of complexity, but the obvious point is that the capacity of MRC to take any degree or type of shape-complexity in its stride is yet another of the important features which make it truly a *general* data-analytic system.

General and Conditional Relationships

Some relationships between Y and some factor C remain the same in regard to degree and form over variation in other factors D, E, F. We will call such relationships *general* or *unconditional*. The definition of a general relationship holds quite apart from how or whether these other factors relate to Y or to C. This might be the case, for example, if Y is a measure of perceptual acuity and C is age. Whatever the form and degree of relationship, if it remains the same under varying conditions of educational level (D), ethnic group (E), and sex (F), then the relationship can be said to be general (insofar as these other factors are concerned). Note that this generality obtains whatever the relationship between acuity and D, E, and F, between age (C) and D, E, F, or among D, E, and F. The *Y–C relationship* can thus be considered unconditional with regard to, or independent of, D, E, and F.

Now consider the same research factors but with Y as a measure of attitude towards racial integration. The form and/or degree of relationship of age to Y is now almost certain to vary as a function of one or more of the other factors: it may be stronger or shaped differently at lower educational levels than higher (D), and/or in one ethnic group than another (E), and/or for men compared to women (F). The relationship of Y to C is now said to be conditional on D and/or E and/or F. In AV contexts, such relationships are called *interactions*, for example, if the *C–Y relationship* is not constant over different values of D, there is said to be a $C \times D$ ("age by educational level") interaction. Greater complexity is possible: the *C–Y* relationship may be constant over levels of D taken by themselves, and over levels of E taken by themselves, yet may be conditional on the *combination* of D and E levels. Such a circumstance would define a "second-order" interaction, represented as $C \times D \times E$ (with, in this case, neither $C \times D$ nor $C \times E$ present). Interactions of even higher order, and thus even more complex forms of conditionality, are also theoretically possible.

To forestall a frequent source of confusion, we emphasize the fact that the existence of a $C \times D$ interaction is an issue quite separate from the relationship of C to D, or D to Y. However age may relate to education, or education to attitude, the existence of $C \times D$ means that the relationship of age to attitude is conditioned by (depends on, varies with) education. Since such interactions are symmetrical, this would also mean that the relationship of education to attitude is conditioned by age.

One facet of the complexity of the behavioral sciences is the frequency with which conditional relationships are encountered. Relationships among variables often change with changes in experimental conditions (treatments, instructions, experimental assistants, etc.), age, sex, social class, ethnicity, diagnosis, religion, geographic area, etc. Causal interpretation of such conditional relationships is even more difficult than it is for general relationships, but it is patently important that conditionality be detected when it exists.

Conditional relationships may be studied directly, but crudely, by partitioning the data into subgroups on the conditioning variable, determining the relationship in each subgroup, and comparing them. However, problems of small sample size and difficulties in the statistical comparison of measures of relationship from subgroup to subgroup are likely to arise. Factorial design AV provides for assessing conditional relationships (interactions), but is constrained to research factors in nominal form and becomes awkward when the research factors are not orthogonal. The versatility of the MRC system obtains here—conditional relationships of any order of complexity, involving research factors with information in any form, and either correlated or uncorrelated, can be routinely handled without difficulty. (See Chapter 8.)

In summary, the generality of the MRC system of data analysis appropriately complements the complexity of the behavioral sciences, where "complexity" is intended to convey simultaneously the ideas of multiplicity and correlation among potential causal influences, the variety of forms in which information is couched, and in the shape and conditionality of relationships. Multiple regression/correlation also provides a full yield of measures of "effect size" with which to quantify various aspects of relationships (proportions of variance, correlation and regression coefficients). Finally, these measures are subject to statistical hypothesis testing, estimation, and power-analytic procedures.

1.2 ORIENTATION

This book was written to serve as a textbook and manual in the application of the MRC system for data analysis by students and practitioners in the diverse areas of inquiry of the behavioral sciences. As its authors, we had to make many decisions about the level, breadth, emphasis, tone, and style of exposition. Its readers may find it useful, at the outset, to have our orientation and the basis for these decisions set forth.

1.2.1 Approach

Nonmathematical

Our presentation of MRC is nonmathematical. Of course, MRC is itself a product of mathematical statistics, based on matrix algebra, the calculus, and probability theory—branches of mathematics familiar only to math majors. There is little question that such a background makes possible a level of insight otherwise difficult to achieve. However, since it is only infrequently found in behavioral scientists, it is bootless to proceed on such a basis, however desirable it may be in theory. Nor do we believe it worthwhile, as is done in some statistical textbooks addressed to this audience, to attempt to provide the necessary mathematical background in condensed form in an introductory chapter or two and then proceed as if it were a functioning part of the reader's intellectual equipment. In our experience, that simply does not work—it serves more as a sop to the author's conscience than as an aid to the reader's comprehension.

We thus abjure mathematical proofs, as well as unnecessary offhand references to mathematical concepts and methods not likely to be understood by the bulk of our audience. In their place, we heavily emphasize detailed and deliberately redundant verbal exposition of concrete examples drawn from the behavioral sciences. Our experience in teaching and consulting convinces us that our audience is richly endowed in the verbal, logical, intuitive kind of intelligence, which makes it possible to understand how the MRC system works, and thus use it effectively. (Dorothy Parker said, "Flattery will get you anywhere.") This kind of understanding is eminently satisfactory (as well as satisfying), since it makes possible the effective use of the system. We note that to drive a car, one does not need to be a physicist, nor an automotive engineer, nor even an auto mechanic, although the latter's skills are useful when you are stuck on the highway, and that is the level we aim for.

Flat assertions, however, provide little intellectual nourishment. We seek to make up for the absence of mathematical proofs by providing demonstrations instead. For example, the regression coefficient for a dichotomous (male–female, yes–no) independent variable equals the difference between the two groups' Y means. Instead of offering the six or seven lines of algebra that would constitute a mathematical proof, we demonstrate that it holds, using a small set of data. True, this proves nothing, since the result may be accidental, but the curious reader can check it out on his own data (and we urge that such checks be made throughout). Whether it is checked out or not, however, we believe that most of our audience would profit more from the demonstration than the proof. If the absence of proof bothers some Missourians, all we can do is pledge our good faith.

Applied

The first word in this book's title is "applied." The heavy stress on illustrations serves not only the function of clarifying and demonstrating the abstract principles being taught, but also that of exemplifying the kinds of applications possible, that is, providing working models. We attend to theory only insofar as sound application makes necessary. The emphasis is on "how to do it." This opens us to the contemptuous charge of writing a "cookbook," a charge we deny, since we do not neglect the whys and wherefores. If the charge is nevertheless pressed, we can only add the observation that in the kitchen, cookbooks are likely to be found more useful than textbooks in organic chemistry.

Data-Analytic

The mathematical statistician proceeds from exactly specified premises (independent random sampling, normality of distribution, homogeneity of variance), and by the exercise of his ingenuity and appropriate mathematical theory, arrives at exact and necessary consequences (F distribution, statistical power functions). He is, of course, fully aware of the fact that no set of real data will exactly conform to the formal premises from which he starts, but this is not properly his responsibility. As all mathematicians, he works with abstractions to produce formal models whose "truth" lies in their self-consistency. Borrowing their language, we might say that inequalities are symmetrical: just as behavioral scientists are not mathematicians, mathematicians are not behavioral scientists.

The behavioral scientist relies very heavily on the fruits of the labors of theoretical statisticians. They provide guides for teasing out meaning from data, limits on inference, discipline in speculation. Unfortunately, in the textbooks addressed to behavioral scientists, statistical methods have often been presented more as harsh straightjackets or Procrustean beds than as benign reference frameworks. Typically, a method is presented with some emphasis on its formal assumptions. Readers are advised that the failure of a set of data to meet these assumptions renders the method invalid. All too often, the discussion ends at this point. Presumably, the offending data are to be thrown away.

Now this is, of course, a perfectly ridiculous idea from the point of view of the working scientist. His task is to contrive situations that yield information about substantive scientific issues—*he must and will analyze his data.* In doing so, he will bring to bear, in addition to the tools of statistical analysis, his knowledge of theory in his field, the past experience he and others have had with similar data, his hunches, and his good sense, both common and uncommon. He would rather risk analyzing his data incorrectly than not at all. For him, data analysis is not an end in itself, but the next-to-last step in a sequence which culminates in providing information about the phenomena. This is by no means to say that he need not be painstaking in his efforts to generate and perform analyses of data from which he can draw unambiguous conclusions. But he must translate these efforts into substantive information.

Most happily, the distinction between "data analysis" and "statistical analysis" has been made and given both rationale and respectability by one of the world's foremost mathematical statisticians, John Tukey. In his seminal *The Future of Data Analysis* (1962), Tukey describes data analysis as the special province of scientists with substantial interest in methodology. Data analysts employ statistical analysis as the most important tool in their craft, but they employ it together with other tools, and in a spirit quite different from that which has come to be associated with it from its origins in mathematical statistics. Data analysis accepts "inadequate" data, and is thus prepared to settle for "indications" rather than "conclusions." It risks a greater frequency of errors in the interest of a greater frequency of occasions when the right answer is *"suggested."* It compensates for cutting some statistical corners by using scientific as well as mathematical judgment, and by relying upon self-consistency and repetition of results. Data analysis operates like a detective searching for clues rather than like a bookkeeper seeking to prove out a balance. In describing data analysis, Tukey has provided insight and rationale into the way good scientists have always related to data.

The spirit of this book is strongly data-analytic, in exactly the above sense. We recognize the limits on inference placed by the failure of real data to meet some of the formal assumptions which underly fixed-model MRC, but are disposed to treat the limits as broad rather than narrow. We justify this by mustering whatever technical evidence there is in the statistical literature (for example, on the "robustness" of statistical tests), and by drawing upon our own and others' practical experience, even upon our intuition, all in the interest of getting on with the task of making data yield their meaning. If we risk error, we are more than compensated by having a system of data analysis which is general, sensitive, and fully capable of reflecting the complexity of the behavioral sciences and thus of meeting the needs of behavioral scientists.

1.2.2 Computation, the Computer, and Numerical Results

Computation

Like all mathematical procedures involving simultaneous attention to multiple variables, MRC makes large computational demands. As the size of the problem increases, the amount of computation required increases enormously; for example, the computational time required on a desk calculator for a problem with $k = 10$ independent variables and $n = 400$ cases is measured in days! With such prodigious amounts of hand calculation, the probability of coming through the process without serious blunders (misreading values, inversion of digits, incorrect substitution, etc.) cannot be far from zero. Rigorous checking procedures can assure accuracy, but at the cost of increasing computational man-days. The only solution is *not* to do the calculation on a desk calculator.

An important reason for the rapid increase during the past two decades in the use of multivariate[4] statistical procedures generally, and for the emergence of MRC as a general data-analytic system in particular, is the computer revolution. During this period, computers have become faster to a degree that strains comprehension, more "user oriented," and, most important of all, more widely available. Computer facilities are increasingly looked upon as being as necessary in academic and scientific settings as are library facilities. And progressive simplification in their utilization ("user orientation") makes the necessary know-how fairly easy to acquire. Fundamentally, then, we assume that MRC computation, in general, will be accomplished by computer.

---

[4] Usage of the term "multivariate" varies. Some authors restrict it to procedures where multiple *dependent* variables are used, by which definition MRC would not be included. However, increasingly and particularly among applied statisticians and behavioral scientists, the term is used to cover all statistical applications wherein "multiple variates are considered in combination" (Cooley & Lohnes, 1971, p. 3), either as dependent or independent variables, or both, or, as in factor analysis, neither; see Tatsuoko (1971) and Van de Geer (1971). See Chapter 11 for a consideration of MRC in relationship to other multivariate methods.

Early in the book, in our exposition of bivariate correlation and regression and MRC with two independent variables, we give the necessary details with worked examples for calculation by desk or pocket calculators (or, in principle, pencil and yellow pad). This is done because the intimate association with the arithmetic details makes plain to the reader the nature of the process: exactly what is being done, with what purpose, and to what result. With two or three variables, where the computation is easy, not only can one see the fundamentals, but there is laid down a basis for generalization to many variables, where the computational demands are great.

With k independent and one dependent variable, MRC computation requires, *to begin with,* $k + 1$ means and standard deviations, and the matrix of $k(k + 1)/2$ correlation coefficients between all pairs of $k + 1$ variables. It is at this point that the serious computation begins, that is, the solution of k simultaneous equations, most readily accomplished by a laborious matrix-arithmetic procedure called "inversion." Appendix 1 describes the mathematical basis of MRC including the role of the centrally important operation of matrix inversion, and the content and meaning of the elements of the inverse matrix. In Appendix 2, we give the actual computational (arithmetic) operations of matrix inversion and multiplication for MRC, suitable for use with a desk or pocket calculator. Although, in principle, the computational scheme given in Appendix 2 may be used with any number of variables, it becomes quite time consuming, and rapidly more onerous, as k increases beyond five or so.[5] The reader without access to computers has our sympathy, but he can manage the computation in small problems by following the procedure in Appendix 2, and may even be rewarded by insights which may accrue from this more intimate contact with the analysis.

But "the human use of human beings" does not include days spent at a desk calculator. As we have noted, we primarily rely on digital computers for MRC computation. Most of our readers will have either direct or on-line access to a computer laboratory, and will either have, or be able quickly to obtain, the modest know-how needed to use the available "canned" programs for MRC. Appendix 3 is devoted to this topic, and includes a description of the characteristics of the most popular and widely available programs, and of the considerations which should enter into one's choice among them. It should be consulted in conjunction with a trip to the computer laboratory to investigate what is available.

We expect that our readers will find the material on "heavy" computation useful, but we have deliberately placed it outside the body of the text to keep it from distracting attention from our central emphasis, which is on understanding

[5] It is difficult to set a value here—who is to say what another will find computationally onerous? Some people find balancing their checkbook a nightmare; others actually enjoy large quantities of arithmetic, particularly when it involves their own data. Five seems a reasonable compromise.

how the MRC system works, so that it may be effectively used in the exploitation of research data.

Our attitude toward computing as such explains the absence of chapter-end problems for the reader to work. Some of the purposes of such exercises can be achieved by carefully following through the details of the many worked illustrative examples in the body of the text. But the highest order of understanding is to be attained when the reader applies the methods of each chapter *to data of his own,* or data with which he is otherwise familiar. There is no more powerful synergism for insight than the application of unfamiliar methods to familiar data.

Numerical Results: Reporting and Rounding

With minor exceptions, the computation of the illustrative problems which fill this book was all accomplished by computer, using various programs and different computers. The numerical results carried in the computer are accurate to at least six significant figures and are printed out to at least four (or four decimal places). We see little point to presenting numerical results to as many places as the computer may provide, since the resulting "accuracy" holds only for the sample data analyzed, and, given the usual level of sampling error, is quite meaningless vis-à-vis the values of the population parameters. We mean nothing more complicated than the proposition, for example, that when the computer mindlessly tells us that, in a sample of the usual size, the product moment correlation between X and Y is .34617952, a guaranteed accurate result, at least the last five digits could be replaced by random numbers with no loss.

In this book, we generally follow the practice of reporting computed correlation and regression coefficients of all kinds and significance test results rounded to three places (or significant figures), and squared correlations (proportions of variance) rounded to four. (Occasional departures from this practice are for specific reasons of expository clarity or emphasis.) Thus, the above r would be reported as .346. But the computer treats it in the calculations in which it is involved as $.34617952\ldots$, and its square as $.34617952\ldots^2$. Thus, when we have occasion to report the square of this r, we do not report $.346^2$, which equals .1197 when rounded to four places, but $.34617952\ldots^2$ rounded to four places, which is .1198. When the reader tries to follow our computations (which he should), he will run across such apparent errors as $.346^2 = .1198$ and others which are consequent on his use in computation of the reported three-digit rounded values. These are, of course, not errors at all, but inevitable rounding discrepancies. Checks which agree within a few points in the fourth decimal place may thus be taken as correct.

Significance Test Results and the Appendix Tables

We employ classical null hypothesis testing, in which the probability of the sample result, P, is compared to a prespecified significance criterion α. If $P <$ (is less than) α, the null hypothesis (usually that the analogous population value is zero) is rejected, and the sample result is deemed statistically "significant" at the α level. In the tests we predominantly use (F and t), the actual value of P is not determined.[6] Instead, the F or t value for the sample result is computed by the appropriate formula, and the result is compared with the value of F or t at the α criterion value found from a table in the Appendix. Then, if the sample value exceeds the criterion value, we conclude that $P < \alpha$, and the null hypothesis is rejected.

We make provision in the Appendix Tables of F and t, and in those used for statistical power analysis, for the significance criteria $\alpha = .01$ and $\alpha = .05$. We see no serious need in routine work for other α values. The $\alpha = .05$ criterion is so widely used as a standard in the behavioral sciences that it has come to be understood to govern when a result is said to be statistically significant in the absence of a specified α value. The more stringent $\alpha = .01$ criterion is used by some investigators routinely as a matter of taste or of tradition in their research area, by others selectively when they believe the higher standard is required for substantive or structural reasons. We are inclined to recommend its use in research involving many variables and hence many hypothesis tests as a control on the incidence of spuriously significant results. The choice of α also depends importantly on considerations of statistical power (the probability of rejecting the null hypothesis), which is discussed in several places and in most detail in Section 4.5.

In reporting the results of significance tests for the many worked examples, we follow the general practice of attaching double asterisks to an F or t value to signify that $P < .01$, and a single asterisk to signify that $P < .05$ (but not .01). No asterisk means that the F or t is not significant, that is, P exceeds .05.

The statistical tables in the Appendix were largely abridged from Owen (1962) and from Cohen (1969), with some values computed by us. The entry values were carefully selected so as to be optimally useful over a wide range of MRC applications. For example, we provide for many values of numerator degrees of freedom (numbers of independent variables) in the F and L tables, and similarly for denominator (error) degrees of freedom in the F and t tables and for n in the power tables for r. On the other hand, we do not cover very low values for n, since they are almost never used. The coverage is sufficiently dense to preclude the need for interpolation in most problems; where needed, linear interpolation is sufficiently accurate for almost all purposes. On very rare occasions more extensive tables may be required, for which Owen (1962) is recommended.

1.2.3 The Spectrum of Behavioral Science

When we address behavioral scientists, we are faced with an exceedingly heterogeneous audience, indeed. We note in passing that our intended audience ranges

---

[6] That is, not by us in this book. Some computer programs compute and print out the actual P for each F or t given (see Appendix 3).

from student to experienced investigator, and from possession of modest to fairly advanced knowledge of statistical methods. With this in mind, we assume a minimum background for the basic exposition of the MRC system, but at some later points and infrequently, we must make some assumptions about background which may not hold for some of our readers, in order that we may usefully address some others. Even then, we try hard to keep everyone on board.

But it is with regard to substantive interests and investigative methods and materials that behavioral scientists are of truly mind-boggling diversity. The rubric "behavioral science" has no exactly delimited reference, but we use it broadly, so that it covers the "social," "biosocial," and even "life" sciences, everything from the physiology of behavior to cultural anthropology, in both their "pure" and "applied" aspects. Were it not for the fact that the methodology of science is inherently more general than its substance, a book of this kind would not be possible. However, our target audience is made up, not of methodologists, but of people whose primary interests lie in a bewildering variety of fields.

We have sought to accommodate to this diversity, even to capitalize upon it. Our illustrative examples are drawn from different areas, assuring the comfort of familiarity for most of our readers at least some of the time. They have been composed with certain ideas in mind: their content is at a level which makes them intellectually accessible to nonspecialists, and they are all fictitious, so they can accomplish their illustrative purposes efficiently and without the distractions and demands for specialized knowledge which would characterize real data. We try to use the discussion of the examples in a way which may promote some cross-fertilization between fields of inquiry—when discussed nontechnically, some problems in a given field turn out to be freshly illuminated by concepts and approaches from other fields. We may even contribute to breaking down some of the methodological stereotypy to be found in some areas, where data are analyzed traditionally, rather than optimally.

1.3 PLAN

1.3.1 Content

The first part of this book (Chapters 1 through 4) develops the basic ideas and methods of multiple correlation and regression. Chapter 2 treats simple linear correlation for two variables, X and Y, and the related linear regression model, with Y as a dependent variable and X as a single independent variable, in both their descriptive and inferential (statistical hypothesis testing and power analysis) aspects. In the first part of Chapter 3, the MRC model is extended to two independent variables, which introduces the important ideas of multiple and partial regression and correlation, and the distinction between simultaneous and hierarchical MRC. In the latter part of Chapter 3, the conceptually straightforward generalization from two to k independent variables is made.

Up to this point, for the most part, the treatment is fairly conventional. Chapter 4, however, introduces a further generalization of MRC, wherein the independent variables are organized into *h sets,* each made up of one or more variables. The utility of this extension arises from the fact that the research factors (presumed causes) being studied are expressed, in general, as sets, to which the ideas of partial relationship and of hierarchical versus simultaneous models are applied. Simple methods for hypothesis testing and statistical power analysis are included. With this chapter, the basic structure of MRC as a general data-analytic system is complete.

The stage having been set, Part II proceeds to detail in a series of chapters how information in any form may be represented as sets of independent variables, and how the resultant MRC yield for sets and their constituent variables is interpreted. In Chapter 5, various methods are described for representing nominal scales (for example, experimental treatment, diagnosis, religion), and the opportunity is grasped to show that the MRC results include those produced by AV, and more. Chapter 6 performs the same task for quantitative (ratio, interval, ordinal) scales, showing how various methods of representation may be used to determine the presence and form of curvilinear relationship with Y. Chapter 7 is concerned with the representation of missing data, a ubiquitous problem in the behavioral sciences, and shows how this property of a research factor may be used as positive information. Chapter 8 presents and generalizes the idea of interactions as conditional relationships, and shows how interactions among research factors of any type may be simply represented as sets, incorporated in analyses, and interpreted.

In Part III ("Applications"), Chapter 9 provides a culmination of the ideas about multiple sets (Chapter 4), implemented by the methods of representing research factors (Chapters 5 through 7) and their interactions (Chapter 8). It shows how conventional ACV may be accomplished by MRC, and then greatly generalizes ACV, first to accommodate multiple, nonlinear, and nominal covariate sets, and then to extend to quantitative research factors. The nature of partialling is closely scrutinized, and the problem of fallible (unreliable) covariates is addressed, as is the use of the system in the study of change.

Chapter 10 extends MRC analysis to repeated measurement and matched-subject research designs. In Chapter 11, canonical correlation analysis and other multivariate methods are surveyed and related to MRC analysis, from which some novel analytic methods emerge.

In the Appendices, we provide the mathematical background for MRC (Appendix 1), the hand computation for k variables together with the interpretation of the results of matrix inversion (Appendix 2), and a discussion of the use of available computer programs to accomplish MRC analyses (Appendix 3). Finally, the necessary statistical tables are provided.

For a more detailed synopsis of the book's contents, the reader is referred to the summaries at the ends of the chapters.

1.3.2 Structure: Numbering of Sections, Tables, and Equations

Each chapter is divided into major sections, identified by the chapter and section numbers, for example, Section 5.3 ("Dummy Variable Coding") is the third major section of Chapter 5. The next lower order of division within each major section is further suffixed by its number, for example, Section 5.3.4 ("Dummy Variable Multiple Regression/Correlation and Analysis of Variance") is the fourth subsection of Section 5.3. Further subdivisions are not numbered, but titled with an italicized heading.

Tables, figures, and equations within the body of the text are numbered consecutively within major sections. Thus, for example, Table 5.3.4 is the fourth table in Section 5.3, and Eq. (5.3.4) is the fourth equation in Section 5.3. (We follow the usual convention of giving equation numbers in parenthesis.) A similar plan is followed in the three Appendices. The reference statistical tables make up a separate appendix and are designated by letters as Appendix Tables A through F.

1.4 SUMMARY

This introductory chapter begins with an overview of MRC as a data-analytic system, emphasizing its generality and superordinate relationship to the analysis of variance/covariance. Multiple regression/correlation is shown to be peculiarly appropriate for the behavioral sciences in its capacity to accomodate the various types of complexity which characterize them: the multiplicity and correlation among causal influences, the varieties of form of information and shape of relationship, and the frequent incidence of conditional (interactive) relationships. (Section 1.1)

The book's exposition of MRC is nonmathematical, and stresses informed application to scientific and technological problems in the behavioral sciences. Its orientation is "data-analytic" rather than statistical-analytic, an important distinction that is discussed. Concrete illustrative examples are heavily relied upon. The means of coping with the computational demands of MRC by desk calculator and computer are briefly described and the details largely relegated to appendices so as not to distract the reader's attention from the conceptual issues. The ground rules for reporting numerical results (including a warning about rounding discrepancies) and those of significance tests are given, and the statistical tables in the appendix are described. Finally, we acknowledge the heterogeneity of background and substantive interests of our intended audience, and discuss how we try to accomodate to it and even exploit it to pedagogical advantage. (Section 1.2)

The chapter ends with a brief outline of the book, and the scheme by which sections, tables, and equations are numbered. (Section 1.3)

2
Bivariate Correlation and Regression

One of the most general meanings of the concept of a relationship between pairs of variables is that knowledge with regard to one of the variables carries information about the other variable. Thus, information about the height of a child in elementary school would have implications for the probable age of the child, and information about the occupation of an adult would lead to more accurate guesses about his income level than could be made in the absence of that information.

2.1 TABULAR AND GRAPHIC REPRESENTATIONS OF RELATIONSHIPS

Whenever data has been gathered on two quantitative variables for a set of cases (for example, individuals), the relationship between the variables may be displayed graphically by means of a scatter plot.

For example, suppose we have scores on a vocabulary test and a digit-symbol substitution task for 15 children (see Table 2.1.1). If these data are plotted by representing each child as a point on a graph with vocabulary scores on the horizontal axis and the number of digit symbols on the vertical axis, we would obtain the scatterplot seen in Figure 2.1.1. The circled dot, for example, represents Child 1, who obtained a score of 5 on the vocabulary test and completed 12 digit-symbol substitutions.

When we inspect this plot, it becomes apparent that the children with higher vocabulary scores tended to complete more digit symbols (d-s) and those low on vocabulary (v) scores were usually low on d-s as well. This can be seen especially well by looking at the average of the d-s scores corresponding to each v score, $\overline{Y}_v$. The child receiving the lowest score, 5, received a d-s score of 12; the children with the next lowest v score, 6, obtained an average d-s score of 14.67, and so on to the highest v scorers, who obtained an average of 19.5 on the d-s

TABLE 2.1.1

Illustrative Set of Data on Vocabulary and
Digit-Symbol Tests

Child (no.)	Vocabulary	Digit symbol
1	5	12
2	8	15
3	7	14
4	9	18
5	10	19
6	8	18
7	6	14
8	6	17
9	10	20
10	9	17
11	7	15
12	7	16
13	9	16
14	6	13
15	8	16

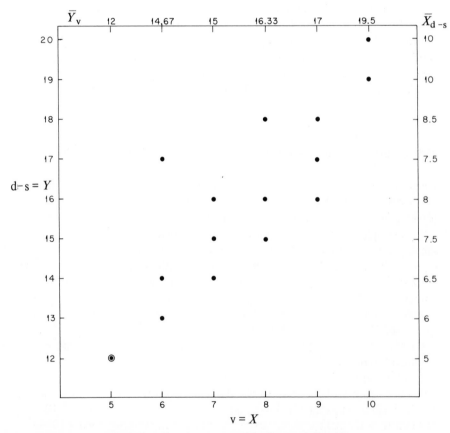

FIGURE 2.1.1 A strong, positive linear relationship.

test. A parallel tendency for vocabulary scores to increase is observed for increases in d-s scores. The form of this relationship is said to be positive, since high values on one variable tend to go with high values on the other variable, and low with low values. It may also be called linear since the tendency for an increase in one variable to be accompanied by an increase in the other variable is consistent throughout the scales. That is, if we were to draw the straight line which best fits the average of the d-s values at each v score (from the lower left-hand corner to the upper right-hand corner) we would be describing the trend or shape of the relationship quite well.

Figure 2.1.2 displays a similar scatter plot for age and the number of seconds needed to complete the digit-symbol task. In this case, low scores on age tend to go with high test time in seconds, and high test times are more common in younger children. In this case the relationship may be said to be negative and linear. It should also be clear at this point that whether a relationship between two variables is positive or negative is a direct consequence of the direction in which the two variables have been scored. If, for example, the vocabulary scores from the first example were taken from a 12-item test, and instead of scoring the number correct a count was made of the number wrong, the relationship with d-s scores would be negative. Since such scoring decisions in many cases may be essentially arbitrary, it should be kept in mind that any positive relationship becomes negative when either (but not both) of the variables is reversed, and

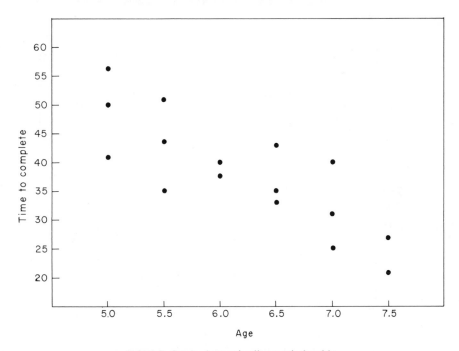

FIGURE 2.1.2 A negative linear relationship.

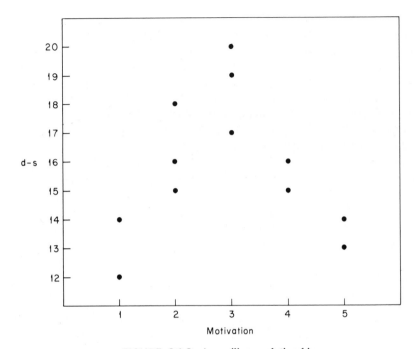

FIGURE 2.1.3 A curvilinear relationship.

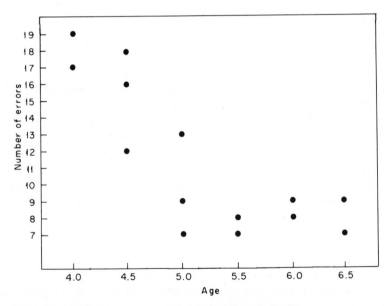

FIGURE 2.1.4 A negative, curved line relationship.

vice versa. Thus, for example, a negative relationship between age of oldest child and income for a group of 30-year-old mothers implies a positive relationship between age of first becoming a mother and income.[1]

Figure 2.1.3 gives the plot of a measure of motivational level and score on a difficult d-s task. It is apparent that the way motivation is associated with performance score depends on whether the motivational level is at the lower end of its scale or near the upper end. Thus, the relationship between these variables is curvilinear. Finally, Figure 2.1.4 presents a scatter plot for age and number of substitution errors. This plot demonstrates a general tendency for higher scores on age to go with fewer errors, indicating that there is, in part, a negative linear relationship. However, it also shows that the decrease in error rate which goes with the same increase in age is greater at the lower end of the age scale than it is at the upper end, a finding which indicates that, although a straight line provides some kind of fit, clearly it is not optimal.

Thus, scatter plots allow visual inspection of the form of the relationship between two variables. These relationships may be linear (negative or positive) or curvilinear; or they may be well described by a straight line, approximated by a straight line, or may require lines with one or more curves to adequately describe them.

Since linear relationships are very common in all sorts of data, we shall concentrate on these in the current discussion, presenting methods of analyzing nonlinear relationships in Chapter 6.

Now suppose that Figure 2.1.1 is compared with Figure 2.1.5. In both cases the relationship between the variables is linear and positive; however, it would appear that vocabulary provides better information with regard to d-s completion than does chronological age. That is, the extent of the relationship with performance seems to be greater for v than for age since one could make more accurate estimates of d-s scores using information about v than using age. In order to compare these two relationships to determine which is greater, we need an index of the size of the relationship between two variables which will be comparable from one pair of variables to another. Looking at the relationship between v and d-s scores, other questions come to mind: Should this be considered a strong or weak association? On the whole, how great an increase in digit-symbol score is found for a given increase in vocabulary score in this group? If v is estimated for d-s in such a way as to minimize the errors, how much error will, nevertheless, be made? If this is a random sample of subjects from a larger population, how much confidence can we have that v and d-s are linearly related in the entire population? These and other questions are answered by correlation and regression methods and their associated tests of significance. In the use and

---

[1] Here we follow the convention of naming a variable for the upper end of the scale. Thus, a variable called income means that high-income numbers indicate high income, whereas a variable called poverty would mean that high numbers indicate much poverty and therefore low income.

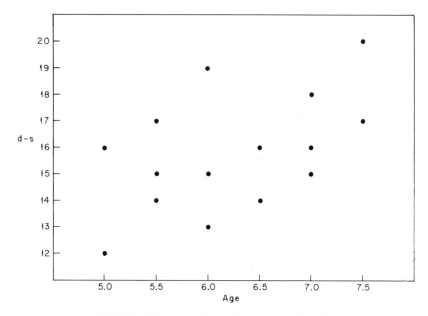

FIGURE 2.1.5 A weak, positive linear relationship.

interpretation of these methods the two variables are literally treated as interval scales, that is, constant differences between scale points on each variable are assumed to represent equal "amounts" of the construct being measured. Although for many or even most psychological scales this assumption is not literally true, empirical work (Baker, Hardyck, & Petrinovich, 1966) indicates that small to moderate inequalities in interval size produce little if any distortion in the validity of conclusions based on the analysis. Further discussion of this issue is found in Chapter 6.

2.2 THE INDEX OF LINEAR CORRELATION BETWEEN TWO VARIABLES: THE PEARSON PRODUCT MOMENT r

2.2.1 Standard Scores: Making Units Comparable

One of the first problems solved by an index of the degree of association between two variables is that of measurement unit. Since the two variables are typically expressed in different units, we need some means of converting the scores to comparable measurement units. It can be readily perceived that any index which would change with an arbitrary change in measurement unit—from inches to centimeters or age in months to age in weeks, for example—could

TABLE 2.2.1

Data on Income and TV Sets Presented in Original Units,
Deviation Units, and z Units

Household	X Income	Y No. TVs	$X - \bar{X} = x$	$Y - \bar{Y} = y$	x^2	y^2
1	7,000	1	−3,500	−1.50	12,250,000	2.25
2	12,000	4	+1,500	+1.50	2,250,000	2.25
3	10,000	2	−500	−.50	250,000	.25
4	13,000	3	+2,500	+.50	6,250,000	.25
Sum	42,000	10	0	0	21,000,000	5.00

$$m \quad \bar{X} = \quad 10,500 \qquad 2.5 = \bar{Y}$$
$$sd_X^2 = \Sigma\, x^2/n \quad = 5,250,000 \qquad 1.25 = sd_Y^2$$
$$sd_X = \sqrt{\Sigma\, x^2/n} \ = \ 2,291.29 \qquad 1.118 = sd_Y$$

	$x/sd_X = z_X$	$y/sd_Y = z_Y$	z_X^2	z_Y^2
1	−1.53	−1.34	2.333	1.800
2	+.65	+1.34	.429	1.800
3	−.22	−.45	.048	.20
4	+1.09	+.45	1.190	.20
Sum	0	0	4.000	4.000
m	0	0		
sd^2	1.00	1.00		
sd	1.00	1.00		

hardly be useful as a general description of the strength of the relationship between height and age.

To illustrate this problem, suppose information has been gathered on the annual income and the number of TV sets of four households (Table 2.2.1).[2] In the effort to measure the degree of relationship between income (X) and the number of TV sets (Y), we are embarrassed by the difference in the nature and size of the units in which the two variables are measured. Although Households 1 and 3 are both below the mean on both variables and Households 2 and 4 are above the mean on both (see x and y, scores expressed as deviations from their means, symbolized as $\bar{X}$ and $\bar{Y}$, respectively), we are still at a loss to assess the correspondence between a difference of \$3,500 from the mean income and a difference of 1.5 TV sets from the mean number of TV sets.

---

[2] In this example as in all examples that follow, the number of cases (n) is kept very small in order to facilitate the reader's following of the computations. In almost any serious research, the n must, of course, be very much larger.

We may attempt to resolve the difference in units by ranking the households on the two variables—1, 3, 2, 4 and 1, 4, 2, 3, respectively—and noting that there seems to be some correspondence between the two ranks. In so doing we have, however, made the difference between Households 1 and 3 ($3,000) equal to the difference between Households 2 and 4 ($1,000); two ranks in each case.

To make the scores comparable we clearly need some way of taking the different variability of the two original sets of scores into account. Since the standard deviation (sd) is an index of variability of scores, we may measure the discrepancy of each score from its mean (m) relative to the variability of all the scores by dividing by the sd:

$$(2.2.1) \qquad sd_X = \sqrt{\frac{\sum x^2}{n}},$$

where $\sum x^2$ means "the sum of the squared deviations from the mean."[3]
The scores thus created are in standard deviation units and are called *standard* or z scores.

$$(2.2.2) \qquad z_X = \frac{X - \bar{X}}{sd_X} = \frac{x}{sd_X}.$$

In Table 2.2.1, the z score for income for Household 1 is −1.53, which indicates that its value ($7,000) falls about $1\frac{1}{2}$ income standard deviations ($2,291) *below* the income mean ($10,500). Although income statistics are expressed in dollar units, the z score is a pure number, that is, it is unit-free. Similarly, Household 1 has a z score for number of TV sets of −1.34, which indicates that its number of TV sets (1 set) is about $1\frac{1}{3}$ standard deviations *below* the mean number of sets (2.5 sets). Note again that −1.34 is not expressed in number of sets, but is also a pure number. Instead of having to compare $7,000 and 1 TV sets for Household 1, we can now make a meaningful comparison of −1.53 (z_X) and −1.34 (z_Y), and note incidentally the similarity of the two values for Household 1.

It should be noted that the rank of the z scores is the same as that of the original scores, and that scores which were above or below the mean on the original variable retain this characteristic in their z scores. In addition, we note that the difference between the incomes of Households 2 and 3 ($X_2 - X_3 = $2,000$) is twice as large, and of opposite direction to the difference between Households 2 and 4 ($X_2 - X_4 = -$1,000$). When we look at the z scores for

---

[3] Note that we distinguish throughout between *sd*, which is a descriptor of the variability of the *sample* at hand and uses n as a divisor, and $\tilde{sd}$, which is a (sample based) estimator of the population variability and uses degrees of freedom as a divisor. The latter will be required for testing hypotheses and finding confidence limits.
   Also note that the summation sign, Σ, is used to indicate summation over all n cases here and elsewhere, unless otherwise specified.

these same households, we find that $z_{X_2} - z_{X_3} = .65 - (-.22) = .87$ is twice as large and of opposite direction to the difference $z_{X_2} - z_{X_4} = .65 - 1.09 = -.44$ ($.87/-.44 = -2$, within rounding error). Such proportionality of differences or distances between scores

(2.2.3)
$$\frac{X_i - X_j}{X_p - X_q} = \frac{z_{X_i} - z_{X_j}}{z_{X_p} - z_{X_q}}$$

is the essential element in what is meant by retaining the original relationship between the scores. This can be seen more concretely in Figure 2.2.1, in which we have plotted the pairs of scores. Whether we plot z scores or raw scores, the points in the scatter plot have the same relationship to each other.

The z transformation of scores is one example of a linear transformation. A linear transformation is one in which every score is changed by adding or subtracting a constant and/or multiplying or dividing by a constant. Changes from inches to centimeters, dollars to francs, and Fahrenheit to Centigrade degrees are examples of linear transformations. Such transformations will, of course, change the ms and sds of the variables upon which they are performed.

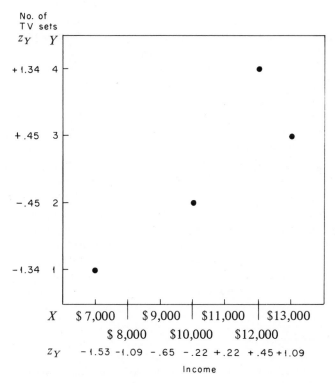

FIGURE 2.2.1 Bivariate distribution of income and TV ownership.

However, since the *sd* will change by exactly the same factor as the original scores (that is, by the constant by which scores have been multiplied or divided) and since *z* scores are created by subtracting scores from their mean, *all linear transformations of scores will yield the same set of z scores.*

Since the properties of *z* scores form the foundation necessary for understanding the correlation coefficient they will be briefly reviewed:

1. The sum of a set of *z* scores ($\Sigma\ z$) and therefore also the mean equal 0.
2. The variance (sd^2) of the set of *z* scores equals 1, as does the standard deviation (*sd*).
3. Neither the shape of the distribution of *X*, nor of its relationship to any other variable, is affected by transforming it to z_X.

2.2.2 The Product Moment *r* as a Function of Differences between *z* Scores

We may now define a perfect (positive) relationship between two variables (*X* and *Y*) as existing when all z_X, z_Y pairs of scores consist of two exactly equal values. Furthermore, the degree of relationship will be a function of the departure from this "perfect" state, that is, a function of the differences between pairs of z_X and z_Y scores. Since the average difference between paired z_X and z_Y is necessarily zero (since $\bar{z}_Y = \bar{z}_X = 0$), the relationship may be indexed by finding the average of the *squared* discrepancies between *z* scores ($\Sigma\,(z_X - z_Y)^2 /n$).

Table 2.2.2 presents the data from Figure 2.1.1. Several things should be noted in this table. Deviation scores (*x* and *y*) sum to zero as do z_X and z_Y. sd_{z_X} and sd_{z_Y} are both 1, $\bar{z}_X$ and $\bar{z}_Y$ are both 0 (which are mathematical necessities), and these equalities reflect the equal footing on which we have placed the two variables.

We find that the squared differences between *z* scores sum to 5.159 which when divided by the number of observations equals .344. How large is this relationship? We have stated that if the two variables were perfectly (positively) related, all *z* score differences would equal zero and necessarily their sum and mean would also be zero. A perfect negative relationship, on the other hand, may be defined as one in which the *z* scores in each pair are equal in absolute value but opposite in sign. Under the latter circumstances, it is demonstrable that the average of the squared discrepancies always equals 4. It can also be proved that under circumstances in which the pairs of *z* scores are on the average equally likely to be consistent with a negative relationship as with a positive relationship, the average squared difference will always equal 2, which is midway between 0 and 4. Under these circumstances, we may say that there is *no* linear relationship between *X* and *Y*.

Although it is clear that this index, ranging from 0 for a perfect positive linear relationship through 2 for no linear relationship to 4 for a perfect negative one, does reflect the relationship between the variables in an intuitively meaningful

TABLE 2.2.2
Calculation of *z* Scores, *z* Score Differences, and *z* Score Products
on Data Example

v scores (X)	d-s scores (Y)	$\frac{X-\bar X}{sd_X}=z_X$	$\frac{Y-\bar Y}{sd_Y}=z_Y$	z_X-z_Y	z_Xz_Y
5	12	−1.79	−1.85	.06	3.32
8	15	.22	−.46	.68	−.10
7	14	−.45	−.93	.48	.42
9	18	.90	.93	−.03	.83
10	19	1.57	1.39	.18	2.18
8	18	.22	.93	−.71	.21
6	14	−1.12	−.93	−.19	1.04
6	17	−1.12	.46	−1.58	−.52
10	20	1.57	1.85	−.28	2.90
9	17	.90	.46	.44	.41
7	15	−.45	−.46	.01	.21
7	16	−.45	0	−.45	0
9	16	.90	0	.90	0
6	13	−1.12	−1.39	.27	1.55
8	16	.22	0	.22	0
Sum 115	240	0	0	0	12.43
SS[a] 915	3910	15	15	5.159	
m 7.67	16	0	0	0	.828
sd² 2.22	4.67	1	1		
sd 1.49	2.16	1	1		

[a]Sum of squared values, that is, ΣX^2, ΣY^2, etc.

way, it is useful to transform the scale somewhat to make its interpretation even more clear. If we divide the average squared discrepancies by 2 and subtract the result from 1, we have

$$(2.2.4) \qquad r = 1 - \frac{1}{2}\left(\frac{\Sigma(z_X-z_Y)^2}{n}\right),$$

which for the data of Table 2.2.2 gives

$$r = 1 - \frac{1}{2}\left(\frac{5.16}{15}\right) = .828.$$

r is the product moment correlation coefficient, invented by Karl Pearson in 1895. This coefficient is the standard measure of the linear relationship between two variables and has the following properties:

1. It is a pure number and independent of the units of measurement.

2. Its absolute value varies between zero, when the variables have no linear relationship, and 1, when each variable is perfectly predicted by the other. The absolute value thus gives the *degree* of relationship.

3. Its sign indicates the direction of the relationship. A positive sign indicates a tendency for high values of one variable to occur with high values of the other, and low values to occur with low. A negative sign indicates a tendency for high values of one variable to be associated with low values of the other. Reversing the direction of measurement of one of the variables will produce a coefficient of the same absolute value but of opposite sign. Coefficients of equal value but opposite sign (for example, +.50 and −.50) thus indicate equally strong linear relationships, but in opposite directions.

2.3 ALTERNATIVE FORMULAS
FOR THE PRODUCT MOMENT r

The formula given above [Eq. (2.2.4)] for the product moment correlation coefficient as a function of squared differences between paired z scores is only one of a number of mathematically equivalent formulas. Some of the other versions provide additional insight into the nature of r, and others facilitate computation. Yet other formulas apply to particular kinds of variables, such as variables for which only two values are possible, or variables which consist of a set of ranks.

2.3.1 r as the Average Product of z Scores

It follows from simple algebraic manipulation of Equation (2.2.4) that

(2.3.1)
$$r = \frac{\sum z_X z_Y}{n} .$$

The product moment correlation is therefore seen to be the mean of the products of the paired z scores. In the case of a perfect positive correlation, since $z_X = z_Y$,

$$r = \frac{\sum z_X z_Y}{n} = \frac{\sum z^2}{n} = 1.$$

For the data presented in Table 2.2.2, these products have been computed and $r = (12.43)/15 = .828$, as before. (Recall our warning in Section 1.2.2 about rounding errors.)

2.3.2 Raw Score Formulas for r

Since z scores can be readily reconverted to the original units, a formula for the correlation coefficient can be written in raw score terms. There are many mathematically equivalent versions of this formula of which the following is a convenient one for computation by computer or desk calculator:

(2.3.2) $$r = \frac{n\sum XY - \sum X \sum Y}{\sqrt{[n\sum X^2 - (\sum X)^2][n\sum Y^2 - (\sum Y)^2]}}.$$

Yet another mathematically equivalent formula for r is

(2.3.3) $$r = \frac{(\sum xy)/n}{sd_X \, sd_Y}.$$

The average of the products of the deviation scores $(\sum xy)/n$ is called the *covariance* and is a measure of the tendency for the two variables to *covary* or go together, expressed in the original units. Thus, we can see that r is an expression of the covariance between standardized variables.

It should be noted that r itself is *not* a function of the number of observations,[4] and that the presence of n in the various formulas only serves to cancel it out of other terms where it is hidden, for example, sd. A formula for r that does not contain n is

(2.3.4) $$r = \frac{\sum xy}{\sqrt{\sum x^2 \sum y^2}},$$

which follows directly from some simple algebraic manipulation of Eq. (2.3.3).

2.3.3 Point Biserial r

When one of the variables to be correlated is a dichotomy (it can take on only two values), the computation of r simplifies. There are many dichotomous variables in the social sciences, such as yes or no responses, left- or right-handedness, and the presence or absence of any trait or attribute. For example, although the variable "sex of subject" does not seem to be a quantitative variable it may be looked on as the presence or absence of the characteristic of being female (or male). As such, we may decide, arbitrarily, to score all females as 1 and all males as 0. Under these circumstances, the sd of the sex variable is determined by the proportion of the total n in each of the two groups; $sd = \sqrt{pq}$, where p is the proportion in one group and $q = 1 - p$, the proportion in the other group. Since r indicates a relationship between two standardized variables, it does not matter whether we choose 0 and 1 as the two values or any other pair of different values.

For example, Table 2.3.1 presents data on the effects of an interfering stimulus on task performance for a group of seven experimental subjects. As can

---

[4] For $n > 2$. The reader may satisfy himself that when $n = 2$, r must equal +1 or −1.

TABLE 2.3.1

An Example of Correlation between a Dichotomous
and a Continuous Variable

Subject no.	Stimulus condition (X)	Task score (Y)	X_A	X_B	z_Y	z_A	z_B	$z_Y z_A$	$z_Y z_B$
1	NONE	67	0	50	−.41	−.88	.88	.36	−.36
2	NONE	72	0	50	1.63	−.88	.88	−1.43	1.43
3	NONE	70	0	50	.81	−.88	.88	−.71	.71
4	NONE	69	0	50	.41	−.88	.88	−.36	.36
5	STIM	66	1	20	−.81	1.16	−1.16	−.94	.94
6	STIM	64	1	20	−1.63	1.16	−1.16	−1.89	1.89
7	STIM	68	1	20	0	1.16	−1.16	0	0
Sum		476	3	260	0	0	0	−4.97	4.97
m		68	.43	37.14	0	0	0	−.707	.707
$\bar{Y}$	NONE	69.5							
$\bar{Y}$	STIM	66.0					$r_{YA} = -.707$		
sd		2.45	.495	14.9			$r_{YB} = .707$		

be seen, the absolute value of the correlation remains the same whether we choose (X_A) 0 and 1 as the values to represent the absence or presence of an interfering stimulus or choose (X_B) 50 and 20 as the values to represent the same dichotomy. The sign of r, however, depends on whether the group with the higher mean on the other (Y) variable, in this case the no stimulus group, has been assigned the higher or lower of the two values. The reader is invited to try other values and observe the constancy of r.

Since the z scores of a dichotomy are a function of the proportion of the total in each of the two groups, the product moment correlation formula simplifies to

$$(2.3.5) \qquad r_{pb} = \frac{(\bar{Y}_1 - \bar{Y}_0)\sqrt{pq}}{sd_Y},$$

where $\bar{Y}_1$ and $\bar{Y}_0$ are the Y means of the two groups of the dichotomy. The simplified formula is called the point biserial r to take note of the fact that it involves one dichotomous or "point" variable (X) and one continuous variable (Y). In the present example

$$r_{pb} = \frac{(66.0 - 69.5)\sqrt{(.428)(.572)}}{2.45} = -.707.$$

The point biserial formula for the product moment correlation displays an interesting and useful property. Under the circumstances in which the two groups on the dichotomous variable are of equal size ($p = q = .5$ and $\sqrt{pq} = .5$), the *r* then equals half the difference between the means of the *z* scores for *Y*, and so 2*r* equals the difference between the means on the standardized variable.

2.3.4 Phi Coefficient

When both *X* and *Y* are dichotomous, the computation of the product moment correlation is even further simplified. The data may be represented by a fourfold table and the correlation computed directly from the frequencies and marginals. For example, suppose a study investigates the difference in preference of homeowners and nonhomeowners for the two candidates in a local election, and the data is presented in Table 2.3.2. The formula for *r* here simplifies to the difference between the product of the diagonals of a fourfold table of frequencies, divided by the square root of the product of the four marginal sums:

$$(2.3.6) \qquad r_\phi = \frac{BC - AD}{\sqrt{(A+B)(C+D)(A+C)(B+D)}}$$

$$= \frac{(54)(60) - (19)(52)}{\sqrt{73 \cdot 112 \cdot 79 \cdot 106}}$$

$$= .272.$$

Once again it may be noted that this is a computing alternative to the *z* score formula, and therefore it does not matter what two values are assigned to the dichotomy, since the standard scores, and hence the *r*, will remain the same. It also follows that unless the division of the group is the same for the two dichotomies ($p_Y = p_X$ or q_X), their *z* scores cannot have the same values and *r* cannot equal 1 or −1. A further discussion of this limit will be found in Section 2.11.1.

2.3.5 Rank Correlation

Yet another simplification in the product moment correlation formula occurs when the data being correlated consist of two sets of ranks. Since the *sd* of a

TABLE 2.3.2
Fourfold Frequencies for Candidate Preference and Homeowning Status

	Candidate U	Candidate V	Total:
Homeowners	*A* 19	*B* 54	73 = *A* + *B*
Nonhomeowners	*C* 60	*D* 52	112 = *C* + *D*
Total:	79 = *A* + *C*	106 = *B* + *D*	185 = *n*

TABLE 2.3.3
Correlation between Two Sets of Ranks

I.D.	X	Y	x	x^2	y	y^2	xy	d	d^2
1	4	2	1	1	−1	1	−1	2	4
2	2	1	−1	1	−2	4	2	1	1
3	3	4	0	0	1	1	0	−1	1
4	5	3	2	4	0	0	0	2	4
5	1	5	−2	4	2	4	−4	−4	16
Sum	15	15	0	10	0	10	−3	0	26
m	3	3							

complete set of ranks is a function only of the number of objects being ranked, some algebraic manipulation yields

$$(2.3.7) \qquad r_{ranks} = 1 - \frac{6 \sum d^2}{n(n^2 - 1)},$$

where d is the difference between the pair of ranks for an object or individual. In Table 2.3.3 a set of 5 ranks are presented with their deviations and differences. Using one of the general formulas (2.3.4) for r,

$$r = \frac{\sum xy}{\sqrt{\sum x^2}\sqrt{\sum y^2}}$$

$$= \frac{-3}{\sqrt{10}\sqrt{10}}$$

$$= -.300.$$

The rank order formula (2.3.7) yields

$$r = 1 - \frac{6 \sum d^2}{n(n^2 - 1)}$$

$$= 1 - \frac{6(26)}{5(24)}$$

$$= 1 - \frac{156}{120}$$

$$= -.300,$$

which checks.

We wish to stress the fact that the formulas for point biserial, phi, and rank correlation are simply computational equivalents of the previously given general formulas for r which result from the mathematical simplicity of dichotomous or rank data. They are of use when computation is done by hand or desk calculator. They are of no significance when computers are used, since whatever formula for r the computer uses will work when variables are scored 0–1 (or any other two values) or are ranked. It is obviously not worth the trouble to write special programs to produce these special-case versions of r when a formula such as Eq. (2.3.2) will produce them.

2.4 REGRESSION COEFFICIENTS: ESTIMATING Y FROM X

To return to one of our original questions, how great an increase in d-s score will we find on the average for a given increase in v score in the group (Tables 2.1.1, and 2.2.2, Figure 2.1.1)? We wish to obtain a single descriptive figure to summarize the rate of change, in a way analogous to the sd as a measure of dispersion. In other words, we want some constant which represents the average change in d-s score per unit increase in v score or, equivalently, allows us to "predict" from v to d-s score. Obviously, if the relationship between vocabulary and d-s were perfect and positive, we could describe the d-s score corresponding to any vocabulary score perfectly simply by adjusting for differences in scale of the two variables. Since, when $r_{XY} = 1$, each individual's estimated $\hat{z}_{Y_j}$ simply equals z_{X_j}, so

$$\frac{\hat{Y}_j - \bar{Y}}{sd_Y} = \frac{X_j - \bar{X}}{sd_X},$$

and solving for $\hat{Y}_j$,

$$\hat{Y}_j = \frac{sd_Y (X_j - \bar{X})}{sd_X} + \bar{Y},$$

and since $\bar{X}$, $\bar{Y}$, sd_X, and sd_Y are known, it remains only to specify X_j and then $\hat{Y}_j$ may be computed.

When, however, the relationship is not perfect, we may, nevertheless, wish to show the $\hat{Y}$ estimate we would obtain by using the best possible "average" conversion or prediction rule from X, in the sense that the computed values will be as close to the actual Y values as is possible with a single linear conversion formula. Larger differences between the actual and predicted scores $(Y_j - \hat{Y}_j)$ are indicative of larger errors. The average error $\Sigma (Y_j - \hat{Y}_j)/n$ will equal zero whenever the overestimation of some scores is balanced by an equal underestimation of other scores. That there be no consistent over- or underestimation is a desirable property, but it may be accomplished by an infinite number of different conversion rules. We therefore define "as close as possible" to correspond to the "least-squares" criterion so common in statistical work—we will choose a conversion rule such that the sum of the squared discrepancies between the actual Y and estimated $\hat{Y}$ will be minimized, that is, as small as the data permit.

It can be proven via calculus (or some rather complicated algebra) that the linear conversion rule which is optimal for converting z_X to an estimate of z_Y is

(2.4.1) $$\hat{z}_Y = r\, z_X.$$

If we wish to convert from our original scores, since $\hat{z}_Y = (\hat{Y}_j - \overline{Y})/sd_Y$ and $z_X = (X_j - \overline{X})/sd_X$,

(2.4.2) $$\hat{Y}_j = r\, sd_Y\, \frac{(X_j - \overline{X})}{sd_X} + \overline{Y}$$

It is useful to simplify and separate the elements of this formula in the following way. Let

(2.4.3) $$B_{YX} = r\frac{sd_Y}{sd_X},$$

(2.4.4) $$A_{YX} = \overline{Y} - B_{YX}\overline{X},$$

from which we may write the regression equation (dropping the j subscript from $\hat{Y}$ and X as understood) as

(2.4.5) $$\hat{Y} = B_{YX}X + A_{YX}.$$

B_{YX} is the regression coefficient for estimating Y from X, and represents the rate of change in Y units per X unit. A_{YX} is called the regression constant or Y intercept and serves to make appropriate adjustments for differences between $\overline{X}$ and $\overline{Y}$. When the line representing the best linear estimation equation (the regression equation) is drawn on the scatter plot of the data in original units, B_{YX} indicates the *slope* of the line and A_{YX} represents the point at which the regression line crosses the Y axis, the estimated $\hat{Y}$ when $X = 0$.

The slope of a regression line is the measure of its steepness, the ratio of how much it rises (or, when negative, falls) to any given amount of increase along the horizontal. Since the "rise" over the "run" is a constant for a straight line, our interpretation of it as number of units of change in Y per unit change in X meets this definition. Taking our running example of 15 pairs of v and d-s scores (Table 2.2.2) for which $\overline{X} = 7.67$, $\overline{Y} = 16$, $sd_X = 1.49$, $sd_Y = 2.16$, and $r = .828$,

$$B_{YX} = .828\, \frac{2.16}{1.49} = 1.20,$$

$$A_{YX} = 16 - 1.20(7.67) = 6.78.$$

The regression coefficient, B_{YX}, indicates that for each unit of increase in X, we will estimate 1.2 units of increase in Y, and that using this rule we will minimize our errors (in the least-squares sense). The A_{YX} term gives us a point for starting this estimation—the point for a zero value of X, which is, of course,

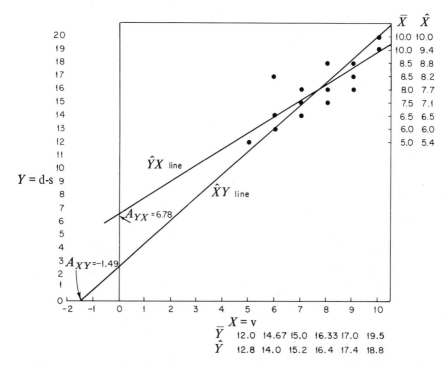

FIGURE 2.4.1 Bivariate distribution of d-s and v scores showing regression lines and intercept.

out of the range for the present set of scores. The equation $\hat{Y} = B_{YX}X + A_{YX}$ produces, when X values are substituted, the $\hat{Y}X$ line in Figure 2.4.1.

We also may, of course, estimate $\hat{X}$ from Y by interchanging X and Y in Eqs. (2.4.3) and (2.4.4):

$$B_{XY} = r(sd_X/sd_Y)$$

$$= .828 \frac{1.49}{2.16} = .572,$$

$$A_{XY} = \bar{X} - B_{XY}\bar{Y}$$

$$= 7.67 - .572\,(16) = -1.49.$$

Substituting Y values in the equation $\hat{X} = B_{XY}Y + A_{XY}$ gives us the $\hat{X}Y$ line in Figure 2.4.1. Note that the two regression lines cross at the point corresponding to the means of the two distributions. As we have already stated, the A constant in the regression equation gives us the $\hat{X}$ *corresponding to* $Y = 0$. When scores are expressed as deviations from their means (but in their original units),

$$\hat{y} = B_{YX}x,$$

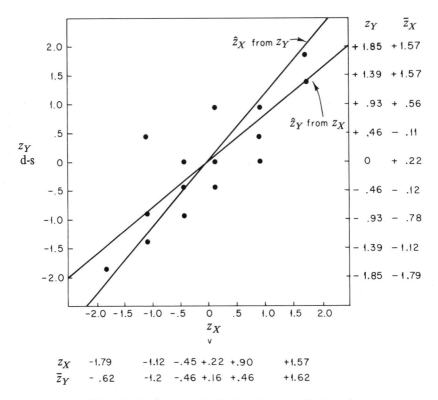

z_Y	z_X
+ 1.85	+ 1.57
+ 1.39	+ 1.57
+ .93	+ .56
+ .46	− .11
0	+ .22
− .46	− .12
− .93	− .78
− 1.39	− 1.12
− 1.85	− 1.79

z_X	−1.79	−1.12	−.45	+.22	+.90	+1.57
$\bar{z}_Y$	− .62	−1.2	−.46	+.16	+.46	+1.62

FIGURE 2.4.2 Bivariate distribution of z scores for d-s and v.

and when $x = 0$ (i.e., $X = \bar{X}$), $\hat{y} = 0$ necessarily. When variables are expressed in standard score form, since $sd_{z_X} = sd_{z_Y} = 1$, then $B_{z_Y z_X} = B_{z_X z_Y} = r$. Since $\bar{z}_Y = \bar{z}_X = 0$, $A = 0$ and the regression lines cross at 0. Plots of the regression lines for the z scores in the previous example can be seen in Figure 2.4.2.

Since regression coefficients are used to estimate one variable from given values of another variable, in the remaining discussion the X variable will be considered the *independent* (or "predictor") variable and the Y variable the *dependent* variable. B and A coefficients will, in general, be presented without subscripts and may be understood to be B_{YX} and A_{YX}. In actual research, of course, designation of dependent and independent variable status implies that Y is hypothesized as dependent upon X in some sense and so is a function of the substantive causal framework of the study.

The meaning of the regression coefficient may be seen especially well in the case in which the independent variable is a dichotomy. If we return to the example from Table 2.3.1 where the point biserial $r = -.707$ and calculate

$$B = -.707 \frac{2.45}{.495} = -3.5,$$

we note that this is exactly the difference between the two group means on Y, $66 - 69.5$. Calculating the intercept

$$A = 68 - (-3.5)(.428) = 69.5,$$

we find it to be equal to the mean of the group coded 0 (the no-stimulus condition). This must be the case since the best (least-squares) estimate of Y for each group is its own mean, and the regression equation for the members of the group represented by the 0 point of the dichotomy is solved as

$$\hat{Y} = B(0) + A = A.$$

2.5 REGRESSION TOWARD THE MEAN

Since $\hat{z}_Y = r z_X$, and the absolute value of $r \leqslant 1.00$, we can see that the estimated $\hat{z}$ score is always closer to its mean of zero than is the z score from which it is estimated (except in the case of perfect correlation, when the z scores must be equal). The fact that the optimal estimation or conversion formula must have this characteristic may be seen by examining the scatter plot of z converted data (Figure 2.4.2). Here we see that indeed, on the whole, the mean of the set of z_Y values corresponding to any given z_X value tends to be closer to the overall $\bar{z}_Y = 0$ than is the z_X value. (The same observation may be made regarding the tendency for $\bar{z}_X$ corresponding to any given z_Y to be closer to zero than is the z_Y value.) Further understanding of this mathematical necessity may be gained by noting that under circumstances of no correlation the $\hat{Y}$ which will minimize the sum of the squared discrepancies is $\bar{Y}$. We may also note that the size of the positive correlation between two variables reflects (among other things) the strength of the tendency for those below (above) the mean on X to also be below (above) the mean on Y. For low positive values of r we will find that, while this is true on the average, there are likely to be many exceptions, with the consequence that the $\bar{Y}$ for any given value of X may be expected to be much closer to the overall $\bar{Y}$ than X is to $\bar{X}$ (when adjustments for differences in unit size have been made).

2.6 ERROR OF ESTIMATE AND MEASURES
OF THE STRENGTH OF ASSOCIATION

In applying the regression equation $\hat{Y} = BX + A$, we have of course only approximately matched the original Y values. How close is the correspondence between the information provided about Y by X (i.e., $\hat{Y}$), and the actual Y values? Or, to put it differently, to what extent is Y associated with X as opposed to being independent of X? How much do the values of Y, as they vary, coincide with their paired X values, as they vary?

As we have noted, variability is indexed in statistical work by the sd or its square, the variance. Since variances are additive, whereas standard deviations are not, it will be more convenient to work with sd^2. What we wish to do is to partition the variance of Y (sd_Y^2) into a portion associated with X, which will be equal to the variance of the estimated $\hat{Y}$ scores, $sd_{\hat{Y}}^2$, and a remainder not associated with X, $sd_{Y-\hat{Y}}^2$, the variance of the discrepancies between the actual and the estimated $\hat{Y}$ scores. (Those readers familiar with other analysis of variance procedures may find themselves in a familiar framework here.) $sd_{\hat{Y}}^2$ and $sd_{Y-\hat{Y}}^2$ will sum to sd_Y^2, provided that $\hat{Y}$ and $Y - \hat{Y}$ are uncorrelated. Intuitively it seems appropriate that they should be uncorrelated since $\hat{Y}$ is computed from X by the optimal rule. Nonzero correlation between $\hat{Y}$ and $Y - \hat{Y}$ would indicate correlation between X (which completely determines $\hat{Y}$) and $Y - \hat{Y}$, and would indicate that our original rule was not optimal. A simple algebraic proof confirms this intuition; therefore,

(2.6.1) $$sd_Y^2 = sd_{\hat{Y}}^2 + sd_{Y-\hat{Y}}^2,$$

and we have partitioned the variance of Y into a portion determined by X and a residual portion not linearly related to X. If no linear correlation exists between X and Y, the optimal rule has us ignore X and minimize our errors of estimation by using $\overline{Y}$ as the best guess for every case. Thus we would be choosing that point about which the squared errors are a minimum and $sd_{Y-\hat{Y}}^2 = sd_Y^2$. More generally we may see that since (by Eq. 2.4.2) $\hat{z}_Y = rz_X$,

$$sd_{\hat{z}_Y}^2 = \frac{\sum (rz_X)^2}{n} = r^2 \frac{\sum z_X^2}{n} = r^2;$$

and since $sd_{z_Y}^2 = 1$, and

(2.6.2) $$sd_{z_Y}^2 = r^2 + sd_{z_Y-\hat{z}_Y}^2,$$

then r^2 is the proportion of the variance of Y linearly associated with X, and $1 - r^2$ is the proportion of the variance of Y not linearly associated with X.

It is often helpful to visualize a relationship by representing each variable as a circle. The area enclosed by the circle represents its variance, and since we have standardized each variable to a variance of 1, we will make the two circles of equal size (see Figure 2.6.1). The degree of linear relationship between the two variables may be represented by the degree of overlap between the circles (the shaded area). Its proportion of either circle's area equals r^2, and $1 - r^2$ equals the area of the nonoverlapping part of either circle. Again, it is useful to note the equality of the variance of the variables once they are standardized: the size of the overlapping and nonoverlapping areas, r^2 and $1 - r^2$, respectively, must be the same for each. If one wishes to think in terms of the variance of the original X and Y, one may define the circles as representing 100% of the variance and the overlap as representing the proportion of each variable's variance associated with the other variable. We may also see that it does not matter in this form of

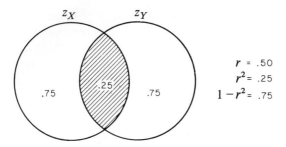

FIGURE 2.6.1 Overlap in variance of correlated variables.

expression whether the correlation is positive or negative since r^2 must be positive.

We will obtain the variance of the residual (nonpredicted) portion when we return to the original units by multiplying by sd_Y^2 to obtain

(2.6.3) $$sd_{Y-\hat{Y}}^2 = sd_Y^2(1 - r^2).$$

The standard deviation of the residuals, that is, of that portion of Y not associated with X is therefore given by

(2.6.4) $$sd_{Y-\hat{Y}} = sd_Y\sqrt{1 - r^2}.$$

For example, when $r = .50$, the proportion of shared variance $= r^2 = .25$ and .75 of sd_Y^2 is not linearly related to X. If the portion of Y linearly associated with X is removed by subtracting $BX + A$ $(= \hat{Y})$ from Y, we have a reduction from the original sd_Y to

$$sd_{Y-\hat{Y}} = sd_Y\sqrt{.75} = .866\,sd_Y.$$

We see that, in this case, although $r = .50$, only 25% of the variance in Y is associated with X, and when the part of Y which is linearly associated with X is removed, the standard deviation of what remains is .866 as large as the original sd_Y.

To make the foregoing more concrete, let us return to our v, d-s score example. The regression coefficient B_{YX} was found to be 1.20 and $A_{YX} = 6.78$ (see Section 2.5). The Y, X, and z_Y values are given in Table 2.6.1. The $Y - \hat{Y}$ values are the residuals of Y estimated from X, or the errors of estimate in the sample. Since $\hat{Y}$ is a linear transformation of X (and $\hat{X}$ is a linear transformation of Y), $r_{Y\hat{Y}}$ (and $r_{X\hat{X}}$) must equal r_{XY} $(= .828)$. The correlation between $Y - \hat{Y}$ (the residual of $\hat{Y}$ estimated from X) and $\hat{Y}$ must, as we have seen, equal zero.

When we turn our attention to the variances of the variables we see that

(2.6.5) $$\frac{sd_{\hat{Y}}^2}{sd_Y^2} = \frac{sd_{\hat{X}}^2}{sd_X^2} = r^2$$

$$= .828^2 = .6862,$$

TABLE 2.6.1

Estimated and Residual Scores for Vocabulary and Digit-Span Example

v X	d–s Y	z_X	z_Y	$\hat{X}$	$\hat{Y}$	$\hat{z}_Y$	$Y - \hat{Y}$	$\hat{Y}_w$	$Y - \hat{Y}_w$
5	12	−1.79	−1.85	5.38	12.80	−1.49	−.80	12.50	−.50
8	15	.22	−.46	7.10	16.40	.18	−1.40	16.43	−1.43
7	14	−.45	−.93	6.53	15.20	−.37	−1.20	15.13	−1.13
9	18	.90	.93	8.82	17.60	.75	.40	17.73	.27
10	19	1.57	1.39	9.39	18.80	1.30	.20	19.03	−.03
8	18	.22	.93	8.82	16.40	.18	1.60	16.43	1.57
6	14	−1.12	−.93	6.52	14.00	−.93	.00	13.83	.17
6	17	−1.12	.46	8.24	14.00	−.93	3.00	13.83	3.17
10	20	1.57	1.85	9.96	18.80	1.30	1.20	19.03	.97
9	17	.90	.46	8.24	17.60	.75	−.60	17.73	−.73
7	15	−.45	−.46	7.10	15.20	−.37	−.20	15.13	−.13
7	16	−.45	0	7.67	15.20	−.37	.80	15.13	.87
9	16	.90	0	7.67	17.60	.75	−1.60	17.73	−1.73
6	13	−1.12	−1.39	5.95	14.00	−.93	−1.00	13.83	−.83
8	16	.22	0	7.67	16.40	.18	−.40	16.43	−.43
Sum 115	240	0	0	115	240	0	0	240	0
m 7.67	16	0	0	7.67	16	0	0	16	0
sd^2 2.22	4.67	1	1	1.53	3.20	.69	1.47		1.49
sd 1.49	2.16	1	1	1.24	1.79	.83	1.21		1.22

$$r_{X z_X} = r_{Y z_Y} = r_{X\hat{Y}} = r_{z_X \hat{Y}} = r_{z_X \hat{z}_Y} = r_{Y\hat{X}} = 1.$$

$$r_{XY} = r_{z_X z_Y} = r_{\hat{X}\hat{Y}} = r_{X\hat{X}} = r_{Y\hat{Y}} = r_{z_Y \hat{z}_Y} = .828,$$

$$r^2_{Y(Y-\hat{Y})} = .3138; \quad r_{(Y-\hat{Y})\hat{Y}} = r_{(Y-\hat{Y})X} = 0.$$

which makes explicit what is meant when we say that (almost) 69% of the variance of each variable is estimated or predicted by the other. Similarly,

$$(2.6.6) \qquad \frac{sd^2_{Y-\hat{Y}}}{sd^2_Y} = r^2_{Y(Y-\hat{Y})}$$

$$= .3138 = 1 - r^2.$$

The ratio $(sd_{Y-\hat{Y}})/sd_Y = \sqrt{1 - r^2} = .560$, which is called the "coefficient of alienation," is the proportion of sd_Y that remains when that part of Y associated with X has been removed. It can also be thought of as the coefficient of noncorrelation, as r is the coefficient of correlation. The standard deviation of the residual scores is given by Eq. (2.6.4) as $sd_{Y-\hat{Y}} = sd_Y \sqrt{1 - r^2} = 2.16(.560) = 1.21$, as shown in Table 2.6.1.

An important distinction must be made here. For purposes of *describing* the sample, we have employed as measures of variability sd and sd^2, which use n as their divisor; for example, Eq. (2.2.1) is applied throughout in Table 2.6.1. When, however, the sample data are used to arrive at an estimate of variability in the *population* (symbolized by $\tilde{sd}$ and $\tilde{sd}^2$), dividing by n would result in a systematic underestimate. Instead, we divide the sum of squared deviations by the number of degrees of freedom (df), a quantity less than n. In correlation and regression analysis, this distinction is particularly important with regard to the variability (hence size) of the errors of estimate or residuals from regression, $Y - \hat{Y}$. In testing for statistical significance and setting confidence limits, it is these population estimates which come into play. For the bivariate case, the estimate of the population error or residual variance for Y is given by

(2.6.7)
$$\tilde{sd}^2_{Y-\hat{Y}} = \frac{\sum(Y-\hat{Y})^2}{n-2} = \frac{(1-r^2)\sum y^2}{n-2}$$

$$= sd^2_{Y-\hat{Y}}\left(\frac{n}{n-2}\right),$$

and its square root, usually called the "standard error of estimate" by

(2.6.8)
$$\tilde{sd}_{Y-\hat{Y}} = \sqrt{\frac{\sum(Y-\hat{Y})^2}{n-2}} = \sqrt{\frac{(1-r^2)\sum y^2}{n-2}}$$

$$= sd_{Y-\hat{Y}}\sqrt{\frac{n}{n-2}}.$$

Here, $df = n - 2$. Thus, while the sd of the *sample* residual Y values was found to be 1.21 by Eq. (2.6.4), the estimated sd of the *population* residuals or standard error of estimate is, by Eq. (2.6.8), $\tilde{sd}_{Y-\hat{Y}} = 1.21\sqrt{15/13} = 1.30$.

Finally, $\hat{Y}_w$ in Table 2.6.1 has been computed to demonstrate what happens when any other regression coefficient or weight is used. In this case we take the slightly incorrect value $B_w = 1.3$ as compared to $B_{YX} = 1.20$ (the A value has been altered to keep $\hat{Y}_w$ centered on $\bar{Y}$). The resulting sd^2 for the sample residuals $Y - \hat{Y}_w = 1.49$, indicating larger errors, that is, greater spread in both directions from the zero mean, than with the optimal least-squares regression coefficient for which the residual variance is 1.47. The reader is invited to try any other value to determine that the residuals will in fact always be larger than with the computed value of B_{YX}.

2.7 SUMMARY OF DEFINITIONS AND INTERPRETATIONS

Product moment r is the rate of change (linear) in z_Y per unit change in z_X (and vice versa) which best fits the data in the sense of minimizing the squared discrepancies between the estimated and actual scores.

r^2 is the proportion of variance in Y associated with X (and vice versa).

B_{YX} is the regression coefficient of Y on X. Using the original raw units, the rate of change in Y per unit change in X, again best fitting in the least squares sense.

Symmetrically, B_{XY} is the rate of change in X per unit change in Y (the slope of the line estimating X from Y). Note that B_{YX} is not equal to B_{XY} unless sd_X = sd_Y; however, $r_{XY} = r_{YX}$.

A is the regression constant which serves to adjust for differences in means, giving the predicted value of the dependent variable when the independent variable's value is zero.

The coefficient of alienation, $\sqrt{1 - r^2}$, is the proportion of sd_Y remaining when that part of Y associated with X has been subtracted; that is, $sd_{Y-\hat{Y}}/sd_Y$.

The standard error of estimate, $\widetilde{sd}_{Y-\hat{Y}}$, is the estimated population sd of the residuals or errors of estimating Y from X.

2.8 SIGNIFICANCE TESTING OF CORRELATION AND REGRESSION COEFFICIENTS

In most circumstances in which r and B are determined, the intention of the investigator is to provide valid inferences from the data at hand to some larger universe of potential data—from the statistics obtained on a sample to the parameters of the population from which it is drawn. Since random samples from a population will not yield values for r and B that exactly equal the population values, statistical methods have been developed to determine the confidence with which such inferences can be accepted. In this section we describe the methods of significance testing—determining the risk that we have rejected a true null hypothesis. In Section 2.9 we present methods of assessing the risk of making the other kind of error—failing to reject a false null hypothesis. In Section 2.10 we provide methods of determining the limits within which we may expect a population value to fall with a specified degree of confidence.

2.8.1 Assumptions Underlying the Significance Tests

It is clear that no assumptions are necessary for the computation of correlation, regression, and other associated coefficients or their interpretation when they are used to describe the available data. However, the most interesting and useful applications of r and B are when they are statistics calculated on a random sample from some population and inferences about the relationship between the variables in the population are desired. As in most circumstances in which statistics are used inferentially, the addition of certain assumptions about the characteristics of the population (when valid) substantially increases the number of useful inferences which can be drawn. Fortunately, the available evidence suggests that even fairly substantial departure from the assumptions will frequently result in little error of inference when the data are treated as if the assumptions were valid.

Probably the most generally useful set of assumptions are those which form what has been called the fixed linear regression model (Binder, 1959). This model assumes that the two variables have been differentiated into an independent variable X and a dependent variable Y. Values of X are assumed to be "fixed" in the analysis of variance sense, that is, selected by the investigator rather than sampled from some population of X values. Values of Y are assumed to be sampled for each of the selected values of X. The residuals from the mean value of Y for each value of X are assumed to be normally distributed with equal variances in the population. It should be noted that no assumptions about the shape of the distribution of X and the total distribution of Y per se are necessary, and that, of course, the assumptions are made about the population and not about the sample. This model, extended to multiple regression, is used throughout the book.

Fortunately, even this rather liberal model can be somewhat violated with typically little risk of error in conclusions about the presence or absence of a linear relationship. A number of studies (Binder, 1959; Boneau, 1960; Cochran, 1947; Donaldson, 1968) have demonstrated the robustness of the t and F tests to failure of distribution and other assumptions, although it must be cautioned that the probabilities (significance) calculated under such circumstances may be somewhat over- or underestimated.

Assumption failure of the heteroscedastic type—circumstances in which the residuals have grossly unequal variances at different values of X—suggests the need for improving the prediction either by an appropriate transformation of X or Y or by including additional variables in the equation.

2.8.2   *t* Test for Significance of *r* and *B*

Usually, the first test of interest is to determine the probability that the population r is different from zero; that some linear relationship exists. To test the null hypothesis that the population r is zero, substitute the sample r and n in the following formula to find Student's

(2.8.1)
$$t = \frac{r\sqrt{n-2}}{\sqrt{1-r^2}}.$$

The resulting t is looked up in a standard table, with $n-2$ degrees of freedom (df) to determine whether the probability (P) is "sufficiently" small that an r as large as the one observed (either positively or negatively, hence "two-tailed") would be obtained from a random sample of size n drawn from a population whose r is zero. If so, the null hypothesis is rejected, and the sample r is deemed "statistically significant." The criterion for "sufficiently" small, α, is conventional; by far the most frequently used are the $\alpha = .05$ and (more stringent) $\alpha = .01$ criteria for statistical significance. Appendix Table A gives, for varying df, the values of t necessary for statistical significance at $\alpha = .01$ and at $\alpha = .05$ (two-tailed). Thus, if the t computed from Eq. (2.8.1) exceeds the tabled value for $df = n-2$ at the prescribed α, then $P <$ (is less than) α, and the null

TABLE 2.8.1

Analysis of Variance of Simple Regression of Y on X

Source	Sum of squares	df	Mean square
Regression	$\Sigma(\hat{Y}-\bar{Y})^2$ $= B\ \Sigma\ xy$ $= r^2\ \Sigma\ y^2$	1	$r^2\ \Sigma\ y^2$
Residual (error)	$\Sigma(Y-\hat{Y})^2$ $= \Sigma\ y^2 - r^2\ \Sigma\ y^2$ $= (1-r^2)\ \Sigma\ y^2$	$n-2$	$\dfrac{(1-r^2)\ \Sigma\ y^2}{n-2}$

$$F = \frac{\text{regression mean square}}{\text{residual mean square}} = r^2\ \Sigma\ y^2 \Big/ \frac{(1-r^2)\ \Sigma\ y^2}{n-2} = \frac{r^2(n-2)}{1-r^2}.$$

hypothesis is rejected "at the α significance level." In our fixed model, the population to which this generalization is confined is strictly the population consisting of the exact values of X as occur in the sample. However, typically, generalization to the population of X values covering the same range as the sample can be made.

Since F for one df in the numerator equals t^2 (with equal error df), it is useful to consider this significance test as carried out by analysis of variance. Since r^2 is the proportion of Y variance associated with X, we may present the data in an analysis of variance format as seen in Table 2.8.1.

It may be deduced from the presence of B_{YX} in one form of the sum of squares for regression that the F test (and consequently the t test) for significance of the correlation coefficient also tests whether the regression coefficient is nonzero in the population. Indeed, B_{YX} can be zero if and only if r equals zero. Also, note that the error or residual mean square is simply the squared standard error of estimate of Eq. (2.6.7).

2.8.3 Fisher's z' Transformation and Comparisons between Independent rs

Although t and F tests are appropriate for testing the significance of the departure of r from 0, a different approach is necessary for estimating the confidence limits of a population r. The reason for this is that the sampling distribution of nonzero correlations is skewed, the more so as the departure from zero increases. To sidestep this problem, R. A. Fisher developed the z' transformation of r with a sampling distribution which is nearly normal and a standard error which depends only on n:

(2.8.2) $z' = \tfrac{1}{2}[\ln(1+r) - \ln(1-r)],$

where ln is the natural (base e) logarithm. Appendix Table B gives the r to z'

transformation directly, with no need for computation. The standard error of z' is given by

(2.8.3)
$$sd_{z'} = \frac{1}{\sqrt{n-3}}.$$

Therefore, in testing the significance of a departure of some obtained r from a hypothetical population r, one need only divide the difference between their z' equivalents by the standard error to obtain a normal curve deviate:

(2.8.4)
$$z = \frac{z'_s - z'_h}{sd_{z'}},$$

where z (not to be confused with z') is the standard normal curve deviate, z'_s the transformed sample r, and z'_h the transformed hypothetical r. The tail area beyond z is used in the usual way to obtain the P value for the significance test (Appendix Table C).

For example, a study determines the correlation between education and income for a sample of 103 adult Black men to be .47. The null hypothesis is that this sample value comes from a population whose r is that of the entire male adult population of the United States, which is known to be .63. We can look up the tabled z' values for $r_s = .47$ and $r_h = .63$ which are .510 and .741, respectively. The standard error is computed as

$$sd_{z'} = \frac{1}{\sqrt{103-3}} = .10,$$

and the normal curve deviate as

$$z = \frac{.510 - .741}{.10} = -2.31.$$

Turning to the normal curve table for the closest tabled value (Appendix Table C), we find the $z = 2.30$ gives us .011 for the area in the tail. Doubling this value gives us approximately .022 as the two-tailed probability (P) level, which would lead us to reject (since P is less than $\alpha = .05$) the null hypothesis that the relationship between income and education is the same for the population of Black men which we sampled as it is for the entire male population of the United States, and we conclude that the correlation between education and income is smaller for Black than for White adult males in the United States.

Similarly, if one wishes to test the significance of the difference between correlation coefficients obtained on two different random samples one may compute the normal curve deviate

(2.8.5)
$$z = \frac{z'_1 - z'_2}{\sqrt{\dfrac{1}{n_1-3} + \dfrac{1}{n_2-3}}}.$$

TABLE 2.8.2

χ^2 Test for the Homogeneity of the Correlations between Two Measures
of Social Participation in 10 Schools

School	r_i	n_i	z'_i	z'^2_i	$n_i - 3$	$(n_i - 3)(z'_i)$	$(n_i - 3)(z'^2_i)$
A	.72	67	.908	.824	64	58.11	52.77
B	.41	93	.436	.190	90	39.24	17.11
C	.57	73	.648	.420	70	45.36	29.39
D	.53	98	.590	.348	95	56.05	33.07
E	.62	82	.725	.526	79	57.28	41.52
F	.21	39	.214	.046	36	7.70	1.65
G	.68	91	.829	.687	88	72.95	60.48
H	.53	27	.590	.348	24	14.16	8.35
I	.49	75	.536	.287	72	38.59	20.69
J	.50	49	.549	.301	46	25.25	13.86
		694	6.025		664	414.70	278.89

$$\chi^2 = 278.89 - \frac{414.70^2}{664} = 19.893*, \quad df = k - 1 = 9$$

$*P < .05.$

The z' transformation is also useful in circumstances in which more than two independent correlation coefficients are to be compared to determine whether they may be considered homogeneous (equal in the population). In this case the test for homogeneity employs the χ^2 (chi-squared) distribution for $k - 1$ degrees of freedom, where k = the number of independent sample coefficients being compared:

$$(2.8.6) \qquad \chi^2 = \sum (n_i - 3)z'^2_i - \frac{[\sum (n_i - 3)(z'_i)]^2}{\sum (n_i - 3)},$$

where the summation is over the k samples. For example, a researcher has calculated correlation coefficients between two measures of social interaction obtained on random samples from 10 elementary schools (see Table 2.8.2). The research question posed is whether this can be considered a homogeneous set of correlation coefficients, that is, whether the values obtained would be expected on the basis of random sampling from a single normal population. We look up the z' values for each of the rs, and find the χ^2 value to be 19.893. Consulting a χ^2 table, we find this value for $df = k - 1 = 9$ meets the .05 α criterion level (16.92). We therefore conclude that the two measures of social participation are not equally correlated among these schools.

2.8.4 The Significance of the Difference between Independent Bs

Since $B_{YX} = r \, sd_Y / sd_X$, it is apparent that we may find circumstances in which rs obtained on two independent samples may be significantly different from each other, yet the regression coefficients may not, or vice versa. This can happen when the larger r occurs in combination with a proportionately larger sd_X, for example. That such findings may be relatively common will be understood better when the problem of restriction of range is considered. At this point, it should be understood that although the test of significance of the departure from 0 is shared by a correlation coefficient and its associated regression coefficients, they do not share the same test in comparisons with their respective counterparts from another sample. In order to test for the significance of the difference between B_E and B_F (that is, B_{YX}s from samples E and F),

$$(2.8.7) \qquad t = \frac{B_E - B_F}{\sqrt{\dfrac{\sum(Y_E - \hat{Y}_E)^2 + \sum(Y_F - \hat{Y}_F)^2}{n_E + n_F - 4} \times \dfrac{\sum x_E^2 + \sum x_F^2}{\sum x_E^2 \cdot \sum x_F^2}}}$$

with $df = n_E + n_F - 4$, and refer to Appendix Table A. Much later (Chapter 8), we will see that this can be accomplished routinely as a test of significance of an interaction.

2.8.5 The Significance of the Difference between Dependent rs

Suppose the question is asked whether some variable X correlates with Y to a significantly greater degree than does another variable V. The three variables may be measured on some sample and the correlation coefficients, r_{XY}, r_{XV}, and r_{VY} obtained. It is not appropriate to determine the significance of the difference between r_{XY} and r_{XV} by way of the Fisher z' and Eq. (2.8.5) since the coefficients have not been determined on independent samples. Just as in the t test for means from dependent or matched samples, it is necessary to take into account the correlation over samples between the coefficients being tested due to the fact that they come from the same sample. The resulting formula yields a t for $n - 3$ degrees of freedom:

$$(2.8.8) \qquad t = \frac{(r_{XY} - r_{VY})\sqrt{(n-3)(1 + r_{XV})}}{\sqrt{2(1 - r_{XY}^2 - r_{VY}^2 - r_{XV}^2 + 2r_{XY}r_{XV}r_{VY})}}.$$

For example, suppose we have investigated the relationship between negative attitude towards urban living (Y), interest in outdoor activities (V), and involvement with young children (X) in a sample of 147 suburban dwellers. We would like to determine whether, in the population, the involvement with young children is more highly related to negative attitude towards urban living (r_{XY} = .39) than is interest in outdoor activities (r_{VY} = .20). We also find r_{XV} = .21, and compute

$$t = \frac{(.39 - .20)\sqrt{(147 - 3)(1 + .21)}}{\sqrt{2\,[1 - .39^2 - .20^2 - .21^2 + 2(.39)(.20)(.21)\,]}}$$

$$= \frac{2.508}{1.262} = 1.987,$$

which, for $n - 3 = 144$ df, is significant at the two-tailed 5% α level (Appendix Table A).

2.9 STATISTICAL POWER

2.9.1 Introduction

In the last section we presented methods of appraising sample data in regard to α, the risk of mistakenly rejecting the null hypothesis when it is true, that is, drawing a spuriously positive conclusion (Type I error). We now turn our attention to methods of determining β, the probability of failing to reject the null hypothesis when it is false (Type II error), and ways in which β can be controlled in planning research.

Any given statistical test of a null hypothesis can be viewed as a complex relationship among the following four parameters:

1. The *power* of the test, defined as $1 - \beta$ (the probability of rejecting the null hypothesis).
2. The region of rejection of the null hypothesis as determined by the α level and whether the test is one-tailed or two-tailed. As α increases, power increases.
3. The sample size n. As n increases, power increases.
4. The magnitude of the effect in the population, or the degree of departure from the null hypothesis. The larger this is, the greater the power.

These four parameters are so related that when any three of them are fixed, the fourth is completely determined. Thus, when an investigator decides for a given research plan his significance criterion α and the sample size he will use, the power of his test is determined. However, the investigator does not in general know what this power is, since he does not know the magnitude of the effect size (ES) in the population.

There are three general strategies for determining the size of the population effect which a research is trying to detect:

1. To the extent that studies which have been carried out by the current investigator or others are closely related to the present investigation, the ESs found in these studies reflect the magnitude which can be expected. Thus, if a review of the relevant literature reveals rs ranging from .32 to .43, the population ES in the current study may be expected to be somewhere in the vicinity of these values. If the investigator wishes to be conservative he may wish to determine the power to detect a population r of .25 or .30.

2. In some research areas an investigator may posit some minimum population effect that would have either practical or theoretical significance. Investigator A determines that unless the population $r = .5$, the importance of the relationship is insufficient to warrant a change in the policy or operations of the relevant institution. Another investigator may decide that a population $r = .10$ would

have a material import for the adequacy of the theory within which the experiment has been designed, and thus would wish to plan the experiment so as to detect such an ES.

3. A third strategy in deciding what ES values to use in determining the power of a study is to use certain suggested conventional definitions of "small", "medium" and "large" effects (Cohen, 1969). These conventional ES values may be used either by choosing one of these values (for example, the conventional "medium" effect size is a population r of .30) or by determining power for all three population ESs. If the latter strategy is taken, the investigator would then make a revision in his research plan according to his estimation of the relevance of the various ESs to his problem.

The point in doing an analysis of the power of a given research plan is that when the power turns out to be insufficient the investigator may decide to revise these plans, or even drop the investigation entirely if such revision is impossible. Obviously, since little or nothing can be done after the investigation is completed, determination of statistical power is of primary value as a preinvestigation procedure. If power is found to be insufficient, the research plan may be revised in ways which will increase it, primarily by increasing n (or possibly, by increasing α). A more complete general discussion of the concepts and strategy in statistical power analysis may be found in Cohen (1965, 1969), and further discussion of power analysis in MRC in Sections 3.8 and 4.5.

2.9.2 Power of Tests of the Significance of r and B

When the null hypothesis to be tested is that the population $r = 0$, r itself is the appropriate measure of ES. Thus the investigator may proceed by determining the population r which he expects (or wishes to allow for), the α criterion he plans to use in the determination of significance, and the sample size included in his plan. Having made such determinations, he may then enter tables with these values and look up the resulting power, $1 - \beta$, or probability that his statistical test will be significant if he has correctly assessed the population value of r. For example, an investigation is planned for which a population $r = .30$ is posited, either on the basis of previous work in the area or because .30 is a conventional definition of a "medium" effect size. It is planned to test the null hypothesis with α (two-tailed) set at .05, and to gather data on 50 subjects. The investigator enters the r power table for $\alpha = .05$ (Appendix Table F.2) with $r = .30$ and $n = 50$ and reads off power = .57. Thus, if the population $r = .30$, the odds are only a little better than 50–50 that the statistical test will be significant and the null hypothesis rejected. If these odds sufficiently distress the investigator (as they should) he may determine that if he increases n to 80 the power will be increased to .78, and that with $n = 84$, power = .80. It has been proposed (Cohen, 1965; 1969) that much as $\alpha = .05$ is used as a convention for significance, power = $1 - \beta = .80$ be used as a convention for power (see Section 4.5.5). Thus a decision may be made to increase the sample size to 84 to effect the necessary increase in power to .80.

The form of power analysis which the above suggests and which is frequently very useful is the direct determination of the necessary sample size (n^*) to attain a given desired power $(1 - \beta)$ to detect a specified population r for a specified α. Appendix Tables G.1 and G.2 are designed for this purpose. For example, with (as before) a population $r = .30$, at $\alpha = .05$, what sample size do we need (n^*) for power to be .80? Entering Table G.2 (for $\alpha = .05$) for column $r = .30$ and (row) desired power = .80, we read out $n^* = 84$ (as before). For power under these conditions to be .90, $n^* = 112$, that is, an increase of 28 cases (or $28/84 = 33\%$) over the sample size necessary for power to be .80 (for these conditions of r and α). If the more stringent $\alpha = .01$ is used in the above problem, for power = .80, $n^* = 124$ and for power = .90, $n^* = 157$ (Appendix Table G.1).

The importance of power analysis as part of research planning cannot be stressed too heavily. The thoughtful use of Appendix Tables F and G will greatly facilitate the related decisions about sample size, α, and the power to which one can aspire during the planning of a research, and to some degree, the interpretation of its results afterwards (Cohen, 1973b).

Conventional magnitudes of r corresponding to small, medium, and large ES that have been suggested as appropriate at least for many areas of psychological investigation are $r = .10, .30,$ and $.50$, respectively. When there is reason to believe that the population ES is small, i.e. $r = .10$, rather large values of n are required—for $\alpha = .05$ and $1 - \beta = .80$, n must be $n^* = 783$ (Appendix Table G.2), and if more stringent (lower) α or β risk levels are desired, for example, $\alpha = .01$ and power = .95, even a sample of 1000 is insufficient to accomplish the goal $(n^* = 1790$, Appendix Table G.1).

Since the statistical test of the H_0 (null hypothesis): $r = 0$ is simultaneously a test of $H_0: B = 0$ (see Section 2.8.2), the power analysis for B may be carried out by means of its associated r.

2.9.3 Power Analysis for Other Statistical Tests Involving r

We have described among other statistical tests involving r, the test that a population r has some specified (nonzero) value, and the test of the hypothesis that two population rs are equal. Both of these tests involve z' and require different definitions of ES and power and n^* tables from those used above. Since neither of these tests arise with much frequency, we conserve the space necessary for their exposition, and instead refer the reader to Cohen (1969), which is a handbook of power analysis that fully treats these and most other cases of power analysis encountered in practice.

2.10 DETERMINING STANDARD ERRORS
AND CONFIDENCE INTERVALS

2.10.1 SE and Confidence Intervals for r and B

If confidence limits for a sample r are desired, the Fisher z' transformation is applied, and the confidence limits for z' determined by applying the appropriate multiple of the standard error of z'. For example, for the 95% confidence interval for an r of .32 in a sample of $n = 67$ cases, we first determine from Appendix Table B that its $z' = .332$. From the normal curve table (Appendix Table C), we find that 1.96 standard units sets off 47.5% of the area on either side of the mean. We compute the standard error of z' of Eq. (2.8.3) as

$$sd_{z'} = \frac{1}{\sqrt{67-3}} = .125.$$

This is multipled by 1.96, and laid off on both sides of z' to give the 95% confidence limits:

$$.332 \pm 1.96(.125) = .332 \pm .245 = .087 \longleftrightarrow .577.$$

These are the 95% limits for z'. Now converting back from z' to r from Appendix Table B gives

95% limits for r: $.09 \longleftrightarrow .52.$

Given $r = .32$ in a random sample of $n = 67$, we can have 95% confidence that the interval .09 to .52 contains the population r, that is, this procedure will include the population r 95% of the time.

Confidence limits may be determined for B by adding and subtracting an appropriate multiple of the standard error of B,

(2.10.1) $$SE_{B_{YX}} = \frac{\tilde{sd}_{Y-\hat{Y}}}{sd_X\sqrt{n}},$$

or, given as a function of the original sds and r:

(2.10.2) $$SE_{B_{YX}} = \frac{sd_Y}{sd_X} \sqrt{\frac{1-r^2}{n-2}}.$$

We return to our example of v and d-s scores, where $r = .828$, $sd_Y = 2.16$, $sd_X = 1.49$, and $n = 15$. Then,

$$SE_{B_{YX}} = \frac{2.16}{1.49} \sqrt{\frac{1-.828^2}{15-2}} = .225.$$

Since SE_B is t distributed with $df = n - 2$, the appropriate multiplier for 95% confidence limits is the t at $\alpha = .05$ for $df = 13$, which is 2.160 (Appendix Table A). Therefore,

$$1.20 \pm 2.160 (.225) = .71 \longleftrightarrow 1.69$$

are the 95% confidence limits for $B_{YX} = 1.20$, indicating that we can be 95% confident that the population regression coefficient falls between .71 and 1.69. Note that no use of the Fisher z' is made here.

Confidence limits can be very useful in data analysis since they give directly the range of probable (for example, 95%) values for the unknown population parameter. They thus provide us with information about the precision of our sample values as point estimates of their analogous population parameters. (Note, incidentally, that for the small sample of the example, precision is rather poor.) Also, they incorporate information about the status of tests of null hypotheses: thus, if the 95% interval does not include zero, the null hypothesis is rejected at α_2 (two-tailed) = .05. Indeed, *all* values outside the interval can be rejected and all values inside the interval can *not* be rejected (at α_2 = .05). We have used the 95% interval as an example, obviously others, more or less stringent, can be and are used, depending on the purpose of the estimate.

It should be noted that such limits, and particularly the direct interpretation of the standard error of estimate ($\widetilde{sd}_{Y-\hat{Y}}$) will be misleading under circumstances of heteroscedasticity and/or curvilinear relationships. Heteroscedasticity exists when the scatter of Y values about the regression line in the population is not equal for all values of X. When this occurs, it will obviously be the case that the expected error in estimating Y from X will vary. The computed standard error of estimate in heteroscedastic regressions is a kind of average which overestimates the error in some regions of X and underestimates it in others.

It is particularly misleading when the trend of a relationship is actually curvilinear and the simple linear (straight-line) regression equation (2.4.5) is used: for some values of X the computed $\hat{Y}$ value will be generally above the bulk of the observed Y values and for other values generally below. This "lack of fit" not only misrepresents the shape of the relationship, but swells the standard error of estimate relative to what it would be for a proper curvilinear fit. If the possibility of curvilinearity exists, it is prudent to inspect the bivariate scatterplot before proceeding to linear correlation and regression. (Curvilinear relationships are the central topic of Chapter 6.)

2.10.2 Confidence Limits on a Single $\hat{Y}_o$ Value

The SE_B should be clearly distinguished from $\widetilde{sd}_{Y-\hat{Y}}$, and each of these standard errors must be further distinguished from the standard error of a given $\hat{Y}_o$, estimated for a new observation on which an X value, X_o, is available.

To understand the difference between this latter value and the standard error of estimate, it is useful to realize that whatever sampling error has occurred in estimating the population regression coefficient by the sample B_{YX} will have more serious consequences for X values which are more distant from $\bar{X}$ than for those near $\bar{X}$.

For example, for the sake of simplicity let us assume that both X and Y have means of zero and sds of 1. Let us further suppose that the B_{YX} value as determined from our sample is .20, whereas the actual population value is .25. For new cases which come to our attention with X_o values = .1, we will estimate $\hat{Y}$ at .02 (= $B_{YX}X_o$) when the actual population mean value of Y for all X_o = .1 is .025, a relatively small error of .005. On the other hand, new values of X_o = 1.0 will yield estimated $\hat{Y}$ values of .20 when the actual mean value of Y for all X = 1 is .25, the error (.05) being ten times as large.

When a newly observed X_o is to be used to estimate $\hat{Y}_o$ we may determine the standard error and thus confidence limits for this $\hat{Y}_o$. The accuracy of such confidence limits is dependent on the validity of the assumption of linearity and equally varying normal distributions of Y values across all values of X. Under such conditions the standard error of an estimated $\hat{Y}_o$ value will be

$$(2.10.3) \qquad \widetilde{sd}_{Y_o - \hat{Y}_o} = \widetilde{sd}_{Y - \hat{Y}} \sqrt{1 + \frac{1}{n} + \frac{(X_o - \bar{X})^2}{n\, sd_X^2}}.$$

Since $\widetilde{sd}_{Y - \hat{Y}}$, the standard error of estimate, is a constant for all values of X, the magnitude of the error is a function of the magnitude of the difference between X_o and $\bar{X}$, that is, the extremeness in either direction of X_o.

The same equation when the X_o value is standardized (z_o) makes the effect of departure from $\bar{X}$ more obvious:

$$(2.10.4) \qquad \widetilde{sd}_{Y_o - \hat{Y}_o} = \widetilde{sd}_{Y - \hat{Y}} \sqrt{1 + \frac{1}{n} + \frac{z_o^2}{n}}.$$

For example, return to the vocabulary-digit span study and assume that we now wish to determine the confidence interval for an estimated d-s score of a newly tested subject with a v score of 9. From Eq. (2.6.8), $\widetilde{sd}_{Y - \hat{Y}} = 1.30, n = 15$ and $z_9 = (9 - 7.67)/1.49 = .90$. Substituting in Eq. (2.10.4),

$$\widetilde{sd}_{Y_9 - \hat{Y}_9} = 1.30 \sqrt{1 + \frac{1}{15} + \frac{.90^2}{15}} = 1.38.$$

Noting that when $X = 9$, $\hat{Y}_9 = 17.60$, we determine the 95% confidence interval by multiplying 1.38 by the t value at $\alpha = .05$ for $df = n - 2 = 13$, which is 2.160 (Appendix Table A), that is, 1.38(2.160) = 2.97. When the necessary assumptions of homoscedasticity and normality are valid, we may be 95% confident that this individual's d-s score will be included within the interval 17.60 ± 2.97, or from 14.63 to 20.57, a rather disappointingly large interval which spans the distance from nearly two-thirds of a sd below the Y mean to the top of the Y range. One can also see that when one is predicting $\hat{Y}$ from an X value which equals $\bar{X}$, the equation simplifies and the standard error of the predicted value becomes $\widetilde{sd}_{Y - \hat{Y}} \sqrt{(n + 1)/n}$.

2.11   FACTORS AFFECTING THE SIZE OF *r*

2.11.1   The Distributions of *X* and *Y*

Since $r = \pm 1.00$ only when each $z_X = z_Y$ or $-z_Y$, it can only occur when the shapes of the frequency distributions of X and Y are exactly the same (or exactly opposite for $r = -1.00$). The greater the departure from distribution similarity, the more severe will the restriction be on the maximum possible r. In addition, as such distribution discrepancy increases, departure from homoscedas-

TABLE 2.11.1
Bivariate Distribution of Experimental and Self-Report Measures
of Conservative Tendency

		Experimental		
		Safe players	Risk takers	Total:
Self-report	Liberal	50	40	90
	Conservative	10	0	10
	Total:	60	40	100

ticity—equal error for different predicted values—must also necessarily increase. The decrease in the maximum possible value of (positive) r is especially noticeable under circumstances in which the two variables are skewed in opposite directions. One such common circumstance occurs when the two variables being correlated are each dichotomies. When the variables have very discrepant proportions, it is not possible to obtain a large positive correlation.

For example, suppose that a group of subjects has been classified into "risk takers" and "safe players" on the basis of behavior in an experiment, resulting in 40 risk takers and 60 safe players. A correlation is computed between this dichotomous variable and self classification as "conservative" versus "liberal" in a political sense, with 10 subjects identifying themselves as conservative (Table 2.11.1). Even if all political conservatives were also safe players in the experimental situation, the correlation will be only [by Eq. (2.3.6)] :

$$r_\phi = \frac{400 - 0}{\sqrt{90 \cdot 10 \cdot 40 \cdot 60}} = .272.$$

Under circumstances in which the dichotomy represents a truly discrete variable, such as male–female, or the presence or absence of some characteristic for which "partial presence" or presence to some degree is not possible, there is no problem in correctly interpreting the correlation coefficient (point biserial or phi) or its associated variance-accounted-for (r^2) estimate. It is, of course, necessary to keep in mind that the population to which we can generalize is one which is characterized by a similar distribution to that of the sample. However, whenever the concept underlying the measure is logically continuous or quantitative[5] —as in the above example of risk taking and liberal–conservative—it is

---

[5] "Continuous" implies a variable on which infinitely small distinctions can be made; "quantitative" is more closely aligned to real measurement practice in the social sciences, implying an ordered variable of many, or at least several, possible values. Theoretical constructs may be taken as continuous, but their measures will be quantitative in the above sense.

highly desirable to measure the variables on a many-valued scale. One effect of this will be to increase the opportunity for valid discrimination of individual differences (see Section 2.11.2). To the extent that the measures are similarly distributed, the risk of underestimating the relationship between the conceptual variables will be reduced.

The Biserial r

When the only available measure of some construct X is a dichotomy, d_X, an investigator may wish to know what the correlation would be between this construct and some other quantitative variable, Y. For example, X may be ability to learn algebra, which we measure by d_X, pass–fail. If one can assume that this "underlying" continuous variable X is normally distributed, and that the relationship with Y is linear, an estimate of the correlation between X and Y can be made, even though only d_X and Y are available. The correlation thus produced is called a biserial correlation coefficient and is given by

$$(2.11.1) \qquad r_{\mathrm{b}} = \frac{(\overline{Y}_p - \overline{Y}_q)pq}{h\,(sd_Y)} = r_{pb}\,\frac{\sqrt{pq}}{h}\,,$$

where $\overline{Y}_p$ and $\overline{Y}_q$ are the Y means for the two points of the dichotomy, p and q ($= 1 - p$) are the proportions of the sample at these two points, and h is the ordinate (height) of the standard unit normal curve at the point at which its area is divided into p and q portions (see Appendix Table C).

For example, we will return to the data presented in Table 2.3.1, where the point biserial r was found to be $-.707$. We now take the dichotomy to represent, not the presence or absence of an experimentally determined stimulus, but rather gross (1) versus minor (0) naturally occurring interfering stimuli as described by the subjects. This dichotomy is assumed to represent a continuous, normally distributed variable. The biserial r between stimulus and task score will be

$$r_{\mathrm{b}} = \frac{(66 - 69.5)(.428)(.572)}{.392\,(2.45)} = -.893,$$

where .392 is the height of the ordinate at the .428, .572 break, found by linear interpolation in Appendix Table C.

The biserial r of $-.893$ may be taken to be an estimate of the product-moment correlation that would have been obtained had X been a normally distributed continuous measure. It will always be larger than the corresponding point biserial r and, in fact, may even nonsensically exceed 1 when the Y variable is not normally distributed. When there is no overlap between the Y scores of the two groups, the biserial r will be at least 1. It will be approximately 25% larger than the corresponding point biserial r when the break on X is $.50 - .50$. The ratio of $r_{\mathrm{b}}/r_{\mathrm{pb}}$ will increase as the break on X is more extreme; for example with a break of $.90 - .10$, the biserial will be about two-thirds larger than r_{pb}.

The significance of r_{b} is best assessed by the t test on the point biserial, or equivalently, the t test on the difference between the Y means corresponding to the two points of d_X.

Tetrachoric r

As we have seen, when the relationship between two dichotomies is investigated, the restriction on the maximum value of r_ϕ when their breaks are very different can be very severe. Once again, we can make an estimate of what the linear correlation would be if the two variables were continuous and normally distributed. Such an estimate is called the tetrachoric correlation. Since the formula for the tetrachoric correlation involves an infinite series, and even a good approximation is a laborious operation, tetrachoric rs are usually obtained by means of diagrams (Chesire, Saffir, & Thurstone, 1933). One enters these nomographs with the 2 X 2 table of proportions and reads off tetrachoric r. Tetrachoric r will be larger than the corresponding phi coefficient and the issues governing their interpretation and use are the same as for biserial and point biserial rs.

Caution should be exercised in the use of biserial and tetrachoric correlations, particularly in multivariate analyses. Remember that they are not observed correlations in the data, but rather hypothetical ones depending on the normality of the distributions underlying the dichotomies.

2.11.2 The Reliability of the Variables

In most research in the behavioral sciences, the concepts which are of ultimate interest and which form the theoretical foundation for the study are only indirectly and imperfectly measured in practice. Thus, typically, interpretations of the correlations between variables as measured should be carefully distinguished from the relationship between the constructs or conceptual variables found in the theory.

The reliability of a variable (r_{XX}) may be defined as the correlation between the variable as measured and another equivalent measure of the same variable. In standard psychometric theory, the square root of the reliability coefficient $(\sqrt{r_{XX}})$ may be interpreted as the correlation between the variable as measured by the instrument or test at hand and the "true"—error-free—score. Since true scores are not themselves observable, a series of techniques has been developed to estimate the correlation between the obtained scores and these (hypothetical) true scores. These techniques may be based on correlations among items, between items and the total score, between other subdivisions of the measuring instrument, or between alternative forms. They yield a reliability coefficient which is an estimate (based on a sample) of the population reliability coefficient.[6] This coefficient may be interpreted as an index of how well the test or

---

[6] Since this is a whole field of study in its own right, no effort will be made here to describe any of its techniques, or even the theory behind the techniques, in detail. Two excellent sources of such information are Lord and Novick (1968) and Nunnally (1967). We return to this topic in Section 9.5.

measurement procedure measures whatever it is that it measures. This issue should be distinguished from the question of the test's *validity,* that is, the question of whether *what* it measures is what the investigator intends that it measure.

The discrepancy between an obtained reliability coefficient and a perfect reliability of 1.00 is an index of the relative amount of measurement error. Each observed score may be thought of as composed of some "true" value plus a certain amount of error:

(2.11.2) $$X = X^* + e.$$

These error components are assumed to have a mean of zero, and to correlate zero with the true scores and with true or error scores on other measures. Measurement errors may come from a variety of sources, such as errors in sampling the domain of content, errors in recording or coding, errors introduced by grouping or an insufficiently fine system of measurement, errors associated with uncontrolled aspects of the conditions under which the test was given, errors due to short or long term fluctuation in individuals' true scores, errors due to the (idiosyncratic) influence of other variables on the individuals' responses, etc.

For the entire set of scores, the reliability coefficient may be seen to equal that proportion of the observed score variance which is true score variance

(2.11.3) $$r_{XX} = \frac{sd_{X^*}^2}{sd_X^2}.$$

Since, as we have stated, error scores are assumed not to correlate with anything, r_{XX} may also be interpreted as that proportion of the measure's variance which is available to correlate with other measures. Therefore, the correlation between the observed scores (X and Y) for any two variables will be numerically smaller than the correlation between their respective unobservable true scores (X^* and Y^*). Specifically:

(2.11.4) $$r_{XY} = r_{X^*Y^*}\sqrt{r_{XX}r_{YY}}.$$

Researchers sometimes wish to estimate the correlation between two theoretical constructs from the correlation obtained between the imperfect observed measures of these constructs. To do so, one corrects for attenuation (unreliability) by dividing r_{XY} by the square root of the product of the reliabilities (the maximum possible correlation between the imperfect measures). From Eq. (2.11.4),

(2.11.5) $$r_{X^*Y^*} = \frac{r_{XY}}{\sqrt{r_{XX}r_{YY}}}.$$

Thus, if two variables, each with a reliability of .80, were found to correlate .44,

$$r_{X^*Y^*} = \frac{.44}{\sqrt{(.80)(.80)}} = .55.$$

As for all other estimated coefficients, extreme caution must be used in interpreting attenuation corrected coefficients, since each of the coefficients used in the equation is subject to sampling error. Indeed, it is even possible to obtain attenuation corrected coefficients larger than 1, when the reliabilities come from different populations than r_{XY} or are underestimated, or when the assumption of uncorrelated error is false. Obviously, since the r_{X*Y*} values are hypothetical rather than based on real data, no significance tests can be computed on their departure from zero or any other value.

The problem of unreliability is likely to be particularly severe when one of the variables is a difference score—that is, when it is obtained by subtracting each person's score on some given variable (A) from that person's score on some other variable (B), where A and B are positively correlated. Such difference scores are common when A and B represent scores before and after some treatment and $A - B$ is intended to represent change. Another common difference score situation is found when A and B are two measures obtained on some inventory and an aspect of inventory *profile* ($A - B$) is being investigated. Under any of these circumstances, the $A - B$ difference score is likely to be less reliable than either original score and may be estimated (assuming $sd_A = sd_B$) by the following formula:

(2.11.6) $$r_{(A-B)(A-B)} = \frac{[(r_{AA} + r_{BB})/2] - r_{AB}}{1 - r_{AB}}.$$

As the correlation between the two variables approaches their average reliability, the reliability of the difference score approaches zero. For example, the reliability of the difference score between two variables each with a reliability of .80 and a correlation of .60 will be only .50. Two variables with reliabilities of .60 and an intercorrelation of .45 would yield a difference score with a reliability of only .27. Reliabilities of .60 are by no means uncommon in the behavioral sciences; in fact, in some circumstances (psychiatric assessment, opinion surveys), they may even be considered reasonably good. Thus, the danger in using difference scores is a real one, since they frequently cannot be expected to correlate very substantially with anything else, being mostly measurement error. See Section 9.6 for the generally superior alternative procedure of working with regressed change.

2.11.3 Restriction of Range

A problem related to the general problem of reliability occurs under conditions when the range on one or both variables is restricted by the sampling procedure. For example, suppose that in the data presented in Table 2.1.1 and analyzed in Table 2.6.1, we had restricted ourselves to the study of children with vocabulary scores of 5–8, rather than the full range of 5–10. If the relationship is well described by a straight line and homoscedastic, we will find that the variance of the Y scores about the regression line, $sd^2_{Y-\hat{Y}}$, remains about the same. Since, when $r \neq 0$, sd^2_Y will be decreased as an incidental result of the reduction in sd^2_X, and since $sd_Y^2 = sd_{\hat{Y}}^2 + sd_{Y-\hat{Y}}^2$, the proportion of sd_Y^2 associated with X, $sd_{\hat{Y}}^2$, will be smaller; therefore, r^2 $(= sd_{\hat{Y}}^2/sd_Y^2)$ and r will be smaller. In the current example, r decreases from .828 to .648. The regression coefficient B_{YX}, on the other hand, will remain approximately the same, since the decrease in r will be offset by an increase in the ratio sd_Y/sd_X. It is 1.10 here, compared with 1.20 before. The fact that B_{YX} tends to remain constant over changes in the variability of X is an important property of the regression coefficient. It will be shown later how this makes them more useful as measures of relationship than correlation coefficients in some analytic contexts.

Suppose that an estimate of the correlation which would be obtained from the full range is desired, when the available data have a curtailed or restricted range for X. If we know the sd_X of the unrestricted X distribution as well as the sd_{X_c} for the curtailed sample and the correlation between Y and X in the curtailed sample $(r_{X_c Y})$, we may estimate r_{XY} by

$$(2.11.7) \qquad \tilde{r}_{XY} = \frac{r_{X_c Y}\,(sd_X/sd_{X_c})}{\sqrt{1 + r_{X_c Y}^2\left(\dfrac{sd_X^2}{sd_{X_c}^2} - 1\right)}}.$$

For example, $r = .25$ is obtained on a sample for which $sd_{X_c} = 5$ while the sd_X of the population in which the investigator is interested is estimated as 12. Situations like this occur, for example, when some selection procedure such as an aptitude test has been used to select personnel and those selected are later assessed on a criterion measure. If the finding on the restricted (employed)

TABLE 2.11.2

Correlation and Regression for a Restricted Range of Vocabulary Scores

Child (no.)	$X = v$	$Y = \text{d-s}$	
1	5	12	
2	8	15	
3	7	14	
6	8	18	
7	6	14	$r_{XY} = .648\ (.828)^a$
8	6	17	$r_{XY}^2 = .3819\ (.6862)$
11	7	15	
12	7	16	$B_{YX} = 1.10\ (1.20)$
14	6	13	
15	8	16	
m	6.80	15.00	
sd	.98	1.73	

[a]Parenthetical values are those for the full-range data of Tables 2.1.1 and 2.6.1.

sample is projected to the whole group originally tested, r_{XY} would be estimated to be

$$\tilde{r}_{XY} = \frac{.25(12/5)}{\sqrt{1 + .25^2 \, [(12/5)^2 - 1]}} = \frac{.60}{\sqrt{1.2975}} = .53$$

It should be emphasized that .53 is an estimate and assumes that the relationship is linear and homoscedastic, which might well not be the case. There is no significance test appropriate for $\tilde{r}_{XY}$; it is significant if $r_{X_c Y}$ is significant.

It is quite possible that restriction of range in either X or Y, or both, may occur as an incidental by-product of the sampling procedure. Therefore, it is important in any study to report the sds of the variables used. Since under conditions of homoscedasticity and linearity, regression coefficients are not affected by range restriction, comparisons of different samples using the same variables should usually be done on the regression coefficients rather than on the correlation coefficients when sds differ. Investigators should be aware, however, that the questions answered by these comparisons are not the same. Comparisons of correlations answer the question "does X account for as much of the variance in Y in group E as in group F?" Comparisons of regression coefficients answer the question "does a change in X make the same amount of score difference in Y in group E as it does in group F?"

2.11.4 Part–Whole Correlations

Occasionally we will find that a correlation has been computed between some variable J and another variable W, which is a sum of variables including J. Under these circumstances, a positive correlation can be expected between J and W due to the fact that W includes J, even when there is no correlation between W and $W - J$. For example, if k test items of equal sd and zero r with each other are added together, each of the items will correlate exactly $1/\sqrt{k}$ with the total score. For the two item case, therefore, each item would correlate .707 with W, when neither correlates with the other. On the same assumption of zero correlation between the variables but with unequal sds, the variables are effectively weighted by their differing sd_i and the correlation of J with W will be equal to $sd_J/\sqrt{\Sigma sd_i^2}$.[7] Obviously, under these circumstances, $r_{J(W-J)} = 0$. In the more common case where the variables or items are positively correlated, the correlation of J with $W - J$ may be obtained by

$$(2.11.8) \qquad r_{J(W-J)} = \frac{r_{JW} sd_W - sd_J}{\sqrt{sd_W^2 + sd_J^2 - 2 r_{JW} sd_W sd_J}}.$$

This is not an estimate and may be tested via the usual t test for the significance of r.

Given these often substantial spurious correlations between elements and totals including the elements, it behooves the investigator to determine $r_{J(W-J)}$,

---

[7] The summation here is over the k items.

or at the very least, determine the expected value when the elements are uncorrelated, before interpreting r_{JW}.

Change Scores

It is not necessary that the parts be literally added in order to produce such spurious correlation. If a subscore is subtracted a spurious negative component in the correlation will also be produced. One common use of such difference scores in the social sciences is in the use of post- minus pretreatment (change) scores. If such change scores are correlated with the pre- and postscores from which they have been obtained, we will typically find that subjects initially low on X will have larger gains than those initially high on X, and that those with the highest final scores will have made greater gains than those with lower final scores. Again, if $sd_{pre} = sd_{post}$ and $r_{pre\ post} = 0$, then $r_{pre\ change} = -.707$ and $r_{post\ change} = +.707$. Although in general we would expect the correlation between pre- and postscores to be some positive value, it will be limited by their respective reliabilities (Section 2.11.2). Suppose that the correlation between two measures would be perfect if they were each perfectly reliable, that is, each subject changed exactly the same amount in true scores as every other. Nevertheless, given the usual measurement error, the pre-post correlation will not equal 1, and it will necessarily appear that some subjects have changed more than have others. Even with fairly good reliability so that the observed $r_{pre\ post}$ is .82, and equal sds,

$$(2.11.9) \qquad r_{pre\ change} = \frac{r_{pre\ post} - 1}{\sqrt{2(1 - r_{pre\ post})}}$$

$$= \frac{.82 - 1}{\sqrt{2(1 - .82)}} = -.30.$$

The investigator may be tempted to interpret this correlation by stating that, for example, subjects with low scores on the initial test were "helped" more than were subjects with high initial scores. However, we have already posited that true changes in all subjects were exactly equal! Obviously, the $r = -.30$ is a necessary consequence of the unreliability of the variables and cannot be otherwise interpreted.

If the post- minus pretreatment variable has been created in order to "control" for differences in pretreatment scores, the resulting negative correlation between pre and change scores may be taken as a failure to remove all influence of prescores from postscores. Since there is an alternative method of adjusting postscores for initial differences (see Section 9.6), such difference scores should probably not be used.

2.11.5 Ratio or Index Variables

Ratio (index or rate) scores are those constructed by dividing one variable by another. When a ratio score is correlated with another variable, or with another ratio score, the resulting correlation depends as much on the denominator of the score as it does on the numerator. Since it is usually the investigator's intent to "take the denominator into account" it may not be immediately obvious that the correlations obtained between ratio scores may be spurious—that is, may be a consequence of mathematical necessities which have no valid interpretive use. Ratio correlations depend, in part, upon the correlations between all numerator and denominator terms, so that $r_{(Y/Z)X}$ is a function of r_{YZ} and r_{XZ} as well as of r_{YX}, and $r_{(Y/Z)(X/W)}$ depends on r_{YW} and r_{XZ} as well as on the other four correlations. Equally problematic is the fact that such correlations also involve the coefficients of variation ($v_X = sd_X/\overline{X}$) of each of the variables. Although the following formula is only a fair approximation of the correlation between ratio scores (requiring normal distributions and homoscedasticity and dropping all terms involving powers of v greater than v^2), it serves to demonstrate the dependence of correlations between ratios on all vs and on rs between all variable pairs:

$$(2.11.10) \quad r_{(Y/Z)(X/W)} \approx \frac{r_{YX}v_Yv_X - r_{YW}v_Yv_W - r_{XZ}v_Xv_Z + r_{ZW}v_Zv_W}{\sqrt{v_Y^2 + v_Z^2 - 2r_{YZ}v_Yv_Z}\sqrt{v_X^2 + v_W^2 - 2r_{XW}v_Xv_W}}.$$

When the two ratios being correlated have a common denominator, the possibility of spurious correlation becomes apparent. Under these circumstances, the approximate formula for the correlation simplifies, since $Z = W$. If all coefficients of variation are equal, when all three variables are *un*correlated, we will find $r_{(Y/Z)(X/Z)} \approx .50$.

Since the coefficient of variation depends on the value of the mean, it is clear that whenever this value is arbitrary, as it is for most psychological scores, the calculated r is also arbitrary. Thus, ratios should not be correlated unless each variable is measured on a "ratio" scale, a scale for which a zero value means literally none of the variable (see Chapter 6). Measures with ratio scale properties are commonly found in the social sciences in the form of counts or frequencies.

At this point it may be useful to distinguish between rates and other ratio variables. Rates may be defined as variables constructed by dividing the number of instances of some phenomenon by the total number of opportunities for the phenomenon to occur; thus, they are literally proportions. Rates or proportions are frequently used in ecological or epidemiological studies, where the units of analysis are aggregates of people or areas such as counties or census tracts. In such studies, the numerator represents the incidence of some phenomenon and the denominator represents the "population at risk." For example, a deliquency rate may be calculated by dividing the number of delinquent boys aged 14–16 in an area by the total number of boys aged 14–16 in the area. This variable may be correlated across areas with the proportion of families whose incomes are below the poverty level, another rate. Since, in general, the denominators will reflect the size of the areas, which may vary greatly, the denominators can be expected to be substantially correlated. In other cases the denominators may actually be the same, as for example, in an investigation of the relationship between delinquency rates and school dropout rates for a given age–sex group. The investigator will typically find that these rates have characteristics which minimize the problem of "spurious" index correlation. In most real data, the coefficients of variation of the numerators will be substantially larger than the coefficients of variation of the denominators, and thus the correlation between rates will be determined substantially by the correlation between the numerators. Even in such data, however, the resulting proportions may not be optimal for the purpose of linear correlation. Section 6.5.4 discusses transformations of proportions, which may be more appropriate for analysis than the raw proportions or rates, themselves.

Experimentally produced rates may be more subject to problems of spurious correlation, especially when there are logically alternative denominators. The investigator should determine that the correlation between the numerator and denominator is very high (and positive), since in general the absence of such a correlation suggests a faulty logic in the study. In the absence of a large correlation, the coefficients of variation of the numerator should be substantially larger than that of the denominator, if the problem of spurious correlation is to be avoided.

Other Ratio Scores

When the numerator does not represent some subclass of the denominator class, the risks involved in using ratios are even more serious, since the likelihood of small or zero correlations between numerators and denominators and relatively similar values of v is greater. If the ratio scale properties of variables are insufficiently "strong" (true zeros and equal intervals), correlations involving ratios should probably be avoided altogether, and an alternative method for removing the influence of Z from Y should be chosen, such as partial correlation (see Chapters 3, 4, and 9).

The difficulties which may be encountered in correlations involving rates and ratios may be illustrated by the following example. An investigator wishes to determine the relationship between visual scanning and errors on a digit-symbol task. All subjects are given four minutes to work on the task. Since subjects who complete more digit symbols have a greater opportunity to make errors, the experimenter decides, reasonably enough, to determine the error rate by dividing the number of errors by the number of d-s completed. Similarly, it is reasoned that subjects completing more d-s substitutions should show more horizontal visual scans and thus visual scanning is measured by dividing the number of visual scans by the number of d-s completed. Table 2.11.3 displays the data for 10 subjects. Contrary to expectation, subjects who completed more d-s did not tend to produce more errors ($r_{ZX} = -.105$) nor did they scan notably more than did low scorers ($r_{ZY} = .106$). Nevertheless, when the two ratio scores are computed, they show a substantial positive correlation (.427) in spite of the fact that the numerators showed slight *negative* correlation ($-.149$), nor is there any

TABLE 2.11.3

Example of Spurious Correlation between Ratios

Subject	No. completed d–s (Z)	No. errors (X)	No. scans (Y)	No. errors / No. completed	No. scans / No. completed
1	25	5	24	.20	.96
2	29	3	30	.10	1.03
3	30	3	27	.10	.90
4	32	4	30	.12	.94
5	37	3	18	.08	.49
6	41	2	33	.05	.80
7	41	3	27	.07	.66
8	42	5	21	.12	.50
9	43	3	24	.07	.56
10	43	5	33	.12	.77

$$r_{ZX} = -.105, \quad r_{ZY} = .106, \quad r_{XY} = -.149$$

$$r_{(X/Z)(Y/Z)} = .427$$

tendency for scanning and errors to be correlated for any given level of d-s completion. Thus, the $r_{(X/Z)(Y/Z)}$ may here be seen as an example of spurious correlation.

2.11.6 Curvilinear Relationships

When the relationship between two variables is only moderately well fitted by a straight line, the correlation coefficient which indicates the degree of *linear* relationship will understate the predictability from one variable to the other. Frequently the relationship, although curvilinear, is monotonic, that is, increases in X are accompanied by increases (or decreases) in Y, although not at a constant rate. Under these circumstances, some (nonlinear) monotonic transformation of X or Y or both may "straighten out" the regression line and provide a better indication of the size of the relationship between the two variables (an absolutely larger r). Since there are several alternative ways of handling (and detecting) curvilinear relationships between variables, the reader is referred to Chapter 6 for a detailed treatment of the issues. Section 6.5 deals with considerations in determining when nonlinear transformations of variables are likely to be appropriate, and methods of selecting among alternative transformations.

2.12 SUMMARY

A linear relationship exists between two quantitative variables when there is an overall tendency for increases in the values of one variable to be accompanied by increases in the other variable (a positive relationship), or for increases in the first to be accompanied by decreases in the second (a negative relationship) (Section 2.1). Efforts to index the degree of linear relationship between two variables must cope with the problem of the different units in which variables are measured. Standard (z) scores are a conversion of scores into distances from their own means, in standard deviation units, and render different scores comparable. The Pearson product moment r is a measure of the degree of relationship between two variables, X and Y, based on the discrepancies of the subjects' paired z scores, $z_X - z_Y$. r varies between -1 and $+1$, which represent perfect negative and perfect positive relationships, respectively. When $r = 0$, there is no linear correlation between the variables. (Section 2.2)

r can be written as a function of z score products, of variances and covariance, or in terms of the original units. Special simplified formulas are available for r when one variable is a dichotomy (point biserial r), when both variables are dichotomies (r_ϕ), or when the data are two sets of complete ranks (rank order correlation). (Section 2.3)

The regression coefficient, B_{YX}, gives the optimal rule for a linear estimate of Y from X, and is the change in Y units per unit change in X, that is, the slope of the regression line. The intercept, A_{YX}, gives the value of Y for a zero value of X. B_{YX} and A_{YX} are optimal in the sense that they provide the smallest squared discrepancies between Y and estimated $\hat{Y}$. r is the regression coefficient for the standardized variables (Section 2.4). Unless $r = 1$, it is a mathematical necessity that the average score for a variable being estimated will be relatively closer to its mean than the value from which it is being estimated will be to its mean, when both are measured in sd units. (Section 2.5)

When Y is estimated from X the sd of the difference between observed scores and the estimated scores (the sample standard error of estimate) can be computed from knowledge of r and sd_Y. The coefficient of alienation represents the error as a proportion of the original sd_Y. r^2 equals the proportion of the variance (sd^2) of each of the variables which is shared with or estimable from the other. (Sections 2.6 and 2.7)

When the observed data represent a sample from some population, various significance tests may be performed on r and B. These are tested against the null hypothesis that they equal zero in the population by the t test. The difference between rs and between Bs determined from independent samples may also be tested for significance. (Section 2.8)

The power of the test of statistical significance may be determined in the planning stage of a research. When such analyses have been done, appropriate changes in the plans may be made, when necessary, to control the probability of making a Type II error—the error of failing to reject a false null hypothesis. (Section 2.9)

When confidence limits are determined for r and B, the investigator will have a concrete picture of the range within which the corresponding population values can be expected to fall with a known probability of error. (Section 2.10)

Confidence limits may also be placed on the $\hat{Y}_o$ predicted from a given X_o value. The size of this confidence interval will increase as the departure of the X value from $\bar{X}$ increases. (Section 2.10.2)

A number of characteristics of the X and Y variables will affect the size of the correlation between them. Among these are differences in the distribution of the X and Y variables (Section 2.11.1), unreliability in one or both variables (Section 2.11.2), and restriction of the range of one or both variables (Section 2.11.3). When one variable is included as a part of the other variable, the correlation between them will reflect this overlap (Section 2.11.4). Errors of inference involving part–whole correlations are particularly common in the examination of difference scores. Scores obtained by dividing one variable by another will produce spurious correlations with other variables under some conditions (Section 2.11.5). The r between two variables will be an underestimate of the magnitude of their relationship when a curved rather than a straight line best fits the bivariate distribution (Section 2.11.6). Under such circumstances, transformations of one or both variables or multiple regression procedures (Chapter 6) will provide a better picture of the relationship between variables.

3

Multiple Regression and Correlation: Two or More Independent Variables

3.1 INTRODUCTION

In Chapter 2 we explored the index of linear correlation between two variables, the Pearson product moment r, and the regression equations for estimating either variable from the other. The regression coefficient B_{YX} is the slope of the line which best estimates Y from X. The regression constant or Y intercept A_{YX} is the value of $\hat{Y}$ for $X = 0$. The complete regression equation, $\hat{Y} = B_{YX}X + A_{YX}$, represents the linear equation (rule) which will minimize the squared discrepancies between the actual Y and $\hat{Y}$ estimated from X.

For example, suppose we wished to study the relationship of the academic rank (Y) of faculty members in some department from information about the number of publications each has authored (X_1). Scores are assigned to academic ranks corresponding to their ordinal position: assistant professor = 1, associate professor = 2, professor = 3.[1] Table 3.1.1 provides the data for the 15 department members and we find the correlation between academic rank and number of publications to equal .463, $B_{Y1} = .070$ by Eq. (2.4.3), and $A_{Y1} = 1.269$ by Eq. (2.4.4).[2] Suppose we decide to determine the improvement in our estima-

---

[1] In assigning these values to the ranks we are assuming that the resulting relationships of Y with the other variables we are studying are adequately described by a straight line. In Chapter 6 we present several methods for dealing with curvilinearity.

[2] In this example, the number of cases has been kept small to enable the reader to follow computations with ease. No advocacy of such small samples is intended (see Sections 3.8 and 4.5 on statistical power).

In this and the remaining chapters the dependent variable is identified as Y and the individual independent variables by X with a numerical subscript, that is, X_1, X_2, etc. This makes it possible to represent independent variables by their numerical subscripts only, for example, B_{YX_1} becomes B_{Y1}.

TABLE 3.1.1
Illustrative Data for a Three Variable Problem

Academic rank (Y)	Number of publications (X_1)	Number of years since Ph.D. (X_2)	
1	2	1	
1	4	2	
1	5	5	
1	12	7	
1	5	10	
1	9	4	$r_{Y1} = .463$ ($r_{Y1}^2 = .2144$)
2	3	3	
2	1	8	$r_{Y2} = .612$ ($r_{Y2}^2 = .3749$)
2	8	4	
2	12	16	$r_{12} = .683$ ($r_{12}^2 = .4667$)
2	9	15	
2	4	19	$R_{Y \cdot 12} = .615$ ($R_{Y \cdot 12}^2 = .3786$)
3	8	8	
3	11	14	
3	21	28	
m 1.80	7.60	9.60	
sd .748	4.96	7.25	
B_{Y_i}	.070	.063	
A_{Y_i}	1.269	1.193	

tion of academic rank which can be accomplished by means of a second independent variable, X_2 = number of years since Ph.D. First, we determine that r_{Y2} = .612, B_{Y2} = .063, and A_{Y2} = 1.193. If X_1 and X_2 were uncorrelated, we could simply use B_{Y1} and B_{Y2} together to estimate Y. However, as might be expected, we find a tendency for those faculty members who have had their degrees longer to have more publications than those who more recently completed their education (r_{12} = .683). Thus, X_1 and X_2 are to some extent redundant, and necessarily their respective estimates, $\hat{Y}_1$ and $\hat{Y}_2$, will also be redundant. What we need in order to estimate Y optimally from both X_1 and X_2 is an equation in which this redundancy (or more generally the relationship between X_1 and X_2) is taken into account. The regression coefficients in such an equation are called *partial regression coefficients* to indicate that they are optimal linear estimates of the dependent variable (Y) when used in combination with specified other independent variables. Thus, $B_{Y1 \cdot 2}$ is the partial regression coefficient for Y on X_1 when X_2 is also in the equation. The full equation is

(3.1.1) $\hat{Y} = B_{Y1 \cdot 2} X_1 + B_{Y2 \cdot 1} X_2 + A_{Y \cdot 12}.$

The partial regression coefficients or B weights in this equation, as well as the regression constant A, are determined in such a way that the sum of the squared differences between (actual) Y and (estimated) $\hat{Y}$ is a minimum. Thus, the multiple regression equation is defined by the same "least-squares" criterion as was the regression equation for a single independent variable. Since the equation *as a whole* satisfies this mathematical criterion, the term partial regression coefficient is used to make clear that it is the weight to be applied to an IV when other specified IVs are also in the equation. Thus $B_{Y1 \cdot 2}$ indicates the weight to be given X_1 when X_2 is also in the equation, $B_{Y2 \cdot 13}$ is the X_2 weight when X_1 and X_3 are in the equation, $B_{Y4 \cdot 123}$ is the X_4 weight when X_1, X_2, and X_3 are also used in the equation for $\hat{Y}$, and so on.

Just as there are partial regression coefficients for multiple regression equations (equations for predicting $\hat{Y}$ from more than one IV) so are there partial and multiple correlation coefficients which answer the same questions answered by the "zero-order" or simple product-moment correlation coefficient in the single IV case. These questions include the following:

1. How well does this group of IVs estimate Y?
2. How much does any single variable add to the estimation of Y already accomplished by other variables?
3. What is the relationship between any given IV and Y when the other IVs have been taken into account?
4. How much change in $\hat{Y}$ is associated with a unit change in each of the IVs when all variables have been standardized to a mean of 0 and an *sd* of 1?

In addition, the tests of statistical significance and power analysis for MRC speak to the questions:

1. How much confidence may be placed in the values of partial and multiple coefficients determined by this method?
2. What is the probability of obtaining a statistically significant result in the study as planned?

In this chapter the methods for determining answers to these questions are presented. For the sake of clarity the special case of two IVs is presented first, with subsequent generalization to the more-than-two IV case.

3.2 REGRESSION WITH TWO INDEPENDENT VARIABLES

To recapitulate, the regression equation provides us with a weight for each IV, that is, each IV is assigned a constant by which every observed value of the IV is to be multipled. When the regression equation is applied to the IV values for any given observation, the result will be an estimated value of the dependent variable ($\hat{Y}$). For any given set of data on which such an equation is determined the resulting set of $\hat{Y}$ values will be as close to the actual Y values as possible, given a single weight for each IV. "As close as possible" is defined by the least-squares principle.

To return to our example of estimating academic rank (Y) from number of publications (X_1) and number of years since Ph.D. (X_2), the full regression equation is

(3.2.1) $\hat{Y}_{12} = .013\,X_1 + .057\,X_2 + 1.15,$

where .013 is the partial regression coefficient $(B_{Y1 \cdot 2})$ for X_1, and .057 is that for X_2 $(B_{Y2 \cdot 1})$. The redundancy of information about Y carried by these two variables is reflected in the fact that the partial regression coefficients (.013 and .057) are smaller in magnitude than their separate zero-order Bs $(B_{Y1} = .070$ and $B_{Y2} = .063)$.[3] We may interpret $B_{Y1 \cdot 2} = .013$ directly by stating that, for any given number of years since Ph.D. (X_2), on the average, each additional publication is associated with an increase of .013 units in academic rank. The $B_{Y2 \cdot 1} = .057$ may be similarly interpreted as indicating that, for faculty members with a given number of publications (X_1), each additional year since Ph.D. is associated with a .057 increase in academic rank. Looked at differently, although when examining the effect of publications alone we found that approximately 14 publications on the average were associated with a unit increase in academic rank $(B_{Y1} = .070 \approx 1/14)$, part of this effect is also attributable to the tendency for faculty members who have published more to have their Ph.D.s longer. "Controlling for" this tendency[4] we find 77 (!) publications necessary for a unit increase in rank (i.e., $B_{Y1 \cdot 2} = .013 \approx 1/77$).

Thus far, we have simply asserted that the regression equation for two or more IVs takes the same form as did the single IV case without demonstrating how its coefficients are obtained. As was the case in presenting correlation and regression with one IV we will initially standardize the variables to eliminate the effects of noncomparable raw (original) units. The regression equation for standardized variables is

(3.2.2) $\hat{z}_Y = \beta_{Y1 \cdot 2} z_1 + \beta_{Y2 \cdot 1} z_2.$

Just as r_{YX} is the standardized regression coefficient for estimating $\hat{z}_Y$ from z_X, $\beta_{Y1 \cdot 2}$ and $\beta_{Y2 \cdot 1}$ are the standardized *partial* regression coefficients for estimating $\hat{z}_Y$ from z_1 and z_2 with minimum squared error. (Do not confuse this use of β with its use as rate of Type II error in power analysis.)

----

[3] Most frequently correlations between independent variables indicate redundancy with regard to Y and therefore smaller (in absolute value) partial than zero-order regression coefficients. Certain patterns of IV correlation do not indicate such redundancy, however, and therefore will not have this effect. See Section 3.4.

[4] The terms "holding constant," or "controlling for," "partialling the effects of," or "residualizing" some other variable(s) indicate a mathematical procedure, of course, rather than an experimental one. Such terms are statisticians' shorthand for describing the average effect of one variable for any given value of these other variables.

The equations for $\beta_{Y1\cdot2}$ and $\beta_{Y2\cdot1}$ can be proved via differential calculus to be

(3.2.3)
$$\beta_{Y1\cdot2} = \frac{r_{Y1} - r_{Y2}r_{12}}{1 - r_{12}^2},$$

$$\beta_{Y2\cdot1} = \frac{r_{Y2} - r_{Y1}r_{12}}{1 - r_{12}^2}.$$

A separation of the elements of this formula may aid understanding. r_{Y1} and r_{Y2} are "validity" coefficients, that is, the zero-order correlations of the IVs with the dependent variable, Y. r_{12}^2 represents the variance in each IV shared with the other IV and reflects their redundancy. Thus, $\beta_{Y1\cdot2}$ and $\beta_{Y2\cdot1}$ are partial coefficients since each has been adjusted to allow for its correlation with the other.

To return to our academic example, the correlations between the variables are $r_{Y1} = .463$, $r_{Y2} = .612$, and $r_{12} = .683$. We determine by Eq. (3.2.3)

$$\beta_{Y1\cdot2} = \frac{.463 - (.612)(.683)}{1 - .683^2} = .084,$$

$$\beta_{Y2\cdot1} = \frac{.612 - (.463)(.683)}{1 - .683^2} = .555,$$

and the full regression equation for the standardized variables is

$$\hat{z}_Y = .084\, z_1 + .555\, z_2.$$

Once $\beta_{Y1\cdot2}$ and $\beta_{Y2\cdot1}$ have been determined, conversion to the original units is readily accomplished by

(3.2.4)
$$B_{Y1\cdot2} = \beta_{Y1\cdot2}\frac{sd_Y}{sd_1},$$

,
$$B_{Y2\cdot1} = \beta_{Y2\cdot1}\frac{sd_Y}{sd_2}.$$

Substituting the sd values for our running example (Table 3.1.1), we find

$$B_{Y1\cdot2} = .084\frac{.748}{4.96} = .013,$$

$$B_{Y2\cdot1} = .555\frac{.748}{7.25} = .057.$$

The constant A which serves to adjust for differences in means, is calculated in the same way as with a single IV:

(3.2.5)
$$A_{Y\cdot12} = \bar{Y} - B_{Y1\cdot2}\bar{X}_1 - B_{Y2\cdot1}\bar{X}_2$$

$$= 1.80 - .013(7.60) - .057(9.60)$$

$$= 1.15.$$

The full (raw score) regression equation for estimating academic rank is therefore

$$\hat{Y}_{12} = .013\,X_1 + .057\,X_2 + 1.15,$$

and the resulting values are provided in the third column of Table 3.3.1 later in the chapter.

3.3   MEASURES OF ASSOCIATION WITH TWO INDEPENDENT VARIABLES

3.3.1   Multiple R and R^2

Just as r is the measure of association between two variables, so the multiple R is the measure of association between a dependent variable and an optimal combination of two or more IVs. Similarly, r^2 is the proportion of each variable's variance shared with the other, and R^2 is the proportion of the dependent variable's variance (sd_Y^2) shared with the optimally weighted IVs. Unlike r, however, R takes on only values between 0 and 1, with the former indicating no relationship with the IVs and the latter indicating a perfect relationship. (The reason that Rs are always positive will become clear shortly.) The formula for the multiple correlation coefficient for two IVs as a function of the original rs is

(3.3.1) $$R_{Y \cdot 12} = \sqrt{\frac{r_{Y1}^2 + r_{Y2}^2 - 2r_{Y1}r_{Y2}r_{12}}{1 - r_{12}^2}}.$$

A similarity between the structure of this formula and the formula for the β coefficients may lead the reader to suspect that R may be written as a function of these coefficients. This is indeed the case; an alternative formula is

(3.3.2) $$R_{Y \cdot 12} = \sqrt{\beta_{Y1 \cdot 2} r_{Y1} + \beta_{Y2 \cdot 1} r_{Y2}}.$$

For the example illustrated in Table 3.1.1 the multiple correlation is thus, by Eq. (3.3.1),

$$R_{Y \cdot 12} = \sqrt{\frac{.2144 + .3749 - 2(.463)(.612)(.683)}{1 - .4667}}$$

$$= \sqrt{.3786} = .615,$$

or by Eq. (3.3.2),

$$R_{Y \cdot 12} = \sqrt{.084(.463) + .555(.612)}$$

$$= \sqrt{.3786} = .615.$$

(We remind the reader who checks the above arithmetic and finds it "wrong" of our warning in Section 1.2.2 about rounding errors.)

We saw in Chapter 2 that the absolute value of the correlation between two variables (r_{YX}) is equal to the correlation between Y and $\hat{Y}_X$. The multiple correlation is actually definable by this property. Thus,

(3.3.3) $$R_{Y \cdot 12} = r_{Y\hat{Y}_{12}},$$

and taking the example values in Table 3.3.1 we see that indeed $r_{Y\hat{Y}_{12}} = .615 = R_{Y \cdot 12}$. That $r_{Y\hat{Y}_{12}}$ and hence $R_{Y \cdot 12}$ can not be negative can be seen from the fact that $\hat{Y}$ is optimally (least-squares) predicted.

The reader will again recall that r_{YX}^2 is the proportion of variance of Y (sd_Y^2) shared with X. In exact parallel, $R_{Y \cdot 12}^2$ is the proportion of sd_Y^2 shared with the optimally weighted composite of X_1 and X_2. These optimal weights are, of course, those provided by the regression equation used to estimate $\hat{Y}$. Thus,

(3.3.4) $$R_{Y \cdot 12}^2 = \frac{sd_{\hat{Y}}^2}{sd_Y^2}$$

$$= \frac{.212}{.560} = .3786,$$

that is, some 38% of the variance in academic rank (Y) is linearly accounted for by number of publications (X_1) and number of years since doctorate (X_2) in this sample.

Again in parallel with simple correlation and regression the variance of the residual, $Y - \hat{Y}_{12}$, is that portion of sd_Y^2 not linearly associated with X_1 and X_2. Therefore (and necessarily),

(3.3.5) $$r_{\hat{Y}(Y-\hat{Y})} = 0,$$

and such variances are additive,

(3.3.6) $$sd_Y^2 = sd_{\hat{Y}_{12}}^2 + sd_{Y-\hat{Y}_{12}}^2.$$

It should also be apparent at this point that a multiple R can never be less than the largest correlation of Y with the IVs and will almost invariably be larger. The optimal estimation of $\hat{Y}_{12}$ under circumstances in which X_2 adds nothing to X_1's estimation of $\hat{Y}$ would involve a 0 weight for X_2 and thus $R_{Y \cdot 12}$ would equal $|r_{Y1}|$, the absolute value of r_{Y1}. Any slight departure of X_2 values from this rare circumstance necessarily leads to some (perhaps trivial) increase in $R_{Y \cdot 12}$ over $|r_{Y1}|$.

3.3.2 Semipartial Correlation Coefficients

One of the important problems that arises in MRC is that of defining the contribution of each IV to the multiple correlation. We shall see that the solution to this problem is not so straightforward as in the case of a single independent variable, the choice of coefficient depending on the substantive reasoning underlying the exact formulation of the research questions. One answer is provided by the semipartial correlation coefficient sr, and its square, sr^2. To understand the meaning of these coefficients it will be useful to consider the "ballantine." Recall that in the diagrammatic representation of Figure 2.6.1,

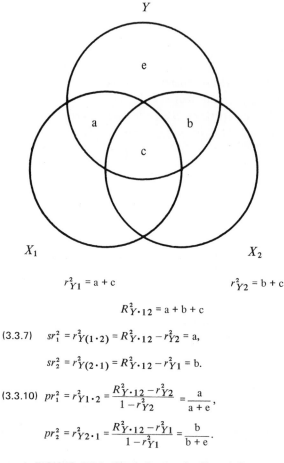

$$r^2_{Y1} = a + c \qquad\qquad r^2_{Y2} = b + c$$

$$R^2_{Y \cdot 12} = a + b + c$$

$$(3.3.7) \quad sr^2_1 = r^2_{Y(1 \cdot 2)} = R^2_{Y \cdot 12} - r^2_{Y2} = a,$$

$$sr^2_2 = r^2_{Y(2 \cdot 1)} = R^2_{Y \cdot 12} - r^2_{Y1} = b.$$

$$(3.3.10) \quad pr^2_1 = r^2_{Y1 \cdot 2} = \frac{R^2_{Y \cdot 12} - r^2_{Y2}}{1 - r^2_{Y2}} = \frac{a}{a + e},$$

$$pr^2_2 = r^2_{Y2 \cdot 1} = \frac{R^2_{Y \cdot 12} - r^2_{Y1}}{1 - r^2_{Y1}} = \frac{b}{b + e}.$$

FIGURE 3.3.1   The ballantine for X_1 and X_2.

the variance of each variable is represented by a circle of unit area. The overlapping area of two circles represents their relationship as r^2. With Y and two IVs represented in this way, the figure is called a ballantine. The total area of Y covered by the X_1 and X_2 areas represents the proportion of Y's variance accounted for by the two IVs, $R^2_{Y \cdot 12}$.

Figure 3.3.1 shows that this area is equal to the sum of areas designated a, b, and c. The areas a and b represent those portions of Y overlapped uniquely by IVs X_1 and X_2, respectively, whereas area c represents their simultaneous overlap with Y. The "unique" areas, expressed as proportions of Y variance, are squared *semipartial* correlation coefficients, and each equals the increase in the squared multiple correlation which occurs when the variable is added to the other IV.[5] Thus,

(3.3.7)
$$a = sr_1^2 = R_{Y \cdot 12}^2 - r_{Y2}^2 \,,$$
$$b = sr_2^2 = R_{Y \cdot 12}^2 - r_{Y1}^2 \,.$$

A formula for sr for the two IV case may be given as a function of zero-order rs as

(3.3.8)
$$sr_1 = \frac{r_{Y1} - r_{Y2}r_{12}}{\sqrt{1 - r_{12}^2}} \,,$$

$$sr_2 = \frac{r_{Y2} - r_{Y1}r_{12}}{\sqrt{1 - r_{12}^2}} \,.$$

For our running example (Table 3.1.1), these values are

$$sr_1 = \frac{.463 - .612(.683)}{\sqrt{1 - .683^2}} = .062,$$

$$sr_1^2 = .0038,$$

or, by Eq. (3.3.7)

$$sr_1^2 = .3786 - .3749 = .0038;$$

and

$$sr_2 = \frac{.612 - .463(.683)}{\sqrt{1 - .683^2}} = .405,$$

$$sr_2^2 = .1642,$$

or, by Eq. (3.3.7),

$$sr_2^2 = .3786 - .2144 = .1642.$$

The semipartial correlation sr_1 is the correlation between all of Y and X_1 from which X_2 has been partialled. It is a *semi*partial correlation since the effects of X_2 have been removed from X_1 but not from Y. Recalling that in this system "removing the effect" is equivalent to subtracting from X_1 the X_1 values estimated from X_2, that is, to working with $X_1 - \hat{X}_{1 \cdot 2}$, we see that another way to write this relationship is

(3.3.9) $sr_1 = r_{Y(X_1 - \hat{X}_{1 \cdot 2})}.$

[5]Throughout the remainder of the book, whenever possible without ambiguity, partial coefficients are subscripted by the relevant independent variable only, it being understood that Y is the dependent variable and that all other IVs have been partialled. Thus, $sr_i = sr_{Y(i \cdot 12...(i)...k)}$, the correlation between Y and X_i from which all other IVs in the set under consideration have been partialled. Similarly, R without subscript refers to $R_{Y \cdot 12...k}$.

Another notational form of sr_1 used is $r_{Y(1\cdot2)}$, the $1\cdot2$ being a shorthand way of expressing "X_1 from which X_2 has been partialled," or $X_1 - \hat{X}_{1\cdot2}$. It is often a great convenience to use this "dot" notation to identify what is being partialled from what, particularly in subscripts, and it will be employed whenever necessary to avoid ambiguity. Thus, $i\cdot j$ means "i from which j is partialled." Note also that in the literature, the term "part" correlation is sometimes used to denote semipartial correlation.

In Table 3.3.1 we present the $X_1 - \hat{X}_{1\cdot2}$ ("residual") values for each case in the example in which academic rank was estimated from publications and years since Ph.D. The correlation between these residual values and Y is seen to equal .062, which is sr_1, and $.062^2 = .0038 = sr_1^2$, as before.

To return to the ballantine (Figure 3.3.1) we see that for our example, area a = .0038, b = .1642, and a + b + c = $R^2_{Y\cdot12}$ = .3786. It is tempting to calculate c (by c = $R^2_{Y\cdot12} - sr_1^2 - sr_2^2$) and interpret it as the proportion of Y variance estimated jointly or redundantly by X_1 and X_2. However, any such interpreta-

TABLE 3.3.1
Actual, Estimated, and Residual Academic Rank Scores

Y	$\hat{Y}_2$	$\hat{Y}_{12}$	$Y - \hat{Y}_{12}$	$\hat{X}_{1\cdot2}$	$X_1 - \hat{X}_{1\cdot2}$	$Y - \hat{Y}_2$
1	1.26	1.24	−.24	3.58	−1.58	−.26
1	1.32	1.32	−.32	4.04	−.04	−.32
1	1.51	1.50	−.50	5.45	−.45	−.51
1	1.64	1.71	−.71	6.38	5.62	−.64
1	1.83	1.79	−.79	7.79	−2.79	−.83
1	1.45	1.50	−.50	4.98	4.02	−.45
2	1.38	1.36	.64	4.51	−1.51	.62
2	1.70	1.62	.38	6.85	−5.85	.30
2	1.45	1.48	.52	4.98	3.02	.55
2	2.20	2.22	−.22	10.60	1.40	−.20
2	2.14	2.13	−.13	10.13	−1.13	−.14
2	2.39	2.29	−.29	12.00	−8.00	−.39
3	1.70	1.71	1.29	6.85	1.15	1.30
3	2.08	2.10	.90	9.66	1.36	.92
3	2.96	3.02	−.02	16.21	4.79	.04
m 1.80	1.80	1.80	.00	7.60	.00	.00
sd .748		.460	.590			
sd^2 .560		.212	.348			

Correlations

	$\hat{Y}_2$	$\hat{Y}_{12}$	$X_1 - \hat{X}_{1\cdot2}$
Y	.612 = r_{Y2}	.615 = $R_{Y\cdot12}$	.062 = sr_1
$Y - \hat{Y}_2$	0 = $r_{(Y\cdot2)2}$	—	.078 = pr_1

tion runs into a serious catch—there is nothing in the mathematics which prevents c from being a negative value and a negative proportion of variance hardly makes sense. A discussion of the circumstances in which this occurs will be found in Section 3.4. On the other hand, a and b can never be negative and *are* appropriately considered proportions of variance; each represents the increase in the proportion of Y variance accounted for by the addition of the corresponding variable to the equation estimating Y.

3.3.3 Partial Correlation Coefficients

Another kind of solution to the problem of describing each IV's participation in determining R is given by the *partial* correlation coefficient pr_1, and its square, pr_1^2. The squared partial correlation may be understood best as that proportion of sd_Y^2 *not* associated with X_2 which *is* associated with X_1. Returning to the ballantine (Figure 3.3.1), we see that

(3.3.10)

$$pr_1^2 = \frac{a}{a+e} = \frac{R_{Y \cdot 12}^2 - r_{Y2}^2}{1 - r_{Y2}^2},$$

$$pr_2^2 = \frac{b}{b+e} = \frac{R_{Y \cdot 12}^2 - r_{Y1}^2}{1 - r_{Y1}^2}.$$

The a area or numerator for pr_1^2 is the squared semipartial correlation coefficient; however the base includes not all of the variance of Y as in sr_1^2, but only that portion of Y variance which is not associated with X_2, that is, $1 - r_{Y2}^2$. Thus, this squared partial r answers the question, "How much of the Y variance which is not estimated by the other IV(s) in the equation is estimated by this variable?" Interchanging X_1 and X_2 (and areas a and b), we similarly interpret pr_2^2. In our academic rank example, we see that by Eqs. (3.3.10)

$$pr_1^2 = \frac{.3786 - .3749}{1 - .3749} = \frac{.0038}{.6251} = .0060,$$

$$pr_2^2 = \frac{.3786 - .2144}{1 - .2144} = \frac{.1642}{.7856} = .2090.$$

Obviously, since the denominator cannot be greater than 1, partial correlations will be larger than semipartial correlations, except in the limiting case when other IVs are correlated 0 with Y, in which case $sr = pr$.

pr may be found more directly as a function of zero-order correlations by

(3.3.11)

$$pr_1 = \frac{r_{Y1} - r_{Y2} r_{12}}{\sqrt{1 - r_{Y2}^2} \sqrt{1 - r_{12}^2}},$$

$$pr_2 = \frac{r_{Y2} - r_{Y1} r_{12}}{\sqrt{1 - r_{Y1}^2} \sqrt{1 - r_{12}^2}}.$$

For our example,

$$pr_1 = \frac{.463 - .612(.683)}{\sqrt{1 - .3749}\,\sqrt{1 - .4667}} = .078,$$

and $pr_1^2 = .078^2 = .0060$, as before;

$$pr_2 = \frac{.612 - .463(.683)}{\sqrt{1 - .2144}\,\sqrt{1 - .4667}} = .4572,$$

and $pr_2^2 = .4572^2 = .2090$, again as before.

In Table 3.3.1, we demonstrate that pr_1 is the correlation between X_1 from which X_2 has been partialled (i.e., $X_1 - \hat{X}_{1 \cdot 2}$) and Y from which X_2 has also been partialled (i.e., $Y - \hat{Y}_2$). Column 6 presents the partialled X_1 values, the residuals from $\hat{X}_{1 \cdot 2}$. Column 7 presents the residuals from $\hat{Y}_2$ (given in column 2). The simple correlation between the residuals in columns 6 and 7 is $.078 = pr_1$ (the computation is left to the reader, as an exercise). We thus see that the partial correlation for X_1 is literally the correlation between Y and X_1, each similarly residualized from X_2. A frequently employed form of notation to express the partial r is $r_{Y1 \cdot 2}$, which conveys that X_2 is being partialled from both Y and X_1, i.e., $r_{(Y \cdot 2)(1 \cdot 2)}$, in contrast to the semipartial r, which is represented as $r_{Y(1 \cdot 2)}$.

Before leaving Table 3.3.1, the other correlations at the bottom are worth noting. The r of Y with $\hat{Y}_2$ of .612 is identically r_{Y2} and necessarily so, since $\hat{Y}_2$ is a linear function of X_2. Similarly, the r of Y with $\hat{Y}_{12}$ of .615 is identically $R_{Y \cdot 12}$ and necessarily so, by definition in Eq. (3.3.3). Also, $Y - \hat{Y}_2$ (that is, $Y \cdot X_2$) correlates zero with $\hat{Y}_2$, since when a variable (here X_2) is partialled from another (here Y), the residual will correlate zero with any linear function of the partialled variable; here, $\hat{Y}_2$ is a linear (regression) function of X_2.

Summarizing the results for the running example, we found $sr_1^2 = .0038$, $pr_1^2 = .0060$ and $sr_2^2 = .1642$, $pr_2^2 = .2090$. Whichever base we use, it is clear that number of publications (X_1) has virtually no *unique* relationship to rank, that is, no relationship beyond what can be accounted for by years since doctorate (X_2). On the other hand, years since doctorate (X_2) is uniquely related to rank and to rank holding publications constant, to a quite substantial degree. We leave further interpretation to the reader . . .

3.4 PATTERNS OF ASSOCIATION BETWEEN Y AND TWO INDEPENDENT VARIABLES

A solid grasp of the implications of all possible relationships among one dependent variable and two independent variables is fundamental to understanding and interpreting the various multiple and partial coefficients encountered in MRC. This section is devoted to an exposition of each of these patterns and its distinctive substantive interpretation in actual research. For simplicity (and without loss of generality) we will define X_1 and X_2 in the following ways:

1. $r_{Y1} \geqslant r_{Y2}$. This simply defines X_1 as whichever of the two IVs has the larger correlation with Y.
2. $r_{Y1} \geqslant 0$. As we saw in Chapter 2, any negative correlation between two variables can be made positive by reversing the direction of scoring of one of the variables, without any fundamental change in the meaning of the association. Therefore, this condition serves only to define the direction in which X_1 is oriented.

3.4.1 Independence: $R^2_{Y \cdot 12} = 0$

This case is perhaps trivial, and occurs only when $r_{Y1} = r_{Y2} = 0$. Figure 3.4.1 presents the ballantine for both the case in which r_{12} is also 0 (Case 1), and the case in which $r_{12} \neq 0$ (Case 2). Since neither X_1 nor X_2 correlates with Y, the value of r_{12} is immaterial to $R^2_{Y \cdot 12}$ —the latter is zero in either circumstance.

3.4.2 Independence: $r_{12} = 0$

In the ballantine of Case 3, we see the case in which $r_{Y1} \neq 0$, while $r_{Y2} = 0$ and $r_{12} = 0$. Here, X_2 is wholly irrelevant to Y and $R^2_{Y \cdot 12} = r^2_{Y1}$. In Case 4, we see the case in which both $r_{Y1} \neq 0$, and $r_{Y2} \neq 0$, but $r_{12} = 0$. Each of the IVs makes a fully independent contribution to $R^2_{Y \cdot 12}$ and therefore $R^2_{Y \cdot 12} = r^2_{Y1} + r^2_{Y2}$. In this case each variable's β and semipartial r will equal its r with Y. The pr_i will be larger than the r_{Yi}, since the denominator will be further reduced by the removal of Y variance associated with the other variable.

3.4.3 Redundancy in Estimating Y

The pattern of correlation between three variables in which all correlations are positive and also $r_{Y2} \geqslant r_{Y1}r_{12}$ is represented as Case 5 in Figure 3.4.1. This is by far the most commonly encountered pattern in nonexperimental research in the behavioral sciences, once the IVs are both oriented so as to produce positive correlations with Y. The area of overlap reflects the redundancy of the IVs with respect to Y. The sr_i and β_i for each IV will be smaller than its r_{Yi} (and will have the same sign), and thus reflect the fact of redundancy: each IV is at least partly carrying information about Y which is being also supplied by the other.

Examples of two-variable redundancy come easily to mind. It occurs when one relates school achievement (Y) to parental income (X_1) and education (X_2), or delinquency (Y) to IQ (X_1) and school achievement (X_2), or fertility (Y) to race (X_1) and income (X_2), or psychiatric prognosis (Y) to rated symptom severity (X_1) and MMPI Schizophrenia score (X_2), or per capital gross national product (Y) to adult literacy rate (X_1) and per capita protein consumption (X_2), or—but the reader can supply many examples of his own. Indeed, redundancy among explanatory variables is the plague of our efforts to understand the causal

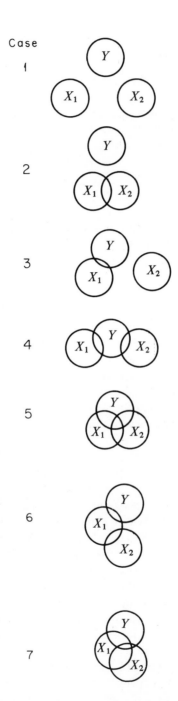

FIGURE 3.4.1 Ballantines representing different relationships of Y and two IVs.

structure that underlies observations in the behavioral and social sciences. This book is replete with examples of redundancy.

The special case when $r_{Y2} = r_{Y1}r_{12}$ (not illustrated in Figure 3.4.1) is of some interest because it means that the information with regard to Y carried by X_2 is *completely* redundant with that carried by X_1. Complete redundancy is equivalent to zero values for sr_2, pr_2, β_2, and B_2 which can only happen when their numerators equal zero, that is, when $r_{Y2} - r_{Y1}r_{12} = 0$. Exactly zero values are virtually nonexistent in real data, but very small and nonsignificant partial correlation and regression coefficients are all too frequently encountered. When this is the case, and r_{Y2} is sizable, the results are consistent with the causal hypothesis that r_{Y2} is "spurious," that is, the observed correlation of Y and X_2 is due to the fact that they are both causally dependent on X_1. Note that we say "consistent with," rather than "proves"—one cannot prove causality on the basis of correlational information alone. (See Section 9.7 for an extended discussion of this and related issues.)

3.4.4 Classical Suppression: $r_{Y2} = 0$

In the ballantine of Case 6 (Figure 3.4.1) we see an example of one type of suppression, called "classical" suppression.[6] Here, while $r_{Y2} = 0$, $r_{Y1} > 0$ and $r_{12} \neq 0$. To see what happens to $R^2_{Y \cdot 12}$ in this case we specialize Eq. (3.3.1) by substituting in it $r_{Y2} = 0$, which results in

$$(3.4.1) \qquad\qquad R^2_{Y \cdot 12} = \frac{r^2_{Y1}}{1 - r^2_{12}}.$$

Since the denominator is less than one, $R^2_{Y \cdot 12}$ must be greater than r^2_{Y1}. In spite of its zero correlation with Y, X_2 increases the variance accounted for in Y by "suppressing" some of the variance in X_1 that is irrelevant to Y.[7] For example, imagine a study of mountain climbing skills of teenagers in which the goal is to establish some paper and pencil test which will serve as an initial screening device for placement in groups of differing levels. A placement test is devised in which responses are required to appropriate questions on the techniques, equipment and situations involved in mountain climbing (X_1). Then field data are gathered in which the actual skills are observed and rated by the instructors to establish the criterion or dependent variable (Y). r_{Y1} is found to equal .32. In recognition of the fact that the questionnaire may also measure (be

---

[6] Conger (1974), publishing after the completion of this chapter, has referred to the three kinds of suppression discussed here as traditional (our classical), negative (our net, Section 3.4.5), and reciprocal (our cooperative, Section 3.4.6).

[7] Clearly the ballantine of Case 6 does not really portray $R^2_{Y \cdot 12}$ accurately, since X_2's unique contribution to Y is not shown. In fact, the existence of suppression of any kind makes the ballantine inaccurate in at least some respect, for example, portraying c as an area when it is negative in Figure 3.3.1.

correlated with) reading ability, which is expected to be irrelevant to climbing skills, a test of reading ability (X_2) is also administered which does indeed correlate .40 with the placement test (X_1) and zero with the field test of climbing. The resulting $R^2_{Y \cdot 12} = .32^2/(1 - .40^2) = .1219$, as opposed to $r^2_{Y1} = .32^2 = .1024$ for X_1 alone. Clearly X_2 adds an additional 2% to the Y variance being accounted for, that is, $sr^2_2 = .1219 - .1024 = .0195$. Equally clearly, X_1 adds .1219 to the zero contribution of X_2 (i.e., $r^2_{Y2} = 0$), so that $sr^2_1 = .1219 - 0 = .1219 = R^2_{Y \cdot 12}$. Thus, the sum of the two sr^2 values (.0195 + .1219 = .1414) exceeds $R^2_{Y \cdot 12}$, a necessary consequence of suppression when there are two IVs (although not necessarily with more than two IVs; see Section 3.6).

It is instructive to observe what happens to the partial correlation under conditions of classical suppression. Since X_1 adds the entire overlapping area, it is not surprising that $sr_1 = pr_1$. By Eq. (3.3.11), when $r_{Y2} = 0$,

$$pr_1 = \frac{r_{Y1} - r_{Y2}r_{12}}{\sqrt{1 - r^2_{Y2}}\sqrt{1 - r^2_{12}}} = \frac{r_{Y1}}{\sqrt{1 - r^2_{12}}} = sr_1$$

$$= \frac{.32}{\sqrt{1 - .40^2}} = \frac{.32}{.9165} = .349.$$

The partial correlation for X_2 when it is a classical suppressor is found by substituting $r_{Y2} = 0$ in Eq. (3.3.11):

$$pr_2 = \frac{0 - r_{Y1}r_{12}}{\sqrt{1 - r^2_{Y1}}\sqrt{1 - r^2_{12}}},$$

$$= \frac{0 - .32(.40)}{\sqrt{1 - .32^2}\sqrt{1 - .40^2}} = \frac{-.1280}{.8683} = -.147.$$

Examination of this formula will satisfy the reader that whatever the sign of r_{12}, pr_2 will be of opposite sign in this case.

It is useful to look at the effects on the standardized partial regression coefficients, β_1 and β_2. (We will henceforth, as with sr and pr, simplify the subscript notation for β and B, dropping Y and the partialled variable(s); for example, $\beta_{Y1 \cdot 2}$ is written β_1 and $\beta_{Y2 \cdot 1}$ is written β_2.)

$$\beta_1 = \frac{r_{Y1} - r_{Y2}r_{12}}{1 - r^2_{12}} = \frac{.32}{1 - .40^2} = .381,$$

$$\beta_2 = \frac{0 - .32(.40)}{1 - .40^2} = -.152.$$

We see that in both cases the standardized *partial* regression coefficients are absolutely larger than the standarized simple regression coefficients (*r*s).

One further word on suppression and β coefficients. Although *r* may never exceed the limits of +1 and −1, under conditions of suppression βs may do so. For example, imagine we have collected data resulting in the following correlations: $r_{Y1} = .70$, $r_{Y2} = .00$, $r_{12} = .60$. (This *is* a mathematically possible result; see Eq. 7.1.1 in Section 7.1.2, and its discussion.) Then,

$$\beta_1 = \frac{.70}{1 - .60^2} = 1.094,$$

$$\beta_2 = \frac{0 - .70(.60)}{1 - .60^2} = -.656.$$

3.4.5 Net Suppression:
Suppression When All rs Are Positive

Case 7 in Figure 3.4.1 presents a pattern of correlation which may be called "net" suppression.[8] This pattern, defined as $r_{Y2} < r_{Y1}r_{12}$ with all correlations positive, may be understood as demonstrating that, in spite of its positive validity coefficient (r_{Y2}), the function of X_2 in the multiple correlation and regression is primarily in suppressing a portion of the variance of X_1 that is irrelevant to (uncorrelated with) Y.

An example: in a study of assessed valuation of homes where both husband and wife are present, the IVs are husband's income and wife's income. The correlation between husband's income and home value, r_{Y1}, equals .70. Wife's income correlates with home value (r_{Y2} =) .20, and the correlation between husband's and wife's income is (r_{12} =) .60. The *pr* coefficients for X_1 and X_2 by Eq. (3.3.11) are

$$pr_1 = \frac{.70 - .20(.60)}{\sqrt{1 - .60^2}\sqrt{1 - .20^2}} = \frac{.5800}{.7838} = .740,$$

$$pr_2 = \frac{.20 - .60(.70)}{\sqrt{1 - .60^2}\sqrt{1 - .70^2}} = \frac{-.2200}{.5713} = -.385.$$

These (fictitious) results provide a good example of how the substantive context of a research must finally determine the interpretive thrust of the findings. It would be consistent to stress the "net suppressing" influence of wife's income in the findings, noting that pr_1 (and sr_1) are larger than r_{Y1}. Thus, we may say that removing the portion of the variance in husband's income associated with wife's income increases the association of husband's income with home value. On the other hand, we may choose to concentrate on the pr_2 = −.385 and say that controlling (partialling) the effects of husband's income (or, equivalently, for subsets of cases where all husbands have the same income), the greater the wife's income, the less the home value. Our theory may predict that for any given husband's income, when the wife has little or no income of her own, the family's interests are more likely to be concentrated on their home and

---

[8] As is the case in classical suppression, the ballantine here represents the raw squared correlations but does not accurately depict $R^2_{Y \cdot 12}$.

therefore its value will be larger, on the average, than when the wife has significant income. The results are also consistent with this hypothesis.

These alternative emphases point up a very important fact about suppression. Suppression is a phenomenon involving two or more IVs, and it is fully symmetrical or mutual. Whenever it can be said that X_2 suppresses X_1, it may also be said that X_1 suppresses X_2. Although the negative partial coefficients will end up attached to the variable with the initially smaller validity coefficient, substantive considerations may lead the researcher to concentrate on its partial relationship to Y rather than that of the other IV. Even in the case of classical suppression when $r_{Y2} = 0$, the emphasis may be placed on the fact that X_1 was "hiding" the true relationship between Y and X_2 as revealed in pr_2. "You pay your money" in the form of a theory, and "you take your choice."

Finally, it is instructive to examine what happens to the β coefficients. In our example involving home value (by Eq. 3.2.3),

$$\beta_1 = \frac{.70 - .20(.60)}{1 - .60^2} = \frac{.5800}{.6400} = .906,$$

$$\beta_2 = \frac{.20 - .70(.60)}{1 - .60^2} = \frac{-.2200}{.6400} = -.344.$$

Notice that each of these coefficients falls outside the boundaries set by its r_{Yi} and 0, that is, $\beta_1 = +.906$ is not in the range 0 to +.70, and $\beta_2 = -.344$ is not in the range 0 to +.20. This will be true in all types of suppression, and may serve as a convenient marker for identifying suppression between two (or among more) IVs.

3.4.6 Cooperative Suppression: $r_{12} < 0$

In his never ending search for high Rs, no circumstance is more attractive to the researcher (although rarely attained) than one in which IVs which correlate positively with Y correlate negatively with each other (or, equivalently, the reverse). r_{12}, being negative, involves a portion of the variance in the IVs all of which is irrelevant to Y; thus, when each variable is partialled from the other, all indices of relationship with Y are enhanced.

For example, imagine that a Director of Personnel is establishing a procedure for selecting sales personnel from among job applicants. A sample of current salesmen is drawn, and each person is rated for overall success in sales performance. These ratings constitute the "criterion" or dependent variable, Y. A series of interviews leads the researcher to suspect that two major components of sales success are social aggressiveness and habits and skills with regard to record keeping. Measures of these two characteristics are devised and administered to the sample of salesmen. The correlation between the measure of social aggressiveness (X_1) and sales success (Y) is found to be .29, the correlation between record keeping (X_2) and Y is .24, and $r_{12} = -.30$, indicating an overall tendency for those high on social aggressiveness to be relatively low on record keeping.

The semipartial correlations are

$$sr_1 = \frac{.29 - .24(-.30)}{\sqrt{1-.30^2}} = \frac{.29 + .0720}{.9539} = .379,$$

$$sr_2 = \frac{.24 - .29(-.30)}{\sqrt{1-.30^2}} = \frac{.24 + .0870}{.9539} = .343,$$

both larger than their respective validity coefficients. $R^2_{Y\cdot12}$ is found by Eq. (3.3.1) to be

$$R^2_{Y\cdot12} = \frac{.29^2 + .24^2 - 2(.29)(.24)(-.30)}{1-(-.30)^2} = .2016,$$

and we note that

$$r^2_{Y1} + r^2_{Y2} < R^2_{Y\cdot12} < sr^2_1 + sr^2_2$$
$$.0841 + .0576 < .2016 < .1440 + .1175.$$

Thus, the independent variables are mutually enhancing under conditions of "cooperative" suppression, and each variable accounts for a *larger* proportion of the *Y* variance in the presence of the other than it does alone.

Computing the βs we find that $\beta_1 = .398$ and $\beta_2 = .359$, both of which retain the sign and exceed in magnitude their respective r_{Yi}, .29 and .24.

Finally, it is important to note that all three kinds of suppression—classical, net, and cooperative—are not frequently found in behavioral science studies. The detailed presentation here is in the interest of enabling the researcher to recognize them when they do occur, and for their value as quasiparadoxical curiosities. A β_i coefficient which falls outside the limits defined by r_{Yi} and 0 signals the presence of suppression. If the X_i in question has a zero (in practice, very small) correlation with *Y*, the situation is one of classical suppression. If its β_i is of opposite sign from its r_{Yi}, it is serving as a net suppressor. If its β_i exceeds its r_{Yi} and is of the same sign, cooperative suppression is indicated.

3.5 MULTIPLE REGRESSION/CORRELATION WITH *k* INDEPENDENT VARIABLES

3.5.1 Introduction

When more than two IVs are related to *Y*, the computation and interpretation of multiple and partial coefficients proceed by direct extension of the two-IV case. The goal is again to produce a regression equation for the *k* IVs of the (raw score) form

(3.5.1) $\hat{Y} = B_{Y1\cdot23...k}X_1 + B_{Y2\cdot13...k}X_2 + B_{Y3\cdot12...k}X_3 + \cdots$

$$+ B_{Yk\cdot123...k-1}X_k + A_{Y\cdot123...k},$$

or, expressed in simpler subscript notation,

$$\hat{Y} = B_1 X_1 + B_2 X_2 + B_3 X_3 + \ldots + B_k X_k + A.$$

When this equation is applied to the data, it yields a set of $\hat{Y}$ values (one for each of the n cases) for which the sum of the $(Y - \hat{Y})^2$ values over all n cases will (again) be a minimum. Obtaining these raw-score partial regression weights, the B_i, involves solving a set of k simultaneous equations in k unknowns. In keeping with the conviction of the authors that carrying out the complex computations involved is not required for a solid working knowledge of MRC, these procedures will not be described here. However, the interested reader is referred to Appendix 2 for a description and worked example of the MRC solution which is feasible for hand (by which we mean desk or pocket calculator) calculation when there are not more than five or six IVs. Readers familiar with matrix algebra may turn to Appendix 1 for a presentation of the general solution for determining the multiple and partial coefficients for the k variable case.

The most frequent method of obtaining these coefficients in the scientific community at large is by means of one of the many widely available computer programs. A discussion of the use of the computer and of the most popular "canned" programs for MRC can be found in Appendix 3. The purpose of this section is to lay down a foundation for understanding the various types of coefficients produced by MRC for the general case of k independent variables, and their relationship to various MRC strategies appropriate to the investigator's research goals.

3.5.2 Partial Regression Coefficients

By direct extension of the one- and two-IV cases, the raw score partial regression coefficient B_i $(= B_{Yi \cdot 123 \ldots (i) \ldots k})$ is the constant weight by which each value of the variable X_i is to be multiplied in the multiple regression equation which includes all k IVs. Thus, B_i is the average or expected change in Y for each unit increase in X_i when the value of each of the $k - 1$ other IVs is held constant.

β_i is the partial regression coefficient when all variables have been standardized. Such standardized coefficients are of interpretive interest when the analysis concerns test scores or indices whose scaling is arbitrary.

For example, let us return to the study in which we seek to account for differences in academic rank in some university department by means of other characteristics of the persons holding these ranks. The two IVs used thus far were the number of publications (X_1) and the number of years since each faculty member had received a doctoral degree (X_2). We now wish to consider two additional independent variables, the sex of the professor, and the number of citations of his (or her) work in the scientific literature in the previous year. These data are presented in Table 3.5.1, where sex (X_3) is coded (scored) 1 for female and 0 for male faculty. The correlation matrix shows us that sex is negatively correlated with academic rank $(r_{Y3} = -.242)$, women $(X_3 = 1)$ being of lower academic rank on the average than are men $(X_3 = 0)$. The number of

TABLE 3.5.1

Illustrative Data: Academic Rank and Four Independent Variables

Subject	Academic rank (Y)	No. of publications (X_1)	Years since Ph.D. (X_2)	Sex (X_3)	No. of citations (X_4)	$\hat{Y}$
1	1	2	1	0	1	1.414
2	1	4	2	0	0	1.365
3	1	5	5	1	1	1.296
4	1	12	7	1	0	1.400
5	1	5	10	0	0	1.679
6	1	9	4	0	1	1.680
7	2	3	3	1	0	1.050
8	2	1	8	0	1	1.643
9	2	8	4	0	0	1.528
10	2	12	16	1	4	2.246
11	2	9	15	0	0	1.951
12	2	4	19	0	3	2.374
13	3	8	8	0	5	2.321
14	3	11	14	0	0	1.959
15	3	21	28	0	3	3.080
m	1.80	7.60	9.60	.267	1.267	1.80
sd^2	.560	24.64	53.5065	.1960	2.5956	.262
sd	.748	4.96	7.25	.443	1.611	.512

rs

	Y	X_1	X_2	X_3	X_4
X_1	.463	1.000	.683	.049	.297
X_2	.612	.683	1.000	−.154	.460
X_3	−.242	.049	−.154	1.000	−.006
X_4	.487	.297	.460	−.006	1.000
$\hat{Y}$	.683				

citations in the current literature (X_4) is positively associated with academic rank ($r_{Y4} = .487$), as well as with the other IVs except sex. Sex correlates very little with the other IVs, except for a slight tendency for the women to be more recent Ph.D.s than the men ($r_{23} = -.154$).

The (raw-score) multiple regression equation for estimating academic rank from these four IVs which may be obtained from computer output (Appendix 3) or by the matrix inversion method of Appendix 2 (where this problem is used illustratively) is

$$\hat{Y} = .0224\,X_1 + .0364\,X_2 - .3266\,X_3 + .1296\,X_4 + 1.2030.$$

These partial B_i coefficients indicate that for any given values of the other IVs: an increase of one in the number of citations is associated with about 13/100 of a unit increase in academic rank (B_4 = .1296); an increase of one unit in X_3, and hence the average difference between the sexes (holding constant the other IVs) is −.3266 of a unit in academic rank (favoring men); and the effects of a unit increase in publication (X_1) and in years since degree (X_2) are .0224 and .0364, respectively. Note also that A = 1.2030 is the estimated rank of a hypothetical male professor fresh from his doctorate with neither publications nor citations, that is, all X_i = 0.

In this problem, the rank estimated by the four IVs for the first faculty member (Table 3.5.1) is

$$
\begin{aligned}
\hat{Y} &= \quad .0224(2) + .0364(1) - .3266(0) + .1296(1) + 1.2030 \\
&= \quad .0448 \quad + .0364 \quad - \quad 0 \quad + .1296 \quad + 1.2030 \\
&= \quad 1.414.
\end{aligned}
$$

The remaining estimated values are given in the last column of Table 3.5.1. (Although we report the B_i and A to more than the usual number of places, if the reader checks these $\hat{Y}$ values he will nevertheless find small rounding discrepancies; recall the warning of Section 1.2.2.)

The regression equation may be written in terms of standardized variables and β coefficients as

$$\hat{z}_Y = .149\, z_1 + .352\, z_2 - .193\, z_3 + .279\, z_4 .$$

The β values may always be found from B values by inverting Eq. (3.2.4):

(3.5.2) $$\beta_i = B_i\, \frac{sd_i}{sd_Y},$$

for example, β_4 = (.1296)(1.611/.748) = .279.

3.5.3 R and R^2

Application of the regression equation to the IVs would yield a set of estimated $\hat{Y}$ values. The *simple* correlation of Y with $\hat{Y}$ equals the multiple correlation; in this example, $r_{Y\hat{Y}} = R$ = .683. As with the one or two independent variable case, R^2 equals the proportion of Y variance accounted for and $R^2 = sd_{\hat{Y}}^2/sd_Y^2$ = .262/.560 = .4671. R^2 may also be written as a function of the original correlations with Y and the β coefficients by extension of Eq. (3.3.2):

(3.5.3) $$R_{Y\cdot 12\ldots k}^2 = \sum \beta_i r_{Yi},$$

where the summation is over the k IVs. Thus in the current example,

$$
\begin{aligned}
R_{Y\cdot 1234}^2 &= .149(.463) + .352(.612) - .193(-.242) + .279(.487) \\
&= .4671
\end{aligned}
$$

as before.

Lest the reader think that this represents a way of apportioning the Y variance accounted for among the IVs (that is, that X_i's proportion is its $\beta_i r_{Yi}$), it is important to recall that β_i and r_{Yi} may be of opposite sign (under conditions of suppression). Thus, the suppressor variable on this interpretation would appear to account for a negative proportion of the Y variance, clearly a conceptual impossibility. The fact that $\beta_i r_{Yi}$ is not necessarily positive precludes the use of Eq. (3.5.3) as a variance-partitioning procedure.

R^2 may also be obtained as a function of the β coefficients and the associations between the IVs as

(3.5.4) $$R^2 = \sum \beta_i^2 + 2 \sum \beta_i \beta_j r_{ij} \quad (i \neq j),$$

where the first summation is over the k IVs, and the second over the $k(k-1)/2$ distinct pairs of IVs. In the current problem,

$$
\begin{aligned}
R^2_{Y \cdot 1234} = \; & .149^2 + .352^2 + (-.193)^2 + .279^2 + 2[(.149)(.352)(.683) \\
& + (.149)(-.193)(.049) + (.149)(.279)(.297) \\
& + (.352)(-.193)(-.154) + (.352)(.279)(.460) + (-.193)(.279)(-.006)] \\
= \; & .4671.
\end{aligned}
$$

This formula *appears* to partition R^2 into portions accounted for by each variable uniquely and portions accounted for jointly by pairs of variables, and some authors so treat it. However, we again note that any of the $k(k-1)/2$ terms $\beta_i \beta_j r_{ij}$ may be negative, as indeed happens for the $\beta_1 \beta_3 r_{13}$ product in this example. Therefore, Eq. (3.5.4), as well as Eq. (3.5.3), can not serve as a variance partitioning scheme. This equation does, however, make clear what happens when all correlations between pairs of IVs equal 0. The triple product terms will all contain $r_{ij} = 0$ and hence drop out, and $R^2 = \Sigma \beta_i^2 = \Sigma r_{Yi}^2$, as was seen for the two-IV case (Section 3.4.2).

3.5.4 sr and sr^2

The semipartial correlation coefficient sr_i and its square sr_i^2 in the general case of k IVs may be interpreted by direct extension of the two IV case. Thus sr_i^2 equals that proportion of the Y variance accounted for by X_i beyond that accounted for by the other $k-1$ IVs, and

(3.5.5) $$sr_i^2 = R^2_{Y \cdot 12 \ldots i \ldots k} - R^2_{Y \cdot 12 \ldots (i) \ldots k}$$

(the parenthetical i signifying its *omission* from the second R^2), or the increase in the squared multiple correlation when X_i is included over the R^2 which includes the other $k-1$ IVs, but excludes X_i. This will be seen to be a very useful coefficient as it may be thought of as the *unique* contribution of X_i to R^2 in the context of the remaining $k-1$ IVs. As in the two-IV case, the semipartial r equals the correlation between that portion of X_i which is uncorrelated with the remaining IV(s) and Y,

(3.5.6) $$sr_i = r_{Y(i \cdot 12 \ldots (i) \ldots k)}$$

$$= r_{Y(X_i - \hat{X}_i \cdot 12 \ldots (i) \ldots k)} \cdot$$

As might be expected, sr_i may also be written as a function of the multiple correlation of the other IVs with X_i,

(3.5.7) $$sr_i = \beta_i \sqrt{1 - R_{i \cdot 12 \ldots (i) \ldots k}^2} \cdot$$

Neither sr_i nor sr_i^2 is provided as routine output by most MRC computer programs. Occasionally, sr_i^2 values will be provided, possibly labeled as the "unique" contribution to R^2. Since $R_{i \cdot 12 \ldots (i) \ldots k}^2$ is also not usually provided by computer output,[9] the above formula will be of less practical use than formulas which express sr_i as a function of typically available values (see Appendix 3). When pr_i is available, sr_i is readily determined by

(3.5.8) $$sr_i^2 = \frac{pr_i^2}{1 - pr_i^2} (1 - R_{Y \cdot 123 \ldots k}^2).$$

3.5.5 pr and pr^2

The partial correlation coefficient pr_i, we recall from the two-IV case, is the correlation between that portion of Y which is independent of the remaining variables $Y - \hat{Y}_{12 \ldots (i) \ldots k}$, and that portion of X_i that is independent of the remaining variables, $X_i - \hat{X}_i \cdot 12 \ldots (i) \ldots k$, that is,

(3.5.9) $$pr_i = r_{Yi \cdot 12 \ldots (i) \ldots k}$$

$$= r_{(Y - \hat{Y}_{12 \ldots (i) \ldots k})(X_i - \hat{X}_i \cdot 12 \ldots (ij) \ldots k)} \cdot$$

pr^2 is thus interpretable as the proportion of that part of the Y variance which is independent of the remaining IVs (i.e., of $1 - R_{Y \cdot 12 \ldots (i) \ldots k}^2$) accounted for *uniquely* by X_i:

(3.5.10) $$pr_i^2 = \frac{sr_i^2}{1 - R_{Y \cdot 12 \ldots (i) \ldots k}^2}.$$

It can be seen that pr_i^2 will virtually always be larger than and can never be smaller than sr_i^2, since sr_i^2 is the unique contribution of X_i expressed as a proportion of the *total* Y variance while pr_i^2 expresses the same unique contribution of X_i as a proportion of that *part* of the Y variance not accounted for by the other IVs.

3.5.6 Illustrative Example

Table 3.5.2 presents the semipartial and partial correlations and their squares for the academic rank example. We see that X_1 accounts uniquely for about 1% of

---

[9] They may, however, be obtained by running additional regression equations using IVs as dependent variables. Alternatively as described in Appendix 1, the quantity $1/(1 - R_{i \cdot 12 \ldots}^2$ $(i) \ldots k)$ is a diagonal element in the inverse of the correlation matrix.

TABLE 3.5.2
Zero-order, Partial and Semipartial Coefficients for Academic Rank
with Four Independent Variables

		r_{Yi}	r^2_{Yi}	sr_i	sr^2_i	pr_i	pr^2_i
X_1	No. of publications	.463	.2144	.106	.0112	.144	.0207
X_2	Years since Ph.D.	.612	.3749	.230	.0531	.301	.0906
X_3	Sex	−.242	.0584	−.186	.0345	−.247	.0608
X_4	No. of citations	.487	.2368	.247	.0609	.320	.1026

$$R^2_{Y \cdot 1234} = .4671, \qquad R_{Y \cdot 1234} = .683$$

the Y variance ($sr^2_1 = .0112$), but about 2% of the Y variance not accounted for by the other three variables ($pr^2_1 = .0207$). Notice that in this example the partial coefficients of the four IVs are ordered differently from the zero-order correlations. For example, although number of publications taken by itself accounts for .2144 (= r^2_{Yi}) of the variance in academic rank, it *uniquely* accounts for only .0112 of this variance, whereas sex with an initially much smaller portion (r^2_{Y3} = .0584) accounts *uniquely* for .0345 (= sr^2_3) of the variance in academic rank. The reason for this may be seen by referring back to the correlations among the IVs in Table 3.5.1, where number of publications is seen to have substantially higher correlations with the other IVs than does sex. Thus, the correlation between X_1 and Y is more redundant with the relationship between the other IVs and Y than is that of sex. It should also be reiterated that none of these coefficients provides a partitioning of R^2 among IVs, unless all correlations between IVs equal zero. In the typical case with correlated IVs, the sr^2_i will sum to less than R^2. Nevertheless, it is still not appropriate even to partition R^2 into these unique (sr^2) and "common" ($R^2 - \Sigma\, sr^2$) portions. As we saw in Section 3.4.3, under conditions of suppression the sum of the sr^2 values for two IVs *must* exceed R^2. When suppression occurs in the more-than-two-IV case, the $\Sigma\, sr^2$ value may be less than, greater than or equal to R^2, depending on the size of the contribution of the variables involved in the suppression and the amount of redundancy among these and other variables. We reject categorically all efforts at partitioning R^2 that can result in negative quantities. (In Section 3.6.2, we present a different approach to this problem.)

3.6 ANALYTIC STRATEGIES

3.6.1 The Simultaneous Model

Thus far, we have been looking at conventional applications of MRC in which all k IVs have been treated simultaneously and on an equal footing. Such a research strategy is clearly most appropriate when we have no logical or theoretical basis for considering any variable to be prior to any other, either in terms of a hypothetical causal structure of the data or in terms of its relevance to the research goals. The resulting regression equation will provide the best linear estimate of the dependent variable for the k IVs and is suitable for use in estimating Y values for other samples from the same population. The partial and semipartial correlations and their squares serve to index the relationship between X_i and Y when all other IVs have been taken into account (held constant by partialling). The values of these partial coefficients, of course, may depend critically on what other IVs are in the equation.

3.6.2 The Hierarchical Model

An alternative analytic strategy to the simultaneous model is one in which the k IVs are entered cumulatively according to some specified hierarchy which is dictated in advance by the purpose and logic of the research. The hierarchical model calls for a determination of R^2 and the partial coefficients of each variable at the point at which it is added to the equation. Note that with the addition of the ith IV, the MRC analysis *at that stage* is simultaneous in i variables. Thus, a full hierarchical procedure consists of a series of k simultaneous MRC analyses, each with one more IV than its predecessor, in a specified order. Since at each stage the R^2 increases, the ordered series of R^2 in hierarchical analysis is called the "cumulative" R^2 series.

A major advantage of the hierarchical MRC analysis of data is that once the order of the IVs has been specified, a unique partitioning of the total Y variance accounted for by the k IVs, $R^2_{Y \cdot 12 \ldots k}$, may be made. Indeed, this is the *only* basis on which variance partitioning can proceed with correlated IVs. Since the sr_i^2 at each stage is the increase in R^2 associated with X_i when all (and only) previously entered variables have been partialled, an *ordered* variance partitioning procedure is made possible by

(3.6.1) $\quad R^2_{Y \cdot 123 \ldots k} = r^2_{Y1} + r^2_{Y(2 \cdot 1)} + r^2_{Y(3 \cdot 12)} + r^2_{Y(4 \cdot 123)} + \cdots + r^2_{Y(k \cdot 123 \ldots k-1)}$

$\qquad\qquad\qquad = r^2_{Y1} + sr^2_{2 \cdot 1} + sr^2_{3 \cdot 12} + sr^2_{4 \cdot 123} + \cdots + sr^2_{k \cdot 123 \ldots k-1}.$

Each of the k terms is found from a simultaneous analysis of IVs in the equation at that point in the hierarchy; each gives the increase in Y variance accounted for by the IV entering at that point *beyond* what has been accounted for by the previously entered IVs. r^2_{Y1} may be thought of as the increment from zero due to the first variable in the hierarchy, an sr^2 with nothing partialled. Summing the terms up to a given stage in the hierarchy gives the cumulative R^2 at that stage, for example, $r^2_{Y1} + sr^2_{2 \cdot 1} + sr^2_{3 \cdot 12} = R^2_{Y \cdot 123}$.

Since the increment attributable to any variable may change considerably depending on where it appears in the hierarchy, and therefore what other variables are partialled from it, let us examine the considerations which may lead to an ordering of the variables. Although several different rationales are offered, these may be combined in determining the order of IVs in actual research.

Causal Priority

Perhaps the most straightforward use of the hierarchical model is when IVs can be ordered with regard to their temporally or logically determined causal priority. Thus, for example, when two IVs are sex (male or female) and an attitudinal variable, sex must be considered the causally prior variable since (at the very least) it antedates attitude. Another common example of causal ordering is in determining the effects of different treatments on some dependent variable when individuals differ with regard to their initial state. For example, a researcher is studying the outcome of marital counseling using as Y a measure of the couple's judgment of how successful they have been in meeting their goals. The IVs include severity of the marital problems as judged at the start of treatment (X_1), the training level of the counselor (X_2), and which of two treatment strategies was used (X_3). The assumed causal priority of the variables is X_1, X_2, X_3. Although a random assignment of couples at differing given levels of severity to the counselor and treatment groups may have been originally planned, scheduling and other problems may have resulted in correlations among the IVs (for example, one of the treatments may have been used more often for more severe problems). Or perhaps, as in many other real situations, the investigator did not wish to lose the information inherent in the actual score values on X_1 and X_2 by arbitrarily chopping the distribution into groups of equal size to make analysis by AV/ACV convenient.

The hierarchical MRC analysis may proceed by entering the IVs in the specified order and determining R^2 after each addition. Thus, suppose that r_{Y1} = −.32, and therefore $R^2_{Y\cdot1}$ $(= r^2_{Y1})$ = .1024, that is, severity of presenting problem (inversely) accounts for slightly over 10% of the variance of counseling success. When we add therapist training level to the equation we find that $R^2_{Y\cdot12}$ = .1432 and may say that, when we have controlled for (partialled) initial severity, training level accounted for about 4% of the Y variance, that is, $sr^2_{2\cdot1} = R^2_{Y\cdot12} - R^2_{Y\cdot1}$ = .0408. The direction of this relationship may be seen from the sign of the partial coefficients (unsquared). Finally, we add the treatment variable to the set and find that $R^2_{Y\cdot123}$ = .1756 $(sr^2_{3\cdot12}$ = .0324). Thus, treatment accounts for 3% of the Y variance after we have controlled for (partialled) severity of presenting problem and therapist training level. Note that since this is the last IV to be entered, the analysis at this stage is identically that which we would have performed had we simply proceeded with the simultaneous model for the $k = 3$ IVs. Also, note that the relevant significance test for the sr^2 contribution of each IV to R^2, the increment due to each IV in turn, is carried out when it is added to the equation (see the following section on significance tests). The final R^2 for the k IVs is the sum of these increments, by Eq. (3.6.1):

$$R^2_{Y\cdot123} = r^2_{Y1} + sr^2_{2\cdot1} + sr^2_{3\cdot12},$$
$$.1756 = .1024 + .0408 + .0324.$$

A special case of the hierarchical model is employed in the analysis of change. Under circumstances in which pre and post values are available on some variable and the researcher wishes to determine whether and to what extent treatment or other variables are associated with change, the postscore may be used as the dependent variable, with prescore entered as the first IV in the hierarchy. Unlike the alternative method involving difference (post- minus pre-) scores, when subsequent IVs are entered into the equation their partial correlations will reflect their relationship with postscores from which prescore influence has been removed. (This is, in fact, an ACV accomplished by means of MRC. Chapter 9 provides a full discussion of ACV, including its use in the study of change in Section 9.6.)

Research Relevance

Not infrequently an investigator gathers data on a number of variables in addition to those IVs that reflect the major goals of the research. Thus, X_1 and X_2 may carry the primary focus of the study but X_3, X_4, and X_5 are also available. The additional IVs may be secondary because they are viewed as having lesser relevance to the dependent variable than do X_1 and X_2, or because hypotheses about their relationships are "weak" or exploratory. Under these circumstances, X_1 and X_2 may be entered into the equation first (perhaps ordered on the basis of a causal model) and then X_3, X_4, and X_5 may follow, ordered on the basis of their presumed relevance and/or priority. Aside from the clarity in interpretation of the influence of X_1 and X_2 which is likely to result from this approach (since the secondary X_3, X_4, and X_5 variables are not partialled from X_1 and X_2), the statistical power of the test of the major hypothesis is likely to be maximal (see Sections 3.8.3 and 4.6.3).

Hierarchical Analysis When Independent Variables Are Highly Correlated

When some or all of the IVs are substantially correlated with each other, the coefficients obtained by the simultaneous model for the entire set may be highly misleading. This situation is sometimes called the problem of multicollinearity. Since all other IVs have been partialled from the relationship between each IV and Y, when two or more IVs have highly redundant associations with Y, none of them may show nontrivial unique relationships, that is, all may show very small sr_i^2.

For example, imagine a study of the influence of attitude toward Traditional Feminine Role (TFR) and the personality trait Ascendance versus Submission (A–S) on the Level of Aspiration (LA) of female high school seniors. The dependent variable (LA) is a scale including items on plans for further education and occupational goals. In addition to the above two IVs, a measure of academic achievement, Grade Point Average (GPA), is included as it is anticipated that aspirations are likely to be tempered by past achievement experience. The correlations and the sr_i, sr_i^2, and β_i of the simultaneous model are presented in Table 3.6.1. We see in the correlation matrix that all three IVs show moderate

TABLE 3.6.1
Illustration of Simultaneous and Hierarchical Models

	LA Y	GPA X_1	TFR X_2	A–S X_3	
GPA X_1	.40	1.00	.20	.15	$R_{Y\cdot123} = .525$
TFR X_2	.40	.20	1.00	.70	$R^2_{Y\cdot123} = .2754$
A–S X_3	.35	.15	.70	1.00	

	Simultaneous model			*Three hierarchical models*		
	sr_i	sr_i^2	β_i	Cum. R^2	Cum. R^2	Cum. R^2
X_1	.325	.1058	.332	X_1 .1600	X_1 .1600	X_3 .1225
X_2	.171	.0294	.242	+ X_2 .2667	+ X_3 .2460	+ X_2 .1696
X_3	.093	.0087	.131	+ X_3 .2754	+ X_2 .2754	+ X_1 .2754

relationships with LA (= Y) (.35 to .40) and that TFR (= X_2) and A–S (= X_3) are very substantially correlated with each other (.70) and much less related to GPA (X_1). The multiple correlation of these IVs with LA is .525, indicating that they together account for 27.54% of its variance. The sr_i^2s determined simultaneously indicate that X_1 accounts *uniquely* for 10.58% of the Y variance and that X_2 and X_3 account *uniquely* for somewhat less than 3% and 1% respectively. These figures might lead a researcher to conclude erroneously that X_2 and X_3 are of only trival importance in accounting for LA.

However, a look at the difference between r^2_{Y1} = .1600 and $R^2_{Y\cdot123}$ = .2754 reveals that together, X_2 and X_3 add 11.54% to the Y variance accounted for, a sizable amount. Suppose we were to specify the order in which variables are hierarchically entered into the equation. Table 3.6.1 presents the cumulative R^2 which would result for three of the six possible orders. (This is done here for illustrative purposes; ordinarily only a single sequence is used.) We have seen that r^2_{Y1} = .1600. When X_2 is added to X_1, it accounts additionally for ($R^2_{Y\cdot12}$ − r^2_{Y1} = $sr^2_{2\cdot1}$ = .2667 − .1600 =) .1067 of Y's variance, and X_3 adds ($R^2_{Y\cdot123}$ − $R^2_{Y\cdot12}$ = $sr^2_{3\cdot12}$ =) .0087 to X_1 and X_2. When X_3 is entered immediately after X_1, it increases R^2 by ($R^2_{Y\cdot13}$ − r^2_{Y1} = $sr^2_{3\cdot1}$ = .2460 − .1600 =) .0860. Although the sequence in which the variables are entered must be determined by the logic of the research, in no sequence can the importance of X_2, X_3, or both be lost when the analysis proceeds in this way.

Two variations on this role of revealing patterns of relationship by hierarchical analysis may also occur in research. Occasionally the goal of a study may be to determine whether, for example, X_4 adds more to X_1, X_2, and X_3 in accounting for Y than does X_5. Such considerations may be especially important in MRC analysis by sets of IVs (see Chapter 4). Under other conditions of high correlation among IVs (and occasionally without it), we will find that in the final equation the β_i of one or more X_i is outside the limits of the interval between r_{Yi} and zero. Under these circumstances, as we saw in Section 3.4, some suppression is taking place. A hierarchical analysis may then be used to determine which IVs are involved in the suppression. In general, if the β_i for X_i moves toward zero when a later X_j is added to the equation, the situation is one of simple redundancy between X_i and X_j in accounting for Y. If, on the other hand, β_i changes sign or increases, the relationship between X_i and X_j is one of suppression.

Hierarchical Analysis Required by Structural Properties

Several types of variables which may be used as IVs in MRC have characteristics which make assessment of their contribution to R^2 meaningful only after related variables have been partialled, thus mandating a specific order. This occurs in the representation of interactions (Chapter 8), curvilinear relationships (Chapter 6), and missing data (Chapter 7). Since the exposition of these methods requires entire chapters, they will not be illustrated here; however, they will be found to entail important applications of the hierarchical model.

In general, the moral of the hierarchical model is that the contribution to R^2 associated with any variable may depend critically upon what else is in the equation. The story told by a single simultaneous analysis for all k variables may, for many purposes, be incomplete. Hierarchical analysis of the variables typically adds to the researcher's understanding of the phenomena being studied, since it requires thoughtful input by the researcher in determining the order of entry of IVs, and yields successive tests of the validity of the hypotheses which define that order.

3.6.3 Stepwise Regression

Although stepwise regression has certain surface similarities with hierarchical MRC, it is considered separately, primarily because it differs in its underlying philosophy, and also because special computer programs and options are available for its computation. As discussed here, these programs are designed to select from a group of IVs the one variable at each stage which has the largest sr^2, and hence makes the largest contribution to R^2. Such programs typically stop admitting IVs into the equation when no IV makes a contribution which is statistically significant at a level specified by the program user.[10] Thus, the stepwise procedure defines an a posteriori order based solely on the relative uniqueness of the variables in the sample at hand.

---

[10] Some stepwise programs operate backwards, that is, by elimination. All k IVs are entered simultaneously and the one making the smallest contribution is dropped. Then the $k - 1$ remaining variables are regressed on Y, and again the one making the smallest contribution is dropped. And so on. The output is given in reverse order of elimination. This order need not agree with that of the accretion method described here.

When an investigator has a large pool of potential IVs, and very little theory to guide selection among them, these programs are a sore temptation. If the computer selects the variables, the investigator is relieved of the responsibility of making decisions about their logical or causal priority or relevance before the analysis, although interpretation of the findings may not be made easier. We take a dim view of the routine use of stepwise regression in explanatory research for various reasons (see below), but mostly because we feel that more orderly advance in the behavioral sciences is likely to occur when researchers armed with theories provide a priori hierarchical ordering which reflects causal hypotheses rather than when computers order IVs post and ad hoc for a given sample.

An option which is available on some computer programs allows for an a priori specification of a hierarchy among *groups* of IVs. An investigator may be clear that some groups of variables are logically, causally, or structurally prior to others, and yet not have a basis of ordering variables within such groups. Under such conditions, variables may be labeled for entering in the equation as one of the first, second, or up to hth group of variables, and the sequence of variables within each group is determined by the computer in the usual stepwise manner. This type of analysis is likely to be primarily hierarchical (between classes of IVs) and only incidentally stepwise (within classes), and computer programs so organized may be effectively used to accomplish hierarchical MRC analysis by *sets* of IVs, as described in the next chapter (Section 4.2.2).

Probably the most serious problem in the use of stepwise regression programs arises when a relatively large number of IVs is used. Since the significance test of an IV's contribution to R^2 proceeds in ignorance of the large number of other such tests being performed at the same time for the other competing IVs, there can be very serious capitalization on chance. The result is that neither the statistical significance tests for each variable nor the overall tests on the multiple R^2 at each step are valid (see the next section). A related problem with the free use of stepwise regression is that in many research problems, the ad hoc order produced from a set of IVs in one sample is likely not to be found in other samples from the same population. When among the variables competing for entry at any given step, there are trivial differences among their partial relationships with Y, the computer will dutifully choose the largest for addition at that step. In other samples, and, more important, in the population, such differences may well be reversed. When the competing IVs are substantially correlated with each other, the problem is likely to be compounded, since the losers in the competition may not make a sufficiently large unique contribution to be entered at any subsequent step before the problem is terminated by "nonsignificance."

Although, in general, stepwise programs are designed to approach the maximum R^2 with a minimum number of IVs for the sample at hand, they may not succeed very well in practice. Sometimes, with a large number of IVs, variables that were entered into the equation early no longer have nontrivial (or "significant") relationships after other variables have been added. Some programs provide for the removal of such variables, but others do not. Also, although it is admittedly not a common phenomenon in practice, when there is suppression between two variables neither may reach the criterion for entrance to the equation, although if both were entered they would make a useful contribution to R^2.

However, our distrust of stepwise regression is not absolute, and decreases to the extent that the following conditions obtain:

1. The research goal is entirely or primarily predictive (technological), and not at all, or only secondarily, explanatory (scientific). The substantive interpretation of stepwise results is made particularly difficult by the problems described above.

2. n is very large, and the original k (that is, before stepwise selection) is not too large; a k/n ratio of one to at least 40 is prudent.

3. Particularly if the results are to be substantively interpreted, a cross-validation of the stepwise analysis in a new sample should be undertaken, and only those conclusions that hold for both samples should be drawn. Alternatively, the original sample may be randomly divided in half, and the two half-samples treated in this manner.

3.7 TESTS OF STATISTICAL SIGNIFICANCE WITH k INDEPENDENT VARIABLES

3.7.1 Significance of the Multiple R

Having obtained a given R, with what confidence may it be asserted that the (linear) relationship between this set of k IVs and Y is not zero in the population? The answer to this question is obtained by an extension of the one independent variable test of significance. In Chapter 2 we saw that the test of the significance of r could be carried out either by the t test, Eq. (2.8.1),

$$t = \frac{r\sqrt{n-2}}{\sqrt{1-r^2}},$$

or by an F test, since for one degree of freedom in the numerator, $F = t^2$. In testing the significance of multiple R there are as many numerator df as there are IVs, and the t test can not be employed. The general equation by which R^2 for any number (k) of IVs may be tested for the statistical significance of its departure from zero is

(3.7.1) $$F = \frac{R^2 (n-k-1)}{(1-R^2) k},$$

with $df = k$ and $n - k - 1$.

F may also be computed (or provided as computer output) as a function of raw scores in the classic analysis of variance format. As we saw in the one-IV case, the total sample variance of Y may be divided into a portion accounted for by the IVs, which is equal to the variance of the estimated $\hat{Y}$ values, $sd_{\hat{Y}}^2$, and a portion not associated with the IVs, the "residual" or "error" variance, $sd_{Y-\hat{Y}}^2$. Similarly, the sum of the squared deviations about the $\bar{Y}$ may be divided into a sum of squares due to the regression and a residual sum of squares. When these two portions of the total are divided by their respective df we have the mean square values necessary for determining the F values, thus[11] :

$$\text{regression MS} = \frac{\text{regression SS}}{k} = \frac{R^2 \sum y^2}{k},$$

(3.7.2)

$$\text{residual or error MS} = \frac{\text{residual SS}}{n-k-1} = \frac{(1-R^2) \sum y^2}{n-k-1},$$

where $\sum y^2 = \sum (Y - \bar{Y})^2$. When F is expressed as the ratio of these two mean squares, we obtain

(3.7.3) $$F = \frac{\text{regression MS}}{\text{residual MS}} = \frac{R^2 \sum y^2 / k}{(1-R^2)(\sum y^2)/(n-k-1)}.$$

Cancelling the $\sum y^2$ term from the numerator and denominator, and simplifying, we obtain Eq. (3.7.1).

Suppose that the study on the LA of female high school seniors presented in Table 3.6.1 was based on $n = 176$ students and the standard deviation of Y was 2.836. When all three IVs were in the equation, we saw that $R^2 = .2754$. Thus, by Eq. (3.7.1)

$$F = \frac{.2754(176 - 3 - 1)}{(1 - .2754)3} = 21.789.$$

Checking the tabled F values (Appendix Tables D.1 and D.2) for 3 numerator and 172 denominator df, we find an F value (by interpolation) of 2.66 as necessary for significance at the $\alpha = .05$ criterion and 3.90 at $\alpha = .01$. Since the obtained F value exceeds the .01 criterion (i.e., $P < .01$), we conclude that the linear relationship between these three IVs and LA is not zero in the population, with a less than one-in-a-hundred chance of doing so erroneously.

The same result is obtained in an analysis of variance format. From the data we would find that the squared deviations of Y about their mean sum to 1415.55 ($= \sum y^2 = n\, sd_Y^2$) and that the sum of the squared deviations of $\hat{Y}$ about the same mean equals 389.81 ($= \sum \hat{y}^2 = R^2 \sum y^2$). The residual SS $[\sum (y - \hat{y})^2$

---

[11] We shift here from $sd_{Y-\hat{Y}}^2$, which is the value of the variance of the residuals or errors of estimate in the sample and thus uses n as the divisor, to $\widetilde{sd}_{Y-\hat{Y}}^2$, which estimates the *population* value of this variance and uses $df = n - k - 1$ as its divisor. The residual or error MS that follows equals $\widetilde{sd}_{Y-\hat{Y}}^2$, the tilde indicating that it is the population estimate rather than the sample value. (See Section 2.6.)

$= \Sigma\ y^2 - \Sigma\ \hat{y}^2] = 1415.55 - 389.81 = 1025.74$. The mean squares and F by Eqs. (3.7.2) and (3.7.3) are thus

$$\text{regression MS} = \frac{389.81}{3} = 129.937,$$

$$\text{residual MS} = \frac{1025.64}{176-3-1} = 5.963,$$

$$F = \frac{129.937}{5.963} = 21.789,$$

as before.

3.7.2 Shrunken $\widetilde{R}^2$

Although we may determine from a sample R^2 that the population R^2 is not likely to be zero, it is nevertheless not true that the sample R^2 is a good estimate of the population R^2. To gain an intuitive understanding of part of the reason for this, imagine the case in which one or more of the IVs account for *no Y* variance in the population, that is, $r^2_{Yi} = 0$ in the population for one or more X_i. Because of random sampling fluctuations we would expect that only very rarely would its r^2 with Y in a sample be *exactly* zero; it will virtually always have some positive value. Thus, in most samples it would make some (possibly trivial) contribution to R^2. The smaller the sample size the larger these positive variations from zero will be, on the average, and thus the greater the inflation of the sample R^2. Similarly, the more IVs we have, the more opportunity for the sample R^2 to be larger than the true population R^2. It is often desirable to have an estimate of the population R^2 and we naturally prefer one which is more accurate than the positively biased sample R^2. Such a realistic estimate of the population R^2 (for the fixed model) is given by

$$(3.7.4) \qquad\qquad \widetilde{R}^2 = 1 - (1 - R^2)\frac{n-1}{n-k-1}.$$

This estimate is necessarily (and appropriately) smaller than the sample R^2 and is thus often referred to as the "shrunken" $\widetilde{R}^2$. The magnitude of the "shrinkage" will be larger for small values of R^2 than for larger values, other things being equal. Shrinkage will also be larger as the ratio of the number of IVs to the number of subjects increases. As an example, consider the shrinkage in R^2 when $n = 200$ and cases where $k = 5, 10$, and 20 IVs, thus yielding k/n ratios of $1/40$, $1/20$, and $1/10$, respectively. When $R^2 = .20$, the $\widetilde{R}^2$ values will equal, respectively, .1794, .1577, and .1106, representing shrinkage of approximately .02, .04, and .09, the last being a shrinkage of almost one-half. When $R^2 = .40$, the comparable $\widetilde{R}^2$ values are, respectively, .3845, .3683, and .3330, smaller shrinkage either as differences from or proportions of R^2. For large ratios of k/n and small R^2, these $\widetilde{R}^2$ may take on negative values; for example, for $R^2 = .10, k = 11, n = 100$, Eq. (3.7.4) gives $\widetilde{R}^2 = -.0125$. Whenever the F value for R^2 of Eq.

(3.7.1) is less than one, a negative $\widetilde{R}^2$ will occur. In such cases, by convention the shrunken $\widetilde{R}^2$ is reported as zero.

It may be of interest to examine the expected value of R^2 for a sample from a population in which all correlations between the dependent and the independent variables are zero. This expected value can be shown to be

$$(3.7.5) \qquad R_E^2 = \frac{k}{n-1},$$

a value almost identical with the k/n ratio which signals the degree of shrinkage to be expected. Thus, in the example of the preceding paragraph, $k/n = .11$, $R_E{}^2 = .111$, so that it is hardly surprising that the observed sample R^2 of .10 shrinks to a negative $\widetilde{R}^2$. (Note that when R_E^2 is substituted for R^2 in the equation for $\widetilde{R}^2$, the resulting value is, as it logically should be, zero.)

It should be clear from this discussion that whenever a subset of IVs has been selected post hoc from a larger set of potential variables on the basis of their relationships with Y, either by the computer performing stepwise regression, or by the experimenter selecting IVs because of their relatively large r_{Yi}s, not only R^2, but also $\widetilde{R}^2$ computed by taking as k the number of IVs *selected*, will be too large because of the inherent tendency of the stepwise procedure to capitalize on chance. A more realistic estimate of shrinkage is obtained by substituting for k in Eq. (3.7.4) the *total* number of IVs from which the selection was made.

3.7.3 Significance Tests of Partial Coefficients

In testing whether a single IV makes a significantly nonzero unique contribution to the multiple R^2 we proceed, as before, to obtain two population variance estimates. The proportion of Y variance uniquely associated with a given X_i, as we have seen, is equal to sr_i^2. The proportion of Y variance associated with "error" (departures from regression), is equal to $1 - R^2$. Dividing these two proportions by their degrees of freedom, 1 and $n - k - 1$ respectively, gives us the F value for the increase in R^2 due to X_i as

$$(3.7.6) \qquad F_i = \frac{sr_i^2/1}{(1-R^2)/(n-k-1)} = \frac{sr_i^2(n-k-1)}{1-R^2}.$$

The conceptual advantage of the F test is that F is a ratio of two independent estimates of the population variance. As we have stated, however, with a numerator $df = 1$, $t = \sqrt{F}$ for the same denominator df. The t test expresses the ratio of the difference between the obtained statistic and the null hypothetical value (zero) to its standard error. In general, in keeping with usual practice, we will report t for statistics with one numerator df, that is, those for a single IV, even though the same information is available from F. Thus,

$$(3.7.7) \qquad t_i = sr_i \bigg/ \sqrt{\frac{1-R^2}{n-k-1}} = sr_i \sqrt{\frac{n-k-1}{1-R^2}}.$$

t_i will carry the same sign as sr_i and will, of course, have exactly the same P value as does the equivalent F test.

As previously noted, sr_i and pr_i differ only with regard to their denominators, sr_i expressing the correlation of Y with X_i from which all other IVs have been partialled, and pr_i expressing the correlation of Y with X_i when all other IVs have been partialled from both X_i and Y. Since neither can equal zero unless the other is also zero, it is not surprising that they must yield the same t_i value for the statistical significance of their departure from zero. It should also be clear that β_i which has the same numerator as sr_i and pr_i, also equals zero only when sr_i and pr_i do, and, since B_i is the product of β_i and the sd_Y/sd_i, it also can equal zero only when they do. Thus, Eq. (3.7.6) and its equivalent Eq. (3.7.7) provide the appropriate F_i and t_i values for the significance of departures of *all* the partial coefficients of X_i from zero. They either are, or are not, *all* significantly different from zero, and to exactly the same degree.

For example, let us return to the data presented in Table 3.6.1, where the obtained R^2 of .2754 was found to be significant ($P < .01$) for $k = 3, n = 176$. The sr_i for the three IVs were respectively .325, .171, and .093. Determining their t values by Eq. (3.7.7) we find

$$t_1 = .325 \sqrt{\frac{176 - 3 - 1}{1 - .2754}} = 5.011,$$

$$t_2 = .171 \sqrt{\frac{176 - 3 - 1}{1 - .2754}} = 2.639,$$

$$t_3 = .093 \sqrt{\frac{176 - 3 - 1}{1 - .2754}} = 1.438.$$

Looking these values up in the t table (Appendix Table A) for 172 df we find that t_1 and t_2 exceed the values required for significance at $\alpha = .01$; however, t_3 is smaller than the value required at $\alpha = .05$. Thus, we conclude that GPA and TFR both make statistically significant unique contributions in accounting for LA. We may *not* reject the null hypothesis that A–S has no linear relationship to LA in the population once the effects of GPA and TFR are taken into account by partialling.

It is quite possible to find examples where R^2 is statistically significant but none of the tests of significance on the individual X_i reaches the significance criterion for rejecting the null hypothesis. This finding occurs when the variables which correlate with Y are so substantially redundant that none of the unique effects is large enough to be significant. On the other hand, it may also happen that one or more of the t tests on individual variables does reach the criterion for significance while the overall R^2 is not significant. The variance estimate for the regression based on k IVs is divided by k to form the numerator of the F test for R^2, making of it an average contribution per IV. Therefore, if most variables do not account for more than a trivial amount of Y variance they may lower this average (the mean square for the regression) to the point of making the overall F not significant in spite of the apparent "significance" of the separate contributions of one or more individual IVs. In such circumstances, we recommend that such IVs *not* be accepted as significant. The reason for this is to avoid spuriously significant results, the probability of whose occurrence is controlled by the requirement that the F for a set of IVs be significant before its constituent IVs are t tested. This, the "protected t test", is part of a general strategy for statistical inference which is considered in detail in the next chapter (Section 4.6).

3.7.4 Significance Tests in Hierarchical Analysis

As we saw in Section 3.6.2, a number of goals in the analysis of data may be served by specifying, a priori, a sequence by which independent variables are to be added into the equation. One of the products of this procedure is a set of squared semipartial correlations representing the unique contribution of each variable when all previously considered variables, and only those, have been partialled from it. As the basic Eq. (3.6.1) of the hierarchical model states, a full set of these hierarchical sr^2 sum to $R^2_{Y\cdot 12\ldots k}$. Each of these may be tested for significance by specializing the standard tests presented above. Two choices of error variance are available for the denominator of the F (or t) ratio. The first of these, called "Model I," is given as the standard output of most computer programs (including stepwise programs). For Model I, the error variance for testing sr^2 is based on the R^2 which includes only those IVs which have entered the equation to that point in the hierarchy. To illustrate: in the example of Table 3.6.1, it was found that when X_1, X_2, and X_3 were treated hierarchically in that order, the resulting cumulative R^2s were respectively $R^2_{Y\cdot 1}$ = .1600, $R^2_{Y\cdot 12}$ = .2667, and $R^2_{Y\cdot 123}$ = .2754. The initial $R^2_{Y\cdot 1}$ = .1600 can also in this context be considered an sr^2 with nothing partialled, and Eq. (3.7.6) with $k = 1$ may be used. Because $R^2_{Y\cdot 1}$ is simply r^2_{Y1} , it can be tested for significance in the usual way, using the basic t test of Eq. (2.8.1). Then, $sr^2_{2\cdot 1}$ = $R^2_{Y\cdot 12} - R^2_{Y\cdot 1}$ = .2667 − .1600 = .1067. This can be tested by using Eq. (3.7.6), with $k = 2$. Since $n = 176$,

$$F = \frac{.1067(176 - 2 - 1)}{1 - .2667} = 25.164 \qquad (df = 1, 173),$$

or, equivalently, we can test $sr_{2\cdot 1}$ = .327 by Eq. (3.7.7) for $t \;(= \sqrt{F})$,

$$t = .327 \sqrt{\frac{176 - 2 - 1}{1 - .2667}} = 5.016 \qquad (df = 173).$$

Referring either F to Appendix Table D.1 for df = 1, 173 or t to Appendix Table A for df = 173, we find $P < .01$.

Note that it is the hierarchical $sr_{2\cdot 1}$ = .327 that concerns us here, *not* the simultaneous $sr_{2\cdot 13}$ = .171 of the previous section. In the simultaneous model, we assessed the unique contribution of X_2 relative to all the other IVs, here X_1 and X_3; in the present hierarchical model, it is the uniqueness of X_2 relative to only the preceding variable, here only X_1.

Since X_3 is the last variable to be entered, its $sr_{3 \cdot 21}$ is exactly the same as in the simultaneous model, with the same significance test, which uses $1 - R^2_{Y \cdot 123}$ = $1 - .2754$ in the denominator, with $n - 3 - 1 = 172$ df. Consider the consequence, however, of a *different* a priori hierarchy being selected, for example, one in which the ordering of the variables is X_3, X_2, X_1 (Table 3.6.1). For this progression we would determine and test for significance the zero-order r_{Y3}, then $sr_{2 \cdot 3}$, and finally $sr_{1 \cdot 23}$. [The reader may determine that each of these tests results in an F (or t) value for which $P < .05$.] It should be clear, however, that the three correlations for this ordering are completely different from those previously tested, which were r_{Y1}, $sr_{2 \cdot 1}$, and $sr_{3 \cdot 12}$, and will necessarily have different significance test values.

An alternative error model, Model II, is available for determining statistical significance in hierarchical MRC. Model II differs from Model I in the definition of the error variance proportion to be used in testing the various partial coefficients. In Model I, as we saw, the denominator of the F ratio changes as each variable is hierarchically entered into the equation. R^2 is determined for the variables up to and including the variable whose partial coefficients are being assessed, and tested with denominator df equal to n minus 1 minus the number of variables then in the equation. Therefore, both the error $(1 - R^2)$ and its degrees of freedom change for each successive significance test. In Model II, neither the error nor its associated df change, being $1 - R^2_{Y \cdot 12 \ldots k}$, and $n - k - 1$ for all k successive partials.

Continuing the previous example, with the variables described in Section 3.6.2 ordered X_1, X_2, X_3, the significance test using Model II error for X_1 would not be the same as using Model I error. In Model I the test is based on

$$F = \frac{r^2_{Y1}}{(1 - r^2_{Y1})/(n - 2)}$$

$$= \frac{.1600}{(1 - .1600)/174} = 33.143 \quad (df = 1, 174),$$

or $t = 5.757$ $(df = 174)$. In Model II it is based on

$$F = \frac{r^2_{Y1}}{(1 - R^2_{Y \cdot 123})/(n - 4)}$$

$$= \frac{.1600}{(1 - .2754)/172} = 37.980 \quad (df = 1, 172),$$

or $t = 6.163$ $(df = 172)$. Similarly, the Model II significance test for $sr^2_{2 \cdot 1}$ would be based on

$$F = \frac{R_{Y \cdot 12}^2 - r_{Y1}^2}{(1 - R_{Y \cdot 123}^2)/(n-4)}$$

$$= \frac{.1067}{(1 - .2754)/172} = 25.328 \qquad (df = 1, 172)$$

(compared to the Model I, $F = 25.164$, $df = 1, 173$). When the F table is entered with 1 and 172 df, we find that $P < .01$ (as was also the case with the test using Model I). The significance test for $sr_{3 \cdot 12}^2$ will be the same with either model since it is the final IV. For a full discussion of the considerations appropriate to a choice of error model, see Section 4.4.2.

3.7.5 Standard Errors and Confidence Intervals for *B* and β

In Chapter 2, we showed how to determine standard error and confidence intervals for r and B in the two-variable case, providing that certain distributional assumptions are made. Similarly, one may determine standard errors for partial regression coefficients; that is, one may estimate the sampling variability of partial coefficients from one random sample to another, using the data from the single sample at hand.

The equation for estimating the standard error of B is particularly useful because it reveals very clearly what conditions lead to large expected sampling variation in the size of B, and hence in the accuracy one can attribute to any given sample B value. A convenient form of the equation for the standard error of B for any X_i is

(3.7.8)
$$SE_{B_i} = \frac{sd_Y}{sd_i} \sqrt{\frac{1 - R_Y^2}{n - k - 1}} \sqrt{\frac{1}{1 - R_i^2}}$$
$$= \frac{\widetilde{sd}_{Y-Y}}{sd_i \sqrt{n}} \sqrt{\frac{1}{1 - R_i^2}}$$

where R_Y^2 is literally $R_{Y \cdot 12 \ldots k}^2$, and R_i^2 is literally $R_{i \cdot 12 \ldots (i) \ldots k}^2$. The ratio of the sds, as always, simply adjusts for the scaling of the units in which X_i and Y are measured. Aside from this, we see from the second term that the size of the SE_B will decrease as the error variance proportion $(1 - R_Y^2)$ decreases and its df $(= n - k - 1)$ increases. (On reflection, this should be obvious.) Note that this term will be constant for all variables in a given regression equation. The third term reveals an especially important characteristic of SE_{B_i}, namely, that it *increases* as a function of the squared multiple correlation of the remaining IVs with X_i, R_i^2. Here we encounter another manifestation of the general problem of multicollinearity, that is, of substantial correlation among IVs. Under conditions of multicollinearity, there will be relatively large values for at least some of the

SE_{B_i}, so that any given sample may yield relatively poor estimates of some of the population regression coefficients, that is, of those whose R_i^2 is large.

In order to show this relationship more clearly it is useful to work with variables in standard score form. B_i expressed as a function of standard scores is β_i. The standard error of β_i drops the first term from (3.7.8) since it equals unity, so that

(3.7.9)
$$SE_{\beta_i} = \sqrt{\frac{1 - R_Y^2}{n - k - 1}} \sqrt{\frac{1}{1 - R_i^2}}.$$

To illustrate the effects of differences in the relationship of a given X_i with the remaining IVs, we return to the example presented in Table 3.6.1. In this example both X_1 and X_2 had zero-order correlations of .40 with Y. Their relationships with X_3, however, differed substantially, with $r_{13} = .15$ and $r_{23} = .70$. Thus, we expect and find that they have substantially different R_i^2, that is, $R_{1 \cdot 23}^2 = .0402$ and $R_{2 \cdot 13}^2 = .4992$. Substituting these values into Eq. (3.7.9) we find

$$SE_{\beta_1} = \sqrt{\frac{1 - .2754}{172}} \sqrt{\frac{1}{1 - .0402}}$$

$$= .0649(1.0207) = .0663,$$

$$SE_{\beta_2} = \sqrt{\frac{1 - .2754}{172}} \sqrt{\frac{1}{1 - .4992}}$$

$$= .0649(1.4131) = .0917.$$

Because of its larger relationship with the remaining independent variable in this example, we find the SE_{β_2} is nearly 40% larger than SE_{β_1}. From Table 3.6.1, the two sample βs are $\beta_1 = .332$ and $\beta_2 = .242$. The discrepancy in their SEs indicates that the $\beta_2 = .242$ is a substantially less reliable estimate of its population value than is the $\beta_1 = .332$.

The t distribution may be used to test the null hypothesis for a β_i, that is, the hypothesis that its population value is zero. It takes the usual form

(3.7.10)
$$t_i = \frac{\beta_i}{SE_{\beta_i}} \quad (df = n - k - 1).$$

Applying this test to β_1 and β_2, we find

$$t_1 = \frac{.332}{.0663} = 5.011,$$

$$t_2 = \frac{.242}{.0917} = 2.639.$$

These necessarily are identical with the ts determined for the corresponding (simultaneous) sr_i (Section 3.7.3); recall that all partial coefficients for X_i (pr_i, sr_i, β_i, and B_i) must share the same t (or F) value.

One may also use the *SE* to determine the interval within which the population value may be expected to fall with a specified level of confidence. For example, we may determine the bounds within which we can assert with 95% confidence that the population β_1 falls. Since the *t* value for $df = n - k - 1 = 172$ for a two-tailed probability of .05 is 1.974 (by interpolation in Appendix Table A), we expect the population value to fall within the interval which extends 1.974 SE_β on either side of β. In the above example, the limits of the 95% confidence interval for β_1 will be from $.332 - 1.974(.0663) = .201$ to $.332 + 1.974(.0663) = .463$.

Confidence intervals may be similarly determined for *B* coefficients. For example, if in the above example $sd_Y = 5$ and $sd_1 = 2$, we can find

$$B_1 = \frac{sd_Y}{sd_1} \beta_1$$

$$= \frac{5}{2}(.332) = .830$$

and

$$SE_{B_1} = \frac{sd_Y}{sd_1} SE_{\beta_1}$$

$$= \frac{5}{2}(.0663) = .166.$$

We can determine the 95% confidence interval to be within the limits $.830 \pm 1.974 (.166)$, that is, from .502 to 1.158.

Confidence limits may also be determined for pr_1 by extension of the procedure described in Section 2.10, by applying the Fisher z' transformation to the *pr* with a standard error of $1/\sqrt{n - k - 2}$. In addition to the determination of confidence limits of *pr*, such standard errors may be used to test the significance of the difference between *pr*s from independent samples, exactly as with zero-order *r*s.

3.7.6   Use of Multiple Regression Equations in Prediction

As we have already noted, the traditional use of MRC in the behavioral sciences has been for prediction, literally forecasting, with only incidental attention to explanation. In this book, the emphasis is reversed, and our almost exclusive concern is with the analytic use of MRC to achieve the scientific goal of explanation. But in its use for purposes of prediction, MRC plays an important role in several behavioral technologies, for example, personnel (including educational) selection, vocational counseling, and psychodiagnosis. In this section we address ourselves to the accuracy of prediction in multiple regression, and some of its problems. The standard error of estimate, $\widetilde{sd}_{Y-\hat{Y}}$, as we have seen, provides us with an estimate of the magnitude of error which we can expect in estimating $\hat{Y}$ values over sets of future $X_1, X_2, \ldots, X_k$ values which correspond to those of the present sample (that is, the fixed-effects model). Suppose, however, we wish to determine the standard error and confidence limits of a *single* estimated $\hat{Y}_o$ from a new set of observed values $X_{1o}, X_{2o}, \ldots, X_{ko}$. In Section 2.10.2, we saw that the expected magnitude of error increases as the X_{io} values depart from their respective means. The reason for this should be clear from the fact that any discrepancy between the sample estimated regression coefficients and the population regression coefficients will result in larger errors in $\hat{Y}_o$ when X_{io} values are far from their means than when they are close.

Estimates of the standard error and confidence limits for $\hat{Y}_o$ predicted from known values $X_{1o}, X_{2o}, \ldots, X_{ko}$ are particularly dependent upon the validity of a rather strong set of assumptions about the nature of the populations from which the sample observations are drawn. The accuracy of the estimate of the standard error of a given $\hat{Y}_o$ value depends on whether the population Y values for any given set of X_i are normally distributed, centered on the population regression surface, and have equal variances from set to set of X_i values. Under these circumstances, the standard error of a $\hat{Y}_o$ predicted from given values of $X_{1o}, X_{2o}, \ldots, X_{ko}$ is given by

$$(3.7.11) \quad \widetilde{sd}_{Y_o - \hat{Y}_o} = \frac{\widetilde{sd}_{Y - \hat{Y}}}{\sqrt{n}} \sqrt{n + 1 + \sum \frac{z_{io}^2}{1 - R_i^2} - 2 \sum \frac{\beta_{ij} z_{io} z_{jo}}{1 - R_i^2}} \, ,$$

where the first summation is over the k IVs, the second over the $k(k-1)/2$ pairs of IVs (i.e., $i < j$) expressed as standard scores, β_{ij} is the β for estimating X_i from X_j, holding constant the remaining $k - 2$ IVs, and R_i^2 is literally $R_{i \cdot 12 \ldots (i) \ldots k}^2$. Although at first glance this formula appears formidable, a closer examination will make clear what elements effect the size of this error. $\widetilde{sd}_{Y - \hat{Y}}$ is the standard error of estimate and as in the case of a single IV, we see that increases in it and/or in the absolute value of the IV (z_{io}) will be associated with larger error. The terms which appear in the multiple IV case which did not appear in the single variable case $(\beta_{ij}$ and $R_i^2)$ are functions of the relationships among the independent variables.[12] When all independent variables are uncorrelated (hence, all β_{ij} and all R_i^2 equal zero), we see that the formula simplifies and $\widetilde{sd}_{Y_o - \hat{Y}_o}$ is minimized (for constant $\widetilde{sd}_{Y - \hat{Y}}$, n, and z_{io} values).

It is worth emphasizing the distinction between the validity of the significance tests performed on partial coefficients and the accuracy of such coefficients when used in prediction. In *analytic* uses of MRC, given the current level of theoretical development in the behavioral and social sciences, the information most typically called upon is the significance of the departure of partial coefficients from zero and the sign of such coefficients. The significance tests for the fixed effects model are relatively robust to assumption failure, particularly so when n is not small. Using the regression equation for prediction, on the other hand, requires applying these typically unstable coefficients to particular X_{io} values, for which the consequence of assumption failure is likely to be much

---

[12] A version of Eq. (3.7.11) using elements of the inverted matrix may be found in Appendix 1.

more serious. Perhaps for this reason a number of investigators have found that in many (probably most) prediction situations, weighting each z_{io} by its r_{Yi} instead of β_i (or equivalently, each X_{io} by r_{Yi}/sd_i instead of B_i), or even using equal weights with the proper sign for the z_{io} (or, equivalently, $1/sd_i$ weights for the X_{io}), yields an estimate which correlates with Y at least as well *on cross validation* as does the regression-produced $\hat{z}_Y$ or $\hat{Y}$.[13]

Further insight may be gained by noting that *regardless* of the *sign, magnitude,* or *significance* of its partial regression coefficient, the correlation between X_i and the $\hat{Y}$ determined from the entire regression equation is

(3.7.12)
$$r_{\hat{Y}i} = \frac{r_{Yi}}{R_{Y \cdot 12 \ldots k}}.$$

Thus it is invariably of the same sign and of larger magnitude than its zero-order r with Y. Reflection on this fact may help the researcher to avoid errors in interpreting data analyses in which variables which correlate materially with Y have partial coefficients which approach zero, or are of opposite sign. When partial coefficients of the X_i approximate zero, whatever linear relationship exists between X_i and Y is accounted for by the remaining independent variables. Since neither its zero-order correlation with Y nor its (larger) correlation with $\hat{Y}$ is thereby denied, the interpretation of this finding is highly dependent on the substantive theory being examined. One theoretical context may lead a researcher to conclude that X_i is only spuriously related to Y, that is, related only because of its relationships with other IVs. Another theoretical context may lead to the conclusion that the true causal effect of X_i on Y operates fully through the other IVs in the equation. Similarly, when the partial coefficients of X_i and r_{Yi} are of opposite sign, X_i and one or more of the remaining IVs are in a suppressor relationship. While it is legitimate and useful to interpret the partialled relationship, it is also important to keep in mind the zero-order correlations of X_i with Y (and hence with $\hat{Y}$).

Without minimizing the utility of the various standard error formulas in the predictive use of MRC, the fact that their use rests on distributional assumptions must be recognized. The surest way of assessing how well a regression equation predicts is to apply it to a new set of data. Particularly when the original n is small, or the scatter of Y values is skewed or unequal for different sets of X_i, or stepwise regression has been used, cross validation to a new sample is strongly indicated. The empirical comparison of the $\hat{Y}$ values generated by applying the old equation to the new sample's X_i values with its actual Y values affords direct and conclusive evidence with regard to the adequacy of prediction.

3.7.7 Multicollinearity

The existence of substantial correlation among a set of IVs creates difficulties, usually referred to as "the problem of multicollinearity." Actually, there are

---

[13] Dawes and Corrigan (1974) provide an excellent analysis of this problem.

three distinct problems—the substantive interpretation of partial coefficients, their sampling stability, and computational accuracy.

Interpretation

We have already seen in Section 3.6.2 that the partial coefficients of highly correlated IVs analyzed simultaneously are reduced. Since the IVs involved lay claim to largely the same portion of the Y variance, by definition, they can not make much by way of unique contributions. Interpretation of the partial coefficients of IVs from the results of a simultaneous regression of such a set of variables which ignores their multicollinearity will necessarily be misleading. Attention to the R_i^2 of the variables may help, but these do not indicate the source of redundancy of each X_i (see Appendix 2).

A superior solution to this problem is the use of the hierarchical rather than the simultaneous model of MRC. The discipline of the a priori ordering of IVs requires the investigator to hypothesize a causal structure which accounts for their correlation. To be sure, the validity of the interpretation depends upon the validity of the causal hypothesis, but this is preferable to the complete anarchy of the simultaneous analysis in which everything is partialled from everything else indiscriminately, including effects from their causes. (See Chapter 4 for *setwise* hierarchical MRC, which extends this strategy for coping with multicollinearity.)

Sampling Stability

The structure of the formulas for SE_{B_i} (Eq. 3.7.8) and SE_{β_i} (Eq. 3.7.9) makes plain that they are directly proportional to $\sqrt{1/(1 - R_i^2)}$—a serious consequence of multicollinearity, therefore, is highly unstable partial coefficients for those IVs which are highly multicollinear. Concomitantly, the trustworthiness of individually predicted $\hat{Y}_O$ is lessened as the R_i^2s for a set of IVs increase, as is evident from the structure of Eq. (3.7.11). Large standard errors mean both a lessened probability of rejecting the null hypothesis (see Section 3.8.3) and wide confidence intervals.

This aspect of the problem of multicollinearity can not be solved, but it can sometimes be prevented from occurring by purging the set of IVs prior to their regression on Y (see Section 4.6.2). This should be done a priori, or at least without knowledge of the r_{Y_i}s, in order to avoid capitalization on chance. The IVs to be purged are those which are of lower causal priority, that is, effects of or surrogates for those which are retained. Thus, this solution is a move halfway in the direction of hierarchical analysis. Alternatively, one can resort to the use of stepwise regression *with cross validation,* the latter being mandatory under conditions of high multicollinearity.

Computation

As the R_i^2s increase, errors associated with rounding in the computation of the inverse of the correlation matrix among IVs (Appendix 1) become potentially serious and may result in partial coefficients which are grossly in error. Fortunately, recent improvements in computer capabilities for carrying many digits in their computations and in their computational recipes (algorithms) are likely to cope adequately with all but the most extensive instances of multicollinearity. When in doubt, seek expert advice at the computer center. (Also see Appendix 3.)

3.8 POWER ANALYSIS

3.8.1 Introduction

Section 2.9.1 explained the purpose and desirability of determining the power of a given research plan to reject at the α significance level a false null hypothesis that the population r equals zero. Thus, given a plan for determining the existence of nonzero correlation between two variables, including n and α, the investigator may enter the table for the selected α with n and the expected (alternate-hypothetical) value of the population r, and read off the power—the probability of finding the sample r to be significant. Alternatively, one may proceed in planning a research by deciding on the significance criterion α and the desired power. Then, having specified the expected population r, a table for the given α is entered with this r and the desired power. The tabled values provide the number of cases necessary (n^*) to have the specified probability of rejecting the null hypothesis (the desired power) at the α level of significance when the population r is as posited. In this section, we extend power analysis beyond simple correlation to the more general MRC for k IVs.

3.8.2 Power Analysis for R^2

Power and n^* can be conveniently tabled for the single IV case. However, the several different coefficients which may be tested in MRC analysis, as well as provision for many possible values of k makes the direct tabling of power and n^* unwieldy. Instead, we provide a table of constants with which one can perform power analysis of tests for the different null hypotheses in MRC with k IVs. These constants are then employed in a simple formula to determine the necessary number of cases (n^*).

To determine n^* for the F test of the significance of R^2, the researcher proceeds with the following steps:

1. Set the significance criterion to be used, α. Provision is made in the Appendix for $\alpha = .01$ and $\alpha = .05$ in the L tables (Appendix Tables E.1 and E.2).

2. Set desired power for the F test. The L tables provide for power values of .10, .30, .50, .60, .70, .75, .80, .85, .90, .95, and .99. (The use of the lower values is illustrated in Chapter 4.)

3. In the L tables (Appendix Tables E.1 and E.2), k_B is used to represent the number of df associated with the source of Y variance being tested. For

$R^2_{Y \cdot 12...k}$, k_B is simply k, the number of IVs. The L tables provide for k_B = 1 (1) 16 (2) 24 (4) 40 (10) 100, that is, for 30 values of k_B between 1 and 100.

4. Look up in the appropriate table (α = .01 or .05) the value of L for the given k_B (row) and specified power (column).

5. Determine the population effect size, ES ($=f^2$, see below) of interest, the expected or alternate-hypothetical value. As was the case for the single IV (where ES = r), the ES may represent a probable population value as indicated by previous work, a minimum value that would be of theoretical or practical significance, or some conventional value as discussed in Section 4.5.4.

The population ES for R^2 is given by

(3.8.1) $$f^2 = \frac{R^2}{1 - R^2}.$$

6. Substitute L (from step 4) and f^2 in

(3.8.2) $$n^* = \frac{L}{f^2} + k + 1.$$

The result is the number of cases necessary to have the specified probability of rejecting the null hypothesis (power) at the α level of significance when f^2 in the population is as posited.

For example, let us return to the research on aspirations of female high school seniors. As part of the planning for the research, the investigator performs a power analysis for R^2 in order to determine the n^* to be used. It is planned to use the α = .05 significance criterion (Step 1), to have a .90 probability of rejecting the null hypothesis (Step 2) and to use three independent variables (Step 3). Checking Appendix Table E.2 (α = .05) for k_B = 3 (row) and power = .90 (column), the L value is found to be 14.17. It is decided that a population R^2 as small as .05 would be of interest and thus the ES is determined to be

$$f^2 = \frac{R^2}{1 - R^2} = \frac{.05}{.95} = .0526.$$

Substituting L and f^2 in Eq. (3.8.2)

$$n^* = \frac{14.17}{.0526} + 3 + 1 = 273.2.$$

Thus, 274 cases are needed to detect (using α = .05) a population R^2 as small as .05 with a 90% probability. If the researcher were content to be able to detect a population R^2 of .10 with the same power and α, only 132 cases would be necessary. Similarly, suppose another investigator feels that .16 is a more realistic value for the population R^2, is content with 80% power, but plans to use the more stringent .01 significance criterion. In this case, the Table E.1 value for L is 15.46 and f^2 = .16/.84 = .1905; thus, from Eq. (3.8.2), only (15.46/ .1905 + 3 + 1 =) 86 cases are needed.

3.8.3 Power Analysis for Partial Correlation and Regression Coefficients

A determination of the number of cases necessary for testing the null hypothesis that any partial correlation or regression coefficient for a given X_i (in a set of k IVs) is zero may proceed in the same manner as for R^2. Having set α and the desired power, the appropriate table is entered now for $k_B = 1$ (since the source of variance is a single X_i), and the L value is read off. The f^2 value for the partial coefficients of a single IV, X_i, is determined by

$$(3.8.3) \qquad\qquad f^2 = \frac{sr_i^2}{1 - R^2}.$$

The L and f^2 values are substituted in Eq. (3.8.2) to determine n^*.

For example, in planning the study on female aspirations, a researcher may expect the population R^2 to be about .10, and decide to determine n^* for the case in which each of the 3 IVs makes a unique contribution of .02 = sr^2. (Note that to the extent that IVs are expected to be somewhat redundant in accounting for Y, the sum of the sr_i^2 will typically be smaller than R^2.)

The researcher then proceeds by deciding that the significance criterion is to be $\alpha = .05$ and that the power desired is .80. Checking Appendix Table E.2 for $k_B = 1$ and power = .80, he finds the L value to be 7.85. Determining from Eq. (3.8.3) that $f^2 = .02/(1 - .10) = .0222$, the L and f^2 values are substituted in Eq. (3.8.2), and

$$n^* = \frac{7.85}{.0222} + 3 + 1 = 357.3.$$

Thus, according to this plan it will take a sample size of 358 to provide an 80% probability of rejecting the null hypothesis at $\alpha = .05$ if the population f^2 is as posited. It is useful to note here the substantial effects of redundancy among the IVs in reducing power (or increasing n^*). If the three IVs were each to account uniquely for one-third of R^2, that is, if they were uncorrelated, the f^2 for each would equal .0370 (= .0333/.90), and $n^* = 216$.

Although the power analysis of partial coefficients proceeds most conveniently by determining f^2 by means of sr^2, the analysis provides the appropriate power to reject the null hypotheses that β, B, and pr are zero as well. Since, as we have seen, these coefficients for a given X_i must have identical significance test results, the power to reject the null hypothesis for any one of them will be the same as for any other for analogous alternative-hypothetical values.

When the research plan calls for hierarchical ordering of the k IVs, the power analysis proceeds as above, except that sr_i^2 is determined for each variable *as of the point of its entry* into the equation and R^2 is posited either for the same point (if Model I error is to be employed), or for the final equation (if Model II error is chosen).

Several other topics in power analysis are presented in Chapter 4, following the exposition of power analysis in the most general form of MRC, where multiple sets of IVs are used. Among the issues discussed there are determination of power for a given n (Section 4.5.8), reconciling different n*s for different hypotheses in a single analysis (Section 4.5.6), and the considerations involved in setting f^2 and power values (Sections 4.5.4 and 4.5.5). Section 4.5.9 discusses some general tactical issues in power analyses.

3.9 SUMMARY

This chapter presents the extension of bivariate linear regression and correlation analysis to the case in which two or more independent variables (IVs), designated X_i ($i = 1, 2, \ldots, k$) are linearly related to a dependent variable Y. As with a single IV, the multiple regression equation with produces the estimated $\hat{Y}$ is that linear function of the k IVs for which the sum over the n cases of the squared discrepancies of $\hat{Y}$ from Y, $\Sigma (Y - \hat{Y})^2$, is a minimum. In addition to the multiple correlation coefficient, MRC yields, for each IV, a variety of *partial* coefficients which describe its unique relationship to Y. (Section 3.1)

The regression equation in both raw and standardized form for two IVs is presented and interpreted. The standardized partial regression coefficients, β_i, are shown to be a function of the correlations among the variables; β_i may be converted to the raw score B_i by multiplying each by sd_Y/sd_i. (Section 3.2)

The measures of correlation in MRC analysis include:

1. R, which expressed the correlation between Y and the best (least-squares) linear function of the k IVs ($\hat{Y}$), and R^2, which is interpretable as the proportion of Y variance accounted for by this function. (Section 3.3.1)

2. Semipartial correlations, sr_i, which express the correlation of X_i from which the other IVs have been partialled with Y. sr_i^2 is thus the proportion of variance in Y uniquely associated with X_i, that is, the increase in R^2 when X_i is added to the other IVs. The ballantine is introduced to provide visual representation of the overlapping of variance with Y of X_1 and X_2. (Section 3.3.2)

3. Partial correlations, pr_i, which give the correlation between that portion of Y not linearly associated with the other IVs and that portion of X_i which is not linearly associated with the other IVs; in contrast with sr_i, it partials the other IVs from both X_i *and* Y. pr_i^2 is the proportion of Y variance *not* associated with the other IVs which *is* associated with X_i. (Section 3.3.3)

Each of these coefficients is exemplified, and shown to be a function of the zero-order correlation coefficients. The reader is cautioned that none of these coefficients provides a basis for a satisfactory Y variance partitioning scheme when the IVs are mutually correlated.

All possible distinctive patterns that can exist in the linear relationships among Y and two IVs are discussed, exemplified, and illustrated with the ballantine. These include zero correlations between the X_i and Y or between X_1 and X_2 (Sections 3.4.1 and 3.4.2), "classical" suppression (Section 3.4.4), redundancy among the IVs in estimating Y (Section 3.4.3), "net" suppression (Section 3.4.5), and "cooperative" suppression (Section 3.4.6).

The case of two IVs is generalized to the case of k IVs in Section 3.5. The use of the various coefficients in the interpretation of research findings is discussed and illustrated with concrete examples. The relationships among the coefficients are given.

The simultaneous consideration of all X_is is an analytic strategy in MRC which is most appropriate when the purpose is prediction, or when no a priori ordering of IVs can be made (Section 3.6.1). When IVs can be so ordered, an alternative strategy, the hierarchical model, is available. The hierarchical or cumulative order of entry of the IVs into the multiple regression results in all earlier entering IVs being partialled from each later entering IV. The order may be determined by hypothetical causal priority, by considerations of the centrality of variables to the purposes of the research, or by structural properties of the IVs. Hierarchical MRC analysis is also useful in uncovering patterns of redundancy and suppression among the IVs. Finally, it provides a valid, although order-dependent, variance partitioning scheme, with successive tests of the hypotheses made explicit by the specific sequence of the IVs. (Section 3.6.2)

An alternative strategy, "stepwise" MRC, in which IVs are entered in a sequence determined by the size of their increment to R^2 is also discussed. Use of this strategy is generally discouraged because of the necessarily post hoc nature of interpretation of the findings and the substantial probability of capitalizing on chance. (Section 3.6.3)

R may be tested for statistical significance by means of an F test (Section 3.7.1). Since the R^2 obtained on any sample uses the optimal linear function of the *sample* X_i values, it follows that it tends to overestimate the population value. The "shrunken" $\widetilde{R}^2$ provides a more realistic estimate of this value. Overestimation of the population R^2 by the sample R^2 is larger as the ratio of k to n increases. When IVs have been selected post hoc from a larger potential set, as in stepwise regression, even this shrunken estimate is too large. (Section 3.7.2)

All partial coefficients share a single t value for the statistical significance of their departures from zero. With two or more IVs, it is possible for R to be significant when no IV yields significant partial coefficients and it is also possible for the partial coefficients of one or more IVs to be significant when R is not. (Section 3.7.3)

In hierarchical MRC, partial coefficients are tested for statistical significance at the point of entry into the equation. Two error models are available in hierarchical analysis. In Model I, the error term and its df are determined from the R^2 produced by only the IVs in the equation at that point. In Model II the R^2 based on all k IVs and its df are used. The consequences of these two models and of different hierarchical ordering are demonstrated on a worked example. (Section 3.7.4)

Standard errors may be determined for B_i and β_i and used for significance testing and to set confidence limits for population values. These standard errors are shown to increase as a function of the multiple correlation of X_i with the other IVs. (Section 3.7.5)

When the regression equation is to be used in estimating a single $\hat{Y}_o$ value from a new set of observed values of X_{1o}, X_{2o}, ..., X_{ko}, the standard error of the $\hat{Y}_o$ value may be determined assuming homoscedasticity and normality. This standard error also increases as a function of the correlations among the IVs, as well as of the departure of the X_{io} values from their respective means. The general issue of the use of MRC in prediction is discussed. (Section 3.7.6)

Large correlations among IVs (multicollinearity) may create problems in interpretation, sampling stability, and computational accuracy. These are discussed, and means of coping with them are suggested. (Section 3.7.7)

In planning a study using MRC it is highly desirable to determine the necessary sample size to attain some given level of statistical power, n^*. Such power analyses may be carried out for R^2 and for the partial coefficients of each of the IVs. Methods of determining n^* and worked examples are provided. (Section 3.8)

4

Sets of Independent Variables

4.1 INTRODUCTION

In the last chapter, we have taken the basic ideas of MRC about as far as they go in the standard textbook treatments. For (in principle) any number k of IVs, we have discussed their joint relationship to Y (R_Y, R_Y^2, the generation of the regression equation for $\hat{Y}$), the various conceptions of the separate relationship of each X_i to Y (r_{Yi}, sr_i, pr_i, B_i, β_i, r_{Yi}^2, sr_i^2, pr_i^2), and significance testing and power analysis for these statistics. In the present chapter, we offer an expansion of these ideas from k single IVs to h *sets* of IVs. It turns out that the basic concepts of proportion of variance accounted for and of correlation (simple, partial, semipartial, multiple) developed in Chapter 3 for single IVs hold as well for sets of IVs. We shall see that this generalization, that is, the use of sets as units of analysis in MRC, proves to be most powerful for the exploitation of data, and is at the core of our expansion of MRC from its limited past role in psychotechnology (for example, "predicting" freshman grade point average) to a truly general data-analytic system.

What is a set? As used here, its meaning is essentially that in common discourse—a group classified as belonging together for some reason. Although we tend to think of a group or set as having plural membership, and this will usually be the case in our applications, the mathematical concept, being general, permits a set to have any number of constituent members, including one or zero. Sets containing only one IV, and sets containing none (empty sets) will be seen to specialize the concepts to cases described in the last chapter. We represent sets by capital letters (e.g., A, B), the number of variables in a set by k subscripted with the letter (e.g., k_A, k_B), and the number of sets by h.

But why organize IVs into sets? There are two quite different kinds of reasons for doing so, the first structural or formal, and the second functional to the substance or logic of the research.

4.1.1 Structural Sets

We use the term "research factor" to identify an influence operating on Y, or more generally an entity whose relationship to Y is under study. The word "factor" is used here in the sense of the AV "factorial design," not that of factor analysis. Thus, sources of Y variance like treatment group, age, IQ, geographic area, diagnostic group, fiscal year, socioeconomic status, kinship system, strength of stimulus, and birth order are all research factors. Note that the examples are quite general and of varied character. Some are inherent properties or characteristics of the research material, while others are the consequences of experimental manipulation. Some are quantitative while others are qualitative or nominal. Although these are important distinctions, the concept of research factor spans them.

Now, when a research factor can be presented by a single IV, no necessity for structural sets arises. This was taken to be the case for the research factors in Chapter 3. But such is not the *general* case. Three formal circumstances make it often necessary to represent a single research factor by a set of multiple IVs, so that its several *"aspects"* are represented:

1. *Nominal or qualitative scales.* When observations are classified by a research factor G as g mutually exclusive and exhaustive qualitative categories, G is defined as a nominal scale (Stevens, 1951) and can be understood as having $g - 1$ aspects and therefore its complete representation requires a *set* made up of $k_G = g - 1$ IVs. Some examples are experimental treatment, religion, diagnosis, ethnicity, kinship system, and geographic area. Thus, the research factor "Religion," a nominal scale made up of the $g = 4$ categories Protestant, Catholic, Jewish and Other, can*not* be represented by a single IV, but requires $k_G = g - 1 = 3$ IVs to represent it. This can be understood as due to the fact that any scheme which will fully represent G, i.e., the distinctions among g groups, must have exactly $g - 1$ aspects. Similarly, in a laboratory experiment in which subjects are randomly assigned to three different experimental groups and two different control groups (hence, $g = 5$), the research factor G of treatment group requires exactly $g - 1 = 4$ IVs to fully represent the aspects of G (that is, the distinctions among the 5 treatment groups). The several different methods for accomplishing this representation (the subject of Chapter 5), can be understood as different systems of aspect representation, but each requires a *set* of $k_G = g - 1$ IVs to fully represent G. For example, in one of these systems (Section 5.5, "Contrast Coding"), *one* of these aspects, represented by *one* IV, might be the distinction between the three experimental groups on the one hand and the two control groups on the other. But the full G information requires a *set*[1] to represent it, and does so for purely structural reasons, that is, the nominal scale form of the research factor.

---

[1] The limiting case here occurs when G is made up of only 2 (= g) groups, for example, male–female, schizophrenic–nonschizophrenic, experimental–control. The G set would then contain only 1 (= $g - 1$) IV, since there is only one aspect of G, the distinction between its

2. *Quantitative scales.* This term is used in this book to convey collectively scales of the kind called ordinal, interval, and ratio by Stevens (1951). For example, interval scales are those whose units of measurement are treated as (more or less) equal, for example, scores of psychological tests and rating scales or sociological indices. These are the conventional variables long used in MRC, and the idea that a quantitative research factor has more than one aspect may seem strange. However, when one wishes to take into account the possibility that such a research factor as age may be related *nonlinearly* to Y (or other research factors), other aspects of age must be considered. Age as such represents only one aspect of the A research factor, its *linear* aspect. Other aspects, which provide for various kinds of nonlinearity in the relationship of A to Y, may be represented by other IVs such as age-squared and age-cubed (see Section 6.2, "Power Polynomials"). Thus, age, broadly conceived as a research factor A, may require a *set* of k_A IVs to represent it for purely formal or structural reasons. Chapter 6 is devoted to several methods of representing aspects of quantitative research factors, each of which requires a set of IVs.[2]

3. *Missing data.* It frequently occurs in research in the social and behavioral sciences that some of the data for a research factor are not available. For example, in a market research survey, some respondents do not answer an item on family income. Thus, the research factor "Income" has not only the obvious aspect of this information when it is given, but also the aspect of whether it is given or not, its "missingness." The latter is often relevant to Y (or other research factors), and can be represented as information. Details of this procedure are the subject of Chapter 7. It is sufficient to point out here simply that the structure of such a research factor defines (at least) two aspects, and thus its representation requires a *set* of IVs.

The above implies that if we are determining the proportion of variance in a dependent variable Y due to a *single* research factor, we will (in general) be finding a squared *multiple* correlation, since the latter will require a *set* of two or more IVs. These are necessary when the research factor has multiple aspects due to any of the three structural reasons just described.

4.1.2 Functional Sets

Quite apart from structural considerations, IVs are grouped into sets for reasons of their substantive content and the function they play in the logic of the research. Thus, if you are studying the relationship between the psychological variable field dependence (Y) and personality (P) and ability (A) characteristics,

---

two constituent groups. The reader may be familiar with the long-standing practice of representing such binary distinctions by assigning as scores zeros to one group and ones to the other for purposes of correlation, already illustrated in Section 2.3.3. This is seen in Chapter 5 to be an instance of the dummy-variable coding method (Section 5.3).

[2] Here the limiting case is that in which the research factor A is represented only by age as such, that is, its linear aspect. This is, of course, the traditional procedure. Another set containing only one IV would result if the investigator was prepared to represent A by only its logarithmic aspect, that is, log age. (See Section 6.5, "Nonlinear Transformations.")

P may contain a *set* of k_P scales from a personality questionnaire and A a *set* of k_A subtests from an intelligence scale. The question of the relative importance of personality and ability (as represented by these variables) in accounting for variance in Y would be assessed by determining $R^2_{Y \cdot P}$, the squared *multiple* correlation of Y with the k_P IVs of P, and $R^2_{Y \cdot A}$ (ditto for the k_A IVs of A), and then comparing them. Similarly, a sociological research that is investigating (among other things) the socioeconomic status (S) of school children, might represent S by occupational index of head of household, family income, mother's education, and father's education, a substantive set of 4 (= k_S) IVs. For simplicity, these illustrations have been of sets of single-IV research factors, but a functional set can be made up of research factors which are themselves sets, for example, a demographic set (D) made up of structural sets to represent ethnicity, marital status, and age. A group of sets is itself a set, and requires no special treatment.

It is often the nature of research that in order to determine the effect of some research factor(s) of interest (a set B), it is necessary to statistically control for (or partial out), the Y variance due to logically antecedent variables in the cases under study. A special case of this is represented by the analysis of covariance (ACV; see Chapter 9). A group of variables deemed antecedent either temporally or logically in terms of the purpose of the research, could be treated as a functional set for the purpose of partialling ("covarying") out of Y's total variance the portion of the variance due to these antecedent conditions. Thus, in a comparative evaluation of compensatory early education programs (B), with school achievement as Y, the set to be partialled might include such factors as family socioeconomic status, ethnicity, number of older siblings and preexperimental reading readiness. This large and diverse group of IVs functions as a single covariate set A in the research described. In another research they might have different functions and be treated separately, or in other combinations.

It is worth noting here that the organization of IVs into sets of whatever kind bears on the interpretation of MRC results, but has no effect on the basic computation. For any Y and k IVs ($X_1, X_2, \ldots, X_k$) in a given analysis, *whatever the set make-up of the IVs*, $R^2_{Y \cdot 12 \ldots k}$ and the array of partial statistics for each X_i (sr_i, pr_i, β_i, B_i) and all relevant F and t values are determined as described in the previous chapter. We shall soon see that sets of IVs may be progressively added in a hierarchical MRC analysis, but for the given group of IVs present at any stage of this analysis, the computation is as described above.

Before leaving the topic of functional sets, an admonitory word is in order. Because it is possible to do so, the temptation exists to assure coverage of a theoretical construct by measuring it in many ways with the resulting large number of IVs then constituted as a set. Such practice is to be strongly discouraged, since it tends to result in reduced statistical power for the sets (see Section 4.5) and an increase in spuriously "significant" single IV results (see Section 4.6), and generally bespeaks muddy thinking. It is far better to sharply reduce the size of such a set, and by almost any means. One way is through a tightened conceptualization of the construct, a priori. In other situations, the large array of measures is understood to cover only a few (or even one) behavioral dimension, in which case their reduction to scores on a few (or even one) factor by means of factor or cluster analysis is likely to be most salutory for the investigation, with little risk of losing Y-relevant information (see Section 4.6.2). Note that such analyses are performed *completely independently* of the values of the r_{Yi} correlations.

4.2 THE SIMULTANEOUS AND HIERARCHICAL
MODELS FOR SETS

4.2.1 The Simultaneous Model for Sets

We saw in the last chapter that, given k IVs, we could regress Y on all of them simultaneously and obtain $R^2_{Y \cdot 12 \ldots k}$, as well as partial statistics for each X_i. Now, these partial statistics are written in shorthand notation (i.e., β_i, B_i, sr_i, pr_i), but it is understood that *all* the IVs (other than X_i) are being partialled. This immediately generalizes to sets of IVs: when sets U, V, W are simultaneously regressed on Y, there are a total of $k_U + k_V + k_W = k$ IVs which together determine $R^2_{Y \cdot UVW}$, and the partial statistics for each IV in the set U has *all* the remaining $k - 1$ IVs partialled: those from V and W (numbering $k_V + k_W$), and also the remaining $(k_U - 1)$ IVs from its own set. It will be shown that, for example, the adjusted Y means of the ACV are functions of the regression coefficients when a covariate set and a set (or sets) of research factors of interest are simultaneously regressed on Y (Chapter 9). We will also see that sets as such may be partialled, so that $U \cdot VW$ may be related to Y by means of a partial or semipartial R.

4.2.2 The Hierarchical Model for Sets

In the preceding chapter (Section 3.6.2), we saw that the k IVs could be entered cumulatively in some specified hierarchy at each stage of which an R^2 is determined. The R^2 for all k variables could thus be analyzed into cumulative increments in the proportion of Y variance due to the addition of each IV to those higher in the hierarchy. These increments in R^2 were noted to be squared semipartial correlation coefficients, and the formula for the hierarchical model for single IVs was given as

$$(3.6.1) \quad R^2_{Y \cdot 12 \ldots k} = r^2_{Y1} + r^2_{Y(2 \cdot 1)} + r^2_{Y(3 \cdot 12)} + r^2_{Y(4 \cdot 123)} + \cdots + r^2_{Y(k \cdot 123 \ldots k-1)}.$$

The hierarchical model is directly generalizable from single IVs to sets of IVs. Replacing k single IVs by h sets of IVs, we can state that these h sets can be entered cumulatively in a specified hierarchical order, and upon the addition of each new set an R^2 is determined. The R^2 for all h sets can thus be analyzed into increments in the proportion of Y variance due to the addition of each new set of IVs to those higher in the hierarchy. These increments in R^2 are, in fact, squared *multiple* semipartial correlation coefficients, and a general hierarchical model equation for sets analogous to Eq. (3.6.1) may be written. To avoid awkwardness of notation, we write it for 4 (= *h*) sets in alphabetical hierarchical order and use the full dot notation; its generalization to any number of sets is intuitively obvious:

(4.2.1) $R^2_{Y \cdot TUVW} = R^2_{Y \cdot T} + R^2_{Y \cdot (U \cdot T)} + R^2_{Y \cdot (V \cdot TU)} + R^2_{Y \cdot (W \cdot TUV)}$.

We will defer a detailed discussion of the multiple semipartial R^2 to the next section. Here it is sufficient to note merely that it is an increment to the proportion of Y variance accounted for by a given set of IVs (of whatever nature) *beyond* what has already been accounted for by prior sets, that is, sets higher up in the hierarchy. Further, the amount of the increment in Y variance accounted for by that set can not be influenced by Y variance associated with subsequent sets, that is, those which are lower in the hierarchy.

Consider an investigation of length of hospital stay (Y) of n = 500 randomly selected psychiatric admissions to eight mental hospitals in a state system for a given period. Assume that data are gathered and organized to make up the following sets of IVs:

1. Set D—Demographic characteristics: age, sex, socio-economic status, ethnicity. Note incidentally that this is a substantive set, itself made up of sets. Assume k_D = 9.

2. Set M—Nine of the scales of the Minnesota Multiphasic Personality Inventory (MMPI). This set is also substantive, but note the necessity for making provision for missing data, a structural feature, and assume k_M = 10.

3. Set H—Hospitals. The hospital to which each patient is admitted is a nominally scaled research factor. With 8 (= g) hospitals contributing data, we will require a (structural) set of k_H = 7 (= g − 1) IVs to represent fully the hospital group membership of the patients.

Although there is a total of 26 (= k_D + k_M + k_H = k) IVs, our analysis may proceed in terms of the 3 (= h) sets, hierarchically ordered in the assumed causal priority of accounting for variance in length of hospital stay as D, M, H.

Suppose that we find that $R^2_{Y \cdot D}$ = .20, that is, the demographic set, made up of 9 IVs, accounts for 20% of the Y variance. Note that this ignores any association with MMPI scores (M) or effects of hospital differences (H). When we add the IVs of the MMPI set, we find that $R^2_{Y \cdot DM}$ = .22; hence, the increment due to M *over and above D*, or with D partialled, is $R^2_{Y \cdot (M \cdot D)}$ = .02. Thus, an additional 2% of the Y variance is accounted for by the MMPI, beyond the demographic set. Finally, the additional of the 7 IVs for hospitals (set H) produces an $R^2_{Y \cdot DMH}$ = .33, and an increment over $R^2_{Y \cdot DM}$ of .11, which equals $R^2_{Y \cdot (H \cdot DM)}$. Thus, we can say that *which* hospital patients enter accounts for 11% of the variance in length of stay, *after* we partial out (or statistically control, or adjust for, or hold constant) the effect of differences in patients' demographic and MMPI characteristics. We have, in fact, performed by MRC an ACV for the research factor "hospitals," using sets D and M as covariates[3] (see Chapter 9). (At the second stage of the analysis, we did the equivalent for the MMPI as the research factor, with the demographic characteristics as covariates, although, since D is not a quantitative scale, this is not within the usual purview of ACV; but see Section 9.4.)

There is much to be said about the hierarchical model and, indeed, it will be said in the next section and throughout the book. For example, as was pointed out in regard to single variables (Section 3.6.2), the increment due to a set may depend critically upon where it appears in the hierarchy, that is, what has been partialled from it, which, in turn, depends on the causal theory underlying the research. The chief point we wish to emphasize here is that the hierarchical model can proceed quite generally with sets of any kind, structural or functional, and this includes sets made up of sets.

4.3 VARIANCE PROPORTIONS FOR SETS

4.3.1 The Ballantine Again

We know of no better way to present the structure of relationships of IVs to a dependent variable Y than the ballantine. It was presented in the last chapter (Figure 3.3.1) for single IVs X_1 and X_2, and we present it as Figure 4.3.1 here for sets A and B. It is changed in no essential regard, and we shall show how the relationships of sets of IVs to Y, expressed as proportions of Y variance, are directly analogous to similarly expressed relationships of single IVs.

A circle in a ballantine represents the total variance of a variable, and the overlap of two such circles represents shared variance or squared correlation. This seems reasonable enough for single variables, but what does it mean when we attach the set designation A to such a circle? What does *the* variance of a *set* of multiple variables mean? Although each of the k_A variables has its own variance, remember that a multiple $R^2_{Y \cdot 12 \ldots k}$ is in fact a simple r^2 between Y and $\hat{Y}$, the latter optimally estimated from the regression equation for variables $X_1, X_2, \ldots, X_k$ [Eq. (3.3.3)]. Thus, what is presented by the circle for set A is the variance of a *single* variable, namely that of $\hat{Y}_A$, the estimated $\hat{Y}$ from the regression equation of the k_A IVs which make up set A. Similarly for set B (i.e., $\hat{Y}_B$) or any other set of IVs. Thus, by treating a set in terms of *how it bears on* Y, we effectively reduce it to a single variable. This lies at the core of the generalizability of the ballantine from single IVs to sets of IVs.

The ballantine in Figure 4.3.1 presents the general case: A and B share variance

---

[3] Omitting the significance test, a detail which will be attended to in Section 4.4. Also, a valid ACV requires that there be no interaction between H and the aggregate M, D covariate set (see Chapters 8 and 9).

with Y, but also with each other.[4] This is, of course, the critical distinction between MRC and the standard orthogonal AV. In an $A \times B$ factorial design AV, the requirement of proportional cell frequencies makes A and B (specifically $\hat{Y}_A$ and $\hat{Y}_B$) uncorrelated with each other, therefore the A and B circles do not overlap each other, and therefore each accounts for a separate and distinguishable (that is, additive) portion of the Y variance. (This is incidentally what makes the computation simpler in AV than in MRC.) Although not fundamental to the distinction between laboratory experiments and field research (which is randomization), it is nevertheless true that field research is characterized by the overlap of research factors while experiments at least make possible their independence or *non*overlap.

The ballantine makes possible putting into direct correspondence proportions of variance, that is, squared correlations of various kinds, to ratios of areas of the circle for Y, as we saw in Section 3.3. The total variance of Y is taken to equal unity (or 100%) and the Y circle is divided into four distinct areas identified by the letters a, b, c, and e. Since overlap represents shared variance or squared correlation, we can see immediately from Figure 4.3.1 that set A overlaps Y in areas a and c; hence

(4.3.2)
$$R^2_{Y \cdot A} = \frac{YA \text{ overlap}}{sd^2_Y} = \frac{a+c}{1} = a + c.$$

The c area arises inevitably from the AB overlap, just as it did in the single IV ballantine in Section 3.3, and is conceptually identical with it. It designates the part of the Y circle jointly overlapped by A and B, since

(4.3.3)
$$R^2_{Y \cdot B} = \frac{YB \text{ overlap}}{sd^2_Y} = \frac{b+c}{1} = b + c.$$

Since the c area is part of both A's and B's overlap with Y, for sets, as for single IVs, it is clear that (for the general case, where $\hat{Y}_A$ and $\hat{Y}_B$ are correlated) the proportion of Y variance accounted for by sets A and B together is not simply the sum of their separate contributions, since area c would then be counted twice, but rather

(4.3.4)
$$R^2_{Y \cdot AB} = \frac{a+b+c}{1} = a + b + c.$$

Thus, the areas a and b represent the proportions of Y variance *uniquely* accounted for respectively by set A and set B. By uniquely we mean relative to

[4] It can be proved that the correlation between A and B, where each is regression weighted to optimally estimate Y, is given by

(4.3.1)
$$r_{\hat{Y}_A \hat{Y}_B} = \frac{\Sigma \beta_i \beta_j r_{ij}}{R_{Y \cdot A} R_{Y \cdot B}},$$

where i indexes an X_i in set A, j indexes an X_j in set B, and the summation is taken over all i, j pairs (of which there are $k_A k_B$).

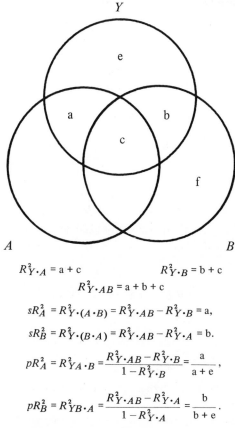

$$R^2_{Y \cdot A} = a + c \qquad\qquad R^2_{Y \cdot B} = b + c$$

$$R^2_{Y \cdot AB} = a + b + c$$

$$sR^2_A = R^2_{Y \cdot (A \cdot B)} = R^2_{Y \cdot AB} - R^2_{Y \cdot B} = a,$$

$$sR^2_B = R^2_{Y \cdot (B \cdot A)} = R^2_{Y \cdot AB} - R^2_{Y \cdot A} = b.$$

$$pR^2_A = R^2_{YA \cdot B} = \frac{R^2_{Y \cdot AB} - R^2_{Y \cdot B}}{1 - R^2_{Y \cdot B}} = \frac{a}{a + e},$$

$$pR^2_B = R^2_{YB \cdot A} = \frac{R^2_{Y \cdot AB} - R^2_{Y \cdot A}}{1 - R^2_{Y \cdot A}} = \frac{b}{b + e}.$$

FIGURE 4.3.1 The ballantine for sets A and B.

the other set, thus area b is Y variance *not* accounted for by set A, but *only* by set B; the reverse is true for area a.

This idea of unique variance in Y for a set is of great importance in MRC, and particularly so in the hierarchical model. It is directly analogous to the unique variance of a single IV discussed in Chapter 3. There we saw that for X_i, the unique variance in Y is the squared semipartial correlation of Y with X_i, which in abbreviated notation we called sr^2_i. It was shown literally to be the r^2 between that part of X_i which could not be estimated from the other IVs and all of Y, the complete cumbersome notation for which is $r^2_{Y(i \cdot 12 \ldots (i) \ldots k)}$, the inner parentheses signifying omission. For a set B, we similarly define its unique variance in Y to be the squared *multiple* semipartial correlation of Y with that part of B which is not estimable from A, or literally that part of $\hat{Y}_B$ which can not be estimated by $\hat{Y}_A$. Its literal notation would be $R^2_{Y \cdot (\hat{Y}_B \cdot \hat{Y}_A)}$ or, somewhat more simply, $R^2_{Y \cdot (B \cdot A)}$, or, even more simply, sR^2_B. In the latter notation, Y is understood, as is the other set (or sets) being partialled. (Obviously, all the above holds when A and B are interchanged.)

4.3.2 The Semipartial R^2

The ballantine is worth a thousand words. "That part of B which is not estimable from A" is represented by the part of the B circle not overlapped by the A circle, that is, the combined area made up of b and f. That area overlaps with the (complete) Y circle only in area b, therefore the proportion of the total Y variance accounted for uniquely by set B is

(4.3.5) $$sR_B^2 = R_{Y \cdot (B \cdot A)}^2 = \frac{b}{1} = b,$$

and, by symmetry, the proportion of Y variance accounted for uniquely by set A is

(4.3.6) $$sR_A^2 = R_{Y \cdot (A \cdot B)}^2 = \frac{a}{1} = a.$$

The ballantine shows how these quantities can be computed. If $R_{Y \cdot AB}^2$ is area a + b + c (Eq. 4.3.4) and $R_{Y \cdot A}^2$ is area a + c (Eq. 4.3.2), then patently

$$b = (a + b + c) - (a + c),$$

(4.3.7) $$sR_B^2 = R_{Y \cdot AB}^2 - R_{Y \cdot A}^2,$$

and, symmetrically,

$$a = (a + b + c) - (b + c),$$

(4.3.8) $$sR_A^2 = R_{Y \cdot AB}^2 - R_{Y \cdot B}^2.$$

The sR^2 can readily be found by subtracting from the R^2 for both sets the R^2 for the set to be partialled.

It is not necessary to provide for the case of more than two sets of IVs in the ballantine,[5] or, indeed, in the above equations. Since the result of the aggregation of any number of sets is itself a set, these equations are self-generalizing. Thus, if the unique variance in Y for set B among a group of sets is of interest, we can simply designate the sets other than B collectively as set A, and find sR_B^2 from Eq. (4.3.7). This principle is applied successively as each set is added in the hierarchical model, each added set being designated B relative to the aggregate of prior sets, designated A. We shall see that this framework neatly accommodates both significance testing and power analysis.

We offer one more bit of notation, which, although not strictly necessary, will be found convenient later on in various applications of the hierarchical model. In the latter, the addition of a new set B (or single IV X_i) results in an increase in R^2 (strictly, a nondecrease). These increases are, of course, the sR_B^2 (or sr_i^2), as already noted. It is a nuisance in presenting such statistics, particularly in tables, to always specify all the prior sets or single IVs which are partialled. Since in

---

[5] A fortunate circumstance, since the complete representation of three sets would require a three-dimensional ballantine, and, generally, the representation of h sets an h-dimensional ballantine.

hierarchical MRC the hierarchy of sets (or single IVs) is explicit, we will have occasion to identify such sR_B^2 (or sr_i^2) as "increments to Y variance at the stage of the hierarchy where B (or X_i) enters," and represent them as I_B (or I_i).

4.3.3 The Partial R^2

We have already identified the overlap of that part of a set circle which is unique (for example, areas b + f of set B) with the total Y circle as a squared multiple *semi*partial correlation (e.g., $sR_B^2 = R_{Y \cdot (B \cdot A)}^2$ = area b). With sets as with single IVs, it is a *semi*partial because we have related the partialled $B \cdot A$ with all of Y. We wrote it as b/1 in Eq. (4.3.5) to make it explicit that we were assessing the unique variance b as a proportion of the *total* Y variance of 1. We can however also relate the partialled $B \cdot A$ to the *partialled* $Y \cdot A$, that is, we can assess the unique b variance as a proportion not of the total Y variance, but of that part of the Y variance not estimable by set A, actually $Y - \hat{Y}_A$. The result is that we have defined the squared multiple *partial* correlation as

$$(4.3.9) \qquad pR_B^2 = R_{YB \cdot A}^2 = \frac{b}{b + e},$$

and symmetrically for set A as

$$(4.3.10) \qquad pR_A^2 = R_{YA \cdot B}^2 = \frac{a}{a + e}.$$

Thus, sR^2 and pR^2 (as sr^2 and pr^2) differ in the base to which they relate the unique variance as a proportion: sR^2 takes as its base the total Y variance while pR^2 takes as its base that proportion of the Y variance not accounted for by the other set(s). Inevitably, with its base smaller than (or at most equal to) 1, pR^2 will be larger than (or at least equal to) sR^2 for any given set.

It is easy enough to compute the pR^2. We have seen how, for example, the b area is found [Eq. (4.3.7)]; the combined areas b + e constitute the Y variance not accounted for by set A, hence $1 - R_{Y \cdot A}^2$. Substituting in Eq. (4.3.9),

$$(4.3.11) \qquad pR_B^2 = \frac{b}{b + e} = \frac{R_{Y \cdot AB}^2 - R_{Y \cdot A}^2}{1 - R_{Y \cdot A}^2},$$

and, symmetrically,

$$(4.3.12) \qquad pR_A^2 = \frac{a}{a + e} = \frac{R_{Y \cdot AB}^2 - R_{Y \cdot B}^2}{1 - R_{Y \cdot B}^2}.$$

To illustrate the distinction between sR^2 and pR^2, we refer to the example of the hierarchy of sets of demographics (D), MMPI (M), and hospitals (H) in relationship to length of hospital stay (Y) of Section 4.2.2. $R_{Y \cdot D}^2 = .20$, and when M is added, $R_{Y \cdot DM}^2 = .22$. The increment was .02; hence, $sR_M^2 = .02$ (or $I_M = .02$), that is, 2% of the *total* Y variance is uniquely (relative to D) accounted for by M. But if we ask "what proportion of the variance of Y not accounted for by D is uniquely accounted for by M?," our base is not the total

Y variance, but only $1 - R^2_{Y \cdot D} = 1 - .20 = .80$ of it, and the answer is $pR^2_M = .02/.80 = .025$. Letting $D = A$ and $M = B$, we have simply substituted in Eqs. (4.3.7) and (4.3.11).

It was also found that the further addition of H resulted in $R^2_{Y \cdot DMH} = .33$. Thus, H accounted for an additional .11 of the *total* Y variance, hence $sR^2_H = .11$ ($= I_H$). But if we shift our base from total Y variance to Y variance not already accounted for by D and M, the relevant proportion is $.11/(1 - .22) = .141$, i.e., pR^2_H. Now letting sets $D + M = A$, and $H = B$, we again have simply substituted in Eqs. (4.3.7) and (4.3.11). [Any desired combination of sets can be effected: if we wished to combine M and H into set B, with $D = A$, we could determine that $sR^2_{MH} = .13$, and $pR^2_{MH} = .13/(1 - .20) = .162$, by the same equations.]

It is worth noting that the pR^2 is rather in the spirit of ACV. In ACV, the variance due to the covariates is removed from Y, and the effects of research factors are assessed with regard to this "adjusted" (partialled) Y variance. Thus in the latter example, D and M may be considered to be covariates whose function is to "equate" the hospitals, so that they may be compared for length of stay, free of any possible hospital differences in the D and M of their patients. In that spirit, we are interested only in the $1 - R^2_{Y \cdot DM}$ portion of the Y variance, and the pR^2_H takes as its base the .78 of the Y variance *not* associated with D and M; hence, $pR^2_H = .11/.78 = .141$ of this adjusted (or partialled, or residual) variance is the quantity of interest.

Partial correlations are sometimes called "net" correlations (for example, as in "net" profit), since they represent the correlation which remains after the effect of other variable(s) have been removed ("debited") from both Y and the set (or single IV) being correlated.

Semipartial R and Partial R

The above development and formulas have been cast in terms of squared correlations, that is, sR^2 and pR^2. For these, as indeed for all correlations, squared values are distinctly lower than unsquared values, for example, a correlation of .20 becomes, when squared, .04. In the behavioral and social sciences, where relationships are not strong, correlations (of whatever kind) only infrequently are as large as .50. Squared correlations are then only infrequently as large as .25. In an effort to keep up morale (or hold his head high among his colleagues in the older sciences), it is understandable when a behavioral scientist prefers to report unsquared rather than squared correlations, for example, sR and pR rather than sR^2 and pR^2. We feel tolerant toward this practice, indeed, even compassionate. But keep in mind the advantage of working with proportions of variance, particularly the additivity of sR^2 in the hierarchical model. There is no inherent advantage in the larger values of sR and pR relative respectively to sR^2 and pR^2, nor of partial relative to semipartial R^2 (or R), since all yield identical statistical significance test and power analysis results, as we shall see below. The important considerations in selecting a descriptive statistic here, as always, are that it correctly represent the research issue and that its meaning be clear to the research consumer.

4.3.4 Area c

Finally, returning once more to the ballantine for sets (Figure 4.3.1), we call the reader's attention to area c, the double overlap of sets A and B in Y. It is conceptually the same as the area c in the ballantine for single IVs (Figure 3.3.1), and shares its problems. Although in the ballantine it occupies an area, unlike the areas a, b, and e, it can *not* be understood to be a proportion of Y variance, because, unlike these other areas, it may take on a negative value, as discussed in Section 3.4 under the heading "suppression." Note that it is *never* properly interpreted as a proportion of variance, whether or not in any given application it is found to be positive, since we cannot alter the fundamental conception of what a statistic means as a function of its algebraic sign in a particular problem. Since variance is sd^2, a negative quantity leads to sd being an imaginary number, for example, $\sqrt{-.10}$, a circumstance we simply cannot tolerate. Better to let area c stand as a useful metaphor which reflects the fact that $R^2_{Y \cdot AB}$ is not equal in general to $R^2_{Y \cdot A} + R^2_{Y \cdot B}$, but may be either smaller (positive c) *or* larger (negative c) than the latter. When area c is negative for sets A and B, we have exactly the same relationship of suppression between the two sets as was described for pairs of single IVs in Section 3.4.

4.4 SIGNIFICANCE TESTING FOR SETS

4.4.1 A General F Test for an Increment (Model I Error)

We have seen that the addition of a set of variables B to a set A results in an increment in the Y variance accounted for, that is, $R^2_{Y \cdot AB} - R^2_{Y \cdot A}$ ($= sR^2_B = I_B$). This is represented by the area b in the ballantine (Figure 4.3.1). This quantity is properly called an increment, because it is not mathematically possible for it to be negative, since $R^2_{Y \cdot AB}$ can not be smaller than $R^2_{Y \cdot A}$.[6]

Our interest, of course, is not in the characteristics of the sample for which these values are determined as such, but rather in those of the population from which it comes. Our mechanism of statistical inference posits the null hypothesis to the effect that *in the population*, there is literally *no* increment in Y variance accounted for when B is added to A, that is, that $R^2_{Y \cdot AB} - R^2_{Y \cdot A} = sR^2_B$ ($= I_B$) $= 0$, or that area b in the ballantine for the population is zero. When this null hypothesis is rejected, we conclude that set B does account for Y variance beyond that accounted for by set A in the population, that is, $B \cdot A$ does, indeed, relate to Y. This null hypothesis may be tested by means of

(4.4.1) $$F = \frac{(R^2_{Y \cdot AB} - R^2_{Y \cdot A})/k_B}{(1 - R^2_{Y \cdot AB})/(n - k_A - k_B - 1)}$$

---

[6] This proposition does not hold for R^2 corrected for shrinkage, that is, $\tilde{R}^2_{Y \cdot AB} - \tilde{R}^2_{Y \cdot A}$ may be negative. This will occur whenever the F of Eq. (4.4.1) is less than one. See Section 6.2.4.

for the source (numerator) $df = k_B$, and the error (denominator) $df = n - k_A - k_B - 1$, and referred to the F tables in the Appendices (Appendix Tables D.1 and D.2).

This formula is applied repeatedly in varying contexts throughout this book, and its structure is worth some comment. Both the numerator and denominator are proportions of Y variance divided by their respective df; thus both are "normalized" mean squares. The numerator is the normalized mean square for unique B variance (area b of the ballantine) and the denominator is the normalized mean square for a particular estimate of "error," i.e., $1 - R^2_{Y \cdot AB}$, which represents Y variation accounted for by neither A nor B (area e of the ballantine). As was done in considering error model alternatives for single IVs in Section 3.7.4, this is designated as Model I error. F is the ratio of these mean squares, which, when the null hypothesis is true, has an expected value of one. When F is sufficiently large to meet the significance criterion, as determined by reference to Appendix Tables D.1 and D.2, the null hypothesis is rejected.[7]

For computational purposes Eq. (4.4.1) can be somewhat simplified:

(4.4.2) $$F = \frac{R^2_{Y \cdot AB} - R^2_{Y \cdot A}}{1 - R^2_{Y \cdot AB}} \times \frac{n - k_A - k_B - 1}{k_B}$$

(for, of course, the same df).

Application to Hierarchical Model

To illustrate the application of this formula, let us return to the study presented in the previous sections on the length of stay of 500 hospital admissions, using demographic (D, $k_D = 9$), MMPI (M, $k_M = 10$) and hospital (H, $k_H = 7$) sets in that hierarchical order. We let A be the set(s) to be partialled and B the set(s) whose unique variance in Y is posited as zero by the null hypothesis. Table 4.4.1 organizes the ingredients of the computation to facilitate the exposition.

The null hypothesis that with D partialled ("holding demographic characteristics constant"), M accounts for no Y variance in the population is appraised as follows (Table 4.4.1, Example 1). It was given that $R^2_{Y \cdot D} = .20$ and $R^2_{Y \cdot DM} = .22$, an increase of .02. To use Eq. (4.4.2), call M set B and D set A. For $n = 500$, $k_B = 10$, $k_A = 9$, we find

$$F = \frac{.22 - .20}{1 - .22} \times \frac{500 - 9 - 10 - 1}{10} = 1.231,$$

which for df of 10 (= k_B) and 480 (= $n - k_A - k_B - 1$) fails to be significant at the $\alpha = .05$ level (the criterion value for $df = 10$, 400 is 1.85, Appendix Table

---

[7] Readers who know AV will find this all familiar. But the reasoning and structure are not merely analogous, but rather mathematically identical, since AV and MRC are applications of the "general linear model." This proposition is pursued in various contexts in this book; cf. Sections 5.3, 5.5, 6.3 and Chapters 8 to 11. But see Section 4.4.2.

D.2). The increase of .02 of the Y variance accounted for by M over D in the sample is thus consistent with there being no increase in the population.

In Example 2 of Table 4.4.1, we test the null hypothesis that the addition of H (which we will now call set B, so $k_B = 7$) to the sets D and M (which *together* we will call set A, so $k_A = 9 + 10 = 19$) results in no increase in Y variance in the population. Since $R^2_{Y \cdot DMH} = .33$ and $R^2_{Y \cdot DM} = .22$ (and hence $sR^2_H = I_H = .11$), substituting the values for sets A and B as redefined, we find

$$F = \frac{.33 - .22}{1 - .33} \times \frac{500 - 19 - 7 - 1}{7} = 11.094,$$

which for $df = 7, 473$ is highly significant, since the criterion F at $\alpha = .01$ for $df = 7, 400$ is 2.69 (Appendix Table D.1).

It was pointed out in Section 4.3.3 that our appraisal of this .11 increment in H over D and M constitutes the equivalent of an ACV with the 19 IVs of the combined D and M sets as covariates. Indeed, hierarchical MRC may be viewed as equivalent to a series of ACV, at each stage of which all prior sets are covariates (since they are partialled) while the set just entered is the research factor whose effects are under scrutiny (see Chapter 9). The "set just entered"

TABLE 4.4.1
Illustrative F Tests Using Model I Error (Eq. 4.4.2)[a]

$R^2_{Y \cdot D} = .20 \quad R^2_{Y \cdot MH} = .18 \quad R^2_{Y \cdot M} = .03$
$R^2_{Y \cdot DM} = .22 \quad R^2_{Y \cdot DH} = .32 \quad R^2_{Y \cdot H} = .17$
$R^2_{Y \cdot DMH} = .33$

xample	Set B	k_B	Set A	k_A	$R^2_{Y \cdot AB}$	$R^2_{Y \cdot A}$	$R^2_{Y \cdot (B \cdot A)} = sR^2_B$	Error $1 - R^2_{Y \cdot AB}$	Source Error df	Error df	F
1	M	10	D	9	.22	.20	.02	.78	10	480	1.231
2	H	7	D, M	19	.33	.22	.11	.67	7	473	11.094**
3	M, H	17	D	9	.33	.20	.13	.67	17	473	5.399**
4	D	9	M, H	17	.33	.18	.15	.67	9	473	11.766**
5	M	10	D, H	16	.33	.32	.01	.67	10	473	.706
6	D	9	M	10	.22	.03	.19	.78	9	480	12.991**
7	D	9	H	7	.32	.17	.15	.68	9	483	11.838**
8	H	7	D	9	.32	.20	.12	.68	7	483	12.176**
9	M	10	H	7	.18	.17	.01	.82	10	482	.588
10	H	7	M	10	.18	.03	.15	.82	7	482	12.596**
11	D	9	empty	0	.20	0	.20	.80	9	490	13.611**
12	M	10	empty	0	.03	0	.03	.97	10	489	1.512
13	H	7	empty	0	.17	0	.17	.83	7	492	14.396**

*$P < .01$.

4.4.2)
$$F = \frac{R^2_{Y \cdot AB} - R^2_{Y \cdot A}}{1 - R^2_{Y \cdot AB}} \times \frac{n - k_A - k_B - 1}{k_B},$$

ource (numerator) $df = k_B$, error I (denominator) $df = n - k_A - k_B - 1$.

may itself be an aggregate of sets. Although it would not likely be of substantive interest in this research, Example 3 of Table 4.4.1 illustrates the F test for the aggregate of M and H (as set B), with D (as set A) partialled.

Application to Simultaneous Model

The F test of Eqs. (4.4.1) and (4.4.2) is also applicable to the simultaneous model. The latter simply means that, given h sets, we are interested in appraising the variance of one of them with *all* the remaining $h - 1$ sets partialled. While in the hierarchical model only higher-order (prior) sets are partialled, in the absence of a concept of hierarchy, it is *all* other sets which are partialled. For this application of the F test we designate B as the unique source of variance under scrutiny, and aggregate the remaining sets which are to be partialled into set A.

Let us reconsider the running example of length of stay (Y) as a function of D, M, and H, but now propose that our interest is one of appraising the unique Y variance accounted for by each set. No hierarchy is intended, so by "unique" to a set we mean relative to all (here, both) other sets, that is, D relative to M and H, M relative to D and H, and H relative to D and M. To proceed, we need some additional R^2 values not previously given in this problem: $R^2_{Y \cdot MH} = .18$ and $R^2_{Y \cdot DH} = .32$.

To determine the unique contribution of D relative to M and H, that is, of $D \cdot MH$, one simply finds $R^2_{Y \cdot DMH} - R^2_{Y \cdot MH} = .33 - .18 = .15 = R^2_{Y \cdot (D \cdot MH)}$, the sR^2_D with both M and H partialled. This quantity might be of focal interest to a sociologist in that it represents the proportion of variance in length of stay of patients associated with differences in their demographic (D) characteristics, the latter freed of any personality differences (M) and differences in admitting hospitals (H) associated with D. This .15 is a sample quantity, and Example 4 (Table 4.4.1) treats D as set B, and aggregates M and H as set A for substitution in Eq. (4.4.2):

$$F = \frac{.33 - .18}{1 - .33} \times \frac{500 - 17 - 9 - 1}{9} = 11.766,$$

which is highly significant since the criterion F at $\alpha = .01$ for $df = 9, 400$ is 2.45 (Appendix Table D.1). Note, incidentally that this example simply reverses the roles of D and M, H of Example 3.

The unique variance contribution of M relative to D and H is tested without further elaboration as Example 5. This might be of particular interest to a clinical psychologist or personality measurement specialist interested in controlling demographic variables and systematic differences between hospitals in assessing the relationship of the MMPI to length of stay.

The last of this series, the Y variance associated with $H \cdot DM$ has already been presented and discussed as Example 2.

Intermediate Models

There is clearly no necessary requirement, given h sets, that *all* $h - 1$ sets be partialled from each in turn. By defining set B as whatever set(s) is under scrutiny, we can designate as set A any of the remaining sets *up to* $h - 1$ of them for partialling, and apply the F test using Model I error given as Eqs. (4.4.1) and (4.4.2). The investigator's choice of what to partial from what is determined by the logic and purpose of his inquiry. For specificity, assume that the h sets are partitioned into *three* groups of sets, as follows: the groups whose unique source is under scrutiny is, as before, designated set B, the group to be partialled from B (again as before) constitutes set A, but now the remaining set(s) constitute a group to be ignored, which we designate set C. All we are doing with this scheme is making explicit the obvious fact that not all sets of IVs on which there are data in an investigation need to be active participants in each phase of the analysis. There is no set C in the Model I Eqs. (4.4.1) and (4.4.2), which makes it evident that it is here being *completely* ignored.[8]

The running example has only three sets, so our "groups," including C will each contain a single set, but this is enough for illustration. There are a total of six different ways of designating three sets as A, B, and C to test the null hypothesis that $B \cdot A$ ignoring C accounts for no Y variance in the population. One of these, $M \cdot D$ (ignoring H) was given as Example 1 in Table 4.4.1. To determine and F test the remaining five, we require two values not previously given, $R^2_{Y \cdot M} = .03$ and $R^2_{Y \cdot H} = .17$. The five tests are given as Examples 6 through 10 in the table, for example, Example 7 gives for $D \cdot H$ (ignoring M)

$$F = \frac{.32 - .17}{1 - .32} \times \frac{500 - 7 - 9 - 1}{9} = 11.838,$$

which is highly significant (F at .01 for $df = 9$, 400 is 2.46; Appendix Table D.1).

In the (fully) simultaneous model, all sets other than the one whose (fully) unique Y variance is under scrutiny are partialled. We have now just seen that we can ignore some sets in some phases of an inquiry. Indeed, the (fully) hierarchical model with h sets is simply a predefined sequence of simultaneous analyses in the first of which a prespecified $h - 1$ sets are ignored, in the second a prespecified $h - 2$ sets are ignored, and, generally, in the jth of which a prespecified $h - j$ sets are ignored, until finally, in the last of which, none is ignored. The analysis *at each stage* is simultaneous—all IVs in the equation *at that stage* are being partialled from each other. Thus, a single simultaneous analysis with all other sets partialled and a strictly hierarchical progression of analyses may be viewed as endpoints of a continuum of analytic possibilities. A

---

[8] This elaboration into a set C which is ignored is also setting the scene for a consideration in Section 4.4.2 of Model II error, in which C is also removed from the error term, i.e., $1 - R^2_{Y \cdot ABC}$.

flexible application of MRC permits the selection of some intermediate possibilities when they are dictated by the theory and logic of the given research investigation.

Partialling an Empty Set

The generality of Eqs. (4.4.1) and (4.4.2) can be seen when they are applied in circumstances when the set A is empty, that is, a null class containing no IVs. Partialling no variables is simply not partialling. With A empty, $R^2_{Y \cdot A} = 0$, $R^2_{Y \cdot AB}$ is simply $R^2_{Y \cdot B}$, and $k_A = 0$. Substituting for this special case, Eq. (4.4.2) becomes

(4.4.3)
$$F = \frac{R^2_{Y \cdot B}}{1 - R^2_{Y \cdot B}} \times \frac{n - k_B - 1}{k_B},$$

with $df = k_B$, $n - k_B - 1$. This is patently the F test of the null hypothesis for any R^2, and, except for a difference in notation, is identical with the equation given for that purpose in Chapter 3 (Eq. 3.7.3). Incidentally, a further specialization to a single IV in set B, renders $k_B = 1$, $R^2_{Y \cdot B} = r^2_{Y1}$, and becomes the equivalent of the standard significance test for a zero-order r, Eq. (2.8.1), since

(4.4.4)
$$F = \frac{r^2_{Y1}}{1 - r^2_{Y1}} \times \frac{n - 1 - 1}{1} = \frac{r^2_{Y1}(n - 2)}{1 - r^2_{Y1}} = t^2,$$

and F for $df = 1$, $n - 2$ is identically distributed as t^2 for $df = n - 2$.

To apply Eq. (4.4.2) specialized as Eq. (4.4.3) or Eq. (3.7.3) to our running example, we need $R^2_{Y \cdot D} = .20$, $R^2_{Y \cdot M} = .03$, and $R^2_{Y \cdot H} = .17$. Examples 11, 12, and 13 give the ingredients and results of testing the null hypothesis using Model I error, respectively for D, M, and H.

Table 4.4.1 contains, for each set, the Y variance it accounts for and the Model I F test of its statistical significance when no other sets are partialled, when each of the other sets is partialled, and when both are partialled. Taking H, for example, we have: H, .17; $H \cdot D$, .12; $H \cdot M$, .15; and $H \cdot DM$, .11—all significant at $P < .01$. Such an accounting may also be organized for D and M. Although this type of exhaustive analysis, in which all possible combinations of the remaining $h - 1$ sets (including none) are partialled from each set may sometimes be profitable, we do not recommend it in general. With h sets, it produces a total of $h2^{h-1}$ distinct sR^2 values, which is a formidable number, even for a few sets: 3 sets yield 12 values in an exhaustive analysis, while 5 sets yield 80, 6 sets 192, and 7 sets 448! Not only do large numbers of significance tests increase the probability of spuriously significant "findings" and thus subvert the credibility of the results of a statistical analysis (see Section 4.6), but it is likely that many (probably most) of the possible partials do not make substantive sense. This will generally be the case when a causally or logically consequent set is partialled from an antecedent set. The strict ordering of h sets as in the hierarchical model selects only h of these $h2^{h-1}$ possibilities, a most desirable procedure when possible. But short of this, careful selection from these many possibilities produces the relatively small number of pertinent hypotheses of a convincing investigation.

Significance of pR^2

In Chapter 3 we saw that the partialled statistics of a single IV, X_i, that is, sr_i, pr_i, B_i, and β_i, all shared equivalent null hypotheses, and hence the same t test for the same $df = n - k - 1$. Conceptually, this can be explained as due to the fact that when any one of these coefficients equals zero, they all must necessarily equal zero.

For any set B, the same identity in significance tests holds for sR_B^2 and pR_B^2 (hence for sR_B and pR_B). Recall that these are both unique proportions of Y variance, the first to the base unity and the second to the base $1 - R_{Y \cdot A}^2$. In terms of areas of the ballantine for sets (Figure 4.3.1), $sR_B^2 = b$, and $pR_B^2 = b/(b + e)$. But the null hypothesis posits that area b is zero, hence $pR_B^2 = 0$. Whether one reports sR_B^2 as was done in Table 4.4.1, or divides it by $1 - R_{Y \cdot A}^2$ and reports pR_B^2, or reports both, since the null hypothesis is the same, the F test of Eq. (4.4.2) tests statistical significance of both sR_B^2 and pR_B^2 .

4.4.2 An Alternative F Test (Model II Error)

An F test is a ratio of two mean squares (or variance estimates), the numerator associated with a source of Y variance being tested, and the denominator providing a reference amount in the form of an estimate of error or residual variance. In the previous section, identifying A and B as sets or set aggregates, the numerator source was $B \cdot A$, and the denominator contained $1 - R_{Y \cdot AB}^2$, area e of the ballantine (Figure 4.3.1), thus treating all Y variance not accounted for by A and B as error in the F test of Eqs. (4.4.1) and (4.4.2). We later introduced the idea of a third set (or set-aggregate) C, whose modest purpose was "to be ignored." Not only was it ignored in that it was not partialled from B in defining $B \cdot A$ as the source for the numerator, but it was ignored in that whatever Y variance it might uniquely contribute was not included in $R_{Y \cdot AB}^2$, and therefore *was* part of the error, $1 - R_{Y \cdot AB}^2$.

These two ways of ignoring C are conceptually quite distinct, and may be considered independently. We obviously have the option of *not* partialling whatever we do not wish to partial from B so that the source of variance in the numerator is precisely what the theory and logic of the investigation dictates it to be, that is, $B \cdot A$ and not $B \cdot AC$. But we may either choose or not choose to ignore C in defining the error term, i.e., either use $1 - R_{Y \cdot AB}^2$ in the F test of Eqs. (4.4.1) and (4.4.2) and thus ignore C, or define an F ratio for $B \cdot A$ which removes whatever additional unique Y variance can be accounted for by C from the error term, resulting in Model II error, that is,

(4.4.5)
$$F = \frac{(R_{Y \cdot AB}^2 - R_{Y \cdot A}^2) / k_B}{(1 - R_{Y \cdot ABC}^2) / (n - k - 1)},$$

where k is the total number of IVs in *all* sets, that is, $k = k_A + k_B + k_C$, or equivalently,

(4.4.6)
$$F = \frac{R^2_{Y \cdot AB} - R^2_{Y \cdot A}}{1 - R^2_{Y \cdot ABC}} \times \frac{n - k - 1}{k_B}$$

with source (numerator) $df = k_B$, and error II (denominator) $df = n - k - 1$. Note that, as with Model I error, this tests both sR^2_B and pR^2_B. The standard F tables (Appendix Tables D.1 and D.2) are used. This is a generalization to sets of Model II error for single IVs (Section 3.7.4).

Which model to choose? It is the prevailing view that, since the removal of additional Y variance associated uniquely with C can only serve to produce a smaller and "purer" error term, one should always prefer Model II error. But this is not necessarily the case: although $1 - R^2_{Y \cdot ABC}$ will always be smaller (strictly, not larger) than $1 - R^2_{Y \cdot AB}$ and hence operate so as to increase F, one must pay the price of the reduction of the error df by k_C, that is, from $n - k_A - k_B - 1$ of Eq. (4.4.2) to $n - k - 1 = n - k_A - k_B - k_C - 1$ of Eq. (4.4.6), which clearly operates to decrease F. In addition, as error df diminish, the criterion F ratio for significance increases and sample estimates become less stable, seriously so when the diminished error df are absolutely small (see Appendix Tables D.1 and D.2). The competing factors of reducing proportion of error variance and reducing error df, depending on their magnitudes, may either increase or decrease the F using Model II error relative to the F using Model I error (see Section 4.6.3).

We can illustrate both possibilities with the running example (Table 4.4.1), comparing Model I F (Eq. 4.4.2) with Model II F (Eq. 4.4.6). If, in testing $M \cdot D$ in Example 1, instead of using Model I error, $1 - R^2_{Y \cdot DM} = .78$ with 480 (= 500 − 9 − 10 − 1) df, we use Model II error, $1 - R^2_{Y \cdot DMH} = .67$ with 473 (= 500 − 9 − 10 − 7 − 1) df, F increases to 1.412 from 1.231 (neither significant). On the other hand, shifting to Model II error in testing $D \cdot H$ in Example 7, brings F down from 11.838 to 11.766 (both significant at $P < .01$).

In those instances in Table 4.4.1 where set C is not empty, and hence where Models I and II are not identical (Examples 1, 6 through 13), the F ratios of the two models differ little and nowhere lead to different decisions about the null hypothesis. (The reader may check these as an exercise.) But before one jumps to the conclusion that the choice makes little or no difference in general, certain characteristics of this example should be noted and discussed, particularly the relatively large n, the fact that there are only three sets, and that two of these (D and H) account uniquely for relatively large proportions of variance. If n were much smaller, the difference of k_C loss in error df in Model II could substantially reduce the size and significance of F, particularly in the case where we let M be set C: the addition of M to D and H results in only a quite small decrease in error variance, specifically from $1 - R^2_{Y \cdot DH} = .68$ to $1 - R^2_{Y \cdot DHM} = .67$. If n were 100, the drop in error df from Model I to Model II would be from 83 to 73. Example 7, which tests $D \cdot H$ would yield a significant Model I $F = 2.034$ ($df = 9, 83, P < .05$), but a nonsignificant Model II $F = 1.816$ ($df = 9, 73$).

Further, consider the consequence of Model II error when the number of sets, and therefore the number of sets in C and (particularly) k_C is large. Many behavioral science investigations, particularly (but by no means solely) of the survey kind, can easily involve upwards of a dozen sets with C containing as many as ten sets collectively including many IVs; hence, k_C can be quite large.[9] The optimal strategy in such circumstances is to order the sets from those judged a priori to contribute most to those judged to contribute least to accounting for Y variance (or most to least confidently judged to account for any Y variance), and use Model I error. Using the latter successively at each level of the hierarchy, the lower-order sets are ignored, and although their (likely small) contribution to reducing the proportion of error variance is given up, their large contribution to error df is retained. (Section 4.6.3 enlarges on this issue.) Although Model II error may be used with the hierarchical model, and has the small advantage of using at all stages the same denominator in Eq. (4.4.5), and a minimum $1 - R^2$ error proportion, it has the disadvantage of using a minimum $df = n - k - 1$, that is, *all* k IVs in the analysis are debited from n, not only the $k_A + k_B$ involved in that stage. Thus, the significance tests of higher order sets where the investigator has "placed his bets" suffer, perhaps fatally, with those of the lower order sets in their use of a weakly determined, low df, error term.

On the other hand, if sets are few and powerful in accounting uniquely for Y variance, Model I error, $1 - R^2_{Y \cdot AB}$, contains important sources of variance due to the ignored C, and may well sharply "negatively bias" (reduce) F at a relatively small gain in error df. The error model of traditional AV generally presented in textbooks for the behavioral and social sciences is Model II error, on the rationale that any potential source of variance provided for by research factors and their interactions should be excluded from the error term.[10] The "within-groups" or "within-cells" error term of AV is a special case of Model II error, since the group or cell membership in multifactor designs is carried by all k IVs.

No simple advice can be offered on the choice between error models in hierarchical or intermediate models of MRC.[11] In general, large n, small h, small k, and sets whose sR^2 are large move us toward a preference for Model II error.

---

[9] This is not as extravagant as it may seem at first glance. For example, in a study using j research factors *and their interactions* (see Chapter 8), there are a total of $h = 2^j - 1$ sets. Four factors thus produce 15 sets, and 5 factors produce 31. Even with few IVs per research factor, the number of IVs mount up and error df accordingly dwindle rapidly. See Section 4.6.2, which recommends that h and k be kept as small as possible.

[10] This is not the case in AV presentations geared to other fields, such as agronomy and engineering. In these it is common practice to "pool" into the error term the sums of squares and df from some sources which are not of interest (usually interactions of high order) with or without tests of significance, treating them as negligible. This is identical with the use of what we are calling Model I error.

[11] Note that the issue can not arise in fully simultaneous MRC, since sets A and B of Eqs. (4.4.1) and (4.4.2) are exhaustive of all h sets and the set C of Eqs. (4.4.5) and (4.4.6) is empty; thus, the two pairs of equations reduce to equality, and the two models are indistinguishable.

One is understandably uneasy with the prospect of not removing from the Model I error the variability due to a set suspected a priori of having a large sR^2 (e.g., Examples 9 through 13 in Table 4.4.1). The overriding consideration in the decision is the maximization of the power of the F test. In the next section we will see how one goes about making this decision during the planning of an investigation, which ideally is when they should be made. At several points later in this book (particularly in Section 4.6.3 and Chapters 6 and 8), the issue of choice between Model I and Model II error comes up again, and its discussion in specific contexts will provide additional guidance.

4.5 POWER ANALYSIS FOR SETS

4.5.1 Introduction

In previous chapters we have presented the basic notions of power analysis (Sections 2.9 and 3.8), and methods for determining:

1. The power of the t test of significance for r_{YX} or B_{YX}, given the sample size (n), the significance criterion (α), and effect size (ES), that is, an alternate hypothetical value of the population r (Section 2.9.2).

2. The necessary sample size (n^*) for the significance test of a partial coefficient of a single X_i in an MRC involving k IVs, that is, its net contribution to R^2 (whether indexed as sr_i^2, pr_i^2, B_i, or β_i), given the desired power, α, and the ES, an alternate hypothetical value for the population sr_i^2 (Section 3.8.3).

3. n^* for the significance test of R^2, given the desired power, α, and an alternate hypothetical value for the population R^2 (Section 3.8.2).

The present section generalizes the previous methods of power analysis to partialled sets, paralleling the significance tests of the preceding Section 4.4. Assume that an investigation is being planned in which at some point the proportion of Y variance accounted for by a set B, over and above that accounted for by a set A, will be determined. We have seen that this critically important quantity is $R^2_{Y \cdot AB} - R^2_{Y \cdot A}$, and has variously and equivalently been identified as the increment due to B (I_B), the squared multiple semipartial correlation for B (sR^2_B or $R^2_{Y \cdot (B \cdot A)}$), and as area b in the ballantine for sets (Figure 4.3.1). This sample quantity will then be tested for significance, that is, the status of the null hypothesis that its analogous value *in the population* is zero will be determined by means of an F test. In planning the investigation, it is highly desirable (if not absolutely necessary) to determine how large a sample is necessary (n^*) to achieve some desired level of probability of rejecting that null hypothesis. As we have seen, this will depend not only on α (usually set at .05 or .01), but also on the actual size of the effect in the population (ES) which must be posited as an alternate to the null hypothesis. Whatever the difficulties of setting a priori specifications for these parameters (particularly ES), there is no alternative to rational research planning—in their absence, one does not know whether one needs to study 10 cases or 10,000 cases in order to achieve the purposes of the research.

4.5.2 Determining $n*$ for an F Test of sR_B^2 with Model I Error

As was the case for determining $n*$ for an F test on $R_{Y \cdot 12...k}^2$ (Section 3.8.2), the procedure for determining $n*$ for an F test on $sR_B^2 = R_{Y \cdot AB}^2 - R_{Y \cdot A}^2$ proceeds with the following steps:

1. Set the significance criterion to be used, α.
2. Set desired power for the F test.
3. Determine the number of source df, i.e., the number of IVs in set B, k_B .
4. Look up the value of L for the given k_B (row) and power (column) in Appendix Tables E.1 (for $\alpha = .01$) or E.2 (for $\alpha = .05$).

Power analysis of significance tests on the F distribution use as their alternate-hypothetical ES index f^2, which is, in general, a ratio of variances.[12] The particular test determines which variances are relevant, and therefore defines the appropriate formula for f^2. Whereas for determining $n*$ for an R^2, f^2 was given as $R^2/(1 - R^2)$ (Eq. 3.8.1), in determining $n*$ to test $R_{Y \cdot AB}^2 - R_{Y \cdot A}^2$ using Model I error,

$$(4.5.1) \qquad f^2 = \frac{R_{Y \cdot AB}^2 - R_{Y \cdot A}^2}{1 - R_{Y \cdot AB}^2} .$$

We stress the fact that these R^2 are alternate-hypothetical values referring to the population, *not* sample values. When the same ratio for *sample* values is combined with the df, the formula for F with Model I error results (Eqs. 4.4.1 and 4.4.2). This occurs after there *is* a sample, whereas in the planning taking place *before* the investigation, the formulation is "*if* f^2 is thus and such in the population, given α and the desired probability of rejecting the null, what $n*$ do I need?" Thus, the next step is

5. Determine a value for f^2. For this, one draws on past experience in the research area, theory, intuition or whatever, to answer the questions "What additional proportion of Y variance do I expect B to account for beyond A?" (the numerator), and "What proportion of Y variance will be accounted for by neither A nor B, that is, the Model I error?" (the denominator). Alternatively, certain conventions discussed below (Section 4.5.4) may be used for setting f^2.

6. Finally, substitute L (from Step 4) and f^2 in

$$(4.5.2) \qquad n* = \frac{L}{f^2} + k_A + k_B + 1.$$

---

[12] By using f^2 instead of some arbitary symbol, we link this treatment to that in Cohen (1969), where power analysis in the AV is treated extensively (Chapter 8), and where are given the relationships of f with noncentrality parameters used in the noncentral F distribution (Section 9.8).

The result is the number of cases necessary to have the specified probability of rejecting the null hypothesis (power) at the α level of significance when f^2 in the population is as posited (for Model I error).

For illustration we return to the running example, where length of stay of psychiatric admissions (Y) is studied as a function of sets of variables representing their demographic characteristics (D), their MMPI scores (M), and the hospitals where they were admitted (H), as described originally in Section 4.2.1. To this point this example has been discussed "after the fact"—results from a sample of 500 cases were presented and used to illustrate significance testing. Now we shift our perspective backwards in time to illustrate the power analysis associated with the planning of this research.

In planning this investigation we know that we will eventually be testing the null hypothesis (among others) that M will account for no variance in Y beyond what is accounted for by D. Thus, M is the set B and D is the set which is partialled from B, set A, and this null hypothesis is that $R^2_{Y \cdot DM} - R^2_{Y \cdot D} = R^2_{Y \cdot AB} - R^2_{Y \cdot A} = 0$, to be tested with Model I error, $1 - R^2_{Y \cdot DM} = 1 - R^2_{Y \cdot AB}$ (that is, the test eventually performed as Example 1, Table 4.4.1). Assume that we intend to use as significance criterion $\alpha = .05$ (Step 1) and that we wish the probability of rejecting this hypothesis (the power of the test) to be .90 (Step 2). There are 10 IVs in set M, so $k_B = 10$ (Step 3). Entering Appendix Table E.2 for $\alpha = .05$, in row $k_B = 10$, column power $= .90$, we find $L = 20.53$ (Step 4). To find f^2, we must posit values which we believe to obtain in the population: assume we judge that the actual population value for $sR^2_M = R^2_{Y \cdot DM} - R^2_{Y \cdot D}$ is .03, in which we judge $R^2_{Y \cdot DM} = .18$ (and hence, necessarily, $R^2_{Y \cdot D}$ is .15). From Eq. (4.5.1),

$$f^2 = \frac{.03}{1 - .18} = \frac{.03}{.82} = .03659$$

(Step 5). Finally (Step 6), we enter $L = 20.53$ and $f^2 = .03569$ to solve (4.5.2) for

$$n^* = \frac{20.53}{.03659} + 9 + 10 + 1 = 581.$$

Thus, if the unique Y variance of $M \cdot D$ in the population (sR^2_M) is .03, and Model I error is $1 - .18 = .82$, then in order to have .90 probability of rejecting the null hypothesis at $\alpha = .05$, the sample should contain 581 cases.

What if we were content to have power of .80 instead of .90? All that changes is the value of L, which (for $\alpha = .05$, power $= .80$) is found to be 16.24 in Appendix Table E.2. Substituting it in Eq. (4.5.2) gives $n^* = 464$; the reduced power demand requires fewer cases.

With many sets and many significance tests to be performed in an investigation, it is probably prudent to use the more stringent $\alpha = .01$ criterion per significance test in order to minimize the number of spuriously significant "results" over the entire analysis (see Section 4.6). If we shift to $\alpha = .01$ (Appendix Table E.1), we find $L = 22.18$ for power $= .80$ and $L = 26.98$ for power $= .90$. Substituting these in turn in Eq. (4.5.2) we find $n^* = 626$ and $n^* = 757$.

Another example: What is the n^* to have .80 power at $\alpha = .01$ in a test of $H \cdot D$ (eventually performed as Example 8, Table 4.4.1). With $k_H = k_B = 7$, from Appendix Table E.1 we determine that $L = 19.79$. Having already posited that $R^2_{Y \cdot D} = .15$, we further suppose as an alternate hypothesis that $R^2_{Y \cdot DH} = .20$. Thus, $R^2_{Y \cdot DH} - R^2_{Y \cdot D} = .05$, $1 - R^2_{Y \cdot DH} = .80$ and f^2 for Model I error is $.05/.80 = .0625$ from Eq. (4.5.1). Substitution this and L in Eq. (4.5.2), and recalling that $k_D = 9$, we have

$$n^* = \frac{19.79}{.0625} + 9 + 7 + 1 = 334.$$

Some other determinations of n^* for this test with different specifications of α and power are

$$\alpha = .01, \text{ power} = .90; L = 24.24; n^* = 405,$$
$$\alpha = .05, \text{ power} = .80; L = 14.35; n^* = 247,$$
$$\alpha = .05, \text{ power} = .90; L = 18.28; n^* = 309.$$

These may be checked by the reader as an exercise.

As a final example in this series, what is the n^* to have .80 power at $\alpha = .01$ in a test of $H \cdot DM$ (eventually performed as Example 2, Table 4.4.1)? Thus, H is set B and the aggregate D, M is set A in the formulas, and $k_H = 7 = k_B$, $k_D + k_M = 19 = k_A$. From Appendix Table E.1, we determine that $L = 19.79$ (as before). We estimate that $R^2_{Y \cdot DMH} = .25$ ($= R^2_{Y \cdot AB}$) in the population, and (as before) $R^2_{Y \cdot DM} = .18$ ($= R^2_{Y \cdot A}$), so from Eq. (4.5.1)

$$f^2 = \frac{.25 - .18}{1 - .25} = \frac{.07}{.75} = .09333.$$

Then, from Eq. (4.5.2) we find

$$n^* = \frac{19.79}{.09333} + 19 + 7 + 1 = 239.$$

Thus, if our sample size is 239, we will have a .80 probability of rejecting the null hypothesis using an $\alpha = .01$ criterion, if the population $f^2 = .09333$. If f^2 is actually larger, either because the unique variance of set B is larger or the error variance is smaller than we have posited, with 239 cases our power will exceed the .80 specified. If we have overestimated f^2, with 239 cases our power will fall short of .80.

It is worth noting that if we treat A as an empty set, the formulas for f^2 (4.5.1) and n^* (4.5.2) for sR^2_B reduce respectively to those given for power analysis of an unpartialled set of variables, that is, for $R^2_{Y \cdot 12 \ldots k}$ (Eqs. 3.8.1 and 3.8.2). Thus, as was the case for significance testing, in power analysis, unpartialled sets can be treated as a special case of partialled sets, $R^2_{Y \cdot B}$ being a special case of $R^2_{Y \cdot (B \cdot A)}$ $(= sR^2_B = I_B)$.

Furthermore, if we treat set B as containing a single IV, and set A as containing the remaining $k - 1$ IVs, then the formulas for f^2, Eq. (4.5.1) and n^*, Eq. (4.5.2), for sR^2_B reduce to those previously given for sr^2_i as Eqs. (3.8.3) and (3.8.2), respectively. Again we see the parallelism of power analysis to significance testing, wherein the same relationship of the general case to the special cases is found.

4.5.3 Determining n^* for an F Test of sR^2_B with Model II Error

Little change occurs in determining n^* when the F test anticipated in the research planning is to use Model II error. The source of variance under scrutiny in the numerator, $R^2_{Y \cdot AB} - R^2_{Y \cdot A}$ $(= sR^2_B = R^2_{Y \cdot (B \cdot A)})$ remains the same. But Model II conceives a third set (or set aggregate) C, whose Y variance is, with that of sets A and B, removed from the error term, which is then $1 - R^2_{Y \cdot ABC}$. The result is that the population ES for Model II error becomes

$$(4.5.3) \qquad f^2 = \frac{R^2_{Y \cdot AB} - R^2_{Y \cdot A}}{1 - R^2_{Y \cdot ABC}} .$$

Note again that this expression for alternate hypothetical population values is identical with the first term of the Model II F ratio (Eq. 4.4.6), where it refers to sample values.

The formula for n^* with Model II simply adds the df for C to those for A and B, and since $k_A + k_B + k_C = k$, the total number of IVs in the analysis, it becomes

$$(4.5.4) \qquad n^* = \frac{L}{f^2} + k + 1,$$

with L to be found as a function of α, desired power and k_B from Appendix Tables E.1 and E.2, exactly as before.

We will use the same examples to illustrate the determination of n^* for Model II error. The first was for $R^2_{Y \cdot DM} - R^2_{Y \cdot D}$ (generally, $R^2_{Y \cdot AB} - R^2_{Y \cdot A}$), which was posited to be .03. $R^2_{Y \cdot DMH}$ (generally, $R^2_{Y \cdot ABC}$) was posited to be .25, so Model II error is $1 - .25 = .75$. Therefore,

$$f^2 = \frac{.03}{1 - .25} = \frac{.03}{.75} = .04$$

from Eq. (4.5.3) (which, of course, can not be smaller than the f^2 for Model I, which was .03659). Since, α $(= .05)$, desired power $(= .90)$, and $k_M = k_B$ $(= 10)$ remain as originally specified, L remains 20.53 (Appendix Table E.2). The total number of IVs, $k = 9 + 10 + 7 = 26$, so from (4.5.4), we find

$$n^* = \frac{20.53}{.04} + 26 + 1 = 540$$

for Model II, compared with 581 for Model I. For changed specifications of α and power, the values of L change, with results for Model II n^* as follows, with Model I n^* in parenthesis for comparison:

$$\alpha = .05, \text{power} = .80; L = 16.24; n^* = 433 \ (464),$$
$$\alpha = .01, \text{power} = .80; L = 22.18; n^* = 582 \ (626),$$
$$\alpha = .01, \text{power} = .90; L = 26.98; n^* = 702 \ (757).$$

The second illustrative example was the n^* for the test for $R^2_{Y \cdot DH} - R^2_{Y \cdot D}$, which was posited to equal .05. Again, Model II error is $1 - R^2_{Y \cdot DMH} = 1 - .25 = .75$. Therefore, for Model II Eq. (4.5.3), $f^2 = .05/.75 = .06667$ (compared to Model I $f^2 = .0625$). L for $\alpha = .01$, power $= .80$, at $k_B \ (= k_H) = 7$ was found from Appendix Table E.1 to be 19.79. Substituting in Eq. (4.5.4) with $k = 26$, we find

$$n^* = \frac{19.79}{.06667} + 26 + 1 = 324$$

for Model II (compared to the Model I $n^* = 334$). For the other specifications and the resulting change in L, the n^* values for Model II are as follows, with Model I n^* in parentheses:

$$\alpha = .01, \text{power} = .90; L = 24.24; n^* = 391 \ (405),$$
$$\alpha = .05, \text{power} = .80; L = 14.35; n^* = 242 \ (247),$$
$$\alpha = .05, \text{power} = .90; L = 18.28; n^* = 301 \ (309).$$

The third illustrative example used above was for a test of $R^2_{Y \cdot DMH} - R^2_{Y \cdot DM}$. Since the numerator contains all the sets which figure in the error term, Model II error is here the same as Model I error. Viewed as Model I, H is set B and the aggregate $D + M$ is set A, so the error term is $1 - R^2_{Y \cdot AB} = 1 - R^2_{Y \cdot DMH}$. Viewed as Model II, since no set is ignored, C is empty, and the error term $1 - R^2_{Y \cdot ABC} = 1 - R^2_{Y \cdot AB}$, that is, the same. Both for power analysis and significance testing, for tests where no sets are ignored in either numerator or denominator, Model II reduces to Model I.

In the two examples where the error models differed, we found that n^* was smaller for Model II than for Model I for the same specifications. These two instances should not be overgeneralized, as we have already argued. The relative size of the n^* of the two models depends on how much reduction in the (alternate-hypothetical) proportion of error variance occurs relative to the cost in df due to the addition of the k_C IVs of set C. The Model II error denominator of f^2 can be written as a function of the Model I error and the additional reduction due to Y variance unique to set C:

(4.5.5) $$1 - R^2_{Y \cdot ABC} = 1 - R^2_{Y \cdot AB} - sR^2_C,$$

so that

$$\text{Model II error} = \text{Model I error} - sR^2_C.$$

The change in df is k_C. Therefore, Model II will require smaller n^* than Model I when sR_C^2 is large relative to k_C, but larger n^* than Model I when sR_C^2 is small relative to k_C. Concretely, in the test of $H \cdot D$ of the second example, $R_{Y \cdot D}^2$ was posited to be .15, $R_{Y \cdot DH}^2$ = .20 (hence, $R_{Y \cdot DH}^2 - R_{Y \cdot D}^2$ = .05), and $R_{Y \cdot DMH}^2$ = .25. The additional reduction of error by Model II over Model I is sR_M^2 = .05 (= sR_C^2), and k_M = 10 (= k_C). If instead, we had posited $R_{Y \cdot DMH}^2$ to be .22, sR_M^2 would be .02 (= sR_C^2). The change results in a reduction in Model II relative to Model I error by .02 (instead of .05); hence, Model II error by Eq. (4.5.6) is .78 (instead of .75); f^2 would then be .05/.78 = .06410 instead of .05/.75 = .06667. The result for L = 19.79 (for k_B = 7, α = .01, power = .80) is that the Model II test of $H \cdot D$ for the revised $R_{Y \cdot DMH}^2$ yields (Eq. 4.5.4)

$$n^* = \frac{19.79}{.06410} + 26 + 1 = 336,$$

which is larger than the Model I (Eq. 4.5.2)

$$n^* = \frac{19.79}{.06667} + 9 + 7 + 1 = 314.$$

The basis for the advice given in Section 4.4.2 should now be clear. When there are many IVs in set C (which may be an aggregate of several sets) and there is no expectation that sR_C^2 is substantial, Model I may require smaller n^* (or for the same n have greater power) than Model II. When k_C is small and/or sR_C^2 is substantial, the reverse holds. The methods of this section, applied to realistically posited values for $R_{Y \cdot A}^2$, $R_{Y \cdot AB}^2$, and $R_{Y \cdot ABC}^2$ in the planning of an investigation, will provide a basis for the a priori choice of error model.

4.5.4 Setting f^2

The key decision required in the power analysis necessary for research planning in MRC, and generally the most difficult, is producing an f^2 for substitution in the equation for n^*. One obviously can not *know* the various population R^2 values which go to make up f^2. But unless some estimates are made in advance, there is no rational basis for planning. Furthermore, unless the f^2 bears some reasonable resemblance to the true state of affairs in the population, sample sizes will be too large or (more often) too small, or, when sample sizes are not under the control of the researcher, the power of the research will be under- or (more often) overestimated.

The best way to proceed is to muster all one's resources of empirical knowledge, both hard and soft, about the substantive field of study and apply them, together with some insight into how magnitudes of phenomenon are translated into proportions of variance, in order to make the estimates of the population R^2 values which are the ingredients of f^2. Some guidance may be obtained from a handbook of power analysis (Cohen, 1969), which proposes operational definitions or conventions which link qualitative adjectives to amounts of correlation broadly appropriate to the behavioral sciences. Translated into proportion of variance terms (r^2 or sr^2), these are "small," .01; "medium," .09; "large," .25. The rationale for these quantities and cautions about their use are given by Cohen (1962, 1969).

Since f^2 is made up of two (Model I) or three (Model II) different R^2 values, it may facilitate the thinking of a research planner to have operational definitions for f^2 itself. With some hesitation, we offer the following: "small," f^2 = .02; "medium," f^2 = .15; "large," f^2 = .35. Our hesitation arises from the following considerations. First, there is the general consideration of obvious diversity of the areas of study covered by the rubric "behavioral and social sciences." For example, what is large for a personality psychologist may well be small for a sociologist. The conventional values offered can only strike a rough average. Secondly, since f^2 is made up of two or three distinct quantities, their confection into a single quantity offers opportunities for judgment to go astray. Thus, what might be thought of as a medium sized expected sR^2 (numerator) may well result in a large or quite modest f^2, depending on whether the expected $1 - R^2$ error (denominator) is small or large. Furthermore, an f^2 = .15 may be appropriately thought of as a "medium" ES in the context of 5 or 10 IVs in a set, but seems too small when k = 15 or more, indicating that, on the average, these variables account for, at most, (.15/15 =) .01 of the Y variance. Nevertheless, conventions have their uses, and the ones modestly offered here should serve to give the reader *some* sense of the f^2 quantity to attach to his verbal formulations, particularly when he cannot cope with estimating the population values themselves. The latter is, as we have said, the preferred route to setting f^2.

4.5.5 Setting Power for n^*

In the form of power analysis discussed thus far, we find the necessary sample size n^* for a given desired power (given also α and f^2). What power do we desire? If we follow our natural inclinations and set it quite large (say at .95 or .99), we quickly discover that except for very large f^2, n^* gets to be very large, often beyond our resources. (For example, in the first example of Section 4.5.2, the test of $M \cdot D$, for α = .05 and power = .99, n^* works out to be 905, about double what is required at power = .80.) If we set power at a low value (say at .50 or .60), n^* is relatively small (for the example, at power = .50, n^* = 271), but we are not likely to be content to have only a 50–50 chance of rejecting the null hypothesis.

The decision as to what power to set is a complex one. It depends upon the result of weighing the "costs" of failing to reject the null hypothesis (Type II error in statistical inference) against the costs of gathering and processing research data. The latter are usually not hard to estimate objectively, while the former include the costs of such imponderables as "failing to advance knowledge," "losing face," and editorial rejections, and of such painful ponderables as not getting continued research support from funding agencies. This weighing of costs is obviously unique to each investigation or even to each null hypothesis to be tested. This having been carefully done, the investigator can then formulate the power value he desires.

Although there will be exceptions in special circumstances, he is likely to choose some value in the .70—.90 range. He may choose a value in the lower part of this range when the dollar cost per case is large and/or when the more intangible cost of a Type II error in inference is not great, that is, when rejecting the null hypothesis in question is of relatively small importance. Conversely, a value at or near the upper end of this range would be chosen when the additional cost of collecting and processing cases is not large and/or when the hypothesis is an important one.

It has been proposed, in the absence of some preference to the contrary, that power be set at .80 (Cohen, 1965, 1969). This value falls in the middle of the .70—.90 range, and is a reasonable one to use as a convention, when such is needed.

4.5.6 Reconciling Different n*s

When more than one hypothesis is to be tested in a given investigation, the application of the methods described above will result in multiple n*s. Since a single investigation will have a single n, these different n*s will require reconciliation.

For concreteness, assume plans to test three null hypotheses (H_i) whose specifications have resulted in $n_1^* = 100$, $n_2^* = 300$, and $n_3^* = 400$. If we decided to use $n = 400$ in the study, we will meet the specifications of H_3 and have much more power than specified for H_1 and more for H_2. This is fine if, in assessing our resources and weighing them against the importance of H_3, we deem it worthwhile. Alternatively, if we proceed with $n = 100$, we will meet the specifications of H_1, but fall short of the power desired for H_2 and H_3. Finally, if we strike an average of these n*s and proceed with $n = 267$, we will have more power than specified for H_1, slightly less for H_2, and much less for H_3.

There is of course no way to have a single n which will simultaneously meet the n* specifications of multiple hypotheses. No problem arises when resources are sufficient to proceed with the largest n*; obviously there is no harm in exceeding the desired power for the other hypotheses. But such is not the usual case, and difficult choices may be posed for the investigator. Some help is afforded if one can determine exactly how much power drops from the desired value when n is to be less than n* for some given hypothesis. Stated more generally, it is useful to be able to estimate the power of a test given some specified n, the inverse of the problem of determining n* given some specified desired power. The next section is devoted to the solution of this problem.

4.5.7 Power as a Function of n

Thus far, we have been pursuing that particular form of statistical power analysis wherein n* is determined for a specified desired power value (for given α and f^2). Although this is probably the most frequently useful form of power analysis, we have just seen the utility of inverting n and power, that is, determining the power which would result for some specified n (for given α and f^2). The latter is not only useful in the reconciliation of different n_i^* (Section 4.5.6), but in other circumstances, for example, when the n available for study is fixed, or when a power analysis is done on an hypothesis post hoc as in a power survey (cf. Cohen, 1962). To find power as a function of n, we rewrite Eq. (4.5.2) for Model I and Eq. (4.5.4) for Model II to yield, respectively,

(4.5.6) $L^* = f^2(n - k_A - k_B - 1)$,

(4.5.7) $L^* = f^2(n - k - 1)$.

Recall that L is itself a function of k_B, α, and power. To find power, one simply uses the L tables (Appendix Tables E.1 and E.2) backwards. Enter the table for the significance criterion α to be used in the row for k_B, and read across to find where the obtained L^* falls. Then read off at the column heading the power values which bracket it.[13]

To illustrate: In Section 4.5.2, we considered a test of $R^2_{Y \cdot DM} - R^2_{Y \cdot D}$ using Model I error at $\alpha = .05$, where $k_M = k_B = 10$, $k_D = k_A = 9$, and $f^2 = .03659$. Instead of positing desired power (e.g., .80) and determining n^* (= 464), let us instead assume that (for whatever reason) we will be using $n = 350$ cases. To determine the power, find, using Eq. (4.5.6),

$$L^* = .03659 (350 - 9 - 10 - 1) = 12.07;$$

entering Appendix Table E.2 (for $\alpha = .05$) at row $k_B = 10$, we find that $L^* = 12.07$ fall between $L = 11.15$ at power = .60 and $L = 13.40$ at power = .70. Thus, with $n = 350$ for these specifications, power is between .60 and .70. (Although usually not necessary, if a specific value is desired, linear interpolation is sufficient for an approximation; here it yields .64.)

For further illustration, what is the power of a test of $R^2_{Y \cdot DH} - R^2_{Y \cdot D}$ using Model II error at $\alpha = .05$, where $k_H = k_B = 7$, $k = 26$, and $f^2 = .06667$ (the second example in Section 4.5.3), for $n = 350$? From Eq. (4.5.7) we find

$$L^* = .06667 (350 - 26 - 1) = 21.53.$$

Entering Appendix Table E.2 (for $\alpha = .05$) at row $k_B = 7$, we find that $L^* = 21.53$ falls just below the $L = 21.84$ for power = .95. (If we wish to consider the alternative $\alpha = .01$, we find in Appendix Table E.1 that $L^* = 21.53$ falls slightly below $L = 21.71$ for power = .85.)

What power does $n = 350$ bring to the test of $R^2_{Y \cdot DMH} - R^2_{Y \cdot DM}$ at $\alpha = .01$, where $k_D = k_B = 7$, $k_D + k_M = k_A = 19$, for $f^2 = .09333$ (the third example in Section 4.5.2)? Since in this case Model II and Model I are indistinguishable, we

---

[13] It is for such applications that the tables provide for low power values (.10 to .60). When a specified n results in low power, it is useful to have some idea of what the power actually is. See the last example in this section.

use Eq. (4.5.6) to find

$$L* = .09333(350 - 19 - 7 - 1) = 30.15,$$

which in the L table for α = .01 (Appendix Table E.1), row k_B = 7, falls above the L value for power of .95, i.e., 28.21. (For α = .05, power is of course even higher: in Appendix Table E.2, $L*$ = 30.15 falls above the L = 29.25 value for power = .99.)

Finally, let us posit for this preceding test that the n is to be 100 (instead of 350). Equation (4.5.6) yields

$$L* = .09333 (100 - 19 - 7 - 1) = 6.81.$$

For α = .01 (for k_B = 7), Table E.1 shows this $L*$ falling between 4.08 for power = .10 and 8.57 for power = .30; for a specific (but approximate) value, linear interpolation yields power about .22. Things are somewhat better for α = .05, where Appendix Table E.2 shows $L*$ = 6.81 bounded by 4.77 for power = .30 and 7.97 for power = .50, with linear interpolation giving approximate power of .43.

4.5.8 Power as a Function of n: The Special Cases of R^2 and sr^2

We have already seen that, as for F tests of statistical significance, the determination of $n*$ for partialled sets is a general case which can be specialized. In Chapter 3, we considered the determination of $n*$ as a function of power (as well as α and f^2) for tests on R^2 and sr_i^2. Treated as special cases, the method of the preceding section, Eq. (4.5.7), can be used to determine the power of these tests as a function of n.

For R^2, f^2 is given in Eq. (3.8.1) as $R^2/(1 - R^2)$. Simply substitute this f^2 in Eq. (4.5.7) to find $L*$, and proceed as before, entering the L table at row $k_B = k$, the total number of IVs.

For sr_i^2, where X_i is one of k IVs, f^2 is given in Eq. (3.8.3) as $sr_i^2/(1 - R^2)$. Again, Eq. (4.5.7) is used to find $L*$, and the L table is entered at row k_B = 1.

4.5.9 Tactics of Power Analysis

We noted earlier that power analysis concerns relationships among four parameters: power, n, α, and ES (indexed by f^2 in these applications). Mathematically, any one of these parameters is determined by the other three. We have considered the cases where n and power are each functions of three others. It may also be useful to exploit the other two possibilities. For example, if one specifies desired power and α for a hypothesis, L is determined. Then, for a specified n, Eqs. (4.5.6) or (4.5.7) can be solved for f^2. This is the *detectible* ES, that is, the population f^2 one can expect to detect using the significance criterion α, with probability given by the specified power desired, in a sample of n cases. One can also (at least, crudely) determine what α one should use, given f^2, desired power, and a sample of size n. Although these are useful forms of power analysis, their detailed consideration is precluded here. The interested reader is referred to Cohen (1965, 1969, 1973b).

It must be understood that these mathematical relationships among the four parameters should serve as tools in the service of the behavioral scientist turned applied statistician, not as formalisms in their own right. We have for the sake of expository simplicity implicitly assumed that when we seek to determine n^*, there is only one possible α, one possible value for desired power, and one possible f^2. Similarly, in seeking to determine power, we have largely operated as if only one value each for α, f^2, and n are to be considered. But the realities of research planning often are such that more than one value for one of these parameters can and indeed must be entertained. Thus, if one finds that for a hypothesis for which $\alpha = .01$, power $= .80$, and $f^2 = .04$, the resulting n^* is 600, and this number far exceeds our resources, it is sensible to see what n^* results when we change α to .05. If that is also too large, we can invert the problem, specify the largest n we can manage, and see what power results for this n at $\alpha = .05$. If this is too low, we might examine our conscience and see if it is reasonable to entertain the possibility that f^2 is larger, perhaps .05 instead of .04. If so, what does that do for power at that given n? At the end of the line of such reasoning, the investigator either has found a combination of parameters that makes sense in his substantive context, or has decided to abandon the research, at least as originally planned. Many examples of such reasoning among a priori alternatives are given in Cohen (1969).

With the multiple hypotheses which generally characterize MRC analysis, the need for exploring such alternatives among combinations of parameters is likely to increase. If H_1 requires $n^* = 300$ for desired power of .80, and 300 cases give power of .50 for H_2 and .60 for H_3, etc., only a consideration of alternate parameters for one or more of these hypotheses may result in a research plan which is worth undertaking.

To conclude this section with an optimistic note, we should point out that we do not always work in an economy of scarcity. It sometimes occurs that an initial set of specifications results in n^* much smaller than our resources permit. Then we may find that when the parameters are made quite conservative (for example, $\alpha = .01$, desired power $= .95$, f^2 at the lower end of our range of reasonable expectation), we still find n^* smaller than our resources permit. We might then use the power analysis to avoid "overkill," and perhaps use our additional resources for obtaining better data, or to test additional hypotheses, or even for additional investigations.

4.6 STATISTICAL INFERENCE STRATEGY IN MULTIPLE REGRESSION/CORRELATION

4.6.1 Controlling and Balancing Type I and Type II Errors in Inference

In the preceding two sections, we have set forth in some detail the methods of hypothesis testing and power analysis for sets. Testing for significance is the procedure of applying criteria designed to control at some rate α, the making of a Type I error in inference, that is, the error of rejecting true null hypotheses, or, less formally, finding things that are not there. Power analysis focusses on the other side of the coin of statistical inference, and seeks to control at some rate β the making of a Type II error, the error of failing to reject false null hypotheses and hence failing to find things that *are* there. The fundamental demand of an effective strategy of statistical inference is the balancing of Type I and Type II errors in a manner consistent with the substantive issues of the research. In practice, this takes the form of seeking to maintain a reasonably low rate of Type I errors, while, at the same time, not allowing Type II errors to become unduly large, or equivalently, maintaining good power against realistic alternatives to the null hypothesis.

For any discrete null hypothesis, given the usual statistical assumptions and the requisite specifications, the procedures for significance testing and power analysis are relatively simple, as we have seen. When one must deal with multiple hypotheses, however, statistical inference becomes exceedingly complex. One dimension of this complexity has to do with whether the Type I error rate is calculated per hypothesis, per group of related hypotheses ("experimentwise") or for even larger units ("investigationwise"). Another is whether α is held constant over the multiple hypotheses or is varied. Still another is whether the hypotheses are mutually independent (orthogonal) or dependent. Yet another is whether the hypotheses are planned in advance or stated after the data have been examined ("post hoc"), the latter being sometimes referred to as "data snooping." And there are yet others. Each of the possible combinations of these alternatives has one or more specific procedures for testing the multiple hypotheses, and each procedure has its own set of implications to the statistical power of the tests its performs.

An example may help clarify the above. Assume an investigator is concerned with hypotheses about the means on a dependent variable across levels of a research factor, G, made up of 6 ($= g$) groups. He may be interested in any of the following kinds of multiple hypotheses, each of which has its own procedure(s):

1. *All simple comparisons between means.* There are $g(g - 1)/2 = 15$ different pairs of means and thus 15 simple comparisons and their null hypotheses. Assume each is t tested at $\alpha = .05$; thus, the Type I error rate *per hypothesis* is controlled at .05. But if, in fact, the population means are all equal, it is intuitively evident that the probability that *at least one* comparison will be "significant," that is, the experimentwise error rate, is greater than .05. The actual rate for $g = 6$ is approximately .40.[14] Thus, the separate αs "mount up," in this case to .40. This error rate may well be unacceptable to the investigator,

---

[14] The calculation requires special tables and the result depends somewhat on sample size. Some other experimentwise error rates for these conditions are (approximately): for $g = 10$, .60 and for $g = 20$, .90. Even for $g = 3$, it is .13. Only for $g = 2$ is it .05.

and almost certainly so to his peers. But each t test at $\alpha = .05$ will be relatively powerful.

There is a large collection of statistical methods designed to cope with the problem of making all simple comparisons among g means. These vary in their definition of the problem, particularly in their conceptualization of Type I error, and they therefore vary in power and in their results. For example, the Tukey HSD test (Winer, 1971, pp. 197–198) controls the experimentwise error rate at α. The Newman–Keuls test and Duncan test both approach Type I error via "protection levels" which are functions of α, but the per-hypothesis Type I error risks for the former are constant and for the latter vary systematically (Winer, 1971, pp. 196–198). Bonferroni (or Dunn) tests employ the principle of dividing an overall α into as many (usually equal) parts as there are hypotheses, and then setting the per-hypothesis significance criterion accordingly; thus, for $\alpha = .05$, each of the 15 comparisons would be tested with the significance criterion set at $\alpha = .05/15 = .0033$ (Miller, 1966, pp. 67–70). The preceding tests of all pairs of means are the most frequently employed, and by no means exhaustive (Games, 1971; Miller, 1966).

One of the oldest and simplest procedures for all pairs of g means is Fisher's "protected t" (or LSD) test (Carmer & Swanson, 1973; Games, 1971; Miller, 1966). First, an ordinary (AV) overall F test is performed on the set of g means $(df = g - 1, n - g)$. If F is not significant, no pairwise comparisons are made. Only if F is significant at the α criterion level are the means compared, this being done by an ordinary t test. Thus, the t tests are protected from large experimentwise Type I error by the requirement that the preliminary F test must meet the α criterion. (In Section 4.6.4, this test is adapted for general use in MRC analysis.)

Note that each of the above tests approaches the control of Type I errors differently, and that therefore each carries different implications to the rate of Type II errors and hence to the test's power.

2. *Some simple comparisons between means.* With g means, only differences between some pairs may be of interest. A frequent instance of this case occurs when $g - 1$ of the groups are to be compared with a single control or reference group, which thus calls for $g - 1$ hypotheses that are simple comparisons. In this special case, the Dunnett test, whose α is controlled experimentwise, applies (Winer, 1971, pp. 201–204). For the more general case where not all pairwise hypotheses are to be tested, protected t and Bonferroni tests (and others) may be used. Again these different tests, with their different strategies of Type I error control, have different power characteristics.

3. *Orthogonal comparisons.* With g groups, it is possible to test up to $g - 1$ null hypotheses on comparisons (linear contrasts) which are orthogonal, that is, independent of each other. These may be simple or complex. A complex comparison is one which involves more than two means, for example, $\bar{Y}_1$ versus the mean of $\bar{Y}_3$, $\bar{Y}_4$, and $\bar{Y}_5$, or the mean of $\bar{Y}_1$ and $\bar{Y}_2$ versus the mean of $\bar{Y}_3$ and $\bar{Y}_5$. These two complex "mean of means" comparisons are however, not

orthogonal. (The criterion for orthogonality of contrasts and some examples are given in Section 5.5.1.) When the maximum possible number of orthogonal contrasts, $g - 1$, are each tested at α, the experimentwise Type I error rate is larger; specifically, it is approximately $1 - (1 - \alpha)^{g-1}$. Thus, for $g = 6$ and the per-contrast hypothesis α set at .05, the experimentwise Type I error rate is $1 - (1 - .05)^5 = .226$. It is common practice, however, not to reduce the per contrast rate α below its customary value in order to reduce the experimentwise rate when orthogonal contrasts are used (Games, 1971).

Planned (a priori) orthogonal comparisons are generally considered the most elegant multiple comparison procedure and have good power characteristics, but, alas, they can only infrequently be employed in behavioral science investigations because the questions to be put to the data are simply not usually independent, for example, those described in paragraphs 1 and 2 above, and in the next paragraph.

4. *Nonorthogonal, many, and post hoc comparisons.* Although only $g - 1$ *orthogonal* contrasts are mathematically possible, the total number of different mean of means contrasts is large (see Chapter 5, footnote 7), and the total number of different contrasts of all kinds is infinite for $g > 2$. An investigator may wish to make more than $g - 1$ (and therefore necessarily nonorthogonal) comparisons, or may wish to make comparisons which were not contemplated in advance of data collection, but rather suggested post hoc by the sample means found in the research. Such "data snooping" is an important part of the research process, but unless Type I error is controlled in accordance with this practice, the experimentwise rate of spuriously significant t values on comparisons becomes unacceptably high. The Scheffé test (Edwards, 1972; Games, 1971; Miller, 1966) is designed for these circumstances. It permits *all possible* comparisons, orthogonal or nonorthogonal, planned or post hoc, to be made, subject to a controlled experimentwise Type I error rate. Because it is so permissive, however, in most applications it results in very conservative tests, that is, in tests of relatively low power (Games, 1971).

The reasons for presenting the above brief survey are twofold. The first is to alert the reader to the fact that for specific and well defined circumstances of hypothesis formulation and Type I error definition, there exist specific statistical test procedures. But even for the "simple" case of a single nominally scaled research factor G made up of g groups, the basis for choice among the alternatives is complex. Indeed, an entire book addressed to mathematical statisticians has been written in this area (Miller, 1966). An excellent article-length exposition for behavioral scientists is provided by Games (1971). Most AV textbooks present some of the more popular multiple comparison procedures for means with worked examples and the necessary tables, for example, Edwards (1972) and Winer (1971).

The second reason for presenting the above survey is to emphasize the fact that, given the variety of approaches to the conception of Type I errors, there are differential consequences to the rate of Type II errors, and thus to the statistical power of these tests. Conventional statistical inference is effectively employed only to the extent that Type I and Type II error risks are appropriately balanced. The investigator can neither afford to make spurious positive claims (Type I) nor fail to find important relationships (Type II). We have seen in Section 4.5 that, all things equal, these two types of errors are inversely related, so that some balance is needed. Yet the complexity which we encountered above when confronted only with the special case of a single nominal scale makes it clear that any effort to treat this problem in comprehensive detail is far outside the bounds of practicality and not in keeping with this book's purpose and data-analytic philosophy, nor with the needs of its intended audience.

What is required instead are some general principles and simple methods which, over the wide range of circumstances in research in the behavioral and social sciences, will serve to provide a practical basis for keeping both types of errors acceptably low and in reasonable balance. The major elements of this approach include parsimony in the number of variables employed, the use of a hierarchical strategy, and the adaptation of the Fisher protected t test to MRC.

4.6.2 Less Is More

A frequent dilemma of the investigator in behavioral science arises in regard to the number of variables he will employ in a given investigation. On the one hand, he is impelled by the need to make sure that the substantive issues are well covered and nothing overlooked, and on the other he needs to keep in bounds the cost in time, money and increased complexity which is incurred with an increase in variables. Unfortunately, the dilemma very often is resolved in favor of having more variables to assure coverage.

In addition to the time and money costs of more variables (which are frequently negligible, hence easily incurred), there are more important "costs" to the validity of statistical inference which are very often overlooked. The more variables, dependent or independent, in an investigation, the more hypotheses are tested (either formally or implicitly). The more hypotheses tested, the greater the probability of occurrence of spurious significance ("investigation-wise" Type I error). Thus, with 5 dependent and 12 independent variables analyzed by 5 MRC analyses (one per dependent variable), there are a total of 60 potential t tests on null hypotheses for partial coefficients alone. At $\alpha = .05$ per hypothesis, if all these null hypotheses were true, the probability that one or more ts would be found "significant" approaches unity. Even at $\alpha = .01$ per hypothesis, the investigationwise rate would be in the vicinity of .50.[15] It is rare in research reports to find their results appraised from this perspective, and many investigations are not reported in sufficient detail to make it possible for a

---

[15] Since the 60 tests are not independent, exact investigationwise error rates can not be given. If they were independent, the two investigationwise Type I error rates would be $1 - .95^{60} = .954$ (for $\alpha = .05$) and $1 - .99^{60} = .453$ (for $\alpha = .01$).

reader to do so—variables that "don't work" may never surface in the final report of a research.

One might think that profligacy in the number of variables would at least increase the probability of finding true effects when they are present in the population, even at the risk of finding spurious ones. But another consideration arises. In each MRC, the greater the number of IVs (= k), the lower the power of the test on each IV (or set of IVs). We have seen this in several ways. First, for any given n, the error $df = n - k - 1$, and are thus diminished as k increases. Second, a glance at the L tables (Appendix Table E) quickly reveals that all things equal, as k_B increases, L increases, and therefore power decreases for any given f^2, α, and n (Section 4.5.7). Also, it is likely that as k increases, the R_is among the IVs increase, which in turn increases the standard errors of partial coefficients (e.g., SE_{B_i}s) and reduces the t_is and hence the power (Section 3.7.5). Thus, having more variables when fewer are possible increases the risks of both finding things that are not so and failing to find things that are. These are serious costs, indeed.

Note that a large n does not solve the difficulties in inference which accompany large numbers of variables. True, the error df will be large, which, taken by itself, increases power. But the investigationwise Type I error rate depends on the number of hypotheses, and not on n. And even potentially high power conferred by large n may be dissipated by large k (as seen in the Appendix E L tables), and by the large R_is that large k may produce.

Within the goals of a research study, the investigator usually has considerable leeway in the number of variables to include, and too frequently the choice is made for more rather than fewer. The probability of this increases with the "softness" of the research area and the degree to which the investigation is exploratory in character, but no area is immune. When a theoretical construct is to be represented in data, a large number of variables may be used to represent it in the interest of "thoroughness" and "just to make sure" that the construct is covered. It is almost always the case, however, that the large number is unnecessary. It may be that a few (or even one) of the variables are really central to the construct and the remainder peripheral and largely redundant. The latter are better excluded. Or the variables may all be about equally related to the construct and define a common factor in the factor-analytic sense, in which case they should be combined into an index, factor score, or sum. The latter not only will represent the construct with greater reliability and validity, but will do so with a single variable. Perhaps more than one common factor is required, but this is still far more effective than a large number of single variables designed to cover (really smother) the construct. These remarks obtain for constructs in both dependent and independent variables.

Other problems in research inference are attendant upon using many variables in the representation of a construct. When used as successive dependent variables, they frequently lead to inconsistent results that, as they stand, are difficult to interpret. When used as a set of IVs, the partialling process highlights their uniqueness, tends to dissipate whatever common factor they share, may produce paradoxical suppression effects, and is thus also likely to create severe difficulties in interpretation.

Insofar as variables and hence hypotheses are concerned, an important general principle in research inference is succinctly stated: "less is more"—more statistical test validity, more power, and more clarity in the meaning of results.

4.6.3 Least Is Last

The hierarchical model can be an important element in an effective strategy of inference. We have already commented briefly on its use when IVs may be classified into levels of research relevance (Section 3.6.2). This type of application is appropriate in investigations which are designed to test a small number of central hypotheses but may have data on some additional research factors which are of exploratory interest, and also in studies which are largely or wholly exploratory in character. In such circumstances, the IVs can be grouped into two or more classes and the classes ordered with regard to their status in centrality or relevance. Each class is made up of one or more research factors, which are generally sets of IVs. Thus, for example, the first group of IVs may represent the research factors whose effects the research was designed to appraise, the second some research factors of distinctly secondary interest, and the third those of the "I wonder if" or "just in case" variety. Depending on the investigator's interest and the internal structure of the research, the levels of the hierarchy may simply be the priority classes or one or more of these may also be internally ordered by research factors or single IVs.

The use of the hierarchical model, particularly when used with Model I error at each priority class level, then prevents variables of lower priority, which are likely to account uniquely for little Y variance, from reducing the power of the tests on those of higher priority by stealing some of their variance, increasing the standard errors of their partial coefficients, and reducing the df for error. In using this stratagem, it is also a good idea to lend less credence to significant results for research factors of low priority, particularly so when many IVs are involved, since the investigationwise Type I error rate over these IVs is likely to be large. We thus avoid diluting the significance of the high priority research factors. This is in keeping with the sound research philosophy which holds that what is properly obtained from exploratory research are not conclusions, but hypotheses to be tested in subsequent investigations.

When hierarchical MRC is used for relevance ordering, it is recommended that Model I error be used at each level of relevance, that is, the first class (U) made up of the centrally relevant research factors uses $1 - R^2_{Y \cdot U}$ (with $df = n - k_U - 1$) as the error term for its F and t tests, the second class (V) made up of more peripheral research factors uses $1 - R^2_{Y \cdot UV}$ (with $df = n - k_U - k_V - 1$), and so on. This tends to make it probable that the tests at each level have minimum error variance per df and thus maximal power. Of course, it is always possible that a test using Model I error is negatively biased by an important source of variance remaining in its error term, but the declining gradient of unique relevance in this type of application makes this rather unlikely.

We summarize this principle, then, as "least is last"—when research factors can be ordered as to their centrality, those of least relevance are appraised last in the hierarchy, and their results taken as indicative rather than conclusive.

4.6.4 A Multiple Regression/Correlation Adaptation of Fisher's Protected t Test

The preceding sections of the chapter have been devoted to the use of sets of IVs as units of analysis in MRC. We have seen how Y variance associated with a set or partialled set can be determined, tested for significance, and power-analyzed. The chapters which follow show how research factors can be represented as sets of IVs. It should thus not come as a surprise to the reader that in formulating a general strategy of statistical inference in MRC, we accord the set a central role.

In Section 4.6.1 above, in our brief review of alternative schemes of testing multiple hypotheses for the special case where the research factor is a nominal scale G, we noted that among the methods available for the comparison of pairs of groups means was a method attributed to R. A. Fisher: the usual AV overall F test over the set of g means is first performed, and if it proves to be significant at the α level specified, the investigator may go on to test any or all pairs at the same α level, using the ordinary t test for this purpose, and interpret his results in the usual way. If F fails to be significant, no t tests are performed—all g population means are taken as equal so that no difference between means (or any other linear contrast function of them) can be asserted, whatever value it may yield. This two-stage procedure combines the good power characteristics of the individual t tests at a conventional level of α with the protection against large experimentwise Type I error afforded by the requirement that the overall F also meet the α significance criterion. For example, in Section 4.6.1, we saw that when all 15 pairwise comparisons among 6 means are performed by t tests using α = .05 per comparison, the probability that one or more will be found "significant" when all 6 population means are equal, that is, the experimentwise Type I error rate, is about .40. But if these tests are performed only if the overall F meets the .05 criterion, the latter prevents us from comparing the sample means 95% of the time when the overall null hypothesis is true. Thus, the t tests are *protected* from the mounting up of small per-comparison α to large experimentwise error rates (but see below).

The virtues of simplicity and practicality of the protected t test procedure are evident. What is surprising is how effective it is in keeping Type I errors low while affording good power. In an extensive investigation of ten pairwise procedures for means compared empirically over a wide variety of conditions, it was unexcelled in its general performance characteristics (Carmer & Swanson, 1973).

To adapt and generalize the protected t test to the MRC system, we use the framework of sets as developed in this chapter and used throughout the book.

We discussed in Section 4.1 the principle that information on research factors of all kinds can be organized into sets of IVs for structural and functional reasons, and in the ensuing sections illustrated how these sets may then serve as the primary units of MRC analysis. Now, the protected t test described above covers only one type of set—a research factor defined by a nominal scale, that is, a collection of g groups. We generalize the protected t procedure, applying it to the functional sets which organize an MRC analysis, whatever their nature. Specifically,

1. The MRC analysis proceeds by sets, using whatever analytic structure (hierarchical, simultaneous, or intermediate) is appropriate.

2. The contribution to Y variance of each set (or partialled set) is tested for significance at the α level by the appropriate standard F test of Eqs. (4.4.2), (4.4.6), or their variants.

3. If the F for a given set is significant, the individual IVs (aspects) which make it up are each tested for significance at α by means of a standard t test (or its equivalent $t^2 = F$ for numerator $df = 1$). It is the *partial* contribution of each X_i which is t-tested, and any of the equivalent tests for its sr_i, pr_i, β_i, or B_i may be used—Eqs. (3.7.7), (3.7.10), or their variants. All standard MRC computer programs provide this significance test, usually for B_i; see Appendix 3.

4. If the setwise F is *not* significant, no tests on the set's constituent IVs are permitted. (The computer program will do them automatically, but the t_is, no matter how large, are ignored.) Overriding this rule removes the protection against large setwise Type I error rates, which is the whole point of the procedure.

This procedure is effective in statistical inference in MRC for several reasons. Since the number of sets is typically small, the investigationwise Type I error rate does not mount up to anywhere nearly as large a value over the tests for sets as it would over the tests for the frequently large total number of IVs. Then, the tests of single IVs are protected against inflated setwise Type I error rates by the requirement that their set's F meet the α significance criterion. Further, with Type I errors under control, both the F and t tests are relatively powerful (for any given n and f^2). Thus, both types of errors in inference are kept relatively low and in good balance.

To illustrate this procedure, we return to the running example of this chapter: length of hospital stay (Y) for $n = 500$ psychiatric patients was regressed hierarchically on 3 sets of IVs, demographic ($D, k_D = 9$), personality ($M, k_M = 10$), and a nominal scale of 8 (= g) hospitals ($H, k_H = g - 1 = 7$), in that order. Using F tests with Model I error and $\alpha = .05$ as the criterion for significance, it was found that set D was significant, set M (actually $M \cdot D$) was not, and set H (actually, $H \cdot DM$) was significant (Section 4.4.1 and Table 4.4.1). Note that the primary focus on sets helps control the investigationwise Type I error risk. Even for $\alpha = .05$, the latter is in the vicinity of .14; the more conservative $\alpha = .01$

for significance per set would put the investigationwise Type I error rate in the vicinity of .03.[16]

Since set D was found to be significant, one may perform a t test (at $\alpha = .05$) on each of the 9 IVs which represent unique aspects of the patients demography.[17] Since these t tests are protected by the significance of F, the mounting up of setwise Type I error is prevented. Without this protection, the setwise error rate for 9 t tests, each at $\alpha = .05$, would be in the vicinity of $1 - .95^9 = .37$.

Set M's increment (over D) to R^2 was found to be nonsignificant, so no t tests on the unique (within M and D) contributions of its 10 IVs are admissible. It would come as no surprise if the computer output showed that one of these 10 t values exceeded the nominal $\alpha = .05$ level. With 10 tests each at $\alpha = .05$, the setwise Type I error rate would be large. Were the tests independent, it would be $1 - .95^{10} = .40$; although they are not, it would be of about that size. In the protected t strategy, the failure of F for the set to be significant is treated so as to mean that *all* IVs in the set have zero population partial coefficients, a conclusion that cannot be reversed by their individual ts.

Finally, the increment of set H (over D and M) to R^2 was found to be significant, and its constituent $k_H = g - 1 = 7$ IVs are t tested, and the significant ones interpreted. These 7 aspects of hospital-group membership may include simple (pairwise) or complex (involving more than two hospitals) comparisons among the 8 hospital Y means, depending on which of several different methods of representing group membership was employed. The latter is the subject matter of the next chapter. The method used in this example was presumably chosen so that the 7 partialled IVs would represent those comparisons (aspects) of central interest to the investigator, and their protected ts test these aspects. Thus far, we have proceeded as with any other set. However, since H is a nominal scale, and is thus made up of g groups, we admit under the protection of the F test any comparisons of interest in addition to the 7 $(= g - 1)$ carried by the IVs. Thus, in full compliance with both the letter and spirit of Fisher's original protected t test, one could t test any of the $8(7)/2 = 28$ pairwise simple comparisons (not already tested) which may be of substantive interest.[18]

We reiterate the generality of our adaptation of the protected t test to MRC. Whether one is dealing with one set or several, whether they are related to Y

---

[16] Again, since the tests are not independent, exact rates can not be determined. The rates given are "ballpark" estimates computed on the assumption of independence, that is, $1 - .95^3$ and $1 - .99^3$, respectively.

[17] A refinement of this procedure would be to test the sR^2s for subsets (for example, the nominal scale for ethnicity) by F, and then perform t tests on the subset's IVs only if F is significant. This gives added protection to the t tests, which is probably a good idea when k for the set is large, as it is here.

[18] In compliance at least with the spirit of Fisher's procedure, one could also test any complex comparisons of interest, but there would usually be few, if any, that had not been included as IVs.

hierarchically or simultaneously, whether error Model I or II is used, and whatever the substantive nature of the set(s), the same procedure is used: the first order of inference is with regard to the set(s), and only when a set's significance is established by its F test are its contents further scrutinized for significance.

In using the protected t procedure, it may happen that after a set's F is found to be significant, none of its IVs yields a significant t. This is apparently an inconsistency, since the significant F's message is that at least one of the IVs has a nonzero population partial coefficient, yet each t finds its null hypothesis tenable. A technically correct interpretation is that collectively (setwise) there is sufficient evidence that there is something there, but individually, not enough evidence to identify what it is. A risky but not unreasonable resolution of this dilemma is to tentatively interpret as significant any IV whose t is almost large enough to meet the significance criterion; whatever is lost by the inflation of the Type I error is likely to be compensated by the reduction of Type II error and the resolution of the apparent inconsistency. Fortunately, the occurrence of this anomaly is rare, and virtually nonexistent when error df are not very small, say greater than 20.

Another difficulty which may arise with the protected t test is best described by a hypothetical example. Assume, for a set made up of many IVs, that one or two of them have large *population* partial coefficients and that the remainder all have population partial coefficients equal to zero. Now when we draw a random sample of reasonable size from this population, we will likely find that the F for the set is statistically significant. This result is quite valid, since this F tests the composite hypothesis that *all* the IVs in the set have zero population coefficients, and we have posited that this is not true. But using the protected t test, the significance of F confers upon us the right to t test *all* the IVs, including those for which the null hypothesis *is* true. For that large group of IVs, the subsetwise Type I error rate will obviously mount up and be high. Of course, we do not know the state of affairs in the population when we analyze the sample. We cannot distinguish between an IV whose t is large because its null hypothesis is false from one (of many) whose t is large because of chance. Obviously, in circumstances such as these our t tests are not as protected as we should like.

Fortunately, a means of coping with this problem is available to us: we invoke the principle of Section 4.6.2—"less is more." By having few rather than many IVs in a set, a significant F protects fewer ts for IVs whose null hypotheses may be true and inhibits the mounting up of Type I error. Moreover, if the investigator's substantive knowledge is used to carefully select or construct these few IVs, fewer still of the ts are likely to be testing true null hypotheses. In this connection, we must acknowledge the possibility that sets D and M in our running example are larger than they need have been; the former may have benefitted from reduction by a priori selection, and the latter by either selection or factor-analytic reduction.

4.7 SUMMARY

This chapter presents one of the core concepts of our application of MRC, that of *sets* of IVs treated as units or fundamental entities in the analysis of data. From this perspective, the single IVs and single group of k IVs treated in Chapter 3 are special cases. Two types of sets are described: sets which come about because of the *structure* of the research factors they represent (for example, Religion is represented as a set of IVs because it is a nominal scale), and sets which have a specific *function* in the logic of the research (for example, a set of covariates which must be adjusted for, or a set of variables which collectively represent "demographic characteristics"). Groups of sets are also described, and such set aggregates are treated simply as sets. (Section 4.1)

The simultaneous and hierarchical models of MRC are shown to apply to sets as units of analysis, and it is shown that Y variance associated with set A may be partialled from that associated with set B, just as with single IVs. For h sets, the simultaneous model appraises Y variance for a given set with the remaining $h - 1$ sets partialled. The hierarchical model orders the sets into an a priori hierarchy, and proceeds sequentially: for each set in hierarchical order of succession, all higher level sets (and no lower level sets) are partialled. The chief quantities of interest are the increments of Y variance accounted for by each set *uniquely*, relative to sets of higher order of priority. These concepts are concretized by an example involving length of stay of psychiatric admissions (Y) as a function of sets of demographic, personality, and hospital variables. (Section 4.2)

The ballantine for sets is presented as a device for visualizing proportions of Y variance associated with sets (A, B) and with partialled sets $(A \cdot B, B \cdot A)$, analogous with those for single IVs. The increments referred to above are squared multiple *semi*partial correlations (e.g., $sR_B^2 = R_{Y \cdot (B \cdot A)}^2$) and represent the proportion of *total* Y variance associated with $B \cdot A$. Similarly, we define the squared multiple *partial* correlations (e.g., $pR_B^2 = R_{YB \cdot A}^2$) as the proportion of the Y variance not accounted for by A (i.e., $1 - R_{Y \cdot A}^2$) which is associated with $B \cdot A$. These two statistics are compared and exemplified. As with single IVs, the troublesome area of overlap of sets A and B with Y, area c, can not be interpreted as a proportion of variance, since it may be negative, in which case we have an instance of suppression between sets. (Section 4.3)

A general test of statistical significance for the Y variance due to a partialled set $B \cdot A$ (that is, of sR_B^2 and, perforce, of pR_B^2) is presented. Two error models for this test are described. Model I error is $1 - R_{Y \cdot AB}^2$, with sets other than A or B (collectively, set C) ignored. Model II error is $1 - R_{Y \cdot ABC}^2$, so that the Y variance unique to C (together with its df) are additionally excluded from error in the significance test of $B \cdot A$'s Y variance. The applicability of the two error models to the hierarchical and simultaneous models of MRC is described, as well as in an intermediate (neither fully hierarchical nor fully simultaneous) MRC model. The two error models are contrasted and exemplified. The concept of partialling an empty set is introduced to demonstrate the generality of the F test. (Section 4.4)

Methods of statistical power analysis for partialled sets, necessary for research planning and assessment, together with the use of the necessary L tables (Appendix Table E), are presented. The determination of n^*, the necessary sample size to attain a desired degree of power to reject the null hypothesis at the α level of significance for a given population effect size (ES), is given for both error models. The ES index f^2, which is a function of two or three distinct population R^2s, is defined, and suggestions for arriving at f^2, including some operational definitions of "small," "medium," and "large" f^2, are offered. Considerations in setting desired power, including the use of a conventional .80 value and the reconciliation of varying n^* for different hypotheses, are discussed. Another form of power analysis, in which one estimates the power that results from a specified n (also α and f^2) is presented for both error models. This is given in general for $B \cdot A$, and then specialized for significance tests of R^2 and sr_i^2. The section concludes with a discussion of the tactics of power analysis in research planning, particularly those of specifying alternative combinations of parameters (α, f^2, power, n) and studying their implications. (Section 4.5)

Finally, the issue of a general strategy of statistical inference in MRC is addressed. By way of illustration, a brief survey is undertaken of the many alternative approaches to testing the multiple null hypotheses which may be generated from a collection of g groups defining a research factor G. The varying conceptions of Type I error and the different types of comparisons have varying consequences to Type II error (hence, to power), and an optimum strategy of statistical inference requires controlling and balancing these types of errors. One element of such a strategy involves minimizing the number of IVs used in representing research factor constructs by judicious selection and the use of composites. Another exploits the hierarchical model of MRC and the use of Model I error in significance testing. A generalization of the protected t test is offered as a simple but effective means of coping with the multiplicity of null hypotheses in MRC. Using sets of IVs as the primary units of analysis, *only* those IVs are t tested for significance whose sets have given rise to significant Fs. This procedure prevents the rapid inflation of setwise and investigationwise Type I error which would occur if the individual ts were not so protected, and at the same time enjoys the good power characteristics of the t test. (Section 4.6)

PART II

THE REPRESENTATION
OF INFORMATION
IN INDEPENDENT VARIABLES

5

Nominal or Qualitative Scales

5.1 INTRODUCTION: THE USES OF MULTIPLICITY

The central feature of MRC is its utility in examining the relationships involved when multiple independent variables (IVs or X_i) are related to a single dependent variable (Y). The multiplicity of MRC can be exploited in three discriminably different ways, namely, by representing multiple research factors, by representing a functional group of research factors, and by representing multiple aspects of a single research factor:

1. The representation of several research factors is the familiar traditional use of the multiplicity of MRC, illustrated in Chapter 3 and in standard textbook treatments, for example, (*a*) Freshman grade point average (Y) versus verbal score (X_1), quantitative score (X_2), and high school class rank (X_3); or (*b*) length of psychiatric hospitalization (Y) versus sex (X_1), marital status (X_2), length of prior hospitalization (X_3), number of prior hospitalizations (X_4), and admission ratings on three symptom rating scales (X_5, X_6, X_7); or (*c*) for census tracts in a large city, infant mortality rate (Y) versus median annual income (X_1), percent nonwhite (X_2), and median education of adults (X_3). In such analyses the multiplicity of IVs makes possible the study of their combined relationships to Y, each IV's overall and unique relationship to Y, and redundancy among the IVs. Note that in these examples each research *factor* is represented by a single IV, and the multiplicity of IVs occurs solely because of the multiplicity of factors, all treated as if they were on the same footing.

2. In the preceding chapter, we saw yet another use for multiplicity, namely for substantive reasons functional to the purpose of the research (Section 4.1.2). Thus, for example, assume that in an educational experiment we wish to control for the socioeconomic status (SES) of the children, e.g., in the sense of the analysis of covariance (Chapter 9). But the concept SES refers to a class of

variables, for example, mother's education, father's education, occupation of head of household, family income. We thus would need a *set* of variables to represent this functional class of SES covariates.

3. We have also seen in the previous chapter how multiplicity may arise because of the need to work with sets of IVs dictated by the form or structure of research factors (Section 4.1.1). A useful conceptualization is that a single research factor (experimental treatment, religion, age, birth order, income, etc.) may have more than one aspect, and each aspect requires an IV to represent it. We noted there three structural circumstances which could produce the necessity for two or more IVs to represent a single research factor, that is, nominal scales, provision for nonlinearity in relationships involving quantitative scales, and the existence of missing data.

Each of the three succeeding chapters takes up one of the latter circumstances. The remainder of this chapter is devoted to an exposition of the several methods of representing and interpreting nominal scales as sets of IVs.

5.2 THE REPRESENTATION OF NOMINAL SCALES

Nominal scales (Stevens, 1951) are those that make qualitative distinctions among the objects they describe, such as religion, experimental group, ethnic group, country of birth, occupation, diagnosis, or marital status. Each such research factor G merely classifies the material under study into one of g groups, where g is some number greater than one. The g groups can not ordinarily be ordered in any way which is generally meaningful. Some authors would not apply the term "measurement" to such qualitative classifications, yet they clearly constitute information. To use nominal scales as IVs in MRC, it is necessary to somehow represent them quantitatively. This problem can be synopsized as "how do you score religion?"

The answer turns out to be that there are several different ways, each useful for a given set of purposes, with none of them requiring the embarrassing (and unscientific) task of assigning higher scores to one religion than another, which is what the question "How do you score religion?" calls to mind. There is no need to score religion on a single IV, as we might do for IQ. Indeed, it would be an error to do so, since no single IV could adequately represent all the information available in the classification of people into g different religious groups. This can only be done by capitalizing on the multiplicity of MRC and using not one but a set of $g - 1$ different IVs, each representing *one* aspect of the distinctions among these g groups. Concretely, if our nominal scale of religion is a classification into the (g =) 4 categories Protestant, Catholic, Jewish, and Other (in which we include "None"), it will take $k_G = g - 1 = 3$ IVs (X_1, X_2, X_3) to fully represent the information in this classification. One might think that it would take (g =) 4 IVs (X_1, X_2, X_3, X_4) to do so, but in our MRC model, the X_4 variable would be fully redundant, that is, provide no additional information beyond what is contained in the set X_1, X_2, X_3. This makes the gth IV not only unnecessary, but mathematically mischievous because it renders the system of underlying simultaneous equations not (uniquely) soluble (see Appendix 1).

Given that religion can be represented by a set of k_G = 3 variables, what are they? We have several choices for the representation (or coding) of this information: coding by dummy variables, by effects, by contrasts, or even by using nonsense coding.

5.3 DUMMY VARIABLE CODING

The idea of dummy variable coding is to render the information of membership in one of g groups by a series of $g - 1$ dichotomies. Thus, one way to score a four-way break is shown in Table 5.3.1. We illustrate this in the table for several different research factors, all involving the assignment of samples of observation units (people, counties, experimental animals, etc.) into four categories. Note that such coding applies whether the categorization comes about as a "natural" property of the research material under study (as religion or region), or as a result of the researcher's experimental manipulation (as in the third example in Table 5.3.1). The only formal requirement is that the observations be assignable to g mutually exclusive and exhaustive categories, that is, that each case be assigned to one (and only one) of the g groups.

For concreteness, take religion and consider X_1. It is a dichotomy in which all Protestants (G_1 cases) are scored 1 and all non-Protestants (non-G_1 cases) are scored 0. In fact, any two different numbers could be used, but 1 and 0 are simple and include the advantage of making it possible to think of X_1 simply as "Protestantness" (or more generally "G_1-ness"). This is a meaningful distinction

TABLE 5.3.1
Dummy-Variable Coding of Illustrative Research Factors
with g = 4 Groups

	Religion	Region	Experimental group	X_1	X_2	X_3
G_1	Protestant	Northeast	Treatment 1	1	0	0
G_2	Catholic	South	Treatment 2	0	1	0
G_3	Jewish	Midwest	Treatment 3	0	0	1
G_4	Other	West	Control	0	0	0
Y	Attitude toward Abortion	Median Family Income	Number of Trials to Criterion			
Units:	People	Counties	Rats			

in the data, and this one variable represents one aspect of religion in the data. Similarly, X_2 and X_3 are each 0, 1 dichotomies that can be understood as representing "Catholicness" ("G_2-ness") and "Jewishness" ("G_3-ness"), respectively. Each is obviously a meaningful IV in its own right and also one aspect of the complete research factor of religion.

Thus, we have $k_G = 3 (= g - 1)$ aspects of our 4 $(= g)$-way break on religion and, although it is not immediately obvious, in our MRC model, that is all that there are: religion is *fully* described by these three IVs. What about Otherness $(G_4$-ness), which we might represent as an X_4 dichotomy? No fourth $(g$th) IV is necessary or desirable, since it would be wholly redundant. No IV is ever permissible that would have an $R = 1$ with the remaining IVs (see Appendix 1), which is exactly what would happen if we included X_4 (or generally X_g) in the equation. One can see intuitively that no X_4 is needed to represent "Otherness" ("G_4-ness"). It is represented implicitly: all cases falling in this category have X_1, X_2, X_3 scores of 0, 0, 0 (see Table 5.3.1), which can be taken to mean non-Protestant, non-Catholic, non-Jewish (non-G_1, non-G_2, non-G_3); since only the Other group (G_4) is left, the scores 0, 0, 0 perforce must designate members of that group. The group that has no explicit representation as an IV (G_g) is not being slighted; indeed, we shall see below that it functions uniquely as a reference group.

The dummy variable system may be clarified for the reader by considering the limiting case, when $g = 2$, as, for example, sex. Clearly, if females are scored 1 and males 0, so that X_1 is defined as femaleness, no second variable X_2 of maleness is required. X_2 would simply be the opposite of X_1, and r_{12} would be -1.00. Correlations of other variables with X_2 would be numerically identical with those with X_1, being only of opposite sign, that is, $r_{Y1} = -r_{Y2}$. The complete redundancy of a second variable to represent the other sex is obvious here. More generally, the gth IV is similarly fully redundant in any g-way break.

Table 5.3.2 is a set of data for $n = 36$ cases, where each is a member of one of four groups and has a Y value. The sample sizes of these groups are, respectively, $n_1 = 13$, $n_2 = 9$, $n_3 = 6$, and $n_4 = 8$. Throughout this chapter we will use these four groups with the Y of Table 5.3.2 to illustrate various forms of coding and their interpretive consequences. The sample sizes are deliberately unequal, since this is the more general case. Where equal sample sizes simplify the results and their interpretation, this will be pointed out. Table 5.3.2 also presents a set of dummy variables X_1, X_2, X_3, given by the coding scheme in Table 5.3.1, which represent group membership. Thus, Cases 3, 7, 13 and 15 (among others) are members of G_1; hence, their X_1, X_2, X_3 scores are 1,0,0. Similarly, Cases 1, 4 and 5 are G_2s and are scored 0, 1, 0 and Cases 18, 22, and 25 are G_3s and are scored 0, 0, 1. Note again that the G_4 group members (for example, 2, 6, and 9) need no X_4 to designate them: they are uniquely scored on the IVs as 0, 0, 0.

With the group membership information thus rendered in quantitative form, the data in Table 5.3.2 can be fully and meaningfully exploited by MRC: means, standard deviations and product moment rs can be computed, also multiple,

TABLE 5.3.2
Illustrative Data for Dummy-Variable Coding
for g = 4 Groups

Case no.	Group	Y	X_1	X_2	X_3
1	G_2	61	0	1	0
2	G_4	78	0	0	0
3	G_1	47	1	0	0
4	G_2	65	0	1	0
5	G_2	45	0	1	0
6	G_4	106	0	0	0
7	G_1	120	1	0	0
8	G_2	49	0	1	0
9	G_4	45	0	0	0
10	G_4	62	0	0	0
11	G_2	79	0	1	0
12	G_4	54	0	0	0
13	G_1	140	1	0	0
14	G_2	52	0	1	0
15	G_1	88	1	0	0
16	G_2	70	0	1	0
17	G_2	56	0	1	0
18	G_3	124	0	0	1
19	G_4	98	0	0	0
20	G_2	69	0	1	0
21	G_1	56	1	0	0
22	G_3	135	0	0	1
23	G_1	64	1	0	0
24	G_1	130	1	0	0
25	G_3	74	0	0	1
26	G_4	58	0	0	0
27	G_1	116	1	0	0
28	G_4	60	0	0	0
29	G_3	84	0	0	1
30	G_1	68	1	0	0
31	G_1	90	1	0	0
32	G_1	112	1	0	0
33	G_3	94	0	0	1
34	G_1	80	1	0	0
35	G_3	110	0	0	1
36	G_1	102	1	0	0

partial and semipartial rs and the ingredients of the MR equation. These values moreover are not only descriptive of the state of affairs in the sample, but can be tested for statistical significance or bounded by confidence limits in order to provide projections to the population. The results of such an analysis of the data of Table 5.3.2 are given in Tables 5.3.3 and 5.3.4. For concreteness and smoothness of exposition, they will be interpreted primarily in terms of religion, assuming Y to be a score on an Attitude toward Abortion scale (ATA), but, of course, the reader may think in terms of the other nominal research factors of Table 5.3.1 with the Ys suggested, or any others of his own choice.

5.3.1 Relationships of Dummy Variables to Y

Since each dummy variable X_i is a dichotomy that expresses one meaningful aspect of group membership, for example, Protestant versus non-Protestant, it yields a meaningful product moment r (incidentally, a point biserial r—see Section 2.3.3) with the dependent variable Y for the sample. These r_{Yi} values are given in the first column of Table 5.3.3. Thus, Protestantness in this sample correlates .318 with ATA, or, equivalently, accounts for $.318^2 = .1011$ of the variance in ATA (column r_{Yi}^2). Similarly, Catholicness correlates $-.442$ and accounts for $(-.442)^2 = .1954$, and Jewishness correlates .355 and accounts for $.355^2 = .1260$, of the ATA variance. The sign of the r_{Yi} indicates, as always, the direction of the relationship, and with dummy variables indicates simply whether the mean of Y in G_i (i.e., $\overline{Y}_i$) is larger (positive) or smaller (negative) than the mean of Y for nonmembers of G_i. The proportion of variance in Y accounted for by each X_i is interpreted as described in Section 2.6, but here there are only two values of X_i: if we were to estimate the Y score of each of the 36 subjects at the grand mean ($m_Y = 81.69$) of the sample, our average squared error would be the variance of Y, $sd_Y^2 = 27.49^2 = 755.70$ (Table 5.3.3). If instead, the $\hat{Y}$ estimate for each member of G_i was the G_i mean, $\overline{Y}_i$, and for each nonmember of G_i was the mean of all nonmembers of G_i, $\overline{Y}_{\text{non-}i}$, now our average squared error over all n cases would be generally smaller. The proportion *by which* (not to which) this error is reduced is what is meant by the "variance accounted for" by a dichotomy. Thus, the operation performed for the X_1 (Protestant) distinction would result in the error from using $\overline{Y}_1$ and $\overline{Y}_{\text{non-}1}$ being .1011 smaller than the error from using the grand mean as the $\hat{Y}$ estimate, that is, smaller by

TABLE 5.3.3

Correlations, Means, and Standard Deviations of the
Illustrative Data for Dummy-Variable Coding

			r				
		Y	X_1	X_2	X_3	r_{Yi}^2	t_i $(df = 34)$
	Y	1.000	.318	−.442	.355	—	—
(P)	X_1	.318	1.000	−.434	−.336	.1011	1.955
(C)	X_2	−.442	−.434	1.000	−.258	.1954	−2.874**
(J)	X_3	.355	−.336	−.258	1.000	.1260	2.214*
	m	81.69	.361	.250	.167	$n = 36$	
	sd	27.49	.480	.433	.373		

$$R_{Y \cdot 123}^2 = .354944. \qquad F = 5.869** \ (df = 3, 32)$$

$$\widetilde{R}_{Y \cdot 123}^2 = .294470.$$

$*P < .05. \qquad **P < .01.$

(.1011)(755.70) = 76.40. When there is no distinction in Y between G_i and the remainder of the sample, i.e., when $\overline{Y}_i = \overline{Y}_{non-i}$, then both of these necessarily equal the grand mean m_Y, there is no difference between the estimated $\hat{Y}$ values, hence no reduction in error, and $r_{Yi}^2 = 0$. At the other extreme, when all G_i members have the same Y score and all nonmembers have some other single Y score, estimation from the respective means is perfect, and the error in estimating $\hat{Y}$ from m_Y is reduced by 100%; hence, $r_{Yi}^2 = 1.00$.

The interpretation of these r_{Yi} and r_{Yi}^2 requires some care. It was pointed out in Chapter 2 that, all things equal, rs change in magnitude directly with changes in the variability of the variables being correlated. If r_{YX} is not zero, and one samples in such a way as to make sd_X larger (smaller) than it would otherwise be, then r_{YX} will be numerically larger (smaller). But the mean of a dummy variable X_i is the proportion of G_i members in the total sample: since the sum of the X_i "scores" for a sample of n cases is n_i, the number of cases in G_i, therefore

(5.3.1) $$\overline{X}_i = \frac{\sum X_i}{n} = \frac{n_i}{n} = p_i ,$$

and the sd of a variable scored 0–1 is (since $\sum X_i^2 = \sum X_i = n_i$)

(5.3.2) $$sd_i = \sqrt{\frac{\sum X_i^2}{n} - \left(\frac{\sum X_i}{n}\right)^2} = \sqrt{p_i - p_i^2} = \sqrt{p_i(1 - p_i)} .$$

Thus, the sd of a dummy variable depends solely on its proportion in the total sample (reaching its maximum of .50 when $p_i = .50$). Since r_{Yi} varies with sd_i, and sd_i depends upon p_i, the magnitude of a correlation with a dummy variable will change with the relative size of that group in the total sample, reaching its maximum when the group is exactly half the total sample, and declining toward zero as the group's proportion in the sample declines toward zero, or increases toward one. Therefore, the interpretation of any given r_{Yi} (or r_{Yi}^2) depends upon the sampling meaning of p_i. (As we will see later, this is true of *any* kind of correlation—multiple, partial, or semipartial, as well.)

Consider two sampling circumstances involving religion and ATA. In the first, we draw a random sample of $n = 36$ cases[1] from some natural population, say a Midwestern college, and determine each case's religion and ATA score. Assume that the data of Table 5.3.2 were so obtained. The p_i for each group reflects, subject to sampling error, its proportion in that population, and so does, indirectly, the magnitude of the r_{Yi}. Thus, for example, the Jewish group (G_3) has $p_3 = .167$ in Table 5.3.3, and $r_{Y3}^2 = .1260$. Were the Jewish group more numerous in this sample (up to $1 - p_3)^2$ as might be the case for some other natural population, then, all things equal, r_{Y3}^2 would be larger.

---

[1] This is an unrealistically small sample size for an actual study. It is used here so that the complete data can be economically presented and, if desired, easily analyzed by desk or pocket calculator.

[2] As long as its proportion falls between p_3 (= .167) and $1 - p_3$ (= .833), its sd will increase; outside these limits it will decrease. See Eq. (5.3.2).

Yet another circumstance resulting in different r_{Y_i}s is a sampling plan where equal numbers of Protestants, Catholics, Jews, and Others are sampled from their respective populations and their ATA scores determined. Here these religions are considered as abstract properties, and their differing numbers in natural populations are ignored. The equal p_is in the resulting data will yield different r_{Y_i}s than those of the previous plan: we would expect a smaller correlation for Protestants and a larger correlation for Jews because of the associated respective decrease and increase in the sds of their dummy variables.

Thus we see that values for r_{Y_i} and $r_{Y_i}^2$ are not solely properties of means on the dependent variable of group samples, but depend also upon the relative sizes of the groups. In the interpretation of such correlations, p_i must be taken into account. Thus, the interpretation from the illustrative data that " 'Jewishness' accounts for 12.6% of the variance in ATA scores at Midwest College" is valid if one understands "Jewishness" to include the property of relatively low frequency. Similarly, in an equal n_i sampling plan that same statement is valid if one understands "Jewishness" to be an abstraction, with the low frequency of Jews in most natural populations ironed out by the fixed equal sample sizes. The latter kind of sampling plan characterizes most manipulative laboratory situations, where no natural population exists. Groups are then intended to represent abstractions (for example, Treatment 1: animals reared from birth in social isolation), and typically are of equal size.

The preceding is an effort to mollify statistical purists who would restrict the use of correlations to the first sampling plan, where a single population is randomly sampled and each case's standing on the two variables (for example, ATA and religion) determined, which is sometimes called the "random model." But we believe that correlations and particularly squared correlations as proportions of Y variance are valuable analytic tools in the second kind of sampling circumstance ("fixed model"), provided only that care is taken in the interpretation of correlational results to keep in mind the sampling plan and its implications.

Keeping the above in mind, we can test any r_{Y_i} for significance (or set up confidence intervals for it) as we can for any r, as described in Chapter 2 (although this test which uses Model I error is not necessarily the most powerful one). Thus, we use Eq. (2.8.1),

$$t = \frac{r\sqrt{n-2}}{\sqrt{1-r^2}}, \quad df = n - 2.$$

For X_2 in the illustrative example, we find

$$t = \frac{-.442\sqrt{34}}{\sqrt{1-(-.442)^2}} = -2.874,$$

which, with $df = n - 2 = 34$, is significant at the 1% level (Table 5.3.3). This means we can conclude that in the natural population sampled, the correlation for Catholicness is nonzero (and negative). It is also exactly the t value we would obtain if we tested for the same data the difference between the Y means of Catholics and non-Catholics, since the two t tests are algebraically identical. We can also set up confidence limits for such r_{Yi}s, and here it is particularly important to keep in mind whether the sampling was of the first or second kind described above, since the two kinds of sampling imply different population distributions of group proportions and therefore, in general, different correlations, as we have seen.

Let us return now to the set of r_{Yi} in Table 5.3.3. We have them for all but the gth group. We could compute r_{Yg} from the raw data, but it is implicit in the other r_{Yi}s and the n_is or p_is:

$$(5.3.3) \qquad r_{Yg} = -\frac{\sum r_{Yi} sd_i}{sd_g} = -\frac{\sum r_{Yi}\sqrt{p_i(1-p_i)}}{\sqrt{p_g(1-p_g)}},$$

where the summation is taken over the $g - 1$ dummy variables.[3] Thus, the correlation of Others versus non-Others with ATA for the illustrative data is (since $p_g = n_g/n = 8/36 = .222$)

$$r_{Yg} = -\frac{.318(.480) + (-.442)(.433) + .355(.373)}{\sqrt{.222(1-.222)}}$$

$$= -\frac{.09367}{.4157} = -.225.$$

It is worth noting, incidentally, that although Y in the present context is a dependent variable in an MRC analysis, Eq. (5.3.3) is valid whatever the nature of Y; it need not even be a real variable—the formula obtains even if Y is a factor in the factor-analytic sense, unrotated or rotated, with the r_{Yi} being factor loadings for a set of dummy variables, which may quite profitably be used in factor analysis.

5.3.2 Correlations among Dummy Variables

We have seen that we can determine the proportion of variance which each of our $g - 1$ dummy variables accounts for separately in Y; they are .1011, .1954, and .1260 (Table 5.3.3). But our primary interest lies not in these three separate aspects of the research factor G, but rather in G taken as a whole. How much variance in ATA does *religion* account for? Is it simply $.1011 + .1954 + .1260 = .4225$? This would be the case if X_1, X_2, X_3 were independent of each other, that is, if $r_{12} = r_{13} = r_{23} = 0$. But that is not the case, as can be seen from Table 5.3.3, nor can it ever be true for dummy variables. These correlations [which are

---

[3] With equal n_i in the g groups, this equation simplifies to $r_{Yg} = -\sum r_{Yi}$.

incidentally r_ϕ coefficients (Section 2.3.4) since dummy variables are dichotomies] give the relationships, in the sample, between such properties as Protestantness and Catholicness. For mutually exclusive categories, such as we are considering, such relationships are necessarily inverse, that is, negative. If a person is Protestant, he is necessarily non-Catholic, and if Catholic, necessarily non-Protestant. The correlation is however never -1.00 because if a person is non-Protestant, he may either be Catholic or non-Catholic, since there are other groups into which he may fall. The size of the correlation between two dummy variables (X_i, X_j) depends on the number of cases in the groups they represent (n_i, n_j) and the total sample n, as follows:

$$(5.3.4)^4 \qquad r_{ij} = -\sqrt{\frac{n_i n_j}{(n - n_i)(n - n_j)}} = -\sqrt{\frac{p_i p_j}{(1 - p_i)(1 - p_j)}}$$

so that the correlation between "Protestantness" and "Catholicness" in our running example (see Table 5.3.3) is

$$r_{12} = -\sqrt{\frac{(.361)(.250)}{(1 - .361)(1 - .250)}} = -.434.$$

Equation (5.3.4), as such, is of only incidental interest. The point worth noting is that the dummy variables, that is, the separate aspects of G, are partially redundant (correlated with each other). Since this is necessarily always true, we cannot find the proportion of variance in Y due to G by simply summing their separate r_{Yi}^2. It is at this point that *multiple* correlation enters the scene. It is designed for just this circumstance: to determine, via R^2, the proportion of variance in Y accounted for by a set of k IVs, taking fully into account whatever correlation (redundancy) there may be among them.

5.3.3 Multiple Regression/Correlation and Partial Relationships for Dummy Variables

Let us review our strategy: we have taken a research factor which is a nominal scale, G, represented by a set of $k_G = g - 1$ dummy variables $(X_1, X_2, \ldots, X_{g-1})$ and can study the relationship of this research factor to Y by running an MRC analysis of Y on the set of X_i representing G. For the illustrative data (Tables 5.3.2 and 5.3.3), we find $R_{Y \cdot G}^2 = R_{Y \cdot 123}^2 = .3549$, and therefore $R_{Y \cdot G} = R_{Y \cdot 123} = .596$. We thus can state that 35.5% of the variance in ATA scores is associated with religion in the sample, or that the R of ATA and religion is .596. Note that R depends on the distribution of the n_i of the four groups; a change in their relative sizes holding their $\overline{Y}_i$ constant would, in general, change the R. This dependence on the p_i is a characteristic of R, as it is of any kind of correlation, and must be kept in mind in interpretation.

---

[4] When all g groups are of the same size, Eq. (5.3.4) simplifies to $r_{ij} = -1/(g - 1)$, so that if our four groups were of equal size, their dummy variables would correlated $-1/(4 - 1) = -.333$.

We can test the R for significance by means of the standard formula

(3.7.1) $$F = \frac{R^2 (n - k - 1)}{(1 - R^2)k},$$

where k is the number of independent variables. For the present application, where $k = k_G = g - 1$,

(5.3.5) $$F = \frac{R^2(n - g)}{(1 - R^2)(g - 1)}$$

$$= \frac{.354944(36 - 4)}{(1 - .354944)(3)} = 5.869.$$

For $df = 3, 32$, the F required for significance at the .01 level is 4.46 (Appendix Table D.1), hence our obtained F is significant. The null hypothesis that religion accounts for no variance in ATA scores in the population sampled can be rejected. Further discussion of the meaning of this test will be given below (Section 5.3.4).

In Chapter 3 it was pointed out that R^2 gives the proportion of Y variance accounted for in the *sample,* but overestimates that proportion in the population. For a better estimate of Y variance accounted for by the IVs in the population, R^2 must be "shrunk" to the $\widetilde{R}^2$ of Eq. (3.7.4):

$$\widetilde{R}^2 = 1 - (1 - R^2)\frac{n - 1}{n - k - 1}.$$

When applied to these data,

$$\widetilde{R}^2 = 1 - (1 - .3549)\frac{35}{32} = .2945,$$

so that our best estimate of the proportion of ATA variance accounted for by religion in the population is 29.4%. Here again it is important to keep in mind how the sampling was carried out, since the population being projected is the one implicit in the sampling procedure.

We turn now to a consideration of the rest of the yield of an MRC analysis with dummy variables, using the illustrative data (Table 5.3.4). In considering the partial coefficients (pr_i, sr_i, β_i, B_i) associated with each X_i, we must first understand the unique role played by the group which is not explicitly coded, the one whose X_i scores are 0, 0, 0. This gth group, G_g, is not only not being slighted, it is on the contrary a reference group, and all the partial coefficients in fact turn upon it.

The Partial Correlation of a Dummy Variable (pr_i)

In MRC generally, a pr_i is the correlation of Y with X_i with all the other IVs held constant, that is, the correlation between Y and X_i for a subset of the data in which the other IVs do not vary. In the specific context of dummy variables, holding the other IVs constant means retaining only the distinction between the ith group and the gth or reference group. Concretely, pr_1 (= .363) is the correlation between ATA and Protestant versus non-Protestant, holding Catholic versus non-Catholic and Jewish versus non-Jewish constant (Table 5.3.4). But the subset of the data for which the latter do not vary includes only Protestants (G_1) and *Others* (G_4), so that the *partialled* $X_1._{23}$ variable represents a new dichotomy of Protestants versus *Others,* and pr_1 = .363 expresses the relationship of the ATA scores to this new dichotomy created by the partialling process. In other words, pr_1 is an expression, in correlational terms, of the difference between the Protestant group and the Other group in ATA scores. Similarly, from Table 5.3.4, pr_2 = −.145 relates ATA to Catholic versus Other (ignoring Protestants and Jews) and pr_3 = .423 relates ATA to Jewish versus Other (ignoring Protestants and Catholics). The interpretation of a given pr_i, as was true for r_{Yi} for dummy variables, must take into account the sampling plan (random or fixed n), since it also depends on the proportions of each group in the sample. Significance testing of pr_i is discussed later.

The Semipartial Correlation of a Dummy Variable (sr_i)

The basic general definition of an sr_i as the relationship of X_i from which all the other IVs have been partialled with an unpartialled Y holds here, but is not particularly helpful. A more useful general property of an sr_i is that sr_i^2 is the amount by which $R^2_{Y.123...k}$ would be reduced if X_i were omitted from the IVs, thus: $sr_i^2 = R^2_{Y.123...k} - R^2_{Y.123...(i)...k}$, the (i) in the subscript symbolizing the omission of X_i. With dummy variables, the omission of X_i has an immediate meaning. Look at the dummy variable coding (Table 5.3.1 above) and consider the consequence of dropping one of the X_i, say X_1. The result would be that G_1 as well as G_4 would have X_2, X_3 values 0, 0, thus the distinction between Protestants and *Others* would be lost. What would remain of our original four-way break would be the three-way break: Catholics, Jews, and Protestant/Others. Thus, sr_1^2 = .0980 (Table 5.3.4) means that the loss of the distinction between Protestants and Others would result in the loss of ATA variance accounted for by religion of 9.8%, or that our R^2 would drop from .3549 to .2569 when we went from $R^2_{Y.123}$ to $R^2_{Y.23}$. Equivalently, we can say that the Protestant−Other distinction accounts for 9.8% of the total Y variance. Again we note that the partialling process changes each X_i from a contrast between G_i and the entire remainder of the sample to one between G_i and G_g, the reference group. In dummy variable coding, sr_i^2 provides in terms of proportion of total Y variance the importance of distinguishing G_i from G_g. Thus, the Catholic−Other distinction accounts for only sr_2^2 = .0139 of the ATA variance, while the Jewish−Other distinction accounts for sr_3^2 = .1406 of the ATA variance. We note again that these values, as all correlations involving dummy variables, are dependent on the proportions of the cases in each group (including the reference group) and should be interpreted with this in mind.

TABLE 5.3.4

Partial and Semipartial Correlations and Regression Coefficients for the
Dummy Variable Coding of the Illustrative Data

X_i	pr_i	pr_i^2	sr_i	sr_i^2	β_i	B_i	$t_i\,(df=32)$
X_1	.363	.1318	.313	.0980	.4051	23.18	2.203*
X_2	−.145	.0210	−.118	.0139	−.1490	−9.46	−0.831
X_3	.423	.1789	.375	.1406	.4525	33.37	2.639*

$$A = 70.13$$

$$\hat{Y} = B_1 X_1 + B_2 X_2 + B_3 X_3 + A$$
$$= 23.18\,X_1 - 9.46\,X_2 + 33.37\,X_3 + 70.13.$$

$$\hat{Y}_1 = 23.18(1) - 9.46(0) + 33.37(0) + 70.13 = 93.31 = \bar{Y}_1.$$
$$\hat{Y}_2 = 23.18(0) - 9.46(1) + 33.37(0) + 70.13 = 60.67 = \bar{Y}_2.$$
$$\hat{Y}_3 = 23.18(0) - 9.46(0) + 33.37(1) + 70.13 = 103.50 = \bar{Y}_3.$$
$$\hat{Y}_4 = 23.18(0) - 9.46(0) + 33.37(0) + 70.13 = A = 70.13 = \bar{Y}_4.$$

$$\widetilde{sd}^2_{Y-\hat{Y}} = sd^2_Y(1 - R^2)\,\frac{n}{n-k-1} = 27.49^2\,(1 - .354944)\,\frac{36}{36-3-1} = 548.41$$

*P < .05.

The Regression Coefficients and the Regression Equation

The general meaning of a raw score partial regression coefficient (B_i) is that it gives the amount and direction of net change in Y, expressed in units of Y, of a change in one unit of X_i. When the X_i are measured in the arbitrary units of typical psychological scales or sociological indices, the B_i and the Y intercept A are frequently not of much interpretive utility. A unique characteristic of dummy variables is that a unit change in an X_i is a change from 0 to 1, for example, from non-Catholic to Catholic. But now remember that B_i is a *partial* regression coefficient, and that the effect of the partialling process on a dummy variable is to relate G_i not to G_{non-i} but rather to G_g. Concretely, then, $B_2 = -9.46$ in Table 5.3.4 means that the expectation for Catholics is $-9.46\,Y$ points relative to *Others,* that is, the ATA sample mean for Catholics is 9.46 points smaller than the ATA mean for Others.

How this comes about can be seen by considering the regression equation in the context of dummy variables. We recall from Chapter 2 that, in general, the regression equation gives a least-squares fit to the data, that is, $\hat{Y}$ is so generated that for all n cases in a sample, the sum of the squared discrepancies between the actual Y and $\hat{Y}$ is a minimum. With nominal IVs, all the cases with the same set of X_i (e.g., 0, 1, 0) are in the same group, so that the regression equation should yield for each group its mean on Y, that is $\bar{Y}_i$, since the mean of any set of observations *is* the best $\hat{Y}$ estimate for G_i in the least-squares sense. In fact it does. In Table 5.3.4, the dummy variable scores of each group given in Table 5.3.1 are substituted in turn in the regression equation, and the meaning of the B_i and A become clear:

Others, the reference group, being coded 0, 0, 0, has as its best-fitted score only 70.13, which is therefore its mean. Generally, then,

(5.3.6) $\bar{Y}_g = A.$

Thus, in dummy variable coded nominal scales, the Y intercept, A (the value fitted for Y when all X_i are zero) is the sample mean of the reference group. In the example, Others have a mean ATA score of 70.13.

Each of the other G_i has added to A only its own B_i, since on its X_i it is coded 1 and on the others 0. Therefore, with dummy-variable coding

(5.3.7) $\bar{Y}_i = A + B_i$

                                            $= \bar{Y}_g + B_i.$

Thus, Protestants' mean ATA score is the Others' mean plus 23.18.

It follows by simple transposition of Eq. (5.3.7) that the regression coefficient B_i of a dummy variable is the difference between the Y means of G_i and G_g :

(5.3.8) $B_i = \bar{Y}_i - \bar{Y}_g.$

Thus, the Jewish B_3 = 33.37 indicates that the mean ATA score for the Jewish group is 33.37 points higher than the mean of the Other group.

It also follows that the differences between B_is for two groups is the difference between their Y means. Since $B_i = \bar{Y}_i - \bar{Y}_g$ and $B_j = \bar{Y}_j - \bar{Y}_g$, then

(5.3.9) $\bar{Y}_i - \bar{Y}_j = B_i - B_j.$

For example, the difference in ATA means between Protestants and Catholics is $23.18 - (-9.46) = 32.64$.

The choice of group to serve as the reference group is, of course, arbitrary from a mathematical viewpoint. If some other group were chosen, the B_i and A would change in numerical value, but the equations above would continue to hold. For example, if G_1 were made the reference group in the illustrative example, the B values for the other three groups would be, in order, −32.64, 10.19, and −23.18, and A would equal 93.31.

A word about the standardized partial regression coefficients: The β_is are dutifully recorded in Table 5.3.4 for the sake of completeness, but in the context of nominal scale coding, they are not particularly useful. Unlike continuous variables, one cannot change the variability of a dichotomy without implicitly changing its mean, as is obvious from Eqs. (5.3.1) and (5.3.2). A particularly useful property of dummy variable *raw* score regression coefficients, that they are differences between means and thus do *not* depend on the relative sizes of the n_i, is lost in β_is, which do vary with changes in relative n_is.

The Statistical Significance of Partial Coefficients

All the partial coefficients of any X_i, namely, pr_i, sr_i, β_i, and B_i, share the same status with regard to statistical significance, all giving the same value of t for the same df (see Section 3.7.3). The test of the null hypothesis that a partial coefficient for X_i equals zero in the population can be written in several alternate ways, one of which we saw in Section 3.7 is

$$(3.7.7) \qquad t = sr_i \sqrt{\frac{n-k-1}{1-R^2}} \qquad (df = n - k - 1).$$

This t test is of singular importance in an MRC dummy variable analysis. Since F for set G was significant, in accordance with the protected t procedure (Section 4.6.4), we may perform and interpret the t tests for the partial coefficients for the constituent IVs of the set. Consider $t = 2.639$ for X_3, the Jewish dichotomy. For 32 df, $t = 2.037$ for $P = .05$ (Appendix Table A) so the string of coefficients is significant. Depending on which coefficient is interpreted, we can make different verbal interpretations, but these are necessarily mathematically equivalent. For example, since pr_3 is significant, we can say that if one considered only the subpopulation of Jews and Others, the population (point-biserial) correlation between this dichotomy and ATA would not be zero. Since sr_3 (and hence sr_3^2) is significant, we can say that if we dropped X_3 from the IVs, thus losing the distinction between Jews and Others, the population R^2 would drop in value from what it is with the Jew–Other distinction maintained. Finally, and perhaps most easily understood, is the interpretation of the statistical significance of $B_3 = 33.37$. Since $B_3 = \bar{Y}_3 - \bar{Y}_4$, its t test is a test of the significance of the difference between the mean ATA score of Jews and that of Others. Thus, the t values of Table 5.3.4 provide significance tests of the difference between the reference group mean ($A = 70.13$) and the mean of each of the other groups.

These tests again call to the fore the centrality of G_g as the reference group. It makes MRC using dummy variables especially appropriate in research where there is a group which is to be compared with other groups. Thus, dummy variable coding would be most appropriate for the third example in Table 5.3.1, where there are three groups receiving different treatments and G_4 is a control group. However, the existence of such a group is not in any sense a requirement for the use of dummy variable coding, since, to the extent to which one is interested in making comparisons between the Y means of pairs of groups, neither of which is G_g, say G_i and G_j, such comparisons are readily accomplished. They are legitimized by the protected t procedure (Section 4.6.4). Since, from Eq. (5.3.9), $\bar{Y}_i - \bar{Y}_j = B_i - B_j$, a test of the difference between the means is equivalent to a test of the difference between the analogous regression coefficients for dummy variables:

$$(5.3.10) \qquad t = \frac{B_i - B_j}{\sqrt{\tilde{sd}_{Y-\hat{Y}}^2 \left(\frac{1}{n_i} + \frac{1}{n_j}\right)}} \qquad (df = n - k - 1),$$

where

(5.3.11) $$s\widetilde{d}^2_{Y-\hat{Y}} = sd^2_Y (1 - R^2) \frac{n}{n - k - 1},$$

the unbiased estimate of the variance error of estimate, or mean square of residuals from regression (see Table 5.3.4). This test uses all the information available in the sample about the Y variability within groups, not only that of G_i and G_j, hence is based on $n - k - 1$ ($= n - g$) df. Thus, the statistical test of the difference in ATA means of Protestants and Catholics is given by

$$t = \frac{23.18 - (-9.46)}{\sqrt{548.41 \left(\frac{1}{13} + \frac{1}{9}\right)}} = \frac{32.64}{10.16} = 3.213,$$

which is significant at the .01 level for $df = 32$ (Appendix Table A).

In summary, MRC with dummy variable coding provides quantitative measures and significance tests for:

1. The relationship between a nominal scale, G (group membership), and Y via R, or the proportion of variance in Y accounted for in the sample by G via R^2.

2. The relationship between each aspect of group membership (e.g., Protestant versus non-Protestant) and Y via r_{Yi}, or the analogous proportion of variance r^2_{Yi}.

3. The relationship between the dichotomies of each coded aspect of group membership versus reference group membership (for example, Protestant versus Other) and Y, via pr_i and sr_i, or the differences between the Y means of the groups, via B_i.

All correlational results are dependent upon the p_i, so the sampling plan (n fixed or random) must be taken into account in interpreting their magnitudes, but not their statistical significance.

5.3.4 Dummy Variable Multiple Regression/Correlation and Analysis of Variance

The reader familiar with the analysis of variance (AV) may have noted some similarities to the above and wondered about its relationship to MRC. It may come as a surprise that they are essentially identical. They are both applications of the same general linear model, and when the independent variables are fixed nominal scales, the identity is exact, although obscured by differences in terminology and traditional usage.

Viewed from the perspective of AV, the data structure considered in this section is that of a set of g "levels" of a factor with n_i observations (Y) in each, that is, a "one-way" AV. For the illustrative data of Table 5.3.2, we would assemble the Y values into the G-designated groups and proceed to find the sums of squares for between groups (B SS), within groups (W SS) and their total (T

TABLE 5.3.5

AV of the Illustrative Data of Table 5.2

Source	SS	df	MS
Between groups	9656.49	$g - 1 = 3$	3218.83
Within groups	17549.15	$n - g = 32$	548.41
Total:	27205.64	$n - 1 = 35$	

G_i	n_i	$\bar{Y}_i$
G_1	13	93.31
G_2	9	60.67
G_3	6	103.50
G_4	8	70.13

$$F = \frac{\text{B SS}/(g-1)}{\text{W SS}/(n-g)} = \frac{\text{B MS}}{\text{W MS}} = \frac{3218.83}{548.41} = 5.869 = F \text{ test of } R^2.$$

$$\eta^2_{Y \cdot G} = \frac{\text{B SS}}{\text{T SS}} = \frac{9656.49}{27205.64} = .354944 = R^2.$$

SS), and then the mean squares (MS) for B and W by division by their respective df (see Table 5.3.5). The F ratio between these MS is a test of the null hypothesis that the population Y means of the g groups are equal.

When this is applied to the illustrative data in Table 5.3.5, we first note that the F of the AV is identically the same as the one computed earlier on the R^2 of these data. This can be understood conceptually in that the null hypothesis of the AV, equality of the g population means, is mathematically equivalent to the null hypothesis of the MRC analysis, which is that no variance in Y is accounted for by group membership. Clearly, if the population means are all equal, Y variance is not reduced by assigning to the members of the population their respective identical means which are necessarily also equal to the grand mean of the combined populations. Each null hypothesis implies and is implied by the other, hence they differ only verbally. The two F ratios are, in fact, algebraically identical. Since B SS = R^2 (T SS) and W SS = $(1 - R^2)$(T SS), substituting these in the AV formula for F in Table 5.3.5 and canceling T SS from numerator and denominator gives Eq. (5.3.5), the F test for R^2.

The proportion of variance in Y accounted for by G is available in AV, although most textbook treatments do not mention it (or relegate it to the narrow usage of tests for curvilinearity of regression). It is the squared "correlation ratio"

(5.3.12) $$\eta^2_{Y \cdot G} = \frac{\text{B SS}}{\text{T SS}}.$$

Note in Table 5.3.5 that the application of this formula gives η^2 (eta squared) =

.354944, identically the same value as was found for $R^2_{Y \cdot 123}$; thus,

(5.3.13) $\eta^2_{Y \cdot G} = R^2_{Y \cdot 123 \ldots (g-1)}.$

Furthermore, just as the shrunken $\widetilde{R}^2$ of Eq. (3.7.4) is an improved estimate of the proportion of variance of Y accounted for in the population, the same improved estimate in AV is called ϵ^2 (epsilon squared), and it is readily proved that

(5.3.14) $\epsilon^2_{Y \cdot G} = \widetilde{R}^2_{Y \cdot 123 \ldots (g-1)}.$

Thus, we see that the yield of an MRC analysis of dummy-variable-coded group membership includes the information which a one-way AV yields, and then some—we also obtain various analytically useful correlational (simple, multiple, partial, semipartial) measures. This is only part of the story of the MRC–AV relationship. First, we will see that other means of coding nominal data (effects and contrast coding) bring out yet other identities between the two systems, and second, that other fixed-AV models (for example, factorial design, covariance analysis) can be duplicated and extended by means of MRC.

5.4 EFFECTS CODING

5.4.1 Introduction

We have seen that in dummy variable coding, the partial coefficients for X_i relate G_i to G_g, one of the g groups selected to serve as a reference group. This is the group which is "omitted" from explicit dummy variable 0, 1 representation among the IVs, the group which would otherwise have produced X_g.

In effects coding, again (as always) we represent a g-way nominal scale research factor G by means of $k_G = g - 1$ IVs, but now the X_is are not dichotomous but trichotomous and the "scores" of G_g on the X_i are not a string of zeros, but a string of minus ones. Table 5.4.1 makes this coding scheme clear. Each X_i aspect of the G research factor contrasts $G_i(1)$ with $G_g(-1)$, the influence of remaining groups being minimized[5] in that contrast by being coded 0, that is, exactly in between. Thus, in the running example, the effects-coded X_1 of Table 5.4.1 primarily distinguishes Protestants from Others and minimizes the influence of Catholics and Jews. X_2 distinguishes Catholics from Others and X_3 distinguishes Jews from Others in the same way. These *raw* variables sound like the *partialled* variables in dummy variable analysis, and for equal n_i they are.

We will study the yield of an MRC with effects coding, again using the illustrative data of Table 5.3.2 but now with the new coding of Table 5.4.1

---

[5] The word "minimized" is used to avoid lengthy discussion of a minor point. The minimum influence of the 0-coded groups on r_{Yi} is literally nil whenever $n_i = n_g$, that is, the sample sizes for the group coded 1 and the group coded −1 are equal. The effect departs from zero as these sample sizes depart from equality to the extent to which the pooled mean of Y for the 0-coded groups departs from the grand mean of all the Y observations.

TABLE 5.4.1

Effects Coding of Illustrative Research Factors
with $g = 4$ Categories

	Religion	Region	Experimental group	X_1	X_2	X_3
G_1	Protestant	Northeast	Treatment 1	1	0	0
G_2	Catholic	South	Treatment 2	0	1	0
G_3	Jewish	Midwest	Treatment 3	0	0	1
G_4	Other	West	Treatment 4	−1	−1	−1

rather than dummy variables. This means merely replacing for all G_4 cases the 0, 0, 0 by −1, −1, −1 in the X_1, X_2, X_3 columns of Table 5.3.2. The results from the MRC analysis are given in Tables 5.4.2 and 5.4.3.

5.4.2 The R^2 and rs

We note first that the $R^2 = .3549$, and its $F = 5.869$, (and $\tilde{R}^2 = .2945$), identically the same values as were obtained by dummy variable coding. We remind the reader (see Section 5.2) that the different kinds of coding are alternative ways of rendering the information as to group membership into quantitative form suitable for correlation. However, taken as *sets* they are equivalent; since each set carries the same information, given the same Y data, they must yield the same R^2 and hence the same F. Indeed we will later see (Section 5.6) that it hardly matters what numbers one puts in a coding table such as Table 5.4.1 insofar as the values found for R^2, F and the group Y means (from the regression equation) are concerned. Results for *individual* aspects (i.e.,

TABLE 5.4.2

Correlations, Means, and Standard Deviations of the
Illustrative Data for Effects Coding

		Y	X_1	X_2	X_3	r^2_{Yi}	$t_i\ (df = 34)$
				r			
	Y	1.000	.328	−.142	.363		
(P)	X_1	.328	1.000	.424	.493	.1076	2.025
(C)	X_2	−.142	.424	1.000	.525	.0202	−.660
(J)	X_3	.363	.493	.525	1.000	.1318	4.858**
	m	81.69	.139	.028	−.056		
	sd	27.49	.751	.687	.621		

$$R^2 = .354944 \qquad F = 5.869**\ (df = 3, 32)$$

$$\tilde{R}^2 = .294470$$

**$P < .01$.

X_is) of a research factor change with changes in coding, but not those from the nominal scale research factor expressed as a set of IVs.

When coded for effects, each X_i aspect of G effects a distinction between its G_i and the G_g, with the effect of the other groups being minimized (see Footnote 5). Thus, $r_{Yi}^2 = .1076$ in Table 5.4.2 means that 11% of the ATA variance is largely accounted for by the Protestant–Others distinction. The r_{Yi} for effects-coded variables equal the sr_i for dummy-coded variables when $n_i = n_g$, and approximate them when the two sample sizes are not grossly unequal. (Compare the r_{Yi} of Table 5.4.2 with the sr_i of Table 5.3.4.) Since the Y data from groups coded 0 in effects-coding may have some effect on the r_{Yi}, it is prudent not to interpret them. It is only when partialled that their interpretation becomes unambiguous.

The three r_{Yi}^2 do not sum to $R_{Y\cdot 123}^2$ as also they did not for dummy variables, and for the same reason: the effects-coded X_i are correlated with each other. We note in passing the correlation between two effects coded variables X_i and X_j:

$$(5.4.1) \qquad r_{ij} = \frac{p_g - (p_i - p_g)(p_j - p_g)}{\sqrt{[p_i + p_g - (p_i - p_g)^2][p_j + p_g - (p_j - p_g)^2]}} .$$

This formula, as such, is of little importance to us. We include it, as we include the r_{ij} values in Table 5.4.2 primarily to emphasize that the r_{ij} are not, in general, zero. When all groups are of the same size, $r_{ij} = .50$ (for any number of groups); it increases above this value when p_g is large relative to the other groups, and falls below it when p_g is small.

5.4.3 The Partial Coefficients in Effects Coding

The Regression Coefficients and the Regression Equation

In dummy variable coding, the consequence of partialling the other IVs from X_i is to make of the partialled X_i a contrast between G_i and G_g. Thus, one of the groups, G_g, is a reference point for all the others. In effects coding, the partialled X_i produces a contrast between G_i and *all* the groups in the sample or, since the total sample includes G_i, one can think of the partialled effects-coded X_i as a contrast between G_i and the remaining groups. In fact, the term "effects" comes from the AV, where the sample effect of an experimental Treatment i on some dependent variable is the difference between that treatment group's mean, $\bar{Y}_i$, and the unweighted mean of *all* the sample means of the observations, $\bar{\bar{Y}}$, that is, the effect of Treatment i on Y is $\bar{Y}_i - \bar{\bar{Y}}$. When the IVs are effects-coded in MRC, the raw regression coefficients *are* the effects for $g - 1$ of the groups

$$(5.4.2) \qquad B_i = \bar{Y}_i - \bar{\bar{Y}},$$

where $\bar{\bar{Y}}$ is the unweighted mean of *all* the g sample means

$$(5.4.3) \qquad \bar{\bar{Y}} = \frac{\sum \bar{Y}_i}{g} .$$

This last quantity is yielded by the analysis as the Y intercept,

(5.4.4) $$A = \overline{\overline{Y}}.$$

When one substitutes a group's effects-coded "scores" in the regression equation (see Table 5.4.3), the group's mean is produced, as with dummy variables. For $g - 1$ of the groups, the regression equation reduces to $\hat{Y}_i = B_i + A = \overline{Y}_i$, that is, the group's effect plus the mean of means, which equals the sample mean. For the gth group, the regression equation becomes

(5.4.5) $$\hat{Y}_g = A - \sum B_i = \overline{Y}_g,$$

the mean of all the means minus the sum of the effects of the $g - 1$ other groups.

Effects coding is therefore particularly appropriate with nominal scales when each group is most conveniently compared with the entire set of groups, rather than with a single reference group, as is facilitated by dummy variable coding. Furthermore, the raw-score regression coefficients of effects coding share with those of dummy variable coding the generally desirable property of independence of the p_i of the groups: since each effect given by B is the difference between a given group's mean and the *unweighted* mean of all the group means (Eqs. 5.4.2 and 5.4.3), changing the relative numbers of cases in the g groups will result in no systematic change of the B_i values or that of A. The significance testing of the B_i will be discussed later.

The standardized coefficients (β_i) share with those of dummy variables dependency on the relative n_i of the groups, since $\beta_i = B_i(sd_i/sd_Y)$, and sd_i is a

TABLE 5.4.3
Partial and Semipartial Correlations and Regression Coefficients for the
Effects Coding of the Illustrative Data

X_i	pr_i	pr_i^2	sr_i	sr_i^2	β_i	B_i	$t_i\,(df = 32)$
X_1	.313	.0980	.264	.0697	.3117	11.41	1.862
X_2	−.481	.2314	−.440	.1936	−.5303	−21.23	−3.100**
X_3	.436	.1901	.389	.1513	.4880	21.60	2.740**

$$A = 81.90 = \overline{\overline{Y}}$$

$$\hat{Y} = B_1X_1 + B_2X_2 + B_3X_3 + A$$
$$= 11.41\,X_1 - 21.23\,X_2 + 21.60\,X_3 + 81.90.$$

$$\hat{Y}_1 = 11.41(1) - 21.23(0) + 21.60(0) + 81.90 = 93.31 = \overline{Y}_1.$$
$$\hat{Y}_2 = 11.41(0) - 21.23(1) + 21.60(0) + 81.90 = 60.67 = \overline{Y}_2.$$
$$\hat{Y}_3 = 11.41(0) - 21.23(0) + 21.60(1) + 81.90 = 103.50 = \overline{Y}_3.$$
$$\hat{Y}_4 = 11.41(-1) - 21.23(-1) + 21.60(-1) + 81.90 = 70.13 = \overline{Y}_4.$$

$$\widetilde{sd}^2_{Y-\hat{Y}} = sd^2_Y(1 - R^2)\,\frac{n}{n - k - 1} = 548.41 \quad \text{(as in Table 5.3.4).}$$

**$P < .01$.

function of the relative sample sizes. They will, in consequence, generally not be analytically useful with nominal scales.

The Semipartial and Partial Correlations of an Effects-Coded Variable

We have seen that an effects-coded IV, when the other effects-coded IVs are partialled from it (e.g., $X_{1 \cdot 23}$) yields a contrast between a given group and all the groups comprising the sample. Since all the groups include the group in question, this is functionally equivalent to contrasting this group with the remaining groups taken collectively. The sr_i here is the correlation between this partialled variable and Y, and sr_i^2 is the proportion of Y variance accounted for by this contrast. Concretely, for our running example on the illustrative data using religion and ATA, $sr_1^2 = .0697$ (see Table 5.4.3) means that 7% of the ATA variance is accounted for by the distinction between Protestants on the one hand and equally weighted Catholics–Jews–Others on the other. Stated slightly differently, for the sample, 7% of the ATA variance is accounted for by the "eccentricity" of Protestants relative to the other groups. Catholics' "eccentricity" on ATA is greater, given that $sr_2^2 = .1936$ (Table 5.4.3), their departure from the other groups, on the average, being towards lower scores as evidenced by the negative sign attached to its partial coefficients.

The partial correlations (pr) relate the partialled effects-coded IV (e.g., $X_{1 \cdot 23}$) not with all of Y (as does sr), but rather with that part of Y left after the other variables have been removed (e.g., $Y \cdot 23$). Thus, pr_1^2 gives

$$\frac{\text{ATA variance due to Protestant group "eccentricity"}}{\text{ATA variance } not \text{ accounted for by remaining groups "eccentricities"}} .$$

As already noted, since the base of the proportion for pr^2 is always smaller (or, at least, not larger) than for sr^2 (where it is the total Y variance), pr_is are numerically larger than sr_is. In a sense, the pr_i values are "fairer" in that they do not depend on the other groups' effects, the latter being removed from Y in its correlation with the partialled X_i.

Since the gth group, the one whose X_i are coded with a string of minus ones, has no X_g variable, the analysis yields no partial coefficients (or t tests) for this group. The analyst may wish to know what the results for this group would have been had some other group replaced it in the coding. They are given in the following equations (specialized from Sections 3.5 and 3.7):

$$(5.4.6) \qquad pr_g^2 = \frac{t_g^2}{t_g^2 + n - g} ,$$

$$(5.4.7) \qquad sr_g^2 = \frac{t_g^2 (1 - R^2)}{n - g} .$$

A formula for and discussion of t_g^2 is given in the next section.

Thus, pr_i and sr_i with effects-coded IVs offer a correlational statement of a group's departure from other groups, the latter treated as equal whatever their relative n_i. This can be most useful in data analysis, since correlations and proportions of variance are pure numbers which express effect sizes in a common metric or *lingua franca* which is not dependent on the size of the unit for Y. (pr_i and sr_i are, however, as are all correlation coefficients, dependent on p_i.) It thus becomes possible to compare effects between different dependent variables. Thus, given another study for the same religious groups (in the same proportions) using another dependent variable, say Political–Economic Conservatism (PEC), it is meaningful to compare the sr_i or pr_i (or their squares) for the same groups across the two studies, while it would not in general be meaningful to compare the B_i since the latter is expressed in units of Y score, and ATA and PEC units are not directly comparable.

It is instructive to note that with equal n_i in the g groups, the sr_i of effects-coded variables are exactly equal to the r_{Yi} of dummy coded variables, thus making clear that the sr_i in effects coding contrasts G_i with the other groups. Further, with equal n_i, the converse also holds: the r_{Yi} of effects-coded variables equal the sr_i of dummy variables.

The Statistical Significance of the Partial Coefficients

Recall that all partial coefficients (pr_i, sr_i, B_i, and β_i) share the same significance tests, one form of which is given in Eq. (3.7.7). Table 5.4.3 gives the t values for each effects-coded IV's set of coefficients. It is simplest to deal with the "effect" of each group, that is, B_i, the Y mean of Group i minus the mean of all the groups' means (Eqs. 5.4.2 and 5.4.3). The null hypothesis tested by the t_i for B_i is that in the population, the Y mean for Group i is equal to the mean of the means of Y of all the groups. Large t_is result in the rejection of the null hypothesis and a conclusion that the population mean of Y for Group i differs from the mean of the population means of all the groups (assuming they are protected—Section 4.6.4). For example, the Jews "eccentricity" on ATA (departure of their mean) is significant, while that of the Protestants is not (at the .05 level for 32 df).[6]

Since the gth group has no X_g variable and hence no t test for its partial coefficients, one cannot determine from the usual MRC computer output whether its $\overline{Y}_g$ departs significantly from $\overline{\overline{Y}}$. Such a test is, however, available from the other results of the MRC analysis:

$$(5.4.8) \qquad t_g = \frac{-g \sum B_i}{\sqrt{\widetilde{sd}_{Y-\hat{Y}}^2 \left[\dfrac{(g-1)^2}{n_g} + \sum \dfrac{1}{n_i} \right]}} \qquad (df = n - k - 1),$$

[6] To further point out identities with AV procedures, we note that these t values are *identically* equal to what would be obtained on the conventional test of a contrast between $\overline{Y}_i$ and the mean of the means of the remaining groups. Obviously, if this difference is zero, the difference between $\overline{Y}_i$ and the mean of the means of *all* the groups will also necessarily be zero, so the null hypotheses are equivalent.

where i runs from 1 to $g - 1$ and the summation is over that range. For the illustrative example,

$$t_4 = \frac{-4\,(11.41 - 21.23 + 21.60)}{\sqrt{548.41\left(\dfrac{3^2}{8} + \dfrac{1}{13} + \dfrac{1}{9} + \dfrac{1}{6}\right)}}$$

$$= \frac{-47.12}{28.49} = -1.654, \quad \text{with} \quad 32\ df.$$

Thus, the data do not warrant concluding at the 5% level (see Appendix Table A.2) that the Other group's population mean departs from the mean of the population means. [It is t_4^2 which is required in the equations for pr_4^2 and sr_4^2 (Eqs. 5.4.6 and 5.4.7) in the previous section.]

Tests of Significance of Differences between Means

Although the orientation of effects coding is one of group versus total sample, the MRC output contains the ingredients to test mean differences between groups, if this should be desired in addition to tests of the group effects, subject to protection (Section 4.6.4). A mean difference between two groups neither of which is G_g is accomplished exactly as for dummy-coded variables, given in Eq. (5.3.10). For the case where $\bar{Y}_i$ is to be compared with $\bar{Y}_g$, the t test is accomplished by

$$(5.4.9) \quad t_{\bar{Y}_i - \bar{Y}_g} = \frac{2B_i + \sum B_j}{\sqrt{\widetilde{sd}_{Y-\hat{Y}}^2\left(\dfrac{1}{n_i} + \dfrac{1}{n_g}\right)}} \quad (j \ne i)\ \ (df = n - k - 1),$$

where the summation is taken over all B coefficients except B_i.

Note that the numerator equals $\bar{Y}_i - \bar{Y}_g$. However the data are coded, one can always write the group means, and therefore functions of the group means, as functions of the regression constants.

Thus, for comparing the Protestant (G_1) mean with the Other (G_4) mean,

$$t_{\bar{Y}_1 - \bar{Y}_4} = \frac{2(11.41) + (-21.23 + 21.60)}{\sqrt{548.41\left(\dfrac{1}{13} + \dfrac{1}{8}\right)}} = \frac{23.19}{10.52} = 2.203,$$

which is significant at $P < .05$ for $df = 32$ (Appendix Table A.2). Note that this is the same value as the t for B_1 in dummy variable coding (see Table 5.3.4), where the regression coefficients are directly the differences between each group mean and the mean of the reference group, i.e., $\bar{Y}_i - \bar{Y}_4$.

In summary, effects coding takes as its point of reference for each of the groups *all* of the groups taken as an equally weighted aggregate. The partial coefficients thus reflect the effect or "eccentricity" of each group relative to the others. Since as a *set* the effects-coded IVs fully represent group membership (as do dummy variables), all results relating to the set will be the same: R^2, its F

and the $\bar{Y}_i$ produced by the regression equation. These, in turn, will agree exactly with the results of the usual AV performed on the data.

5.5 CONTRAST CODING

5.5.1 Introduction

In statistical parlance, a contrast for a set of g sample means (or other statistics) is any linear function of them of the form

$$(5.5.1) \qquad C = a_1 \bar{Y}_1 + a_2 \bar{Y}_2 + \cdots + a_g \bar{Y}_g,$$

subject to the condition that

$$(5.5.2) \qquad a_1 + a_2 + \cdots + a_g = 0.$$

This is a very general formulation for the kinds of comparisons that we are always making, explicitly or implicitly, in the analysis of data. The null hypothesis is that the contrast applied to the g population means equals zero, and may be tested for significance by means of t, given the usual assumptions.

Indeed, although not written in the form of Eq. (5.5.1), we have already tested for significance contrasts in both dummy variable and effects coding, in the former between the Y mean of one of the groups and that of the reference group (5.3.8), and in the latter between the Y mean of one of the groups and the mean of the means of all the groups (5.4.2). Contrast coding differs from other coding primarily in its generality: any single contrast within a set of *independent* contrasts can be represented and tested by means of contrast coding.

To see how the form of Eq. (5.5.1) expresses a contrast, consider any set of g groups comprising G. We wish to compare (on Y) the mean of the u means of a subset U with the mean of the v means of another subset V, omitting from this contrast the remaining q groups in subset Q, if any. For concreteness, imagine a G set of $g = 9$ groups defined by differing responses to a survey interview item. We wish to compare a subset U made up of 4 $(= u)$ of these groups with another subset V made up of another 3 $(= v)$ groups, ignoring the remaining 2 $(= q)$ groups of subset Q (for example, "did not answer" and "not applicable"). This comparison of means of means is accomplished for this example by

$$C = \frac{\bar{Y}_1 + \bar{Y}_2 + \bar{Y}_3 + \bar{Y}_4}{4} - \frac{\bar{Y}_5 + \bar{Y}_6 + \bar{Y}_7}{3},$$

or restated in the form of Eq. (5.5.1) as

$$C = \frac{1}{4} \bar{Y}_1 + \frac{1}{4} \bar{Y}_2 + \frac{1}{4} \bar{Y}_3 + \frac{1}{4} \bar{Y}_4 - \frac{1}{3} \bar{Y}_5 - \frac{1}{3} \bar{Y}_6 - \frac{1}{3} \bar{Y}_7 + 0 \, \bar{Y}_8 + 0 \, \bar{Y}_9.$$

Note that the full set of a coefficients satisfy the condition (5.5.2):

$$\sum a_i = \frac{1}{4} + \frac{1}{4} + \frac{1}{4} + \frac{1}{4} - \frac{1}{3} - \frac{1}{3} - \frac{1}{3} + 0 + 0 = 0.$$

Thus, we can state generally how the contrast coefficients are written for a simple contrast between U and V subsets, ignoring Q: the a values for each of the U groups is $1/u$, for each of the V groups $-1/v$, and for each of the Q groups, if any, 0.

The importance of these contrast coefficients lies in the fact that a contrast is effected by an IV whose coded values are the a coefficients. Thus, each subject is "scored" on the contrast IV with the a value which gives the role or weight which his group has in it.

The above conditions to accomplish means of means contrasts by MRC are necessary, but not sufficient. We have seen that the full representation of the information entailed in membership in one of g groups requires a set of $g - 1$ IVs. The contrast described above is only one member of such a set. For this to function as a contrast between means of means in MRC, two further conditions must obtain: all $g - 1$ contrasts must be represented in the IV set (that is, 8 in the example), and the contrasts must be independent in the specific sense that the *coefficients* for any pair of contrasts must be linearly independent. What this means is best illustrated by again using our running example of $g = 4$ groups.

Table 5.5.1 presents two research factors, religion, as before, and a new one of experimental groups, to which we will refer later, as well as two sets of contrast codes. These are *alternative* sets, the first more appropriate to religion and the second to a set of experimental groups which will be taken up in the next section.

Consider again the example of the four religious groups, with Attitude toward Abortion (ATA) as Y. Assume we are interested in, among other things, the difference between the majority religious groups, Protestants and Catholics, on the one hand, and the minority groups, Jews and Others, on the other. Thus, $u = 2$, $v = 2$ (and $q = 0$), and our a coefficients for the contrast are $1/u$ for G_1 and G_2 and $-1/v$ for G_3 and G_4, in order: $\frac{1}{2}, \frac{1}{2}, -\frac{1}{2}$, and $-\frac{1}{2}$ (which sum to zero). These are entered as X_1 for Set I in Table 5.5.1; they are the coding coefficients for this, the first contrast.

Note that if we were to multiply these a values by any nonzero constant, w, we would still have a contrast in that the condition of Eq. (5.5.2) would be satisfied. Furthermore, no change (other than in sign if w is negative) would

TABLE 5.5.1

Two Alternative Sets of Contrast Codes for Illustrative
Research Factors with $g = 4$ Categories

Set I					Set II			
G_i	Religion	X_1	X_2	X_3	G_i	Experimental group	X_1	X_2	X_3
G_1	Protestant	$\frac{1}{2}$	1	0	G_1	Drug–Frontal Lesion	$\frac{1}{2}$	$\frac{1}{2}$	$\frac{1}{4}$
G_2	Catholic	$\frac{1}{2}$	-1	0	G_2	Drug–Control Lesion	$\frac{1}{2}$	$-\frac{1}{2}$	$-\frac{1}{4}$
G_3	Jewish	$-\frac{1}{2}$	0	1	G_3	Placebo–Frontal Lesion	$-\frac{1}{2}$	$\frac{1}{2}$	$-\frac{1}{4}$
G_4	Other	$-\frac{1}{2}$	0	-1	G_4	Placebo–Control Lesion	$-\frac{1}{2}$	$-\frac{1}{2}$	$\frac{1}{4}$

occur in correlations involving X_1 (zero order, partial, semipartial), nor in its β coefficient, nor, most important of all, would the usual t tests involving X_1 change. All that would change is B_1; it would simply be divided by w. Thus, only the *scale* of a contrast is affected by a constant multiplier w of the a values. In this example, multiplication by $w = 2$ would give the convenient a' coefficients 1, 1, −1, and −1, leave all correlations and β_1 intact, but make B'_1 one-half of B_1. In the interests of simplicity, we will not generally adopt the option of multiplying the $1/u$ and $-1/v$ values by a constant, but simply allow for it as a possibility.

We now proceed to construct two further contrasts that are independent of each other and of the first contrast. Assume that a second issue of interest is the difference between Protestants ($u = 1$) and Catholics ($v = 1$), ignoring the other 2 ($= q$) groups. Such a simple difference is rendered by $(1/u =) 1$ and $(-1/v =) -1$ for the groups compared, with the two groups irrelevant to the contrast coded 0, that is, as in X_2 of Set I in Table 5.5.1: $C = \bar{Y}_1 - \bar{Y}_2$. Similarly, X_3 codes the analogous contrast between G_3 and G_4, the two minority groups as 0, 0, 1, −1. (The resulting structure is called a "nested" design.)

Obviously there are many other contrasts possible among four groups.[7] Recall, however, that *all* the information about membership in one of g groups is contained in $g - 1$ variables, however we may choose to code them. We can write many different sets, each of $g - 1 = 3$ contrasts (for example, Set II in Table 5.5.1), but once $g - 1$ different contrasts are written, they exhaustively represent group membership, and all other possible contrasts can be written as a function of these. The latter can be seen in the fact that we can, with any kind of coding, produce the Y mean of each group by employing the regression equation, and once the means can be produced, we can contrast them in any way we like.

There is one further property that our contrast codes must have: they should represent wholly different or independent issues, that is, they should be orthogonal, as in the "orthogonal comparisons" of AV (cf. Edwards, 1972, pp. 130−152). Mathematically, this demand is satisfied when, for each pair of contrasts, the sum of the products of the coding coefficients (a) is zero. To illustrate, refer to the Set I a values in Table 5.5.1. For the a coefficients of X_1 and X_2: $(\frac{1}{2})(1) + \frac{1}{2}(-1) + (-\frac{1}{2})(0) + (-\frac{1}{2})(0) = 0$. For the as of X_1 and X_3: $(\frac{1}{2})(0) + (\frac{1}{2})(0) + (-\frac{1}{2})(1) + (-\frac{1}{2})(-1) = 0$. Also for the coding of X_2 and X_3: $(1)(0) + (-1)(0) + (0)(1) + (0)(-1) = 0$. (A check will show that Set II also has this property.)

Thus, the strategy of contrast coding is to express the hypotheses of interest in the form of $g - 1$ *different* (orthogonal) contrasts, using means of means coding. The MRC analysis then directly yields functions of contrast values and also their significance tests.

Now assume that we use the coding of Set I (Table 5.5.1) for the illustrative

---

[7] The total number of different means of means contrasts among g groups is $1 + (3^g - 1)/2 - 2^g$, a rather large number for even relatively few groups. For $g = 4$, it is 25. (For $g = 6$, it is 301, and for $g = 10$, it is 28, 501!)

data of Table 5.3.2), replacing the dummy variable coding given there. Thus, all cases in G_1 are coded $\frac{1}{2}$, 1, 0 for X_1, X_2, and X_3 (instead of 1, 0, 0), all G_2 cases are coded $\frac{1}{2}$, -1, 0, etc. We present the results of a full MRC analysis using the Set I contrast codes in Tables 5.5.2 and 5.5.3.

5.5.2 The R^2 and rs

We note from Table 5.5.2 that R^2 = .3549 (and $\tilde{R}^2$ = .2945) and its F = 5.869, exactly as with any other coding of the nominal scale of group membership for these data; changes in coding have effects on results from individual X_i, but the set *as a set* represents the group membership information and will therefore account for the same amount of Y variance as any other set for the same data.

Now note the correlations among the contrast-coded IVs, X_1, X_2, and X_3, .114, .112, and .013 (Table 5.5.2). None of them is zero, despite the fact that the a coefficients with which the 36 observations were contrast-coded for group membership are orthogonal. The r_{ij} would all be zero if (and only if) the groups were of equal size, and they are not. (It may be of interest to note that when the n_i are equal and hence all r_{ij} = 0, $R^2_{Y \cdot 123}$ would simply be the sum of the r^2_{Yi}.) We stress that the requirement is that the *contrast coefficients* be mutually orthogonal, not that the X_i be mutually orthogonal (or uncorrelated). The condition of "unequal cell frequencies" is a nuisance in the analysis of contrasts in AV, but is automatically handled in MRC.

Some ambiguity may accompany the interpretation of the r^2_{Yi}. None occurs when only two a values are used in coding an X_i. It is then a simple dichotomy, and its r^2_{Yi} is clearly the proportion of Y variance associated with the dichotomous distinction. X_1 is such a variable, distinguishing G_1 and G_2, coded $\frac{1}{2}$, from G_3 and G_4, coded $-\frac{1}{2}$. Note that *no* distinction is made between G_1 and G_2 or between G_3 and G_4; they are treated as groups of $n_1 + n_2 = 22$ and $n_3 + n_4 = 14$ cases. Thus, X_1 as such is a function of the *weighted* means, so that, for

TABLE 5.5.2
Correlations, Means, and Standard Deviations of the
Illustrative Data for Contrast Coding of Set I (Table 5.5.1)

		Y	X_1	X_2	X_3	r^2_{Yi}	t_i (df = 34)
Maj–Min	X_1	−.079	1.000	.114	.112	.0062	−.461
P–C	X_2	.444	.114	1.000	.013	.1971	2.889**
J–O	X_3	.363	.112	.013	1.000	.1321	2.275*
	m	81.69	.111	.111	−.056		
	sd	27.49	.488	.774	.621		

R^2 = .354944.

$\tilde{R}^2$ = .294470.

F = 5.869**   (df = 3, 32)

**P < .01.

example, the Protestants figure more strongly than the Catholics in the pooled majority religion group, in the ratio of their n_is, 13:9. The unweighted means of means contrast, where all groups are treated as equal is carried by the partialled contrast variable, $X_{1 \cdot 23}$, and will be discussed below. With this in mind, the zero-order $r_{Y1}^2 = .0062$ indicates that less than 1% of ATA variance is accounted for by the pooled majority versus pooled minority distinction, and is not significant ($t = -.461$).

In X_2 and X_3, we have a trichotomy: 1, −1, and 0 coded values are used. In the *partialled* results (see below), the 0-coded groups are omitted as we intend, but they play a role in determining the zero-order r_{Yi} (as was also the case in effects coding). We would therefore, in general, not wish to interpret the unpartialled correlations with Y of such trichotomous contrast variables.

5.5.3 The Partial Coefficients in Contrast Coding

The Regression Coefficients and the Regression Equation

It is the partialling process which makes our contrast-coded variables yield unambiguously interpretable results about the contrasts or "issues" which have been coded into group membership.

Specifically, the B_i coefficients in contrast coding are functions of the values of the contrast between unweighted means of means, and when groups are coded 0, they are in fact omitted. With the u means of subset U each coded $1/u$, the v means of subset V each coded $-1/v$, and irrelevant groups coded 0, the value of C, the contrast in the sample as defined by Eq. (5.5.1), is given by[8]

(5.5.3)
$$C = B \left(\frac{u + v}{uv} \right).$$

Applying this to the B_i of Table 5.5.3, we obtain the contrast values

$$C_1 = -9.82 \left(\frac{2 + 2}{(2)(2)} \right) = -9.82 \quad \left(= \frac{\bar{Y}_1 + \bar{Y}_2}{2} - \frac{\bar{Y}_3 + \bar{Y}_4}{2} \right),$$

$$C_2 = 16.32 \left(\frac{1 + 1}{(1)(1)} \right) = 32.64 \quad (= \bar{Y}_1 - \bar{Y}_2),$$

$$C_3 = 16.69 \left(\frac{1 + 1}{(1)(1)} \right) = 33.38 \quad (= \bar{Y}_3 - \bar{Y}_4).$$

Thus, C_1 indicates that the ATA mean of means of the two majority groups is 9.82 points below the mean of means of the two minority religious groups. Note that here it is unweighted means of means that are being compared; all groups, regardless of n_i count equally. Implicitly, such contrasts treat each of the groups

---

[8] If the option of multiplying the $1/u$, $-1/v$, 0 codes by a nonzero constant w is taken, the right-hand term of Eq. (5.5.3) is simply multiplied by w.

as characterized by an abstract property, and ignore differences in the group sizes. Thus, "majority religion" here is conceived as the direct average of Protestant and Catholic means, *not* as the overall average of Protestants and Catholics combined, in which Protestants figure more heavily, which is as the unpartialled X_1 represents them. C_1 might be thought of as the *net* "effect" of the abstract property "being in a majority rather than a minority religious group" (although, of course, causality would need to be supported on other than statistical grounds). Similarly, $C_2 = 32.64$ is the net "effect" on ATA of being Protestant rather than Catholic, and $C_3 = 33.38$ is the net "effect" of being a Jew rather than an Other.[9] Unlike the results with the unpartialled X_2 and X_3, the groups coded 0 are effectively omitted from the contrast.

In keeping with the theme of ignoring differences in group size, the Y intercept A represents the same reference value as in effects coding—the unweighted mean of all the group means as in Eqs. (5.4.3) and (5.4.4).

Perhaps of even greater interest than the size of the contrast as given by Eq. (5.5.3) is its statistical significance. The formal null hypothesis for a contrast is that when population means are substituted, the population contrast is zero, that is, the difference between population means or means of population means is zero. The t_i value associated with the B_i provides exactly the significance test of this null hypothesis. From Table 5.5.3, we note that the majority—minority contrast is not significant, but that both of the others are.

When these B_i values and A are combined into the regression equation and the contrast-coded values for each group substituted, they yield, as for all other nonredundant coding, the group means, as shown in Table 5.5.3. The contrast model can be viewed as yielding the $\overline{Y}_i$ of a group by adding to the mean of the means $\overline{\overline{Y}}$ ($= A$) the "effect" provided by the group's role in each contrast. Thus, the Protestant $\overline{Y}_1$ comes about, by adding to the mean of means, 81.90, one-half of the value of the majority—minority contrast $[(-9.82)(\frac{1}{2}) =]$ -4.91, and the value of the Protestant versus Catholic contrast $[(16.32)(1) =]$ 16.32, but none of the irrelevant value of the Jewish versus Other contrast $[(16.69)(0) = 0]$, thus: $81.90 - 4.91 + 16.32$ to $= 93.31 = \overline{Y}_1 . B_1$ is an ingredient in all the means, since all groups figure in that contrast, whereas B_2 and B_3 each figures in only the two means which it compares.

As was the case for dummy variable and effects coding, it is noted again that the contrast-coded B_i values are not affected by varying sample sizes. Since they are a function of means or unweighted means of means, the expected value of a contrast is invariant over changes in relative group size. This is generally (although not universally) a desirable property of B_i values in nominal coding, and

---

[9] Again, we disavow the causal implication of the term "effect," which is not logically defensible with nonexperimental IVs. The reader is invited to take it as a mathematical metaphor and may prefer to substitute the more neutral word "difference." Causal interpretations are never warranted by statistical results, but require logical and substantive bases. (see Section 9.7.)

TABLE 5.5.3

Partial and Semipartial Correlations and Regression Coefficients for the
Set I Contrast Coding of the Illustrative Data

X_i	pr_i	pr_i^2	sr_i	sr_i^2	β_i	B_i	t_i $(df = 32)$
X_1	−.209	.0437	−.172	.0296	−.1742	−9.82	−1.212
X_2	.494	.2440	.452	.2045	.4593	16.32	3.213**
X_3	.423	.1789	.375	.1404	.3770	16.69	2.639*

$$A = 81.90 = \bar{\bar{Y}}_i$$

$$\hat{Y} = B_1 X_1 + B_2 X_2 + B_3 X_3 + A$$
$$= -9.82\, X_1 + 16.32\, X_2 + 16.69\, X_3 + 81.90.$$

$$\hat{Y}_1 = -9.82(\tfrac{1}{2}) + 16.32(1) + 16.69(0) + 81.90 = 93.31 = \bar{Y}_1.$$
$$\hat{Y}_2 = -9.82(\tfrac{1}{2}) + 16.32(-1) + 16.69(0) + 81.90 = 60.67 = \bar{Y}_2.$$
$$\hat{Y}_3 = -9.82(-\tfrac{1}{2}) + 16.32(0) + 16.69(1) + 81.90 = 103.50 = \bar{Y}_3.$$
$$\hat{Y}_4 = -9.82(-\tfrac{1}{2}) + 16.32(0) + 16.69(-1) + 81.90 = 70.12 = \bar{Y}_4.$$

$*P < .05.$ $**P < .01.$

it is the lack of this property which renders the standardized β_i coefficients of generally little use, as already noted.

The Semipartial and Partial Correlations of a Contrast-Coded Variable

Whereas the B_i values express contrasts in units of Y, the semipartial and partial correlation coefficients and their squares are "pure" or dimensionless numbers. The use of such measures of effect size has the advantage of working with a constant absolute scale from 0 to 1, which then also makes possible the comparison of contrasts across different dependent variables and even different investigations (provided that the p_i are comparable). This may not be an important feature when Y is expressed in some palpable familiar unit, like dollars, or bushels of wheat, or even IQ points. But many dependent variables in behavioral sciences are expressed in arbitrary units of limited familiarity to the investigator, to say nothing of the reader. Measures of correlation and proportion of variance are most illuminating in almost all investigations, even when the unit in which Y is measured is directly comprehensible. However, since all correlations of nominally coded data depend on the distribution of p_i, the nature of the sampling (n fixed or random, see Section 5.3.1) must be taken into account.

With a contrast-coded X_i (where all $g - 1$ IVs have been included), sr_i^2 is the proportion of the *total* Y variance accounted for in the sample by contrast i. Thus, from Table 5.5.3, .0296 of the ATA variance is accounted for by X_1 (literally $X_{1 \cdot 23}$), the majority–minority religion contrast, where the religions *within* categories count *equally,* irrespective of their relative sample sizes (but the relative size of the two *categories* is influential, as always, r being maximal when the latter are equal). The sr_1 of $-.172$ expresses the effect in correlational terms, the negative sign indicating that the positively coded majority religious groups had, on the average, the lower ATA mean. Similarly, $.2045$ $(= sr_2^2)$ of the ATA variance is accounted for by the Protestant–Catholic distinction, or $sr_2 = +.452$, the positive sign in this instance indicating that $\bar{Y}_1 > \bar{Y}_2$. The Jewish–Other contrast gives $sr_3^2 = .1404$, $sr_3 = .375$. Note that the sr_i^2s do *not* sum to $R_{Y \cdot 123}^2$, since, given that the n_is are not equal, the $r_{ij} \neq 0$.

This leaves for consideration the partial correlation for contrast-coded X_i. The pr_i^2 is the proportion of that part of the Y variance *not accounted for by the other contrasts* which is accounted for by contrast i. Thus, $pr_2^2 = .2440$ indicates that contrast 2 (the Protestant–Catholic distinction) accounts for 24.4% of the Y variance remaining after the variance due to contrasts 1 and 3 has been removed. This is necessarily larger than $sr_2^2 = .2045$, since the base of the proportion of the latter is larger, namely *all* the Y variance. (Recall that in all MRC, $pr_i^2 \geqslant sr_i^2$, the equality holding when the other IVs account for no variance.) The choice between sr and pr depends, as always, on what seems to be the more appropriate interpretive framework, the total Y variance or the residual Y variance after the effects of the other variables have been removed. In the absence of other considerations, it seems more reasonable to use the former base when group membership is defined by a naturally varying ("organismic") variable, that is, other sources of variance are always present, and the latter base when it is defined by experimental manipulation, that is, other sources of variance need not be present. Thus, with G defining religion, sr might be the preferred measure. The statistical significance of sr_i and pr_i is the same as that of B_i, as always.

In summary, when contrast codes are written as a function of differences between means or between unweighted means of means, a simple function of B_i yields the value of contrast i in units of Y, sr_i^2 and pr_i^2 express the contrast in terms of proportion of Y variance, and sr_i and pr_i in terms of correlation with Y.

5.5.4   Contrast Set II: A 2 × 2 Factorial Design

We can use our running example of four groups to illustrate the contrast coding which yields the results of a 2 × 2 factorial design in AV. We continue to use exactly the same (artifical) data as throughout, but now reinterpret our four groups as experimental animals subjected to the Treatments D: Drug versus Placebo, and F: Frontal Lesion versus Control Lesion, the dependent variable Y being a measure of retention error (see Table 5.5.1 above). As throughout this chapter, we retain the same unequal ns, in order, 13, 9, 6, and 8. Although such an experiment would ordinarily have been planned for equal ns, we assume for the sake of generality and realism that data have been lost due to animal mortality or other reasons.

With the groups designated as in Table 5.5.1, the first "main effect" contrast is for Drug versus Placebo, a contrast between the mean of means of the ($u =$) 2 Drug groups (G_1 and G_2) and the mean of the means of the ($v =$) 2 Placebo groups. This can be stated in the form of Eq. (5.5.1), and results in the $1/u$, $-1/v$ contrast coefficients (a) as described in the previous section: $\frac{1}{2}, \frac{1}{2}, -\frac{1}{2}, -\frac{1}{2}$. These are the coded values which constitute X_1 for Set II in Table 5.5.1. (This is, coincidentally, the same as the first contrast for Set I.)

Similarly, the second main effect contrast is for Frontal versus Control Lesion. For this contrast, too, $u = v = 2$, but the combination is G_1 and G_3 versus G_2 and G_4, and the a coefficients are, in order, $\frac{1}{2}, -\frac{1}{2}, \frac{1}{2}, -\frac{1}{2}$. These are entered as the coded values for X_2 (Table 5.5.1, Set II). Note that the orthogonality condition for the coded values of X_1 and X_2 is satisfied: $(\frac{1}{2})(\frac{1}{2}) + (\frac{1}{2})(-\frac{1}{2}) + (-\frac{1}{2})(\frac{1}{2}) + (-\frac{1}{2})(-\frac{1}{2}) = \frac{1}{4} - \frac{1}{4} - \frac{1}{4} + \frac{1}{4} = 0$.

We have coded the two main effects in what is now clearly a 2×2 factorial design AV (with unequal, in fact, disproportionate cell n_is). The AV correctly leads us to expect that the remaining single df (of "between cells") carries the Drug $\times$ Lesion interaction. The multiplication sign in the conventional symbols for interactions is neither arbitrary nor accidental. Throughout MRC, whether we are dealing with nominal or quantitative IVs or combinations of these, *interactions are carried by products of variables* (or variable sets).[10] We accordingly code X_3, the interaction contrast, by computing for each group the *product* of its X_1 and X_2 codes: $(\frac{1}{2})(\frac{1}{2}) = \frac{1}{4}$, $(\frac{1}{2})(-\frac{1}{2}) = -\frac{1}{4}$, $(-\frac{1}{2})(\frac{1}{2}) = -\frac{1}{4}$, and $(-\frac{1}{2})(-\frac{1}{2}) = \frac{1}{4}$ (Table 5.5.1, Set II, X_3). Applied to the four means, these are necessarily a coefficients satisfying Eq. (5.5.2) and define a contrast in the sense of Eq. (5.5.1), as do those of X_1 and X_2, the two main effects. Moreover, given that the main effects coefficients are orthogonal, they will necessarily be orthogonal to their interaction products, for example, X_1 with X_3: $(\frac{1}{2})(\frac{1}{4}) + (\frac{1}{2})(-\frac{1}{4}) + (-\frac{1}{2})(-\frac{1}{4}) + (-\frac{1}{2})(\frac{1}{4}) = 0$.

We have, then, as given for Set II in Table 5.5.1, a full ($k_G = g - 1 = 3$) set of coded mutually orthogonal contrasts, representing the two main effects and the interaction of a 2×2 factorial design. Now, assume that we use this coding for the illustrative data of Table 5.3.2, replacing the dummy variable coding of the X_i given there by the Table 5.5.1 Set II coding, for example, G_3 cases are coded $-\frac{1}{2}, \frac{1}{2}, -\frac{1}{4}$ instead of the 0, 0, 1 dummy variable codes. The results of a full MRC analysis using the 2×2 factorial contrasts are given in Tables 5.5.4 and 5.5.5.

Table 5.5.4 shows, exactly as in each previous coding for the same data, that $R^2_{Y \cdot 123} = .3549$ and its $F = 5.869$ (df 3, 32, $P < .01$). Again we reiterate that coding variations effect the results from individual X_i, not those of the *set* of X_i taken as a whole. A little over one-third of the variance in retention error is accounted for by experimental group membership, an amount significantly

---

[10] The implications for data analysis using MRC of this statement are far-reaching. See Chapter 8, which is devoted to interaction.

TABLE 5.5.4

Correlations, Means and Standard Deviations of the Illustrative Data
for 2 × 2 Factorial Design Contrast Coding (Set II, Table 5.5.1)

Contrast		Y	X_1	X_2	X_3	r^2_{Yi}	$t_i\ (df = 34)$
Drug–Placebo	X_1	−.079	1.000	.158	.019	.0062	−.461
Frontal–Control	X_2	.570	.158	1.000	.216	.3249	4.045**
D × F	X_3	.120	.019	.216	1.000	.0144	.705
	m	81.69	.111	.028	.042		
	sd	27.49	.488	.499	.246		

$$R^2 = .354944 \qquad F = 5.869** \ (df = 3, 32)$$

$$\tilde{R}^2 = .294470$$

**P < .01.

different from zero. As in the previous section we note the orthogonality of our contrast *coefficients* (Table 5.5.1) does not result in the 36 coded cases' intercorrelations of the X_i being zero, since the n_is are not equal.

The zero-order correlations of the contrast variables with Y in Table 5.5.4 implicitly define contrasts differently than do the partialled values in Table 5.5.5 below. The former do not distinguish in X_1 between the 13 animals in the Drug–Frontal Lesion group and the 9 in the Drug–Control Lesion group, but simply treat them as a single Drug group of 22 cases coded $\frac{1}{2}$, contrasted with a single Placebo group of 14 cases coded $-\frac{1}{2}$. Any effect due to Frontal–Control, or any interaction effect contaminates the zero-order correlation r_{Yi}, since the cells which are combined are unequal. This holds for all the X_is: the unpartialled r_{Yi} is at least potentially contaminated by the other effects unless all n_is are equal. From Table 5.5.4, we see that, for example, the "gross" effect of the Frontal–Control Lesion contrast is significant, accounting in the sample for .3249 of the retention variance.

Table 5.5.5 assembles the partialled results which, being partialled, are now for uncontaminated and hence unambiguous contrasts. The raw score regression equation is given, and the coded values are explicitly substituted to yield the group means. At the foot of the table, these means are organized in a 2 × 2 table to facilitate relating the B_i and A coefficients to the factorial structure of the contrast set.

As in the results for the contrast coding of Set I, it is the partialling process that gives us the intended contrasts. For the first main effect contrast, Drug versus Placebo, since $u = v = 2$, the right hand term of Eq. (5.5.3) = 1, and $C_1 = B_1\ (1) = -9.82$, that is, the mean of the two Drug group means (76.99) is 9.82 retention error units below the mean of the two Placebo means (86.81), a difference which is not significant ($t = -1.212$). For the second main effect contrast, again $u = v = 2$, so $C_2 = B_2\ (1) = 33.00$—the mean of the Frontal groups' means (98.40) is 33.00 retention error units above the mean of the Control groups means (65.40), a highly significant difference ($t = 4.070$).

TABLE 5.5.5
Partial and Semipartial Correlations and Regression Coefficients for the
2 X 2 Factorial Design Contrast Coding (Set II, Table 5.5.1)

X_i	pr_i	pr_i^2	sr_i	sr_i^2	β_i	B_i	$t_i\,(df=32)$
X_1	−.209	.0437	−.172	.0296	−.1742	−9.82	−1.212
X_2	.584	.3411	.578	.3341	.5994	33.00	4.070**
X_3	−.008	.0001	−.006	.0000	−.0066	−.74	−.045

$$A = 81.90 = \bar{\bar{Y}}$$

$$\hat{Y} = B_1X_1 + B_2X_2 + B_3X_3 + A$$
$$= -9.82\,X_1 + 33.00\,X_2 - .74\,X_3 + 81.90.$$

$$\hat{Y}_1 = -9.82(\tfrac{1}{2}) + 33.00(\tfrac{1}{2}) - .74(\tfrac{1}{4}) + 81.90 = 93.31 = \bar{Y}_1.$$
$$\hat{Y}_2 = -9.82(\tfrac{1}{2}) + 33.00(-\tfrac{1}{2}) - .74(-\tfrac{1}{4}) + 81.90 = 60.67 = \bar{Y}_2.$$
$$\hat{Y}_3 = -9.82(-\tfrac{1}{2}) + 33.00(\tfrac{1}{2}) - .74(-\tfrac{1}{4}) + 81.90 = 103.50 = \bar{Y}_3.$$
$$\hat{Y}_4 = -9.82(-\tfrac{1}{2}) + 33.00(-\tfrac{1}{2}) - .74(\tfrac{1}{4}) + 81.90 = 70.13 = \bar{Y}_4.$$

Means in 2 X 2 Form

	Frontal	Control	Means of means	Difference
Drug	93.31	60.67	76.99	−9.82 $(=B_1)$
Placebo	103.50	70.13	86.81	
Means of means	98.40	65.40	$81.90 = \bar{\bar{Y}} = A$	
Difference	33.00 $(= B_2)$			

Interaction effect $= (\bar{Y}_1 - \bar{Y}_2) - (\bar{Y}_3 - \bar{Y}_4)$
$$= (93.31 - 60.67) - (103.50 - 70.13)$$
$$= 32.64 - 33.38 = -.74\ (= B_3).$$

**$P < .01$.

An interaction contrast for nominally scaled research factors can always be interpreted as the difference between differences of means (or with multifactorial designs, the difference between differences of any contrasts of means (see Chapter 8). When interaction contrasts are coded as products of $1/u$, $-1/v$, 0 codes, their B coefficients give the contrast value directly,[11]

(5.5.4) $$C_I = B_I.$$

In our running example, the value of the interaction $B_3 = -.74$. This is the difference between the difference between Frontal and Control means for the

---

[11] If the $1/u$ and $-1/v$ values of the two main effect contrasts had been multiplied respectively by constants w and x, perhaps to make them simple integer values, the value of the interaction contrast becomes $B_I wx$. (See Footnote 8.)

Drug groups ($93.31 - 60.67 = 32.64$) and the difference between the Frontal and Control means for the Placebo groups ($103.50 - 70.13 = 33.38$); $32.64 - 33.38 = -.74$ (Table 5.5.5). The interaction effect in the sample is negligible, and its t virtually zero. Thus, the data provide no reason to suppose that the Frontal–Control effect operates differentially in the Drug and Placebo conditions. Since interactions are symmetrical, we could say equivalently that the data provide no reason to suppose that the drug effect operates differentially in the Frontal and Control conditions.

The regression equation bares the anatomy of the groups' means. For example, the retention mean of 93.31 for the Drug–Frontal Lesion group (G_1) is composed of a component due to Drug as opposed to Placebo [$(-9.82)(\frac{1}{2}) = -4.91$], another due to Frontal as opposed to Control [$(33.00)(\frac{1}{2}) = 16.50$], another due to the interaction combination of Drug and Frontal [$(-.74)(\frac{1}{4}) = -.18$], and finally another due to factors constant in the experiment, reflected in the mean of all the means (81.90); thus, $-4.91 + 16.50 + -.18 + 81.90 = 93.31$.

No detailed discussion of the other partialled statistics is necessary here, since the principles governing their interpretation are the same as discussed in the previous section of the Set I contrasts. Only four points need be made:

1. We reiterate that the contrasted subsets conceived as abstract qualities (or as means of means) are expressed by the partialled variables, not the unpartialled ones, for example, by $X_{1 \cdot 23}$, not by X_1. $X_{1 \cdot 23}$ is the Drug–Placebo distinction, where Drug is represented by the average of the Drug–Frontal and Drug–Control groups, irrespective of how many of each there are in the sample; similarly for Placebo. X_1, on the other hand, represents Drugs as the overall mean of the combined Drug–Frontal and Drug–Control cases, wherein the former, containing more cases, plays a larger role in this pooled groups mean (and similarly for Placebo). The partialled results in Table 5.5.5 represent correlations and regression coefficients for the partialled variables: literally, for $X_{1 \cdot 23}$, $X_{2 \cdot 13}$, and $X_{3 \cdot 12}$. Thus, the sr^2 for X_2 with Y is $r^2_{Y(2 \cdot 13)}$, which equals .3341 (Table 5.5.5). The analogous squared correlation of the unpartialled X_2 with Y is $r^2_{Y2} = .3249$ (Table 5.5.4), a different quantity. The difference is not great in this particular instance, but may be when n_is are markedly disparate; the two rs may even be of opposite sign. Incidentally, it would be interpretively important to note in this problem that not only are sr^2_1 and sr^2_3 not significant (ts of -1.212 and $-.045$), but that the sr^2_2 of .3341 virtually accounts for all of the $R^2_{Y \cdot 123}$ of .3549—almost all the variance in retention due to the four treatment combinations is accounted for by the Frontal–Control contrast.

2. The AV has made interactions familiar to data analysts, but its neglect of measures of effect size makes the idea of correlations with interaction contrasts an unfamiliar one. Multiple regression/correlation and its representation of sources of variance as independent variables facilitates thinking about interactions as "first-class citizens," fully on an equal footing with "main" effects as sources of variability. If a distinction such as Drug–Placebo can be meaningfully

correlated with Y, and its squared correlation interpreted as a proportion of Y variance associated with this distinction, then it is equally meaningful to interpret correlations or squared correlations of interaction contrasts with Y. Thus, for the sample in this example, it is as meaningful to state that the D $\times$ F interaction correlates with retention $-.006$ (= sr_3) as it is to state that the Drug–Placebo distinction correlates $-.172$ (= sr_1). (See Table 5.5.5–neither is significant.) Provided that one keeps in mind their dependence on the details of the experimental conditions (for example, level of drug dosage, method of administration) and the p_i, such correlations and their squares can be treated as any others generated from data (see Section 1.1.1).

3. The results of the MRC analysis of this 2 $\times$ 2 data structure are not an approximation to those obtained from an *exact* (or "fixed constants") factorial design AV–they are *identical* with those results (when Model II error is used; see Section 4.4.2). They will, of course, not agree with the results of the more popular (and frequently inappropriate) approximate AV procedures for disproportionate n_is such as "unweighted means" or "weighted means". The MRC analysis provides not only an exact solution in regard to statistical significance of effects, but also, as always, the useful additional unit-free information given by various types of correlation coefficients and their squares.

4. As throughout this chapter, the presentation and illustration for contrast coding has been for unequal (in fact, disproportionate) n_is, for the sake of generality. With equal n_i, the results simplify: the X_i become uncorrelated (r_{ij} = 0), and R^2 is the sum (over the $g - 1$ IVs) of the r_{Yi}^2. Also $r_{Yi} = sr_i$ (= β_i); there is no difference between a mean of means and a combined group mean. Another consequence of equal n_i is that *all* correlations with Y (r_{Yi}, sr_i, pr_i, and R) reach their absolute maxima for given means. And, of course, the results are identical with those of the standard factorial design AV.

Summarizing this section, we have shown how main effect and interaction contrasts in a 2 $\times$ 2 factorial design can be coded as IVs, how the resulting regression coefficients can be interpreted as contrast values and appraised for statistical significance, and how to interpret the regression equation and correlations. The difference between the unpartialled and partialled contrast variables was stressed, as was the fact that the MRC gives exact results in the general case of disproportionate sample sizes. (See Sections 8.1 and 8.2, which provide further discussion of the 2 $\times$ 2 design).

5.6 NONSENSE CODING

A challenging name and a contraintuitive idea. We have noted that all the coding methods yield exactly the same R^2, F, and regression equations which solve for $\overline{Y}_i$. We have also repeatedly stated that the $k_G = g - 1$ IVs which are used to carry information of membership in one of g groups of G must be nonredundant, that is, the R of each of them with the remaining $g - 2$ IVs must not equal 1.00. This turns out to be not merely a necessary, but surprisingly also a *sufficient* condition for the set of $g - 1$ IVs to represent group membership. To state this more plainly, *any* nonredundant set of $g - 1$ IVs will, for the same data, give the same R^2, F, and a regression equation which solves for the Y_i.

Concretely, consider the coding of X_1 in Set I of Table 5.6.1. The coded values of 5, 0, −4, and 6 were obtained by entering a table of random numbers at random, and reading off the first four different digits. A coin was then flipped to decide the sign. For X_2 of Set I, the X_1 values are squared, and for X_3 they are cubed. (If there were more groups, we would simply continue powering: in general, $X_i = X_1^i$, so that one would have $g - 1$ IVs, the last of which would be $X_{g-1} = X_1^{g-1}$.)

Now, when these patently nonsensical "scores" for religion or experimental group (or whatever) are used to code group membership in place of the dummy variable coding in our running example of data in Table 5.3.2, and an MRC performed, the results are rather startling: $R^2_{Y \cdot 123} = .3549$, and its $F = 5.869$, exactly as with the various sensible forms of coding previously employed with these data. Moreover, the raw-score regression equation that results,

$$\hat{Y} = 10.6863\, X_1 + 2.6019\, X_2 - .6867\, X_3 + 60.67,$$

will correctly yield the group Y means when the "scores" for each group are substituted. The rest of the results are gibberish.

This trick will, of course, work with X_1 as any four different real numbers, positive or negative, integral or fractional. The squaring and cubing is meaningless, except that it assures the satisfaction of the condition of (linear) nonredundancy among the X_i—it is not mathematically possible to *perfectly* estimate by *linear* equations a power function of a set of numbers by other power functions of these numbers. The r_{ij} among data coded by powered variables can be quite high and their multiple R_i even higher, but not as high as 1.00. (For these data, for example, $r_{13} = .96$, and $R_{1 \cdot 23} = .97$.)

Indeed, although convenient, it is not at all necessary to proceed by powering. By whatever means it is achieved, the only requirement is that of linear nonredundancy. Sets II and III of Table 5.6.1 were constructed by a haphazard assignment of any old values with an eyeball check for nonredundancy. They yield the same $R^2_{Y \cdot 123} = .3549$ and $F = 5.869$, and regression equations which produce the group means.

A proof or even detailed explanation of this peculiar phenomenon would take us into matrix algebra and n-dimensional geometry. Perhaps it can be understood by extension from correlation with a dichotomy, such as sex. The r of any variable with sex (X) traditionally scores X as 0 for one sex and 1 for the other, but it actually makes no difference in the resulting r which two values are used (Section 2.3.3): recall from Chapter 2 that X is standardized in the process of determining r (as is the other variable). Whether X is rendered as 0, 1 or −261, −12 or −.46, 72, the same two values will result for z_X. Another way to see this

TABLE 5.6.1

Some Examples of Nonsense Coding for $g = 4$ Groups

	Set I			Set II			Set III		
	X_1	X_2	X_3	X_1	X_2	X_3	X_1	X_2	X_3
G_1	5	25	125	1	-7	0	0	0	.71
G_2	0	0	0	-1	-1	0	0	0	.04
G_3	-4	16	-64	-4	$\frac{1}{2}$	24	1	4	1
G_4	6	36	216	1	6	-1	2	108	2

is that however we scale X, the Y on X best-fitting regression line will have the same amount of "fuzz," or "errors of estimate" of Y values about it, and the proportion of this "fuzz" to total Y variance, which is $1 - r^2$, will remain the same; hence, r^2 will remain the same. By extension, in multiple correlation of a set of $g - 1$ coded values X_i, however these X_i values are scaled, the best-fitting regression "surface" (a hyperplane of $g - 1$ dimensions) will have the same amount of Y dispersion about it, hence the same $1 - R^2$, hence the same R^2. The F ratio also remains invariant since it depends only on R^2, n, and g, which are all invariant.

The point is that despite the nonsensical character of the coding, any full set of $g - 1$ codes carries complete information about group membership, and any result which depends upon the set as a whole will yield correct and meaningful results: R^2, $\widetilde{R}^2$, F, and the group means from the regression equation. On the other hand, results from single variables will be as nonsensical as each variable is when taken by itself. In other words, one's intuition about the meaninglessness of nonsense-coded data is correct about the discrete results from any one of them—there is no sense, in general, to be had from an r_{Yi}, pr_i, sr_i, β_i, B_i, nor is A meaningful, in general.

Other than as a bit of statistical curiosa, one might well ask: Why settle for nonsense coding which only gives R^2 and F (and the $\overline{Y}_i$, which can certainly more easily be found than by performing an MRC analysis) when other coding alternatives additionally yield a substantial harvest of interpretable material? The answer is obvious that in general one should not, but there are circumstances in which there are practical compensations in data handling for this loss. Imagine a set of interview data from a survey (for example, market research, public opinion poll) in which the responses to some question are to be studied in regard to their relationship to one or more dependent variables by means of MRC. Now assume that these responses have been "coded," not in the sense in which we have been using the term, but as it is used in survey research, that is, each response has been classified as falling into one of a set of mutually exclusive and exhaustive categories that is, a nominal scale. These are numbered, ordinarily in no particular order, and although frequently with consecutive digits, some researchers use a convention of assigning the digit 9 to "not applicable" and/or 0 to "did not respond," or some other, so that the g response categories are not necessarily coded from 1 to g. No matter. In any case, such a coding scheme partitions the total group of interview respondents into a set of g categories, to each of which a unique number has been assigned. These numbers may be written on the form or already punched on cards. Now if one takes these numbers to represent a nonsense coded X_1 variable, one can use a transgeneration computer routine (see Appendix 3) to square them for X_2, cube them for X_3, etc. until they have been raised to the $g - 1$ power for X_{g-1}.[12] The set of X_i now carry the information as to response-group membership for the MRC, and this has been accomplished by standard computer routines. On the other hand, to use other than nonsense coding might require much clerical effort or special ad hoc computer programming. Although much analytic yield may be lost by nonsense coding, it must be pointed out that a typical data analyst following typical textbooks and doing an AV of the Y variable in g response groups would obtain no more than is yielded by the MRC with nonsense coding: R^2, F, and the means. Indeed, it is a fair guess that he would obtain less in that he would not determine R^2, the proportion of Y variance accounted for by response-group membership [although it is readily available to him in the equivalent η^2 (Eq. 5.3.12)]. The compleat data analyst should therefore find room for nonsense coding in his tool kit.

5.7 GENERAL IMPORTANCE

The detailed examination of nominal scale coding of this chapter must not be allowed to obscure what is probably its most important feature—the ease with which a nominal scale G represented by a set of $g - 1 = k_G$ IVs can be combined with other research factors. To the G scale discussed in this chapter one can readily add other variables: for example, another nominal scale H (expressed as a set of $h - 1 = k_H$ IVs) and/or a quantitative scale V with k_v IVs either linearly or with provision for curvilinearity (see Chapter 6), with or without provision for missing data (see Chapter 7). Further, interactions of G with other such sets ($G \times H$, $G \times V$, $G \times H \times V$) are readily represented and analyzed (see Chapter 8). This possibility, in turn, when combined with the hierarchical model, makes it relatively simple to analyze covariance-type designs which would be of a complexity that would defy analysis by the conventional ACV (Chapter 9). Thus, when the technique of nominal scale coding and analysis by sets of IVs (Chapter 4) is incorporated in the analytically rich MRC analysis, new vistas in

---

[12] If g is so large as to result in values which exceed the maximum digit length provided for by the MRC program, right hand digits can be dropped, or a linear transformation applied to scale the data down, again by means of the transgeneration program. Also when g is large (say, greater than 5), it would probably be wise to use double- or extended-precision computing (see Section 6.2.4, Appendix 3, and Wampler, 1970).

data analysis open up. Herein lies the fundamental importance of nominal scale coding.

5.8 SUMMARY

A nominal scale (or qualitative variable) G that partitions observations into g groups can be represented as $g - 1$ independent variables in MRC by various forms of coding. All these coding forms yield identical R_Y^2, $\widetilde{R}_Y^2$, F for R_Y^2, and via the regression equations, the group means on Y, results which are identical with those of analysis of variance. The MRC analysis, in its results for individual X_i, provides additional useful interpretive material: zero-order correlations of these variables with Y, partial and semipartial correlations (and their squares as proportions of variance), regression coefficients, and the status of these in regard to statistical significance. The various alternative coding methods facilitate the interpretation of the results for single X_i for different frames of reference or desired comparisons:

1. *Dummy-variable coding.* One group is selected as a reference group, and the partial coefficients compare each of the remaining groups with it. This is particularly appropriate for research in which one group is to be a control group and the others are to be compared with it. (Section 5.3)

2. *Effects coding.* The reference point here is the aggregate of all the groups, their means being equally weighted. The term "effect" is taken from the analysis of variance, where it is the difference between a group's mean and the (unweighted) mean of all the groups; the raw score regression coefficients yield these values directly. (Section 5.4)

3. *Contrast coding.* The $g - 1$ variables are constructed so as to provide a set of independent comparisons between means or (unweighted) means of means, selected in accordance with the issues of the research. Not only main effects, but comparisons of groups within sets and interactions can be written in contrast form, and, in addition to correlational information, the MRC analysis provides values for the contrasts and tests of their statistical significance. (Section 5.5)

4. *Nonsense coding.* Here the coding values are arbitrary, and individual results are meaningless, although not those for the set as a whole. Its utility lies in its convenience in certain circumstances arising in data analysis. (Section 5.6)

Finally, it was stressed that the fundamental importance of nominal scale coding lies in the possibilities of incorporating as IVs in MRC nominal scales together with quantitative data, variables with missing data, and interactions of these so as to readily represent the issues under study, however complex, far beyond what is possible by other data-analytic means. (Section 5.7)

6

Quantitative Scales

6.1 INTRODUCTION

In Stevens' (1951) familiar classification of measurement scales, above nominal scales, the hierarchy goes: ordinal, interval, and ratio. We treat these scale types in this chapter under the collective rubric "quantitative," and note that they share the minimum property that the scale numbers assigned to the objects order them with regard to the measured attribute. Ordinal (rank order) scales have only this property, interval scales add the property of equal units (intervals), and ratio scales have both equal intervals and equal ratios, hence a true scale zero (Section 1.1.2). Despite these differences in the amount of information they yield, we find it convenient to treat them together. This chapter presents several alternative methods of representing (coding) quantitative scales, and although the choice among these methods is not completely unrelated to the level of scaling, the relationship is neither strong nor simple: most methods may be used for all three types of quantitative scales, some are more appropriate or useful for one level of scaling than another. We will not focus on these relationships in the course of presenting the alternative methods (Sections 6.2–6.5), but view them from the perspective of scale type afterwards (Section 6.6), and conclude with a summary (Section 6.7).

The introduction to the preceding chapter (Section 5.1) presented the notion of a research factor having different aspects, each represented by an IV. Chapter 5 presented several ways in which this idea was employed for nominal data, for example, one aspect of Religion is majority versus minority religious group membership (Section 5.5), another (from another coding method) is Protestant versus Other (Section 5.3), etc. One does not spontaneously think of aspects of quantitative scales, which have been traditionally employed in behavioral science "as is." Thus, research factors like age, IQ, MMPI Sc score, median annual income (for census tracts), luminance, trials to criterion, symptom ratings, percentile rank in high school graduating class, and birth order have been used to represent their respective constructs in correlation and regression analysis. We would argue that each such variable presents only *one* aspect of the construct, its *linear* aspect, and in some research problems there may be other relevant aspects which should be used in addition to, or in place of, this linear aspect. This chapter thus addresses itself to the question of how *curvilinear* relationships can be detected, represented, and studied within the confines of *linear* MRC by the use of multiple and/or nonlinear aspects of research factors. In keeping with the preceding chapter, these methods can be viewed as alternative means of *coding* quantitative research factors.

To effect the necessary distinction between a research factor construct and its representation as IVs, we will use a capital letter to designate the construct as represented by a set of k IVs, where k is one or more, and designate each aspect (as usual) as X_i (i = 1 to k). Because linear aspects are pivotal, we will use the lower case letter to represent the linear aspect of the construct, or the construct as measured. Thus, V is age as a research factor, made up of k variables. Those might be, for example, $X_1 = v$, $X_2 = v^2$, $X_3 = v^3$, hence, $k = 3$ (Section 6.2); or, V might simply contain $X_1 = \log v$, hence $k = 1$ (Section 6.5). In this notation, traditional MRC has been largely concerned with quantitative variables, where each V has been constrained to its linear aspect v.

6.2 POWER POLYNOMIALS

6.2.1 Method

It is a most useful mathematical fact that in a graph of q data points relating y to v (where the values of v are all different), an equation of the following form will define a function which fits these points *exactly:*

(6.2.1)
$$y = A + Bv + Cv^2 + Dv^3 + \cdots + Qv^{q-1},$$

$A, B, \ldots, Q$ being constants.

Now this is an equation which relates *one* variable to y, and since that variable v is raised to successive integer powers beyond the first, the equation is clearly *non*linear. Now, consider the standard regression equation

(3.5.1)
$$\hat{Y} = B_1 X_1 + B_2 X_2 + B_3 X_3 + \cdots + B_k X_k + A.$$

This equation relates k variables (X_1 to X_k) to Y, and since the X_i are all to the first power, and there are no product terms such as $X_i X_j$ this is a *multiple linear* equation. Now, since there are no practical restraints on how we define our X_i, we use the trick of letting $X_1 = v$, $X_2 = v^2$, $X_3 = v^3$, $\ldots$, $X_k = v^{q-1}$, and now what had been a nonlinear equation for one independent variable has become a linear equation for k independent variables, that is, a multiple *linear* regression equation of the appropriate form for our system.

This bit of legerdemain is not a mere verbal trick. In order to accomplish *linearity*, we have undertaken *multiplicity* of the X_i, but the price is cheap. By bringing nonlinear relationships into the fold in this way, we make it possible to determine *whether* and specifically *how* a relationship is nonlinear and to write an equation which *describes* this relationship.

Now let us back up and consider the number of data points we wish to fit. For such a V as IQ, there may be more than 50 different values of V; hence, the complete polynomial (6.2.1) would have over 50 terms. We would be reluctant to spend that many *df* (one for each X_i) to relate Y to V, and, fortunately, it is neither necessary nor desirable. Indeed, we do not seek to fit each squiggle of the curve due to random sampling or measurement error; we wish to fit a smooth curve with as few X_i as are needed to approximate the true function. For most behavioral science data, certainly for the data of the "soft" behavioral sciences, the first two or three powers will suffice: in the terminology introduced in the preceding section, we can represent in polynomial form most quantitative variables where nonlinearity is to be allowed for by a set V made up of

(6.2.2) $X_1 = v$ (linear), $X_2 = v^2$ (quadratic), $X_3 = v^3$ (cubic),

a total of $k = 3$ aspects of V. Very often it will be sufficient to use only the first two terms; rarely, one might wish to include more than three.

Using the X_i aspects representing V, one analyzes the data by MRC using the *hierarchical model* (Section 3.7.4) for significance testing, and the *simultaneous* model (Section 3.6.1) for plotting the curve. The process is made clear by two concrete examples.

6.2.2 An Example: A Quadratic Fit

Figure 6.2.1 shows a bivariate plot of variables Y and v for $n = 36$ cases. Ignore for the moment the line and curve—simply consider the gestalt of the 36 points. They suggest a curve, possibly due to an asymptote or "ceiling effect" in Y—increasing v beyond approximately 100 does not seem to be associated with further increases in Y. When we compute r_{Yv}, we find it to equal .767, with $t = 6.970$ ($P < .01$). The linear correlation is high, but the fact that it is significant means only that the linear correlation in the population is very likely not zero, and *not* that the relationship in the population is necessarily adequately described by a straight line. Thus, the best-fitting straight line, or linear aspect of V makes it possible to account for .5883 (= r^2_{Yv}) of the Y variance. The resulting linear regression equation turns out to be

$$\hat{Y}_1 = .8398\, v + 14.49.$$

This line is easily plotted by substituting one high and one low value of v, finding their respective $\hat{Y}$s from this equation, plotting the two v, $\hat{Y}$ points and connecting them with a straight line (for $v = 40$, $\hat{Y}_1 = 48.1$, and for $v = 140$, $\hat{Y}_1$

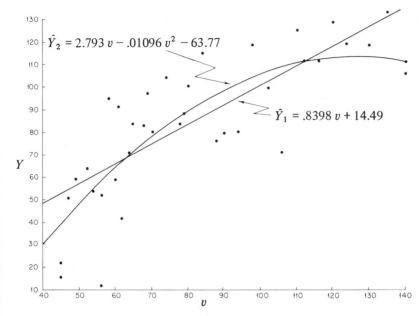

FIGURE 6.2.1 Polynomial regression of Y on V.

= 132.1); see Figure 6.2.1. Despite the large r, there are quite a few unbalanced points in the middle of the v scale above the line and several more points with low v below it.

Can matters be improved by taking into account the quadratic aspect of V? This can be determined by means of the hierarchical model: with $X_2 = v^2$ added to $X_1 = v$, a second MRC analysis is performed, and the increase in $R^2_{Y\cdot12}$ over $R^2_{Y\cdot1}$ $(= r^2_{Y1})$ is noted. This is literally sr^2_2 (the squared semipartial correlation for X_2 partialling X_1) but, with the hierarchical model it is convenient to use the general symbol I with subscript to denote an *increment* in R^2 due to the addition of one or more variables. This I_2 value is determined and tested for significance to determine whether, in the population, the inclusion of this term increases R^2.

In the upper section of Table 6.2.1, both the ingredients and results of this operation are given. First, however, note that these are most certainly not orthogonal variables: v and v^2 are very highly correlated with each other (.990), and also with v^3 (.965 and .992). This is a feature of polynomial regression using powers of v: when, as is usually the case, v is a typical variable made up of positive values, the powers of v used as polynomial terms are always highly correlated. It would be a mistake to conclude from $r_{12} = .990$ that X_1 and X_2 are functionally equivalent. To be sure, $.990^2 = .9801$ of the variance in X_2 is linearly accounted for by X_1, and only $1 - .9801 = .0199$ of the X_2 variance is

TABLE 6.2.1
Polynomial MRC Analysis of Regression of Y on V

		r					Hierarchical model				
	Y	X_1	X_2	X_3	IVs	$cum\ R^2$	F	df	I	F_I	d
$(v)\ X_1$	.767	1.000	.990	.965	X_1	.58828	48.580**	1, 34	.58828	48.580**	1,
$(v^2)\ X_2$	.725	.990	1.000	.992	X_1, X_2	.65076	30.745**	2, 33	.06248	5.904*	1,
$(v^3)\ X_3$	.675	.965	.992	1.000	X_1, X_2, X_3	.65079	19.878**	3, 32	.00003	.003	1,

m 83.09   81.69   7422.   741389.

$$n = 36$$

sd 30.10   27.49   4948.   718223.

Simultaneous model
Linear equation:

$$\hat{Y}_1 = .8398\ X_1 + 14.49$$

t_{B_1} 6.970** $df = 34$

Quadratic equation:

$$\hat{Y}_2 = 2.793\ X_1 - .01096\ X_2 - 63.77$$

t_{B_i} (3.441**) $-2.430*$ $df = 33$

Cubic equation:

$$\hat{Y}_3 = 2.596\ X_1 - .008654\ X_2 - .000008476\ X_3 - 58.48$$

t_{B_i} (.650) $(-.188)$ $-.050$ $df = 32$

*P < .05.   **P < .01.

not, but what is operating in the hierarchical model when X_2 is brought in after X_1 is this latter *un*correlated variance; in other words (and as always), it is the *partialled* $X_{2 \cdot 1}$, that is, v^2 from which v is (linearly) removed, that represents the pure quadratic variable. It may be only 2% of X_2, but it is 100% of $X_{2 \cdot 1}$.

Table 6.2.1 shows the process. First, X_1 by itself accounts for .58828 of the Y variance, as we have seen. In conformity with the plan, it appears in the *cum* R^2 column (cumulating from zero variance accounted for by no prior IVs) and is significant, as we also have seen, the $F = 48.580$ simply being the square of the already noted $t = 6.970$. When X_2 is added to X_1, the *cum* R^2 (now $R^2_{Y \cdot 12}$) is .65076, which, when tested by the standard test for the significance of an R^2 (Eq. 3.7.1), gives $F = 30.745$ $(P < .01)$. This merely means that in the population, the use of X_1 and X_2 will result in a nonzero R^2. Our interest rather focuses on the increment in R^2 due to the addition of X_2, that is, $I_2 = R^2_{Y \cdot 12} - R^2_{Y \cdot 1} = .65076 - .58828 = .06248$. Thus, an additional 6.25% of Y variance has been accounted for by the quadratic term $X_{2 \cdot 1}$. To test whether this increment is significant, we specialize the test for the significance of the increase in R^2 of an added set B to that of a set A, using Model I error, already given:

(4.4.2) $F = \dfrac{R^2_{Y \cdot AB} - R^2_{Y \cdot A}}{1 - R^2_{Y \cdot AB}} \times \dfrac{n - k_A - k_B - 1}{k_B}$ $(df = k_B, n - k_A - k_B - 1)$.

In this general formula, k_A is the number of IVs in set A and k_B is the number of IVs in set B. We specialize it for the hierarchical polynomial by noting that $R^2_{Y \cdot AB} - R^2_{Y \cdot A} = I_B = I_i$ for some specific X_i added term, $R^2_{Y \cdot AB}$ is the cumulative $R^2_{Y \cdot 12 \dots i}$ up through that term, and $k_B = 1$. Thus, specializing Eq. (4.4.2), the significance of the increment due to a single added polynomial term X_i is

(6.2.3)[1] $F = \dfrac{I_i(n - i - 1)}{1 - R^2_{Y \cdot 12 \dots i}}$ $(df = 1, n - i - 1)$.

Since, as has been pointed out, I_i is sr_i^2 with all lower-order terms partialled, this F is simply the t^2 for the partialled coefficients for X_i. Most computer programs provide t (or F) for each B, including that for the highest order, B_i, making hand computation of Eq. (6.2.3) unnecessary. Applying Eq. (6.2.3) to the increment, we find

$$F = \frac{.06248 \, (36 - 2 - 1)}{1 - .65076} = 5.904 \qquad (df = 1, 33),$$

which is significant $(P < .05)$. Thus, F indicates that the population R^2 using v and v^2 is larger than the one using v alone; or that the quadratic aspect of V accounts for some Y variance (over and above the linear aspect). This in turn means that the relationship in the population is not straight line, but has (at least) one bend in it. Equivalently, we note that $B_2 = -.01096$ and its $t = 2.430$ $(= \sqrt{5.904}), P < .05$, as before.[2]

Now, using the Bs and A from the computer output, we can write the quadratic regression equation (Table 6.2.1)

$$\hat{Y}_2 = 2.793 \, X_1 - .01096 \, X_2 - 63.77.$$

---

[1] Since this test is for error Model I (see Section 4.4), it is negatively biased (conservative) when terms of *higher* order than X_i are making real contributions to the population R^2. But when the latter is not the case, this test is statistically more powerful than other tests (Section 4.5). See discussion of alternative error models for polynomials in Section 6.3.2. Also, note that this equation is equivalent to that of the test for sr_i^2 (Eq. 3.7.6).

[2] Do not be surprised that although B_2 is so small, it is nevertheless significant. Remember that $X_2 = v^2$, a four or five digit number: even so small a weight results in a large component in $\hat{Y}$. For example, when $v = 100$, $X_2 = v^2 = 10,000$, and $B_2 X_2 = -.01096$ $(10,000) = -109.6$, a sizable contribution to Y. Reference to the standardized β coefficients will help avoid being misled: for this problem, for example, $\beta_2 = -1.80$ and $\beta_1 = 2.55$.

This equation gives the quadratic *curve* which best fits the 36 points in Figure 6.2.1. If a plot is desired (it need not be necessary—see Section 6.2.4), it can be accomplished by substituting the v values 40, 50, . . . , 140 into this equation with v for X_1 and v^2 for X_2, solving each time for $\hat{Y}_2$, and drawing a smooth curve through the resulting v, $\hat{Y}_2$ points. For example, at $v = 40$, $\hat{Y}_2 = 2.793$ $(40) - .01096 (40^2) - 63.77 = 30.4$; at $v = 50$, $\hat{Y}_2 = 2.793 (50) - .01096 (50^2) - 63.77 = 48.5$; . . . ; at $v = 140$, $\hat{Y}_2 = 112.4$. This curve is plotted in Figure 6.2.1 and permits comparison with the fit provided by the straight line of the linear equation. It follows the track of the points which invites the eye, clearly doing a better job at low and middle values of v than the straight line.

Note that in order to write the regression equation we have shifted from the hierarchical to the simultaneous model of MRC. We do not take the B_1 from the linear equation (.8398) with the B_2 from the quadratic equation ($-.01096$) to produce the best-fitting equation. If the B_1 of 2.793 of the quadratic equation is the best weight for X_1, as guaranteed by the mathematics of MRC, then the B_1 of .8398 cannot be. On the other hand, the t test of B_1 in the quadratic equation ($t = 3.441$) is, at most, of academic interest; it is certainly *not* a test of whether there is a significant linear component in the regression. That was obtained in testing the $B_1 = .8398$ of the linear equation ($t = 6.970$), or, equivalently, in testing the $I_1 = .5883$ ($F = 48.580 = 6.970^2 = t^2$). Stated generally, the statistical significance of a polynomial term is that of the increment in R^2 it produces *upon entry into the equation* in the hierarchical model, or, equivalently, that of its *B at the point of entry*, not its B after higher order terms have been entered, and hence partialled from it.

What happens when the cubic term, $X_3 = v^3$ is introduced into the equation? The increase in R^2 is trivial ($I_3 = .00003$) and, applying Eq. (6.2.3), nonsignificant ($F = .003$); equivalently, the $B_{3.12} = .000008476$ has a nonsignificant $t = .050$ ($= \sqrt{F}$). Table 6.2.1 gives the former in the X_3 row at the top of the table, and the latter in the cubic equation at the bottom. Again the two null hypotheses are mathematically equivalent and $F = t^2$ (within rounding discrepancy). The results of this test indicate quite emphatically that the cubic aspect of V bears no material relationship to Y. Some features of the rejected cubic polynomial deserve comment. It is rejected because it does not provide a significantly better fit than the quadratic, *not* because its fit in absolute terms is not good: its $R^2_{Y.123}$ of .65079 is highly significant with $F = 19.878$ for $df = 3, 32$. Were we to plot the curve of the cubic equation in Figure 6.2.1, it would be virtually indistinguishable from that of the quadratic. This is because B_3 of the cubic equation is small (more relevantly, β_3, which is standardized for the very large sd_3, is small—only $-.20$) and its B_1, B_2, and A do not differ greatly from those of the quadratic.

The cubic equation dramatizes the already noted irrelevance of the significance test for lower order B_j in the polynomial of ith order: the ts for B_1 and B_2 become very small (.650 and $-.188$) even though both the linear and quadratic terms were each found to be significant upon entry into the equation cumulatively. This is due to the partialling from them of the highly linearly correlated X_3, as well as each from the other, in the cubic equation. In more explicit notation, the coefficients are $B_{1 \cdot 23}$, $B_{2 \cdot 13}$, and $B_{3 \cdot 12}$. Only the significance test of the last is meaningful. The partialling process of these highly correlated IVs is so debilitating to them individually (see Section 3.7.5) that none of them yields a t as large as one, yet collectively they yield an R^2 of .65 (so $R = .81$) with F of 19.878 $(P < .001)$!

We could continue the process by adding higher polynomial terms, but they are rarely useful with behavioral science data. (For these data, $I_4 = .039$ and $I_5 = .009$, neither significant nor large; we discuss in Section 6.2.4 the question of how high it pays to go.)

In summary, we have found that the regression of Y on V is curvilinear (although a straight line provides some fit), and more particularly, that it is quadratic. In other words, both the linear and quadratic aspects of V are related to Y, but not the cubic aspect.

6.2.3 Another Example: A Cubic Fit

We offer another example of polynomial regression to demonstrate its operation for a more complex relationship, and to further exemplify the general method. Consider the plot of points in Figure 6.2.2, relating Y to w. W is of a different

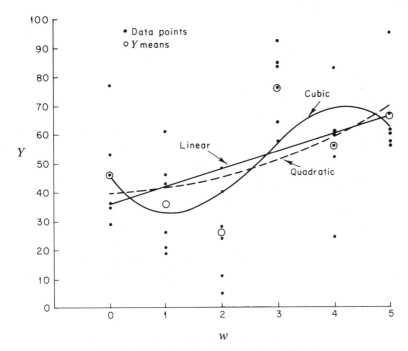

FIGURE 6.2.2 Polynomial regression of Y on W.

nature than V: it has only 6 values, they are equally spaced, and for each of these values there is the same number of points. These features suggest that W is a variable produced by experimental manipulation (e.g., number of exposures to a stimulus, or drug dosage level), and that the data structure is the product of a laboratory experiment rather than a field study. These features are important to the substantive interpretation of the results but not relevant to the present analysis; we would proceed as we do here with data at unequal intervals and/or unequal ns at each level of W. (But see Section 6.3.2.)

The correlation of Y with w (the linear aspect of W) is found to be .452, so that 20.4% of the Y variance can be accounted for by w, a highly significant amount ($F = 8.713**$, $df = 1$, 34; see Table 6.2.2). Again we caution the reader that this only means that in the population (as well as in the sample) a straight line accounts for some variance and not necessarily that it provides an optimal fit. The equation for this line is given in the middle of the table and we note that t for B_1 is 2.952** (redundantly, since it is simply $\sqrt{F} = \sqrt{8.713}$). The linear

TABLE 6.2.2
Power Polynomial MRC Analysis of Regression of Y on W

	Y	X_1	X_2	X_3	IVs	cum.R^2	F	df	I	F_I	df
			r				Hierarchical model				
$(w)\ X_1$	.452	1.000	.960	.906	X_1	.20399	8.713**	1, 34	.20399	8.713**	1, 3
$(w^2)\ X_2$	.466	.960	1.000	.986	X_1, X_2	.21787	4.596*	2, 33	.01388	.586	1, 3
$(w^3)\ X_3$	.435	.906	.986	1.000	X_1, X_2, X_3	.32148	5.054**	3, 32	.10361	4.887*	1, 3
m	51.00	2.50	9.17	37.50							
sd	22.69	1.71	8.90	44.87							

$$n = 36$$

Simultaneous model

Linear equation:

$$\hat{Y}_1 = 5.993\, X_1 + 36.02$$

t_{B_1} 2.952** $df = 34$

Quadratic equation:

$$\hat{Y}_2 = .6428\, X_1 + 1.071\, X_2 + 39.57$$

t_{B_i} (.310)   .765 $df = 33$

Cubic equation:

$$\hat{Y}_3 = -29.80\, X_1 + 17.74\, X_2 - 2.222\, X_3 + 46.24$$

t_{B_i} (−1.935)   (2.317*)   −2.211* $df = 32$

*$P < .05$.   **$P < .01$.

equation is plotted in Figure 6.2.2. Although the straight line accounts for a statistically significant and by no means trivial amount of Y variance, the fit is not impressive.

When the quadratic term $X_2 = w^2$ is added into the equation, R^2 increases by only .01388 $(= I_2)$, which yields by Eq. (6.2.3) $F = .586$ (equivalently, t for B_2 $= \sqrt{F} = .765$), a nonsignificant result. The data provide no warrant for the relevance of the quadratic aspect of W to Y. Note that it is not curvilinearity that is being rejected, but quadratic curvilinearity, that is, a tendency for a parabolic arc to be at least partially descriptive of the regression; a higher order, necessarily curvilinear, aspect of W may be significant. To show what this small (1.4%) and nonsignificant quadratic addition means graphically, the quadratic equation (Table 6.2.2) is plotted with a dashed line in Figure 6.2.2. The line is curved, but only slightly so since B_2 is small, and lies close to the straight line. Although the fit is better (it can never be worse), it is only slightly and nonsignificantly so.

The addition of $X_3 = w^3$ to the equation, in contrast, does make a substantial difference. It increases R^2 to .32148 $(F = 5.054^{**}, df = 3, 32)$ an increment beyond the quadratic of $I_3 = .10361$, which yields by Eq. (6.2.3) an $F = 4.887^*$ for $df = 1, 32$ (or equivalently, $t = -2.211^*, df = 32$). Table 6.2.2 also gives the cubic equation, which is also plotted in Figure 6.2.2. The cubic always gives a two-bended function (although the bends need not appear within the part of the range of the independent variable under study—see Section 6.2.4), and the net fit is visibly improved, particularly at $w = 0, 2$, and 3. We conclude that this regression is curvilinear and, more particularly, that it is cubic. By this we mean that the cubic aspect of W relates to Y, and also that the best-fitting equation utilizing w, w^2, and w^3 will account for more Y variance in the population than one which has only w and w^2, or (necessarily) only w. Note, however, that although I_3 is both substantial and significant, and the cubic curve has both a maximum and a minimum as does the trend of Y means for the values of w, the maximum and minimum of the curve do not coincide with those of the trend. This is discussed in Section 6.2.4.

The population cubic equation is estimated, without bias, by the cubic equation produced for our sample. This best-fitting cubic equation provides a particularly good example of a trap for the unwary who try to interpret its lower-order coefficients, B_1 and B_2. In full notation, these are $B_{1.23}$ and $B_{2.13}$, and there's the rub. We have already stressed the fact that the significance tests of these do not provide proper bases for decisions about the shape of the function: for example, the nonsignificant t of $B_{1.23}$ (-1.935) and the significant t of $B_{2.13}$ (2.317) are easily misinterpreted to yield exactly opposite conclusions from the valid ones we have drawn above. Moreover, even the signs of these coefficients may mislead; $B_{1.23}$ is -29.80, even though we know that the slope of the best-fitting line is positive, since $B_1 = +5.993$ $(r_{Y1} = +.452)$. The mutual partialling of higher-order by lower-order polynomial terms, given their high correlations, results in extreme suppression effects which render the magnitude, sign, and statistical significance of lower order regression coefficients in the simultaneous model analytically useless for practical purposes.

6.2.4 Interpretation, Strategy, and Limitations

Interpretation of Polynomial Regression Results

The polynomial in v of Eq. (6.2.1) defines a function which contains powers of v up to (let us say) the kth. Those of our readers with knowledge of the differential calculus will note that when the first derivative of this equation is set equal to zero and solved for v, the results will be $k - 1$ values of v at which Y is either a maximum or minimum (or point of inflection) in that region. This means that a kth order polynomial is an equation which gives rise to a curve with $k - 1$ bends. It follows that if the data-points we seek to fit have $u (= k - 1)$ real (nonchance) bends, they will not be well fitted by a polynomial of lower order than $u + 1 (= k)$.[3] Thus a straight line is defined by an equation of order one and has $1 - 1 = 0$ bends, a quadratic has $2 - 1 = 1$ bend, a cubic $3 - 1 = 2$ bends, etc.

The differential calculus makes it possible to determine where v achieves a maximum or minimum Y value in the equation. For the quadratic equation, this point is given as a function of its constants as

(6.2.4)
$$v_M = \frac{-B_{1 \cdot 2}}{2B_{2 \cdot 1}},$$

which comes at

(6.2.5)
$$\hat{Y}_M = \frac{4B_{2 \cdot 1}A - B_{1 \cdot 2}^2}{4B_{2 \cdot 1}}.$$

(We employ the full subscript notation to avoid confusion.) For the quadratic example (Section 6.2.2), these values are $v_M = 127$, where $\hat{Y}_M = 114$, consistent with the plot in Figure 6.2.1. Note that v_M happens to fall within the range of v of the data in this example.

Just as the sign of B_1 of the linear equation indicates whether the line generally rises (+) or falls (−), the sign of $B_{2 \cdot 1}$ in the quadratic equation indicates whether the quadratic curve is concave upward (+) or concave downward (−). If the former, v_M gives a minimum, if the latter, a maximum.

---

[3] Although for the typical data of behavioral science, a polynomial of the kth order will be sufficient as well as necessary, if the data points being fitted are relatively free of sampling and measurement error, a polynomial of order higher than k may materially and significantly improve the fit. This is because the additional bends it provides will occur outside the range of v we are studying (mathematically, v stretches from minus to plus infinity), while within that range the proper number $(k - 1)$ of bends are provided for, and the function fits the points better than that of the kth order equation.

Similarly, we can find the minimum and maximum of w_M for the cubic equation by substituting its constants in

$$(6.2.6) \qquad w_M = \frac{-B_{2 \cdot 13} \pm \sqrt{B_{2 \cdot 13}^2 - 3B_{1 \cdot 23}\, B_{3 \cdot 21}}}{3B_{3 \cdot 21}}.$$

The values for Y_M may then be found by substituting the w_M values in the cubic regression equation. For the cubic example (Section 6.2.3), the w_M values are 1.05 and 4.28, where the Y_M are, respectively, 31.9 and 69.5. The sign of $B_{3 \cdot 12}$, gives the general regression trend: if negative, as in the case here, the lower w_M gives a minimum and the higher a maximum, and if positive, the reverse. Since for this example B_1 was positive and significant, $B_{2 \cdot 1}$ not significant, and $B_{3 \cdot 12}$ negative and significant, the function can be described (without plotting) as generally rising with a minimum followed by a maximum—the shape of a forward-tilted **S**.

Maxima and minima of polynomial regression equations are of interest in some applications, but they need not be routinely determined. They also require some caution in interpretation. In the quadratic regression problem (Section 6.2.2 and Figure 6.2.1), for example, we found the $\hat{Y}$ maximum to come at $v = 127$. It would however be an error to conclude that *in the population,* increases in v beyond 127 are accompanied by decreases in Y. Were the single data point $v = 140$, $Y = 116$ omitted, the value of v_M would be displaced beyond the upper limit of our range of v, where it may be completely fictive (for example, an IQ of 200). In the cubic regression example (Section 6.2.3 and Figure 6.2.2), the observed displacement of the maximum and minimum certainly limits our confidence in the projectibility of such values to the population. We are unable to offer any general guidance about the size of the error in maxima and minima due to sampling and inadequacy of fit which is to be anticipated; one must rely on one's substantive knowledge of the field of application, leavened with some common sense.

The possible wobble in v_M and accompanying $\hat{Y}_M$ values should not be allowed to detract from the areas of solidity of the method:

1. A data plot with $k - 1$ real bends will require a polynomial of (certainly no less than and probably no more than) the kth order.

2. By noting the size and statistical significance of each I_i, and the sign of B_i when it enters the equation, the general shape of the regression can be known. For many applications, this makes physically plotting the curve unnecessary.

How Many Terms?

It is difficult to offer general advice as to how high k is to be in the representation of V by a polynomial of the kth order. There are several interrelated reasons for this difficulty. Behavioral science is an exceedingly heterogeneous domain, covering areas as diverse as brain biochemistry and cultural anthropology. This heterogeneity, in turn, is accompanied by diversity in the metric quality of the data and in the precision of research formulations and hypotheses, which then will be related to the purpose of an investigator when he uses the polynomial method. It hardly needs to be pointed out that numbers issuing from paper chromotography of a substance found in the brain and those from attitude surveys of welfare mothers are of quite different character, yet computer programs do not distinguish between them. Finally, purely statistical considerations (n, number of observed values of V) which vary widely from research to research also play a role in setting or finding k.

Remember that with n observations of Y, each associated with one of q different values of V, a polynomial in $q - 1$ terms will exactly fit the Y means of all these V values. Concretely, if we have 100 (= n) paired observations of grade point average (Y) and IQ (V), with 36 (= q) distinct values of IQ, the 36 Y means of these IQ values will be perfectly fitted by a polynomial equation in powers of v up to v^{35}. Do we want to fit this equation? Most assuredly *not*, and for several reasons.

One obvious reason is that this perfect fit will hold for only this sample. Consider that with 100 cases distributed over 36 points, the Y means for each V value are based on an average of $100/36 = 2.8$ cases. Many will be based on only one case. Obviously such means will be wildly unstable as estimates of their population values. Of what use is it to fit them exactly? The 35 B_i and A of the 35th-order polynomial equation will similarly be grossly unstable estimates of their population values. For the population equation, almost all of the 35 B_i values will be zero or negligible.

If our n were 10,000, and, assuming for the sake of simplicity that q is still 36, the average number of cases on which the 36 Y means are based would be 278, and only a few would be based on such a small number as to render them unstable. Do we want to fit the 35th-order polynomial now? Still not. Even if we go further and assume that these 36 means coincide perfectly with their population values, so that the sample equation coincides perfectly with the population equation, would it be desirable to have it? It would not, because the curve through the means would still not represent the relationship between the two constructs under study, owing to the inevitable departures, however small, of the Y and V scale units from perfect equality.

Quite apart from questions of sample size and adequacy of scaling, consider: is there more information in the 36 constants we determine from the equation than in the 36 means which they fit? No, if anything, less: the means are easier to interpret. Underlying this argument is the traditional scientific strategy of parsimony. If it takes 36 numbers to account for 36 other numbers, no explanation (order, theory) has been accomplished.

What we want, then, is that the order of the polynomial k be smaller, usually much smaller, than $q - 1$. Given that curvilinearity is to be provided for or investigated, it is difficult to think of circumstances where k should exceed four, and frequently two is sufficient. Within this range, a large number is likely to be employed (*a*) when one's purpose is the precise fitting of the curve, or (*b*) when the data are relatively free of error (that is, small dispersions about the *Y* means of the *V* values relative to the overall *Y* variance), or (*c*) when *n* is very large relative to the total number of IVs to be studied (say, a ratio of more than 20 to 1). In contrast, a *k* of 2 or maybe 3 will suffice (*a*) when one's purpose is to allow for or detect curvilinearity in a relationship rather than to closely specify it, or (*b*) when measurement error is substantial (rating scales, interview responses, most psychological tests and sociological indices), or (*c*) when *n* is not large relative to the total number of IVs (say, a ratio of less than 10 to 1). In general, the "less is more" principle of Section 4.6.2 applies.[4]

In our exposition and examples, we have used $k = 3$ because it is in the middle of our range, and the strategy was that *k* was set in advance of the analysis, implicitly as a hypothesis that no higher order was needed. An alternative strategy may be employed in which one proceeds hierarchically and evaluates each *I* until a statistical decision rule halts the procedure. The following rules or combinations thereof may be employed:

1. *Statistical significance.* Starting with the linear term one proceeds cumulatively until an I_j is reached which fails to meet the significance criterion which is being employed (usually $P < .05$). The process is terminated, and $k = j - 1$.[5]

2. *Change in* $\widetilde{R}^2$. Thus far in this chapter, we have been concerned only with the observed R^2 and increments to it. However, we know that the addition of any *IV* to a set, even a random, nonpredictive variable, will result in an increase in R^2 —a decrease is mathematically impossible, and *I* exactly zero exceedingly rare. This fact of MRC should not be lost sight of in its application to the polynomial method. It can be instructive, while cumulating, to track the size and particularly the change in the $\widetilde{R}^2$, the estimated proportion of variance in *Y* accounted for in the population by a polynomial of that order. Unlike *I*, these changes may be positive *or* negative. For example, the $\widetilde{R}^2$s of the quadratic example (Section 6.2.2) are, successively, .576, .630, and .618: $\widetilde{R}^2_{Y \cdot 123}$ is smaller than $\widetilde{R}^2_{Y \cdot 12}$. Two stopping criteria use changes in $\widetilde{R}^2$:

2a. *No increase in* $\widetilde{R}^2$. It can be proven that whenever the *F* ratio for an I_j is less than one, $\widetilde{R}^2$ for the *j*th order will be smaller than $\widetilde{R}^2$ for the $(j - 1)$th order, that is, a decrement will occur. If one is tolerant of relatively large *k* in the interest of a close fit, one may take successive terms until $\widetilde{R}^2$ drops or fails to

---

[4] Yet another practical consideration is the degree of accuracy provided by the computer program used. Large *k,* even $k = 5$, requires a degree of computing accuracy not attained by many standard MRC programs (Wampler, 1970). See "Computing Problems," and Appendix 3.

[5] In the unlikely circumstance that one is dealing with a symmetrical U- or inverted U-shaped curve, one should not terminate after the linear term; similarly, and even less likely, if one is dealing with a symmetrical trigonometric wavelike function with one maximum and one minimum, one should not terminate with $k = 2$. It happens that our cubic example (Table 6.2.2) is such a case: I_2 is not significant, but I_3 is.

increase. However, this may easily lead to too many terms, since under conditions of no increase in population R^2, F will exceed one with a probability greater than .32. This procedure, then, is appropriate when one is seeking a maximum true fit, and is willing to pay the price of large k and the risk of overfitting.

2b. *Minimum increase in $\widetilde{R}^2$.* When maximum fit is not the goal, a reasonable criterion for stopping is failure for $\widetilde{R}^2$ to increase by some arbitrary minimum consonant with the purpose of the investigation, ranging between, say, .02 and .05. This criterion would ordinarily be used in conjunction with a significance test on I, that is, one stops either when I fails to be significant, *or* when the increase in $\widetilde{R}^2$ fails to meet the criterion minimum.

In much of the research done in the large social science sector of the behavioral sciences, the actual fitting of curves is not necessary, or even appropriate. In these areas, a multiplicity of causal factors not represented in the equation plus measurement and scaling errors typically operate so as to preclude accounting for more than about one-third of the Y variance, if that much. Departures from linearity here are likely due to "ceiling" or "floor" effects. These may be real, or due to lack of scaling fidelity. Or, the latter may produce other departures from linearity. In any case, the nonlinearity is simple, and V will be well represented using only v and v^2 (and barely possibly, v^3 may make a nonnegligible contribution). In the "harder" sectors of behavioral science (for example, psychophysiology, sensory psychology, learning), if a polynomial fitting function is to be used, v^3 will more often be needed additionally (and barely possibly, v^4 may make a nonnegligible contribution).

There are other considerations which lead to a preference for k to be small. With reasonably large n, two or three additional IVs due to larger k in V may not be serious, but, as will be seen in Chapter 8, when we are interested in interactions of V with other sets representing quantitative or nominal scales, each interaction requires IVs equaling k times the number of IVs in the other set(s). A generous value for k easily results in a substantial increase in the total number of IVs in the analysis, with the attendant increase in instability and loss of power (Section 4.6). Also, problems of computation accuracy may arise with a polynomial where k is as much as 4 or 5 (see below).

In many investigations, our major purpose is to adequately represent the constructs in our IVs as they relate to Y, rather than fit a function. Thus, when for a variable–construct like age, length of hospitalization, number of siblings, or socioeconomic status, when we say that it accounts for 12% of the Y variance, we want it to mean that the construct has that given degree of effect or association. If it should actually be the case that this 12% is accounted for linearly, while the addition of the quadratic aspect of the variable would raise the Y variance accounted for to 18%, we will have understated the association of our construct and cheated ourselves and our readers of a complete understanding of the import of our data. Such blunders are easily avoided by the use of polynomials of low order. This should not be taken as a blanket recommendation to adopt the slogan "No v without its v^2!," but rather to be prepared to cope with nonlinear relationships when they arise or when there is good reason to suspect them. No research prescription can replace, let alone override, the experience and judgment of the investigator.

Computational Problems

We have seen that the correlations among the powers of v run high. With more than two terms, the multiple Rs of any power with the others run even higher. Theoretically, no matter how high they get, as long as such Rs do not reach 1.000 (and, theoretically they cannot), the matrix of IVs is not, in matrix-algebraic parlance, singular, and, in theory, the MRC can proceed.

In practice, however, there is a limit on the accuracy provided by the given computing algorithm (recipe) used by the program and realized on the computer. As the Rs among the X_i get very large and very close to one, the matrix of r_{ij} is said to be "highly multicollinear" (or "ill conditioned" or "near singular"), its determinant approaches zero (see Appendix 1), and the computing goes haywire. The program may "bomb," or produce garbage, or worse—give results that look reasonable but actually are not even correct to one significant digit (Wampler, 1970).

This constitutes by itself a very good reason not to climb too high up the polynomial tree, but stay securely on its lower branches. This is not, however, a sufficient prescription. Before one has reached an order of the polynomial which is justifiable in terms of the purpose of the analysis, serious computational inaccuracy may occur. What to do about this problem?

First, and obviously, consult with the technical staff attached to the computer laboratory with regard to your problem and their resources. Most labs have "double-precision" or "extended-precision" options, and several different MRC programs that differ, among other features, in their numerical accuracy (see Appendix 3). The fruits of such consultation may make it unnecessary to take any other measures.

The Rs among the X_i can be reduced, with no loss of information, by expressing v as a deviation from its mean, $v' = v - \bar{v}$. With $X_1 = v'$ instead of v, simply square, cube, etc. the v' to get higher-order terms. This will sharply reduce those r_{ij} where i is an even-numbered (2, 4) and j an odd-numbered (1, 3) power,[6] hence reduce their Rs, and thus improve the accuracy of the output.

A minor nuisance that may arise is exceeding the number of digits per datum provided for by the MRC program. When v has d digits, v^i has di or $di - 1$ digits. Concretely, if v is a 3-digit variable, v^4 will have up to $3 \times 4 = 12$ digits, which may exceed the program's maximum. This problem can be handled by dropping as many right-hand digits in v^i as necessary to avoid exceeding the maximum. The problem may well be mitigated by using $v' = v - \bar{v}$, where, if v is a variable

---

[6] When v is symmetrically distributed about its mean, these r_{ij}s become zero, but even with readily observable departures from symmetry, they are quite low.

subject to measurement error, $\bar{v}$ is carried to no more places than v. Then, higher powers of v' may also have right-hand digits dropped, as necessary.

Finally, the problems of high multicollinearity and excess digits may generally be solved by going from power polynomials to coding by means of orthogonal polynomials, as described in Section 6.3.3.

Scaling

Quite deliberately, no attention was given to the nature of the quantitative scales used in the illustrative examples. The issue of the level of scaling and measurement precision required of quantitative variables in MRC is complex and controversial. We take the position that, in practice, almost anything goes.

Formally, fixed model regression analysis demands that the quantitative independent variables be scaled at truly equal intervals and measured without error. Meeting this demand would rule out the use of all psychological tests, sociological indices, rating scales, and interview responses; excepting some experimentally manipulated "treatments," this eliminates virtually all the kinds of quantitative variables on which the behavioral sciences depend. Such variables have no better than approximately equal intervals and at the very least 5 or 10% of their variance in measurement error.

Regression models which permit measurement error in the IVs exist, but they are complex and have even less satisfiable requirements, for example, multivariate normal distributions. The general conviction among data analysts is that the assumption failure entailed by the use of fallible IVs in the fixed-regression model, which we use throughout, does not materially affect any positive conclusions which are drawn.[7] Naturally, if the construct V is measured by a test with a reliability of .20 (hence, with 80% measurement error), a conclusion that Y variance is not being significantly accounted for by V, whatever the order of polynomial we use from first to kth, may mislead us. But this is not a problem unique to regression analysis—reaching negative conclusions about constructs from unreliable or otherwise invalid measures is an elementary error in research inference.

The inequality of intervals in V demands further scrutiny. We must first distinguish between randomly unequal intervals and systematically unequal intervals. Ordinary crudeness of measurement in interval or ratio scales results in randomly unequal intervals. For example, for a given psychological test which yields an IQ score, the unit 106–107 may be slightly larger or slightly smaller than either the 105–106 or 108–109 units. There is no systematic relationship in the size of adjacent unit intervals, and by extension, intervals made up of adjacent groups of units, for example, 95–99, 100–104, 105–109. Such units or ranges of units may not reflect identically equal intervals on a true measure of

---

[7] This statement is false if taken out of the present context of the regression on Y of one research factor V. Measurement error in multifactor MRC may lead to seriously misleading partial coefficients. See Section 9.5.

the construct (which, of course, we cannot know), but there is no reason to believe that the intervals are more or less equal in one part of the scale than another. Another way to say this is that the serial correlation (Harris, 1963) of each interval size with the next over the entire scale will be about zero.

Such random lack of fidelity in scaling characterizes the indirect approach to measurement of psychometrics and sociometrics: test and factor scores and indices—generally, measurement accomplished by combining elements. It is a tolerable condition when, as seems to be generally the case, the inequality of interval size is small when expressed proportionately, say when the mean interval size is three or four times the standard deviation of interval sizes. The effect that it has is to produce jaggedness in the observed function which is additional to that produced by sampling error. Since, as argued above, we do not seek to fit such jags with polynomials of high order, they merely contribute to the size of the proportion of residual variance, $1 - R^2$. Compared to other factors, this contribution is typically a relatively small one. We conclude that, in general, the random inequality of intervals found in quantitative variables produces no dramatic invalidation of MRC, but merely makes a minor contribution to the residual (error) variance.

Systematic inequality of intervals may arise at any intended level of quantitative scaling, ordinal, interval, or ratio. If the measurement process wherein numbers are assigned to phenomena is such that, relative to hypothetical "true" measures, the intervals at one end of the scale are larger than those at the other, or larger at both ends than in the middle, or more generally, different in different regions of the scale, adjacent interval sizes will be correlated (the serial correlation will be positive). This may arise in many ways in a domain as diverse as behavioral sciences. For example, in ordinal scaling of responses to an attitude item from an interview, an investigator may inadvertently provide more scale points in the vicinity of his own attitude position than in other regions. As another example, difficulty in writing hard items for an intended interval scale of reasoning ability may make higher intervals smaller than lower ones. A familiar example is that the use of percentile values from unimodal, approximately symmetrical distributions makes middle intervals large relative to extreme ones.

The effect of such systematic interval inequality in a variable is to alter the shape of its observed relationship with other variables from what it would be if measured in equal units. True linear relationships become curvilinear (or the reverse), bends in curves may increase (or decrease) in number or degree of curvature. But rather than invalidate MRC analysis, it is explicitly to fit functions irrespective of their shape that the polynomial method is used. So whether the shape of the function Y of V is truly reflected in the data, or is due to systematic interval inequality, or both, a polynomial fit to the actual data can be obtained.

The scaling requirements for the dependent variable are at least as modest. Measurement error does not even theoretically violate the model, and inequality of interval size, random and systematic, operate as for the IVs, producing noise and curvature change respectively. It should also be noted that dichotomous dependent variables (employed–unemployed, married–single, pass–fail) may be coded 1–0, and used as dependent variables. With this coding, the B_i (and A and $\hat{Y}$) are simply interpreted as proportions, which is very convenient. This practice is in formal violation of the model, which demands that for any given combination of X_i values, the Ys be normally distributed (and of constant variance), a patent impossibility for a variable which takes on only two values. Yet in practice, and with support from the central limit theorem and empirical studies, dichotomous dependent variables are profitably employed in MRC.

In summary, then, neither measurement error nor inequality of intervals precludes the use of polynomial MRC, despite some formal assumption violation of the fixed regression model. In practice, ordinal scales, as well as those which seek (not necessarily successfully) to yield interval or ratio level measurement, may be profitably employed.

The Polynomial as a Fitting Function

We have seen in the examples that the use of polynomial MRC correctly indicates the fact of nonlinearity of regression and its general nature. But the fit it provides with a few terms need not necessarily be a good one. In the quadratic example (Figure 6.2.1), the fit is reasonably good (provided that the maximum is not taken too seriously). But in the cubic example (Figure 6.2.2), although the polynomial function yields a maximum and minimum within the range studied, these are displaced from where the Y means of the data to be fitted have their respective maximum and minimum by one point in w (hardly a negligible amount in a six-point scale). Thus, although the cubic term significantly improves the fit over that provided by the quadratic polynomial, and properly mirrors the fact that the shape of the regression is two bended, the bends do not come in the proper place; hence, the fit, in absolute terms, is not very good. The cubic polynomial does the best it can, but simply cannot manage the steep climb between $w = 2$ and $w = 3$ necessary for a really good fit. Of course, a fifth-order polynomial can, since, given that there are only six values of w, it will go through all the means exactly. But, we have argued against the empirical fitting of high-order polynomials as being unparsimonious, hence unlikely to contribute to scientific understanding. This argument has its maximum force, when, for q points, the polynomial is of the order $q - 1$.

Our claim for the polynomial of low order is modest: it is a good *general* fitting function for most behavioral science data when curvilinearity exists, particularly for data of the "soft" kind. The close fitting of curves with few parameters requires either strong theory which takes the form of an explicit mathematical model (of which examples can be found in such fields as mathematical learning theory, sensory psychology and psychophysics, and econometric theory—see Section 6.5.2), or for purely empirical fitting, a collection of fitting functions of which the polynomial is only one. Curve fitting is a branch of applied mathematics in its own right, and any detailed attention to it is beyond our scope (and competence). The interested reader might wish to pursue the possibilities of other general fitting functions: trigonometric, Bessel, and Chebyshev functions, in particular. These, like the polynomial, may be used as coding procedures in MRC; they are other functions of v than positive integer powers. To take a simple example, one such fitting function uses as aspects of v:

(6.2.7) $X_1 = \sin v,$ $X_2 = \sin 2v,$ $X_3 = \sin 3v,$ etc.,

the angle v being expressed in radians. The application of this function to the cubic problem (Table 6.2.3) results in a better fit at $k = 1, 2,$ and 3 than that obtained for the power polynomial, the successive cumulative R^2s being, respectively, .266, .274, and .374. But it is the fit which is improved, not our understanding of the relationship. In closing, we reiterate: the polynomial of low order has the virtues of simplicity, flexibility, and general descriptive accuracy. It will work well in most behavioral science applications, and particularly for those in which we wish to represent a quantitative research factor using two or three terms, rather than to achieve a maximal fit.

6.3 ORTHOGONAL POLYNOMIALS

6.3.1 Method

The terms in the polynomials of the last section were hardly uncorrelated. We saw that for typical (positive) scores, the correlations among the first few powers of v are in the nineties. What made these nonorthogonal IVs work was the use of the hierarchical model. When v^2 ($= X_2$) is entered into an equation already containing v ($= X_1$), the partialling process makes of it effectively $v^2 \cdot v$ or ($X_{2 \cdot 1}$), that is, v^2 from which v has been (linearly) partialled. Now, by the very nature of the system (as was seen in Chapter 2) $X_{2 \cdot 1}$ is necessarily orthogonal to (correlated zero with) X_1. More generally, at any stage of cumulation, $X_{i \cdot 12 \ldots i-1}$ is orthogonal to $X_1, X_2, \ldots,$ and X_{i-1}, and whatever portion of the Y variance it accounts for (I_i) is different from (not overlapped with) that accounted for by the latter individually and collectively ($I_1, I_2, \ldots, I_{i-1}$).

The use of k positive-integer powers of v is one way of coding V. Orthogonal polynomials constitute a means of coding V into k IVs that not only represent linear, quadratic, cubic, etc. aspects of V, but do so in such a way that these coded values are orthogonal *to each other*. Recall that this is exactly the same demand that was set for contrast coding of nominal scales (Section 5.5). In fact, it is purely a matter of terminology whether the orthogonal polynomials are called curve (or trend) components or contrasts.

The reader is probably familiar with orthogonal polynomials from "trend analysis" in AV. We will see here, as before, that MRC can produce identically the same results as AV, and beyond that, can use orthogonal polynomial coding more flexibly and for other purposes.

TABLE 6.3.1

Orthogonal Polynomial Coding for u-Point Scales; First-, Second-,
and Third-Order Polynomials for u = 3 to 12[a]

$u = 3$		$u = 4$			$u = 5$			$u = 6$			$u = 7$		
X_1	X_2	X_1	X_2	X_3	X_1	X_2	X_3	X_1	X_2	X_3	X_1	X_2	X_3
1	1	-3	1	-1	-2	2	-1	-5	5	-5	-3	5	-1
0	-2	-1	-1	3	-1	-1	2	-3	-1	7	-2	0	1
-1	1	1	-1	-3	0	-2	0	-1	-4	4	-1	-3	1
		3	1	1	1	-1	-2	1	-4	-4	0	-4	0
					2	2	1	3	-1	-7	1	-3	-1
								5	5	5	2	0	-1
											3	5	1

$u = 8$			$u = 9$			$u = 10$			$u = 11$			$u = 12$		
X_1	X_2	X_3	X_1	X_2	X_3	X_1	X_2	X_3	X_1	X_2	X_3	X_1	X_2	X_3
-7	7	-7	-4	28	-14	-9	6	-42	-5	15	-30	-11	55	-33
-5	1	5	-3	7	7	-7	2	14	-4	6	6	-9	25	3
-3	-3	7	-2	-8	13	-5	-1	35	-3	-1	22	-7	1	21
-1	-5	3	-1	-17	9	-3	-3	31	-2	-6	23	-5	-17	25
1	-5	-3	0	-20	0	-1	-4	12	-1	-9	14	-3	-29	19
3	-3	-7	1	-17	-9	1	-4	-12	0	-10	0	-1	-35	7
5	1	-5	2	-8	-13	3	-3	-31	1	-9	-14	1	-35	-7
7	7	7	3	7	-7	5	-1	-35	2	-6	-23	3	-29	-19
			4	28	14	7	2	-14	3	-1	-22	5	-17	-25
						9	6	42	4	6	-6	7	1	-21
									5	15	30	9	25	-3
												11	55	33

[a]This table is abridged from Table 20.1 in Owen (1962). Reproduced with the permission of the publishers. (Courtesy of the U.S. Atomic Energy Commission.)

Table 6.3.1 presents orthogonal polynomial coding (a coefficients as in Section 5.5) for quantitative scales having from (u =) 3 to 12 scale points at equal intervals. X_1, X_2, and X_3 give, respectively, the linear, quadratic, and cubic coefficients. Although there exist for u points orthogonal polynomials up to the order $u - 1$, only the first three are given here, since higher orders as not often useful (see the discussion in Section 6.2.4).[8] The coefficients of each polynomial for a given u sum to zero, and their products with the coefficients of any other order for that u also sum to zero, the latter constituting the orthogonality property. For example, consider the quadratic (X_2) values for a five-point scale: its five values sum to zero; its orthogonality with X_3 is demonstrated by $(2)(-1) + (-1)(2) + (-2)(0) + (-1)(-2) + (2)(1) = 0$, and with X_1 by $(2)(-2) + (-1)(-1) +$

[8] Higher-order polynomials and larger numbers of points are available elsewhere. The most extensive are those of Anderson and Houseman (1942), which go up to the fifth order and to u = 104. Pearson and Hartley (1954) give up to the sixth order and u to 52.

$(-2)(0) + (-1)(1) + (2)(2) = 0$. The values are all integers and exact. Each column of coefficients, when plotted against an equal-interval scale from low to high has the distinctive shape of its order (subject to discreteness): the linear coefficients define a rising straight line, the quadratic coefficients define a concave upward parabola, and the cubic coefficients define a curve that rises to a maximum, falls to a minimum, and then rises again. We can think of these as pure (because they are mutually uncorrelated) renditions of aspects of shape into which we transform or code our equal-interval u-point scale.

The feature of orthogonal polynomials which is important for MRC is that the coefficients are zero correlated. The property that the coefficients of each polynomial sum to zero is important to AV, where it greatly simplifies the computation. It is also in the interests of computational simplicity in AV that the requirements that there be equal numbers of observations at each scale point and that the scale points be at equal intervals are made in textbook presentations of AV trend analysis. We shall see that these conditions are not necessary in MRC applications, although our worked example satisfies them.

6.3.2 The Cubic Example Revisited

Consider again the data presented graphically in Figure 6.2.2: at each of 6 ($= u$) equally spaced values of W (0, 1, 2, 3, 4, and 5) there are 6 values of Y, for a total of $n = 36$ observations. Instead of the powers, w, w^2, and w^3, W is represented by the orthogonal polynomials under $u = 6$ in Table 6.3.1. Assume that, as in Section 6.2.3, we decide to provide for all three orders given in Table 6.3.1. We thus code W (ranging from 0 to 5) as the linear aspect $X_1 = -5, -3, -1, 1, 3$, and 5, respectively; as the quadratic aspect $X_2 = 5, -1, -4, -4, -1$, and 5, respectively; similarly for X_3. The results of a simultaneous MRC analysis are given in Table 6.3.2.

TABLE 6.3.2
Simultaneous Orthogonal Polynomial MRC Analysis of Y on W

	Y	X_1	X_2	X_3	$r_{Yi}^2 = sr_i^2 = I_i$	t $(df = 34)$
X_1	.45165	1.000	.000	.000	.20399	2.9518**
X_2	.11780	.000	1.000	.000	.01388	.6917
X_3	-.32189	.000	.000	1.000	.10361	1.9824
m	51.00	.000	.000	.000	$n = 36$	
sd	22.69	3.416	3.742	5.477		

$R_{Y \cdot 123}^2 = .32148$; $F = 5.054**$ $(df = 3, 32)$; $\tilde{R}_{Y \cdot 123}^2 = .25787$

$\hat{z}_Y = .452 Z_1 + .118 Z_2 - .322 Z_3$

$\hat{Y} = 3.000 X_1 + .714 X_2 - 1.333 X_3$

$**P < .01$.

Note first that the correlations among the X_i are all zero. This occurs because the orthogonal polynomials have been applied in a problem where the numbers of observations at each of the u scale points are equal. Were this not the case, despite the orthogonality of the polynomial *coefficients*, their unequal representation in the X_i would render the r_{ij} not generally zero (as was the case with contrast coefficients in Section 5.5).

Given that the r_{ij} are all zero, it follows that for each X_i, r_{Yi} will equal sr_i and β_i, since there is nothing from other IVs to partial. This therefore means that the proportion of Y variance accounted for by each curve-function aspect (or "trend component") is given directly by the zero-order r_{Yi}^2. Thus, in the problem, the linear aspect of W accounts for .20399 of the Y variance, the quadratic aspect .01388, and the cubic .10361 (Table 6.3.2). But reference to the previous analysis of these data by hierarchical MRC using nonorthogonal integer power polynomials (Table 6.2.2) shows that these are *identically* the respective I values obtained there. What was there accomplished by the hierarchical procedure with its partialling out of lower-order terms, is done here by making the IVs uncorrelated, thus rendering partialling unnecessary. Remember, however, that for this simple state of affairs, equal ns at each of the u values are necessary.

Since the X_i are uncorrelated, $R_{Y \cdot 123}^2$ is simply the sum of the three r_{Yi}^2, .32148 with $F = 5.054^{**}$ $(df = 3, 32)$ as before. This property also results in the standardized regression equation being very easily written, since $\beta_i = r_{Yi}$. These β_i are exactly those of the hierarchical model (not given): β_1, $\beta_{2 \cdot 1}$, and $\beta_{3 \cdot 12}$, respectively. The (unstandardized) regression equation given in Table 6.3.2 naturally differs in its constants from the cubic equation for W represented by powers given in Table 6.2.2, because W is here coded by the orthogonal polynomial coefficients (note, for example, that $A = 0$); when the latter are substituted as "scores" for the X_i representing W, the equation yields exactly the same cubic function plotted in Figure 6.2.2.

When we test the r_{Yi} for significance, we do not get exactly the same results as before, because although the proportions of variance accounted for are the same, different models for the error (residual) terms are used in the two analyses. One can, in fact, identify four versions of the two error models discussed in Chapter 4, leading potentially to four different F (or t) tests of proportions of variance, the results of whose applications are given in Table 6.3.3:

1. *Model I(0): $1 - r_{Yi}^2$*. The routine t test of a simple r uses as error the total residual from its regression (Eq. 2.8.1), with $df = n - 2$. These are the values given above in Table 6.3.2. Equivalently, this can be viewed as a special case of the general F test using Model I error (Eq. 4.4.1), where A is an empty set, and $k_B = 1$. Hence, $df = 1, n - 2$ as in Eq. (4.4.4), which is simply the square of the above t. We give the latter in Table 6.3.3 to facilitate comparison among the error models. Note that this is truly a version of error Model I: where there is variance being accounted for by *any* curve components other than X_i, it is part of $1 - r_{Yi}^2$ and negatively biases the F. This model is not recommended for orthogonal polynomials.

TABLE 6.3.3
The Fs of Alternative Models for Testing Orthogonal
Polynomials in the Cubic Example

ject of W	I_i	Model I(0) Error: $1-r^2_{Yi}$		I(i) $1-R^2_{Y\cdot12...i}$		II(k) $1-R^2_{Y\cdot12...k}$		II($u-1$) $1-R^2_{Y\cdot12...(u-1)}$	
		F	df	F	df	F	df	F	df
inear	.20399	8.713**	1, 34	8.713**	1, 34	9.620**	1, 32	14.122**	1, 30
uadratic	.01388	.478	1, 34	.586	1, 33	.654	1, 32	.961	1, 30
ubic	.10361	3.930	1, 34	4.887*	1, 32	4.887*	1, 32	7.173*	1, 30

$P < .05.$ **$P < .01.$

2. *Model I(i): $1 - R^2_{Y\cdot12...i}$* . As noted in Section 6.2.3, and as is our general practice, the error for each I_i is the proportion of variance remaining after the first i terms have been removed, with $df = n - i - 1$. Thus, this is an instance of the general error Model I, $1 - R^2_{Y\cdot AB}$ (Eq. 4.4.1), with set A containing the first $i - 1$ terms and set B, the single IV X_i term being tested. F is negatively biased when there is variance due to curve components of order higher than i, but this possibility is compensated for by increased df (relative to Model II error), and is only justified when n is relatively small, and the analyst's outlook is that progressively higher order variance contributions are progressively unlikely.

3. *Model II(k): $1 - R^2_{Y\cdot12...k}$* . Although there are $u - 1$ possible polynomial terms for u different values of the quantitative variable, we most often take only the first k, assuming that the other $u - k - 1$ terms are making no real contribution. This is then a version of Model II error, $1 - R^2_{Y\cdot ABC}$ with sets A and B as before, and set C made up of the polynomial terms $i + 1$ through k. With Model II(k), the same error term is used for all the polynomials, $1 - R^2_{Y\cdot12...k}$, with $df = n - k - 1$. This error includes the unexamined higher-order terms as well as the variation about the u means of Y. Specifically, then, we specialize the equation for F with Model II error (Eq. 4.4.6) for this case:

$$(6.3.1) \qquad F = \frac{I_i(n-k-1)}{1-R^2_{Y\cdot12...k}} \qquad (df = 1, n-k-1).$$

Unlike the analogous specialized F test for Model I error (Eq. 6.2.3), all the I_i are multiplied by a constant and referred to the same F distribution for any given problem. Note also that this F ratio (or the $t = \sqrt{F}$) will be supplied by standard MRC computer programs as the test on each partial coefficient when all k terms are in the equation (for equal $n_i = n/u$). For the above problem, $k = 3$, $n = 36$, and $R^2_{Y\cdot123} = .32148$, so the Fs of Model II(k) in Table 6.3.3 are found by multiplying each I_i by $(n - k - 1)/(1 - R^2_{Y\cdot123}) = 32/.67852$, and each F is for $df = 1, 32$.

4. *Model II(u − 1):* $1 - R^2_{Y \cdot 12 \ldots (u-1)}$. This error residual is made up only of the variation about the Y means for each of the u points, which is then used to test each of the k polynomials. Since no aspect of the shape of the curve up to order $u - 1$ is included in the error, no risk is run that F will be negatively biased. This, then, is the purest version of Model II error, with set C of Eqs. (4.4.5) or (4.4.6) containing the polynomial terms $i + 1$ through $u - 1$. As noted in Sections 4.4 and 4.5, this avoidance of the risk of negative bias in F pays a price in statistical power because the df for this error is $n - u$. This is the error model in AV "trend analysis" using orthogonal polynomials, where it is called "within groups" or "pure" error (cf. Edwards, 1972; Hays, 1973); the equivalence is exact—the F ratios of Table 6.3.3 for Model II($u - 1$) are identically those which would be found using the AV computing procedure. We specialize the general F test for Model II (Eq. 4.4.6) for the present purpose as

$$(6.3.2) \qquad F = \frac{I_i(n - u)}{1 - R^2_{Y \cdot 12 \ldots (u-1)}} \qquad (df = 1, n - u).$$

This model requires finding $R^2_{Y \cdot 12 \ldots (u-1)}$. It is not necessary to represent the orthogonal polynomials between k and $u - 1$ in order to do so—merely treat the u scale points as if they were levels of a nominal scale, code them by $u - 1$ dummy variables, and run them as such (see Sections 5.3 and 6.5). For the present artificial problem it happens that the curve has a large quintic component, $I_5 = .22666$, so that $R^2_{Y \cdot 12345} = .56665$, which greatly reduces the residual and makes the Fs considerably larger than in the other models. This should not be considered characteristic—with real data, the Fs would most likely be no larger, and are based on fewer df for error.

An advantage which accrues to the determination of $1 - R^2_{Y \cdot 12 \ldots (u-1)}$ ("pure" error) is the possibility of testing the "goodness of fit" of the kth order polynomial. This is determined by testing whether the polynomial terms between $k + 1$ and $u - 1$, inclusive, significantly increment the $R^2_{Y \cdot 12 \ldots k}$. This test is accomplished by specializing the general Model I error F test (Eq. 4.4.2), letting set A include the first k terms and set B the terms $k + 1$ through $u - 1$, so that

$$(6.3.3) \quad F = \frac{(R^2_{Y \cdot 12 \ldots (u-1)} - R^2_{Y \cdot 12 \ldots k})(n - u)}{(1 - R^2_{Y \cdot 12 \ldots (u-1)})(u - k - 1)} \qquad (df = u - k - 1, n - u).$$

To illustrate this with the cubic example, as noted, $R^2_{Y \cdot 12345} = .56665$, $R^2_{Y \cdot 123} = .32148$ (Table 6.3.2), so

$$F = \frac{(.56665 - .32148)(36 - 6)}{(1 - .56665)(6 - 3 - 1)} = 8.486 \qquad (df = 2, 30),$$

which is highly significant ($P < .01$). Thus, quartic and quintic terms combined add significantly to $R^2_{Y \cdot 123}$ (we have already noted that this is due to the quintic term) so the fit provided by the third order polynomial is not adequate.

Nevertheless, Model I(i) and II(k) both lead to the correct conclusion about the linear and cubic contributions to the fit.

In the example on power polynomials, Model I(i) was used for the test of I_i. Note, however, that Model II(k) and in principle Model II($u - 1$) can also be used with power polynomials, where they have exactly the same properties as described here for orthogonal polynomials. Model I(0), however, is clearly unsuitable, since, without partialling, the r_{Yi} for $i > 1$ are heavily contaminated with effects of lower-order terms.

Which model to choose? The decision requires the resolution of the competing demands of maximizing statistical power and minimizing the risk of negatively biasing F, given the context of substantive and strategic considerations, as discussed for the general Models I and II in Sections 4.4 and 4.5.

We find little to commend Model I(0), even for orthogonal polynomials. Most data have sturdy linear components, so that when a frail quadratic or cubic component is tested, the linear component in the error term may seriously debilitate its F. This has, in fact, occurred to the cubic component in the example.

The hierarchical Model I(i) may bias tests on lower-order components because of substantial higher-order components. We are inclined to favor it, however, particularly when n is small, say less than 40, because in real data, as we have argued, the likelihood of real contributions diminishes rapidly as order increases beyond the first or second. We favor the hierarchical strategy with error Model I in applied MRC generally: when likely to be real effects are put in early, and "I wonder if . . ." and likely nonexistent effects, later, the risk of negatively biasing the Fs of the former is more than compensated for by the increased statistical power that accrues from the greater error df (see Section 4.6.3).

Model II(k) commends itself when k is small (e.g., three) and n large. It is generally sensible to exclude from error the k terms in which one is seriously interested, and when n is sizable, $n - k - 1$ are enough df to maintain power. It may be checked for "goodness of fit" by Eq. (6.3.3).

Model II($u - 1$) is safe in its avoidance of underestimating the Fs. It is, however, a luxury in those applications of orthogonal polynomials where u is large relative to n. Since real polynomial components between $k + 1$ and $u - 1$ (inclusive) are unlikely, such safety can be afforded only when $n - u$ is large, say, at least 30. But if the F test of Eq. (6.3.3) is significant, this model *must* be used. If not significant, some justification is provided for the use of Models II(k) and (partly) I(i). Model II($u - 1$) works well in the present example, the somewhat small $n - u = 30$ being compensated for by the large I_5.

It is worth giving thought to the selection of the model, even though in many problems where there is only V to be regressed on Y, all models lead to the same conclusions. In more complex problems, however, where interactions of other research factors with V are to be included (see Chapter 8), careful attention to power loss through biasing F and loss of error df becomes more important. For example, in such situations, Model II($u - 1$), which, as extended, discards not only the df from the higher-order terms, but all those from interactions of these terms with other research factors, would be grossly wasteful of error information and hence of statistical power.

6.3.3 Unequal n and Unequal Intervals

The previous example had equal n at each of the u points of V. This property is required in the familiar AV use of orthogonal polynomials in order that the computational simplicity of the method be maintained. It is, however, not necessary to meet this condition in MRC applications of orthogonal polynomials.

When the ns are not equal, the r_{ij} among the coded aspects of V are generally not zero, since the orthogonal polynomial coefficients are unequally weighted (as with contrast coefficients—see Section 5.5). With the X_is not independent, equality among r_{Yi}^2s and sr_i^2s in the simultaneous model is lost and r_{Yi}^2 no longer equals the amount of variance accounted for purely by the ith trend component, I_i. This, however, constitutes no problem in analyzing the data—we simply revert to the hierarchical model, and, as in Section 6.2, find I_i as an increment in R^2 when X_i is added to the IVs. We can use the error Model I(i) term to test I_i, as in Section 6.2, or that of either Model II(k) or II($u-1$), the considerations of the last section applying to the decision.

Another circumstance which defeats the simplicity of the AV use of orthogonal polynomials but is easily handled in MRC is inequality of the given intervals in V.[9] A scale with unequal given intervals can be conceived as one with equal intervals some of whose scale points have no data. For example, data may have been obtained for the values of V: 1, 2, 4, 6, 9. We can code these as if we had a 9-point scale, using coefficients under $u = 9$ in Table 6.4.1, but of course, omitting the coded values for points 3, 5, 7, and 8. The effect on the r_{ij} among the coded X_i is as before: they take on generally nonzero values, and the analysis proceeds by finding I_i by means of hierarchical MRC. Designating the number of points on which there are observations q, the number of polynomial terms which provide a perfect fit of the Y means is $q-1$ (instead of $u-1$). Models I(i), II(k), or II($u-1$) can be used for significance testing, as before, but q replaces u in the discussion of these models and in Eq. (6.3.2).

Finally, using orthogonal polynomial coefficients and hierarchical MRC, we can analyze problems in which neither the given intervals nor the numbers of observations per scale point are equal. Simply proceed as in the preceding paragraph.

6.3.4 Applications and Discussion

Three kinds of circumstances of research design and some already noted problems in computation particularly invite the use of orthogonal polynomials.

---

[9] We do not refer here to inequality of measured intervals relative to a true underlying scale, as discussed above in Section 6.2.4, but inequality of the numerical intervals *as given* in the problem. The points made in Section 6.2.4 continue to apply.

Experiments

The greatest simplicity in the application of orthogonal polynomials occurs under equal n, equally spaced conditions, V. Such data are produced by experiments, where V is some manipulated variable (number of rewards, level of illumination, size of discussion group) and Y is causally dependent on this input. Typically, such data sets are characterized by relatively small n and small n/u (that is, numbers of observations per scale point). These are the circumstances in which one would prefer to use Model II(k) error, $1 - R^2_{Y \cdot 12 \ldots k}$, or when k is large, Model I(i) error. Unequal n in such experiments, usually the result of data loss, requires the hierarchical model to find the I_i, but does not affect the decision about the error model to be used.

Note that throughout this chapter, the n observations are taken to be independent. The frequently occurring case where n subjects or matched sets of subjects yield observations for each of the u points is not analyzed as described above.[10]

Developmental Studies and Continuous V

Cross-sectional studies of subjects at different age levels (V) are another type which lend themselves to orthogonal polynomial MRC. These are not true experiments, age being an organismic variable rather than a treatment applied to randomly selected subjects, but their analysis proceeds in the same way. Again, cases per age level may or may not be equal, these two situations being handled as described above. When n is large, as in surveys, the choice of error models for significance testing [other than I(0)] is not likely to be critical.

V in the examples thus far given has been made up of u discrete points. This is not required—V may be measured quasi-continuously (for example, age in months, or even days, over a ten-year span) and then broken into u equal class intervals, each represented by its midpoint. This is equivalent to measuring V with error, since there are obvious discrepancies between each midpoint and the actual values of V in its interval. Although this technically violates our assumptions, if u is large (say 10 or 12), the effects of this violation on the validity of the conclusions is likely to be small. The discussion under *Scaling* in Section 6.2.4 on random errors of measurement in V is relevant here.

Sometimes one or both ends of V are "open," as, for example, age given as: under 20, 20–30, 30–40, . . . , 70 and over. Such variables may be treated provisionally as if they represented equal intervals ($u = 7$ is implied here), provided that one keeps in mind the possibility that the end intervals produce lateral displacements of the true curve. If the true relationship is (or is assumed to be) smooth, distortion due to inequality of the end intervals may be detected by the polynomial.

---

[10] The analysis by MRC of such "mixed-model" designs is discussed in Chapter 10. In such designs, the observations are not independent, and the approach to the analysis must be quite different.

Serial Data without Replication

Thus far we have been assuming that at each of the u points of V, there are multiple Y observations or "replications," the means of which define the function to be fitted. We now consider instances where there is only one observation at each of the u points; hence, $n = u$. Usually, such data are produced by a single subject, and V is time, or learning trials, or, generally, serial position. A diverse array of substantive research areas, ranging from economics to psychophysiology to psychotherapy, yield data of this kind. To exemplify the latter, Y may be ratings by a psychotherapist or semantic counts from transcripts of each of a series of 50 consecutive thrice-weekly (Monday, Wednesday, Friday) sessions; thus $n = u = 50$ (see Footnote 8). The purpose of analysis might be to test propositions or answer questions about the nature of the trend over time of Y for this patient, for example, is it generally rising (positive linear), does it wax and wane (negative quadratic with a maximum within the series)? Another purpose might be descriptive; each of a group of patients with such data might be described, hence measured, by a relevant facet of his series for use in further analyses, for example, the regression coefficient for the linear polynomial term as rate of change or $R_{Y \cdot 123}$ as an index of "orderliness" of the trend.

Since the number of observations per point on V is equal (namely, one), the $r_{ij} = 0$, hence the r_{Yi}^2 sum to $R_{Y \cdot 12 \ldots k}^2$, and the simultaneous MRC form of analysis is appropriate. But since $n = u$, Model II($u - 1$) for testing the I_i becomes impossible, since error $df = n - u = 0$. Pooled higher-order polynomial terms must constitute the error term; hence, either Model I(i) or II(k) would be used for testing hypotheses about the trend of the series. (For purely descriptive purposes, no error model is needed.) A reasonable analysis of the above example for most purposes would be to let $k = 3$ or 4 and use error Model II(k), which has an adequate number (45 or 46) of df.

It is unnecessary to be overly rigid about the equality of time intervals, when, as here, the degree of inequality (weekend versus midweek intervals) is not great.[11] The likely small effect of such imprecision in the intervals of V has already been commented on above. In some situations, one might be able to give up the idea of V as a time scale and finesse the problem by conceiving of it as simply ordinal position, adjusting the interpretation accordingly.

Computational Advantages

We have seen that with equal numbers of observations at equal intervals of v, the $r_{ij} = 0$ for orthogonal polynomial coding. When these numbers are unequal,

[11] In this particular instance, the extra day aside, there might be a sociopsychological "weekend" effect" as such. If this is suspected, it can be allowed for by adding beyond the k polynomial terms, a contrast $IV\ X_c$ coded 1 for Fridays, -1 for Mondays, and 0 for Wednesdays. Since this contrast is not in general orthogonal to the polynomials, its variance contribution is given by sr_c^2.

the r_{ij} are not generally zero, but, for almost any set of real data, the r_{ij} will be small, rarely attaining .50 among the coded X_i of the first few polynomial terms. This fact, plus the possibility of breaking a quasi-continuous V into equal class intervals discussed above, may be used to solve the problems of high multicollinearity and/or an excessive number of digits which may be encountered with integer power polynomials (see Section 6.2.4, "Computational Problems"). The method is obvious: with two- or three-digit scorelike data ranging over many points and an interest in third- or higher-order polynomials, one can break v into a fairly large number (say 10 or more) equal intervals, and code them by orthogonal polynomials. This will result in unequal ns per interval, but the resulting r_{ij} will be small enough so as to preclude the problems in numerical accuracy which could well occur if the powers of v were used: also the number of digits per X_i will remain small, usually not exceeding two.

Orthogonal Polynomials and Power Polynomials

Finally, it will be salutary to keep in mind that orthogonal polynomials are simply integer power polynomials coded so as to be orthogonal (and have mean zero), and thus partake of the fundamental properties of the latter. The discussion in Section 6.2.4 under the headings "How Many Terms?," "Scaling," and "The Polynomial as a Fitting Function" is thus equally relevant to orthogonal polynomials. In addition, the property of orthogonality is useful in simplifying some analyses and/or improving the numerical accuracy of the results.

6.4 NOMINALIZATION

When the primary interest in the regression of Y on some quantitative construct V is descriptive rather than analytic, one can assure adequate representation of V by treating it as if it were a nominal scale and coding it by dummy variables, as described in Section 5.3 (or by other coding methods for nominal scales). This method has the virtue of simplicity both in execution and communication of the results, which in some circumstances may be reason enough to employ it.

Proceed as follows: If V takes on many values, break it into g equal intervals, where g is a relatively small number between 3 and 10. The decision as to the number depends on the degree of descriptive detail which is desired. Instead of u as in the preceding section, we use g to denote the number of intervals (and also to designate the highest interval), in conformity with the notation used with nominal scales—we are, in fact, treating the cases in each interval as if they constituted a group in the sense used in Chapter 5. Each group may then be represented by the value of the midpoint of its range. Thus, for example if V is IQ with a range from 73 to 134, we could set up the intervals as 70–79, 80–89, . . . , 130–139 so that $g = 7$ and the "groups" are represented as 74.5, 84.5, . . . , 134.5. If a smaller g and, hence, a coarser grain is preferred, one could work with the intervals 60–79, 80–99, 100–119, 120–139; hence, $g = 4$.

If V takes on relatively few values, no breaking into class intervals is necessary; the number of groups is simply the number of values of V for which we have data.

From this point, simply proceed as described under Dummy Variable Coding (Section 5.3): code these g groups by means of $g - 1$ dummy variables, and do a (simultaneous) MRC analysis. Since V is represented by $g - 1$ IVs, it will be well represented—as well, in fact as if we were using $g - 1$ IVs coded by orthogonal polynomial coefficients: recall from Chapter 5 that *any* nonredundant coding into $g - 1$ variables will yield the same R^2 as any other. This procedure necessarily produces a "perfect" fit between Y means and V as broken into g intervals, no matter what the shape of the line drawn through the Y means of the g groups. These Y means are found from the B_i and A, as given in Section 5.3 in Eqs. (5.3.6) and (5.3.7). With the Y means plotted against the group (interval) midpoints, the relationship between Y and V may readily be graphically portrayed.

The yield from this analysis, then, is the proportion of Y variance accounted for by V (R^2 and $\tilde{R}^2$), its statistical significance (F test of R^2), and the Y means for the successive intervals of V (via the regression constants), which facilitates graphic portrayal of the Y on V regression. The remaining yield (r_{Yi}, sr_i, pr_i, β_i) is not generally of much interpretive utility.

What is lost and what is gained by nominalization, relative to either of the polynomial methods? What is lost, chiefly, is the analytic yield of the latter, for example, statements about whether and how the regression departs from linearity. Secondarily, $g - 1$ is usually larger than the few k terms of polynomial regression; this both leads to "overfitting" and a loss in df from the residual which may be of consequence when n is not large and/or when interactions with other variables are also to be included. What is gained is simplicity with adequate description, and, of lesser importance, the avoidance of the necessity for hierarchical regression (and possibly of computational problems due to multicollinearity of the IVs). The simplicity of method and communication of results may be advantageous in applications in social survey, market, and advertising research. In such research, too, the quantitative research factors often come from interviews and questionnaires as precoded items which are already broken down into (more or less) equal intervals, for example, age, income, size of family, length of residence.

6.5 NONLINEAR TRANSFORMATIONS

6.5.1 Introduction

Thus far in this chapter, we have sought to assure the proper representation of V, a quantitatively expressed research factor, by coding it multiply as a set of k IVs, each representing a single aspect of V: as a set of integer powers of v

(Section 6.2), or orthogonal polynomials (Section 6.3), or of dummy variables (Section 6.4). In the present section we consider the representation of V by a single nonlinear (monotonic) transformation of v, where v is used to denote the scale on which V is measured, or the linear aspect of V (Section 6.1).

The effect of a *linear* transformation of a variable is to stretch or contract it *uniformly* and/or to shift it up or down the numerical scale. Because of the nature of product-moment correlation, particularly its standardization of the variables, linear transformation of X or Y has no effect on correlation coefficients of any order, or on the proportions of variance which their squares yield. Not so those transformations or functions which stretch or contract v nonuniformly, the *non*linear transformations such as log v, or a^v, or 2 arcsin $\sqrt{v}$. These relationships are also strictly *monotonic,* that is, as v increases, the function either steadily increases or steadily decreases.

Whenever a constant additive change in one variable is associated with other than a constant additive change in another, the need for nonlinear transformations may arise. Certain kinds of variables and certain circumstances are prone to monotonic nonlinearity. For example, in learning experiments, increases in the number of trials do not generally produce uniform (linear) increases in the amount learned. As another example, it is a fundamental law of psychophysics that constant increases in the size of a physical stimulus are not associated with constant increases in the subjective sensation. As age increases during childhood, measured skills do not change uniformly. As total length of prior hospitalization of mental patients increases, scores of psychological tests and rating scales do not generally change uniformly. Certain variables are more prone to give rise to nonlinear relationships than others: time-based variables such as age, length of exposure, response latency; money-based variables such as annual income, savings; variables based on counts, such as number of errors, size of family, number of hospital beds; and proportions of all kinds.

The application of nonlinear transformations arises from the utilization of a simple mathematical trick. If Y is a logarithmic function of v, then being nonlinear, the $Y - v$ relationship is not optimally fitted by linear correlation and regression. But then the relationship between Y and *log v is* linear. Similarly, if Y and v are reciprocally and hence nonlinearly related, Y and $1/v$ *are* linearly related. Thus, by taking nonlinear functions of v or Y which represent specific nonlinear aspects, we can "linearize" some relationships and bring them into our system.

In addition to the linearization of relationships, nonlinear transformation is of importance in connection with the formal statistical assumptions of regression analysis—that Y is normally distributed and of constant variance (homoscedastic) over sets of values of the IVs. If a variable w does not meet these conditions, then it may be possible to find a nonlinear transformation of w which will better meet them. It is in this connection that they are found in textbook presentations of AV. Although our emphasis in this chapter is on the goal of linearization, the variance stabilizing and normalizing function of nonlinear transformation is of some importance, particularly for statistical tests when n is small, and for assessing the degree of accuracy of estimation in different regions of the regression surface. Fortunately, it has been the general experience of applied statisticians that when regressions are linearized, the assumptions of normality and constant variance tend to be satisfied, and vice versa (Acton, 1959, p. 221).

Given that some nonlinear relationship exists, how does one determine which (if any) is appropriate when? As already noted, curve fitting is a complex mathematical art far beyond the scope of this presentation. Further, for some purposes, its successful application depends on the theoretical structure of the substantive area to which it is applied. All that can be attempted here is a sketch of some of the considerations involved, and a general guide to practice.

It is useful to organize the presentation around the question of the strength of the theoretical model which underlies the data. Strong models dictate the mathematical relationship, and the task is one of transforming the variables into a form amenable to linear MRC analysis. Weaker models imply certain features or aspects of variables which are likely to linearize relationships. In the absence of any model, one may attempt to cope with observed nonlinearity by nonlinear transformation as a matter of practical convenience.

6.5.2 Strong Theoretical Models

In such fields as mathematical psychology and sociology, neuropsychology, and econometrics, relatively strong theories have been developed which result in the postulation of (generally nonlinear) relationships between dependent and independent variables. The adequacy of these models is then assessed by observing how well the equations specifying the relationships fit suitably gathered data. We emphasize that the equations are not arbitrary, but hypothetically descriptive of "how things work." The independent and dependent variables are observables, the form of the equation is a statement about a process, and the values of constants of the equation are estimates of parameters that are constrained or even predicted by the model.

Assume that a model leads to an equation of the following form:

(6.5.1) $$c^w = dv,$$

where c and d are constants. It asserts that changes in v are associated with changes in w as a power of a constant. The relationship between v and w is clearly nonlinear. But if we take logarithms of both sides, we obtain

(6.5.2) $$w \log c = \log d + \log v,$$

$$w = \frac{1}{\log c} \log v + \frac{\log d}{\log c}.$$

This is a linear equation, that is, w is a *linear* function of $\log v$. If we let $\hat{Y} = w$

and $X_1 = \log v$, we can express it in the familar form of a simple regression equation

$$\hat{Y} = B_1 X_1 + A,$$

where

(6.5.3)
$$B_1 = \frac{1}{\log c},$$

so that

(6.5.4)
$$c = \text{antilog}(1/B_1),$$

and since

(6.5.5)
$$A = \frac{\log d}{\log c},$$

(6.5.6)
$$d = \text{antilog}(A/B_1).$$

Thus, we see that Eq. (6.5.1) implies that w is a logarithmic function of v (or a linear function of $\log v$). Applying this now to suitably generated data, we regress w (= Y) on the logarithm of v (= X_1), find B_1 and A, and from Eqs. (6.5.4) and (6.5.6), estimate the constants c and d for substitution in the equation of the model (Eq. 6.5.1). These values will be of interest since they estimate parameters in the process being modeled, as will r^2 as a measure of the fit of the model (see below).

Consider another model, expressed by the equation

(6.5.7)
$$w = cd^v.$$

As before, c and d are constants, but now the independent variable v is an exponent (power) of a constant. Thus, this nonlinear relationship is specifically one in which w is an *exponential* function of v. Again, the relationship can be linearized by taking logarithms:

(6.5.8)
$$\log w = \log c + v \log d,$$
$$\log w = (\log d)v + \log c.$$

The latter is in the form of our standard linear regression equation, with $\hat{Y} = \log w$ as the dependent variable, and $v = X_1$. Log w is now linear in v, with

(6.5.9)
$$B_1 = \log d,$$

(6.5.10)
$$d = \text{antilog } B_1$$

and

(6.5.11)
$$A = \log c,$$

(6.5.12)
$$c = \text{antilog } A.$$

As yet another example of a nonlinear relationship defined by a strong theoretical model,

(6.5.13) $w = cv^d$,

where, again, c and d are constants, and w is a power function of v. This, too, can be linearized by taking logarithms of both sides:

(6.5.14) $\log w = d \log v + \log c.$

Log w (= $\hat{Y}$) is a linear function of log v (= X_1), with

(6.5.15) $B_1 = d,$

and

(6.5.16) $A = \log c,$

(6.5.17) $c = \text{antilog } A.$

This theoretical model has been offered by Stevens (1961) as the psychophysical law which relates v, the energy of the physical stimulus (light, sound, weight, electric shock) measured in appropriate units, to the perceived magnitude of the sensation, w. In this model, the exponent or power d estimates a parameter that characterizes the specific sensory function and conditions, and is not dependent on the units of measurement, whereas c does depend on the units in which w and v are measured. We stress that Stevens' law is not merely one of finding an equation which fits data—it is rather an attempt at a parsimonious description of how human discrimination proceeds. It seeks to replace Fechner's law, which posits a different fundamental equation, one of the form of Eq. (6.5.1).

We do not specify the base of the logarithm in the above transformations. Either common (base 10), or natural (base e), or any other logarithmic system can be used, since logarithms to different bases are proportional to each other and hence perfectly linearly related.

Although logarithms provide the most generally useful procedure for effecting linearity, other functions of variables are sometimes useful. Take, for example, the hyperbolic function

(6.5.18) $w = \dfrac{v}{c + dv}.$

By algebraic manipulation, it can be written as a linear function of the reciprocals of w and v:

(6.5.19) $\dfrac{1}{w} = c \, \dfrac{1}{v} + d.$

With $\hat{Y} = 1/w$, and $X_1 = 1/v$, the equation is linear in $\hat{Y}$ and X_1 with, in terms of our standard representation of the regression equation, $B_1 = c$ and $A = d$.

The transformation of nonlinear equations into linear form may sometimes be accomplished by transformation of variables other than logarithmic or reciprocal. For example, some geometric transformation or taking a square root may turn the trick. On the other hand, not all equations lend themselves to linear transformation—they are "intrinsically nonlinear" (Draper & Smith, 1966, p. 264).

Even when we accomplish the transformation into linear form, a problem exists which is worth mentioning. The constants B_i and A computed for the transformed equation are those which minimize the squared deviations of Y (as always). When Y is a transformed variable (for example, $\log w$ or $1/w$), minimization of the squared Y deviations does not generally result in the minimization of the deviations in w; hence, the constants are not least-squares estimates in w. (For a treatment of this and other problems in handling nonlinear relationships, see Draper & Smith, 1966, Chapters 5 and 10, and references; Ezekiel & Fox, 1959.)

In addition to the interest in estimating the parameters, an obvious question about a model is how well it fits the data. This will be appraised by R^2 (or r^2), keeping in mind that aside from all other considerations its value depends on the range of v employed, but the details require some attention. In the typical "strong theory" situation, the independent variable (v) will be represented by a (usually small) number of values, u, at each of which a number of independent observations (n_i) of w have been obtained. For concreteness, assume the experiment to be in psychophysics and on a single subject, with 8 ($= u$) points along the dimension of physical energy (v) at each of which he gives 12 ($= n_i$) subjective magnitude responses (w), a total of 96 ($= n$) observations. The model used is that of Eq. (6.5.13), which, through logarithmic transformations of both variables, becomes linear in $\hat{Y}$ ($= \log w$) and X_1 ($= \log v$).

Now the question, "How well does the model fit the data over the range of v employed?" needs closer specification. Each of the 8 points along X_1 has a mean for its 12 Y observations. The adequacy of the fit can be appraised either with regard to these 8 means or with regard to the 96 observations. Thus, as the AV has made familiar, the total sum of squares of Y can be partitioned into a between stimulus level portion reflecting the variation of the 8 ($= u$) means with 7 ($= u - 1$) df, and a within stimulus level portion reflecting the variation in response to a given stimulus level with $df = n - u = 88$. The variation accounted for by the model is, in turn, a part of the variation of the stimulus level means, and its proportion of the latter can be taken as a measure of the model's fit over the v range studied. From this perspective, the within stimulus variation is considered "pure" error, and not the business of the model. An alternative basis on which to appraise the model is the total variation in Y including the within stimulus variation. Which of these is deemed appropriate is purely a matter of the substantive theory involved. We will show how both can be determined by a single hierarchical MRC analysis.

The proportion of the total Y ($= \log w$) variance accounted for by the model (6.5.14), which posits a linear relationship with X_1 ($= \log v$) is given by r^2_{Y1} .

Assume it to be .70. Now the proportion of Y variance accounted for by stimulus levels (that is, by the V construct over the given range without any scaling assumptions) can be determined by nominalization: we can simply treat the 8 $(= u)$ groups of observations as a nominal scale, coding them by 7 $(= u - 1)$ dummy variables, and starting afresh, determine $R^2_{Y \cdot 12 \ldots 7}$ for this set, as was described in Section 6.4. Assume it to be .73. However, we can find this same value by proceeding hierarchically. Recall from Chapter 5, and particularly from the discussion of nonsense coding of nominal scales (Section 5.6) that *any* nonredundant coding of the u levels into $u - 1$ IVs will represent the scale. Instead of separately running R^2 for 7 dummy variables after determining r^2_{Y1} for $X_1 = \log v$, we can proceed cumulatively after $X_1 = \log v$, by adding into the equation *any* $u - 2 = 6$ dummy variables X_2, X_3, X_4, X_5, X_6, X_7. The R^2 obtained for these seven IVs, $X_1 = \log v$ plus the six dummy variables, will exactly equal the R^2 obtained from seven dummy variables, .73. (In AV terms, the model variance is part of the between stimulus level variance.) We thus have the following results:

$$R^2_{Y \cdot 1} \ (= r^2_{Y1}) = .70; \quad R^2_{Y \cdot 12 \ldots 7} = .73; \quad 1 - R^2_{Y \cdot 12 \ldots 7} = .27.$$

The model accounts for .70 of the total Y variance, the stimulus levels (Y means of the 8 values of V) account for .73 of the total Y variance, hence .27 of it is "pure" error or within stimulus level variance. The increment in total Y variance accounted for by stimulus level means over that accounted for by the model is a sR^2, which here reflects the "lack of fit to the means,"

(6.5.20) $I_{\text{LFM}} = R^2_{Y \cdot 12 \ldots 7} - R^2_{Y \cdot 1} = .73 - .70 = .03.$

An obvious question to pose is whether this increment is significant. We put to work the general Model I F test for an increment in R^2 of Eq. (4.4.2), where A is the set representing the model, here simply X_1 with $k_A = 1$, and B is the set representing the additional dummy variables necessary to make up the full between stimulus levels variance, here X_2 through X_7, with $k_B = 6$. Thus,

$$F = \frac{.73 - .70}{1 - .73} \times \frac{96 - 1 - 6 - 1}{6} = \frac{.03}{.27} \times \frac{88}{6} = 1.630 \quad (df = 6, 88)$$

and is not significant.

Before interpreting this result, we should raise questions about a descriptive formulation of the model's adequacy. As noted, one alternative is to use as a base the variance of all 96 observations, of which the model over the range of v employed accounts for .70 $= r^2_{Y1}$. Is it large enough? That is a substantive issue. Is it significant? It is highly significant ($F = 219!$), but equally trivial: all it says is that the model certainly accounts for a nonzero amount of variance. But then, so would almost any model which might be posited other than facetiously. For example, if r^2 between $\log w$ ($= Y$) and $\log v$ ($= X_1$) is .70, then r^2 between w and v would likely be in the vicinity of .65, which would also be highly significant ($F = 175!$), and this linear model would not be seriously entertained.

Indeed, it is foregone that any experiment of this kind would yield a statistically significant r^2 or R^2 for the model.

But aside from its significance, is the total Y variance the appropriate base for a proportion due to the model? It might be argued that the theory which yields the model does not attempt to account for "pure" error, but only seeks to fit the Y means for the stimulus values. In this case, the base is not properly unity, but the between stimulus variance, or nonerror variance, that is,

(6.5.21) $$\frac{R^2 \quad \text{for model}}{R^2 \quad \text{for stimulus values}} = \frac{R^2_{Y\cdot 1}}{R^2_{Y\cdot 12\ldots 7}},$$

or, for this example, $.70/.73 = .96$. This is rather more impressive than $.70$, and assuming the substantive validity of removing the "pure" error from the base, suggests that 96% of the variance of the response magnitude means was accounted for by the model for the given range of v studied. Also, the finding that the $.03$ value for lack of fit to the means is not significant, directly implying that the departure of $.96$ from 1.00 is not significant, have we then "proved" the model?

Unfortunately, not quite. This would involve the classical statistical fallacy of proving the null hypothesis, specifically, concluding from the finding that the sample $I_{LFM} = .03$ is not significantly different from zero, that the population I_{LFM} is in fact zero. This need not be the case—increasing the number of observations, and hence the statistical power of the test, might result in a significant departure from zero. The nonsignificance of the $.03$ value is properly interpreted to mean that our results are consistent with the model. But our model is not thus proven, since other models, perhaps implying markedly different processes of sensory discrimination, might be offered, now or later, with which the data are equally consistent, or more consistent. Although low, nonsignificant I_{LFM}s are encouraging, the adequacy of a model is appraised in competition with other models. Nor is the decision purely statistical—models come from theoretical networks that must satisfy other demands, such as plausibility and parsimony.

To avoid undue complexity in this brief survey, we have considered models with only one independent variable. If the theoretical model equation $w = f(v, t)$ is capable of transformation into linear form, the generalization of the previous material is straightforward. The B_i and A will estimate interpretable parameters of the theory. To assess the adequacy of the fit of the model by hierarchical MRC, since there are two independent variables, we find the between stimulus (or whatever) level variance (or nonerror variance) by adding to them any $u - 3$ dummy variables. In Eq. (6.5.20), we replace $R^2_{Y\cdot 1}$ by $R^2_{Y\cdot 12}$ to find I_{LFM}, and for the F test, the set A represents the model $X_1 = f(v)$ and $X_2 = f(t)$ with (therefore) $k_A = 2$, and the set B is made up of the $u - 3 = k_B$ dummy variables representing (after the variance of the model is removed) lack of fit.

The interested reader will wish to refer to more detailed and more specialized sources on the building and testing of theoretical models. For psychology, we particularly recommend Lewis (1960) and Luce, Bush, and Galanter (1963); in sociology, Coleman (1964) and Blalock and Blalock (1968); in economics, Johnston (1963).

6.5.3 Weak Theoretical Models

Under this rubric we intend to discuss linearizing transformations in relationships between variables which are well below the level of exact mathematical specification of the strong theoretical models discussed above, yet above the level of purely empirical cut-and-try. If, when we measure some construct V using units of some particular scale v, we know or suspect a priori that additive changes in v are not associated with additive changes in w, we know that a linear relationship does not obtain. Now, given the context of this a priori knowledge, we may be able to specify a particular nonlinear transformation of v (nonlinear aspect of V), or of the dependent variable w, which will likely linearize the relationship. Certain variables invite certain expectations about the effective linearizing transformation. It is these expectations that we characterize as weak theory.

Logarithms and Proportional Change

When, as v changes by a constant *proportion*, w changes by a constant additive amount, then w is a logarithmic function of v; hence, w is a linear function of log v:

v	8	12	18	27	40.5
$\log_{10} v$	.90	1.08	1.26	1.43	1.61
w	5	8	11	14	17

The constant proportionate increase (by one-half) in v results in unequal additive increases (by respectively 4, 6, 9, and 13.5), while w increases additively by a constant of 3 units; log v, however increases additively by a constant of .18 (within rounding error).[12] Thus, $w = Y$ and log $v = X_1$, and the relationship is cast into linear form.

Conversely, if constant additive changes in v are associated with proportional changes in w, then log w (= Y) is a linear function of v (= X_1), and again the linear regression model correctly represents the relationship.

Finally, proportionate changes in v may be associated with proportionate changes in w, for example,

v	8	12	18	27	40.5
w	2	4	8	16	32

[12] These are common (base 10) logarithms. Logarithms to any other base would also increase by a constant amount, albeit by a different constant: for natural (base e) logarithms, log v would increase by .69 for proportional increases in v of .5.

If logarithms of both variables are taken, then

$$X_1 = \log v \qquad .90 \quad 1.08 \quad 1.26 \quad 1.43 \quad 1.61$$

$$Y = \log w \qquad .30 \quad .60 \quad .90 \quad 1.20 \quad 1.51$$

Each proceeds by constant additive changes and again Y is a linear function of X_1 after logarithmic transformation.

We note, in passing, that the relationships of the last three paragraphs may be formally stated by the equations of the mathematical models given in the last section, Eqs. (6.5.1), (6.5.7), and (6.5.13). However, our present "weak theory" framework is more modest; we are here not generally interested in estimating model parameters (c and d) since we do not have a theory which generated equations in the first place. All we have is a notion that when we measure a certain construct V by means of a scale v, it changes (approximately) proportionately in association with changes in other variables. This is sufficient to represent V not by its raw scale value v, but rather by $\log v$, in order to make its relationships to other variables linear (or more nearly linear).

Consider, for example, the construct of "chronicity" in mental hospital patients (V). Its raw measure is years of hospitalization (v). We would be content to let v represent V if we believed it generally true that the difference between 1 and 2 years of hospitalization represented the same "amount" of chronicity as the difference between 9 and 10 years. This seems unlikely; we would probably prefer to suppose that chronicity accrues at a diminishing rate as years of hospitalization increases. Log v has this property,[13] more specifically, the difference between log 1 and log 2 is equal to the difference between log 5 and log 10. In a study which relates chronicity (with or without other IVs) to intellectual deterioration, we might therefore represent V by $\log v$.

The question as to what function of v represents equal "amounts" of V is difficult to discuss in the abstract, dependent as it is upon substantive issues in the area under investigation, past experience, and intuition (not to mention those of philosophy of science). But we do know that the logarithmic transformation of certain kinds of vs are more likely to be more linearly related to other variables than the variables themselves. This tends generally to be the case for measures of time. This is particularly true when the variable is indexing growth, since biological growth tends to be proportional to size. Thus, such variables as age, or time-related ordinal variables like learning trials or blocks of trials are frequently effectively log-transformed. Even when physical growth is not involved, the logarithmic transformation of time units may linearize relationships, as in the psychiatric chronicity example above, or for time required to complete

---

[13] So do many other nonlinear transformations, for example, $\sqrt{v}$. The difference between $\sqrt{1}$ and $\sqrt{2}$ is equal to the difference between $\sqrt{7.56}$ and $\sqrt{10}$. The rate of increase is greater for the square root than for the logarithmic transformation. We presume above that from past experience with this variable, we prefer the logarithmic rate.

a standard task, or reaction time (but see below). This is also frequently the case for physical variables, as for example energy (intensity) measures of light, sound, electric shock or chemical concentration of stimuli in psychophysical or neuro-psychological studies, or physical measures of biological response. At the other end of the behavioral science spectrum, it has been noted that variables like family size and counts of populations as occur in vital statistics or census data are frequently made more tractable by taking logarithms. So, often, are variables expressed in units of money, for example, annual income or gross national product.

By logarithmic transformation, we intend to convey not only log v, but such functions as log $(v - K)$ or log $(K - v)$, where K is a nonarbitrary constant. Note that such functions are not linearly related to log v so that when they are appropriate, log v will not be. K, for example, may be a sensory threshold or some asymptotic value. Examples of the use of such transformations in psychology are given by Lewis (1960).

The Square Root Transformation

The most likely use of a square root transformation occurs for variables which arise in conformity with a "Poisson process." The Poisson distribution describes the distribution of the number of times per sample in which a rare event occurs within either a restricted period of time or space (or both), when occurrences are independent of each other. For example, the number of first mental hospital admissions in 1970 for each of New York City's Health Areas, or this number for a given Health Area for each of 30 years are variables which may yield a Poisson distribution. Counts of errors in cognitive or perceptual tasks may also be so distributed.

In a Poisson distribution, the mean and variance are equal, thus the distribution is positively skewed (that is, have a long, right-hand tail). If w is a dependent variable, which depending upon the values of a set of IVs gives rise to different Poisson distributions, then changes in the general (mean) level of w will necessarily be accompanied by similar changes in the variance of w, thus violating the assumptions of constant dependent variable variance and normality in the fixed-regression model. Further, and at least as serious, there is the likelihood that w will not be linearly related to other variables.

Such data are effectively handled by taking $\sqrt{w}$ as Y (or X_i). This will likely operate so as to equalize the variance, reduce the skew, and linearize relationships to other variables. Several refinements of this transformation have been proposed, namely $\sqrt{w + .5}$ (Bartlett, 1936) and $\sqrt{w} + \sqrt{w + 1}$ (Freeman & Tukey, 1950).

Note that in the spirit of this section, a square root transformation is applied not because a strong mathematical model dictates it, but rather because the data are counts of events which are rare in the sense that for any given condition, the mean of such a set is much smaller than the largest value, and which for varying conditions, means are equal (or proportional) to variances. Many count measures are of this nature, even Rorschach subscores.

The Reciprocal Transformation

Reciprocals arise quite naturally in the consideration of rate data. Imagine a perceptual–motor or learning task presented in time limit form—all subjects are given a constant amount of time (T), during which they complete a varying number of units (u). One might express the scores in the form of rates as u/T, but since T is a constant, we may ignore T and simply use u as the score. Now, consider the same task, but presented in work limit form—subjects are given a constant number of units to complete (U), and are scored as to the varying amounts of time (t) they take. Now if we express their performance as rates, it is U/t, and if we ignore the constant U, we are left with $1/t$, not t. If rate is linearly related to some other variable v, then for the time limit task, v will be linearly related to u, but for the work limit task, v will be linearly related not to t, but to $1/t$. There are other advantages to working with $1/t$. Often, as a practical matter in a work limit task, a time cutoff is used that a few subjects reach without completing the task. Their exact t scores are not known, but they are known to be very large. This embarassment is avoided by taking reciprocals, since the reciprocals of very large numbers are all very close to zero and the variance due to the "error" of using the cutoff $1/t$ rather than the unknown true value of $1/t$ is negligible relative to the total variance of the observations.

Reciprocals of variables are of use in other circumstances (Lewis, 1960, pp. 92–97), but we can offer no helpful easy generalizations. We note, however, that hyperbolas involve the reciprocals of either or both variables (e.g., Eq. 6.5.19 above), and are characterized by asymptotes parallel to either or both axes of a graph. If one's weak theory suggests an asymptotic curve, recourse to reciprocalization should be considered.

6.5.4   No Theoretical Model

For the sake of completeness and symmetry (but without great enthusiasm), we offer the possibility of nonlinear transformations in the absence of any prior theoretical model. With special graph paper one or both of whose axes are scaled in logarithmic or other nonlinearly transformed unit of the original measurement scale, one can simply seek a good fit. Or, with plots on ordinary graph paper and some experience or guidance from what different nonlinear functions look like (e.g., Ezekiel & Fox, 1959, pp. 69–74), one can find linearizing transformations of one or both variables. However, the question arises as to why such a task is undertaken and we will turn to its consideration later.

First, it must be noted that in such "empirical" fitting, there will often be two or more different nonlinear transformations which do about equally well. In the absence of any theoretical basis, the choice is arbitrary. The choice of the better (or best) gives of course no assurance that it is better in the population from which the data were sampled or when the transformed variable is correlated with a new variable. Indeed, when the better or best is chosen to describe the relationship in the sample, one is likely to be overstating it for the population because of capitalization on chance.

Consider again the example used to illustrate a quadratic polynomial fit (Section 6.2.2), analyzed in Table 6.2.1 and graphed in Figure 6.2.1. The squared correlation of v with Y was .5883 (= $.767^2$, see Table 6.2.1), and the relationship both looks (Figure 6.2.1) and turns out upon significance testing of I_2 to be curvilinear. Now consider that the r^2s of v with $1/Y$ is .6717, with log Y .6405, and with $\sqrt{Y}$ .6167, all larger than .5883. Now the largest and smallest of these do not differ significantly ($t = 1.53$, $df = 33$),[14] and it would be difficult to justify on statistical grounds the use of the reciprocal rather than the log or square root to describe the population relationship. This example is not atypical. The similarity of r^2 for different monotonic nonlinear transformations of scorelike data is a consequence of the high r^2 among such transformations. We have already seen this phenomenon for power polynomials (e.g., Section 6.2.2). For these data, the r^2s among $1/Y$, log Y and $\sqrt{Y}$ are .9773, .9497, and .9938.

If the pitfalls described above are understood, the use of a nonlinear transformation lacking a theoretical basis because it tidies up the data is a (marginally) legitimate procedure. Tidiness, after all, is an integral part of the business of science. What is dubious is such use for the purpose of conferring an aura of capital-S Science on modest data, a temptation that frequently besets the soft behavioral scientist. It strikes us that the use of the frankly atheoretical polynomial (power or orthogonal) as described above is to be preferred in such situations.

On the other hand, certain uses of nonlinear transformations are clearly legitimate. If a transformation found to "work" in one sample is then used in another sample, or in relationships involving other variables, no capitalization on chance occurs. The latter bridges over to the happiest use of nonlinear transformations in the absence of prior theory—the suggestion of some theory provided by the effective nonlinear transformation. This is now a suggestion of how something works, which can then serve as a hypothesis to be tested with new data.

6.5.5 Transformations of Proportions[15]

Counted fractions (u/v), where both the numerator and demoninator vary, are a ubiquitous mode of expression in behavioral science. They are variously further

[14] The test of the significance of the difference between two correlation coefficients which share an array in common (here v) is due to Hotelling (1940). It was given in Chapter 2 as Eq. (2.8.8).
[15] A strictly logical organization would subsume the material in this section (and also the next) under the section on weak theoretical models (Eq. 6.5.3). The distinctions of this area of data representation, however, make it more convenient to treat it by itself.

expressed numerically as proportions, percentages, per thousand or per some other power of ten, the choice being purely a matter of convenience or number of digits desired. Thus, such variables as percent human responses on the Rorschach for psychiatric patients ($100\ u/v$), proportion of women employees in a sample of light manufacturing companies (u/v), and number of newborn of low birth weight per 10,000 live births in New York City Health Areas ($10,000\ u/v$) all illustrate this mode of measurement. The denominator v is used to qualify the numerator u, usually to adjust for amount or size in some sense.

The transparency and straightforwardness of proportions is unfortunately deceptive, for several reasons (additional to those discussed in Section 2.11.5):

1. For division by the denominator to be an appropriate method for qualifying the numerator makes an implicit assumption that the correlation of u with v is perfectly linear, or nearly so, and that the regression line goes through the origin ($u = v = 0$). Otherwise, one is either over- or underadjusting or adjusting nonuniformly over the range. We will pursue this problem no further here, but in Chapter 9 we take up the general problem of adjusting variables.

2. When the observations (Rorschach responses, employees, live births) are viewed as samples, we note that the proportions they yield are not of constant variance. Recall that the variance of a proportion equals $p(1 - p)$, and thus varies with p, and further, the sampling distribution of p is not generally normal. When used as a dependent variable, these properties formally violate the assumptions of the underlying fixed model sampling theory (which violation may or may not be serious).

3. The unit of measurement for proportions is almost never constant over the scale. It is intuitively evident (at least to some) that the .04 difference between .01 and .05 is much more important than the .04 difference between .48 and .52 "for almost all purposes except winning elections," as Tukey[16] says. In ratio terms, for example, from .01 to .05 represents a fivefold increase, while from .48 to .52 an increase merely of 8%. Similarly, and quite symmetrically, a change from .95 to .99 is much more important than the .48 to .52 change, the former can be thought of as going 80% of the way toward the maximum, that is, $.04/(1 - .95)$, the latter not quite 8%, that is, $.04/(1 - .48)$. Thus, the .00–1.00 scale of proportions or linear transformations thereof is bunched up at its extremes relative to its center. In most circumstances, one needs a nonlinear transformation which opens it up and "stretches the tails" in order to achieve a unit of measurement which is more nearly linearly related to other variables.

Three tail-stretching transformations of proportions have been found particularly useful in linearizing relationships: the arcsine, the probit, and the logit transformations. All three progressively stretch the scale as one proceeds from $p = .50$ in both directions symmetrically toward .00 and 1.00. For none of them is there a material amount of stretching in the .25–.75 range of p, so that if all or

---

[16] The treatment in this section has gained greatly from our reading of an unpublished manuscript by John Tukey, "Data Analysis and Behavioral Science," June 1962.

almost all of one's observations fall in this range, there is little point in using any of these transformations. All are available in commercial graph paper ruled according to the transformation. In the absence of an exact mathematical model, no a priori basis for choosing among them can be offered, but possibly with the aid of graphic plotting or previous experience with a given measure in a given substantive area, a choice can be made. However, the choice may not be crucial; it is generally the case that if one of these transformations of proportions makes relationships for a given variable more nearly linear, the others will also. Any are likely to perform better than the untransformed proportion.

The three transformations stretch the tails to varying degrees, and will be discussed in order of increasing degree of stretching, that is, arcsine, probit, and logit. Table 6.5.1 presents the three transformations.

The Arcsine Transformation

The arcsine (or angular, or "angit") is the most familiar of the transformations used with proportions, and often the only one presented in textbooks oriented toward the behavioral sciences. Specifically, it is

$$(6.5.22) \qquad A = 2 \arcsin \sqrt{p},$$

that is, twice the angle (measured in radians) whose trigonometric sine equals the square root of the proportion being transformed.

Table 6.5.1 gives the A values for p up to .50. For A values for $p > .50$, let $p' = 1 - p$, find A_p from the table, and then compute

$$(6.5.23) \qquad A_{p'} = 3.14 - A_p.$$

For example, for the arcsine transformation of .64, find A for .36 (= 1 − .64) which equals 1.29, then find 3.14 − 1.29 = 1.85. Table 6.5.1 will be sufficient for almost all purposes, but for a more exact statement of the transformation of $p = 0$ and 1, and for a denser argument of p with A to four decimal places, see Owen (1962, pp. 293–303).

The amount of tail stretching effected by a transformation may be indexed by the ratio of the length of scale on the transformation of the p interval from .01 to .11 to that of the interval from .40 to .50. For A, this index is 2.4 (compared with 4.0 for the probit and 6.2 for the logit).

Apart from its tail-stretching and hence linearizing property, the arcsine transformation has a unique property which commends it particularly for use as a dependent variable, namely, constant variance. Unlike p itself, or the other transformations, the sampling distribution of A under the binomial model is approximately normal, and the variance of A does not change with its value, but remains constant at approximately $1/v$, where v is the denominator of the counted fraction giving rise to the proportion. Note, however, that constancy is lost if the denominators vary across the observations.

TABLE 6.5.1
Arcsine (A), Probit (P), and Logit (L) Transformations
for Proportions $(p)^a$

p	A	P	L	p	A	P	L
.000	.00	$-^b$	$-^b$	.16	.82	4.01	−.83
.002	.09	2.12	−3.11	.17	.85	4.05	−.79
.004	.13	2.35	−2.76	.18	.88	4.08	−.76
.006	.16	2.49	−2.56	.19	.90	4.12	−.72
.008	.18	2.59	−2.41	.20	.93	4.16	−.69
.010	.20	2.67	−2.30	.21	.95	4.19	−.66
.012	.22	2.74	−2.21	.22	.98	4.23	−.63
.014	.24	2.80	−2.13	.23	1.00	4.26	−.60
.016	.25	2.86	−2.06	.24	1.02	4.29	−.58
.018	.27	2.90	−2.00	.25	1.05	4.33	−.55
.020	.28	2.95	−1.96	.26	1.07	4.36	−.52
.022	.30	2.99	−1.90	.27	1.09	4.39	−.50
.024	.31	3.02	−1.85	.28	1.12	4.42	−.47
.026	.32	3.06	−1.81	.29	1.14	4.45	−.45
.028	.34	3.09	−1.77	.30	1.16	4.48	−.42
.030	.35	3.12	−1.74	.31	1.18	4.50	−.40
.035	.38	3.19	−1.66	.32	1.20	4.53	−.38
.040	.40	3.25	−1.59	.33	1.22	4.56	−.35
.045	.43	3.30	−1.53	.34	1.25	4.59	−.33
.050	.45	3.36	−1.47	.35	1.27	4.61	−.31
.055	.47	3.40	−1.42	.36	1.29	4.64	−.29
.060	.49	3.45	−1.38	.37	1.31	4.67	−.27
.065	.52	3.49	−1.33	.38	1.33	4.69	−.24
.070	.54	3.52	−1.29	.39	1.35	4.72	−.22
.075	.55	3.56	−1.26	.40	1.37	4.75	−.20
.080	.57	3.59	−1.22	.41	1.39	4.77	−.18
.085	.59	3.63	−1.19	.42	1.41	4.80	−.16
.090	.61	3.66	−1.16	.43	1.43	4.82	−.14
.095	.63	3.69	−1.13	.44	1.45	4.85	−.12
.100	.64	3.72	−1.00	.45	1.47	4.87	−.10
.11	.68	3.77	−1.05	.46	1.49	4.90	−.08
.12	.71	3.83	−1.00	.47	1.51	4.92	−.06
.13	.74	3.87	−.95	.48	1.53	4.95	−.04
.14	.77	3.92	−.91	.49	1.55	4.97	−.02
.15	.80	3.96	−.87	$.50^a$	1.57	5.00	.00

^{a}See text for values when $p > .50$.
^{b}See text for transformation when $p = 0$ or 1.

The Probit Transformation

This transformation is variously called "probit," "normit," or, most descriptively, the "normalizing transformation of proportions," a specific instance of the general normalizing transformation (see Section 6.5.6). We use the term "probit" in recognition of its wide use in bioassay, where it is so designated.

Its rationale is straightforward. Consider p to be the cumulative proportion of a unit normal curve (that is, a normal curve "percentile"), determine its base-line value, z_p, which is expressed in sd departures from a mean of zero, and add 5 to assure that the value is positive, that is,

$$(6.5.24) \qquad P = z_p + 5.$$

Table 6.5.1 gives P as a function of p for the lower half of the scale. When $p = 0$ and 1, P is at minus and plus infinity, respectively, something of an embarrassment for numerical calculation. We recommend that for $p = 0$ and 1, they be revised to

$$(6.5.25) \qquad p_0 = \frac{1}{2\,v},$$

and

$$(6.5.26) \qquad p_1 = \frac{2\,v - 1}{2\,v},$$

where v is (as throughout), the denominator of the counted fraction. This is arbitrary, but usually reasonable. If in such circumstances, these transformations make a critical difference, prudence suggests that this transformation be avoided.

For P values for $p > .50$, as before, let $p' = 1 - p$, find P_p from Table 6.5.1, and then find

$$(6.5.27) \qquad P_{p'} = 10 - P_p.$$

For example, the P for $p = .83$ is found by looking up P for .17 ($= 1 - .83$) which equals 4.05, and then finding $10 - 4.05 = 5.95$. For a denser argument for probits, which may be desirable in the tails, see Fisher and Yates (1963, pp. 68–71), but any good table of the inverse of the normal probability distribution will provide the necessary z_p values (Owen, 1962, p. 12).

The probit transformation is intermediate in its degree of tail stretching—as noted in the previous subsection, its index is 4.0, compared with 2.4 for the arcsine and 6.2 for the logit.

The probit transformation is intuitively appealing whenever we conceive that some construct, if measurable directly, would be normally distributed, but our available measure is instead a cumulative proportion (or frequency). Thus, we go from the ordinate of such an assumedly normal cumulative distribution to its baseline measure. Such a transformation will often seem plausible in some areas of experimental and physiological psychology, psychometrics, and epidemiology.

The Logit Transformation

This transformation is related to the logistic curve, which is similar in shape to the normal curve, but generally more mathematically tractable. The logit transformation is

(6.5.28)
$$L = \frac{1}{2} \ln \frac{p}{1-p},$$

where ln is the natural logarithm (base e). As with probits, the logits for $p = 0$ and 1 are at minus and plus infinity, and the same device for coping with this problem (Eqs. 6.5.25 and 6.5.26) is recommended: replace $p = 0$ by $p = 1/(2v)$ and $p = 1$ by $(2v - 1)/(2v)$, and find the logits of the revised values. As before, Table 6.5.1 gives the L for p up to .50; for $p > .50$, let $p' = 1 - p$, find L_p and change its sign to positive for $L_{p'}$, that is,

(6.5.29)
$$L_{p'} = -L_p.$$

For $p = .98$, for example, find L for .02 (= 1 − .98), which equals −1.96, and change its sign, thus L for .98 is +1.96.

The logit stretches the tails of the p distribution the most of the three transformations. The tail-stretching index (described above) for the logit is 6.2, compared with 4.0 for the probit and 2.4 for the arcsine.

The quantity $p/(1 - p)$ is the odds related to p, for example, when $p = .75$, the odds are .75/.25, or 3:1, or simply 3. The logit, then, is simply half the natural logarithm of the odds. Therefore logits have the property that for equal intervals on the logit scale, the odds are changed by a constant multiple; for example an increase of .35 on the logit scale represents a doubling of the odds, since .35 is $\frac{1}{2} \ln 2$.

We also note the close relationship between the logit transformation of p and Fisher's z' transformation of the product moment r (see Section 2.8.3 and Appendix Table B). If we let $r = 2p - 1$, then the z' transformation of r is the logit of p. We can take advantage of this fact if we wish a denser argument in the tails than is provided in Table 6.5.1, since r to z' transformation tables are more readily available than p to L tables. A useful logit transformation table is given by Fisher and Yates (1963, p. 78).

Note that all three transformations are given in the form most frequently used or most conveniently tabled. They may be further transformed linearly if it is found convenient by the user to do so. For example, if the use of negative values is awkward, one can add a constant to L of 5, as is done for the same purpose in probits. Neither the 2 in the arcsine transformation (Eq. 6.5.22), nor the $\frac{1}{2}$ in the logit transformation (Eq. 6.5.28) is necessary for purposes of correlation, but they do no harm, and are tabled with these constants as part of them, in accordance with their conventional definitions.

6.5.6 Normalization of Scores and Ranks

Normalization is more widely applicable as a nonlinear monotonic transformation than was described above in connection with proportions. Whenever it seems reasonable to suppose that a construct being measured by a variable v is represented with greater fidelity by rescaling it so that it yields a frequency distribution of normal form, this transformation may be applied. More specifically, as with other transformations, the usual goal is linearization of relationships with other variables.

A frequent candidate for normalization is ranked data. When a third grade teacher characterizes the aggressiveness of her 30 pupils by ranking them from 1 (most) to 30 (least) or vice versa, the resulting 30 values may occasion difficulties when they are treated numerically as measures. Ranks are necessarily rectangularly distributed, that is, there is one score of 1, one score of 2, . . . , one score of 30. If, as is likely, the difference in aggressiveness between the most and next-most (or the least and next-least) aggressive child is greater than between two adjacent children in the middle (for example, those ranked 14 and 15), then the scale provided by the ranks is not likely to produce linear relationships with other variables. This need to "stretch the tails" is the same phenomenon encountered with proportions; it presupposes that the distribution of the construct to be represented *has* tails, that is, is bell shaped, or normal. Since individual differences for many well measured biological and behavioral phenomena seem to approximate this distribution, in the face of ranked data, it is a reasonable transformation to apply in the absence of specific notions to the contrary. Even if the normalized scale is not optimal, it is likely to be superior to the original ranks.

The method for accomplishing this is simple. Following the procedure described in elementary statistics textbooks for finding centiles ("percentiles"), express the ranks as cumulative proportions, and refer these either to a unit normal curve table (Appendix Table C) to read off z_p, or use the P column of Table 6.5.1, where 5 has been added to z_p to yield probits.[17]

Having come this far, it is apparent that the original scale need not be made up of ranks at all, but may be any scale whose unit of measurement is believed not to be equal or even changing in size regularly from one end to the other. The basis for this suspicion may be a grossly irregular shaped frequency distribution, multimodal or strangely skewed. Such distributions may be rescaled into normality by dint of force: write the cumulative frequency distribution on the original scale, grouping into a dozen or so intervals if and as needed, and then go from the cumulative p values to their normal curve deviates, z_p (or P), as above. The result is a monotonic transformation to a normal distribution, however irregular the original scale.

Normalization should be used judiciously, but whenever an original scaling yields a wild distribution, and particularly where there is some basis for belief that the construct being assessed is usefully conceived as being more or less

---

[17] An alternative, slightly superior model for the normal transformation is tabled as "expected values of order statistics from the normal distribution" by Owen (1962, pp. 151–154), and goes up to $n = 50$. The difference between the models would be very slight unless n is quite small, say less than 10.

normally distributed in the population being sampled, it is worth the effort to normalize.

6.5.7 The Fisher z' Transformation of r

It is sometimes the case that a variable is measured and expressed in terms of r. The most common instance of this generally rare circumstance is in "Q sorting" (Stephenson, 1953), where items are sorted into rating categories of prescribed size (usually defining a quasi-normal distribution) so as to describe a complex phenomenon, such as personality. The similarity of two such Q-sort descriptions, for example, actual self and ideal self, is then indexed by the r between ratings over the set of items. Such a scaling is more likely to relate linearly to other variables if the rs are transformed by the Fisher z' transformation (as described in Sections 2.8 and 2.10), and are more likely to satisfy the normality and equal variance assumptions when used as a dependent variable. The z' transformation of r is given in Appendix Table B.

6.6 LEVEL OF QUANTITATIVE SCALE
AND ALTERNATIVE METHODS OF REPRESENTATION

The alternative methods of representing quantitative scales described in this chapter do not relate closely to the ordinal-interval-ratio distinction of Stevens (1951), but rather to the analytic goal and content of the research area. A given level of scaling may be represented by all, or almost all, of the methods of representation described above, although some combinations of level and method will occur more frequently than others. If one's purpose is primarily descriptive, or one wishes simply to assure that a construct is being represented in the IVs whatever the shape of the relationship, then simple fitting functions with multiple IVs such as polynomials (powers or orthogonal) or nominalization are appropriate, and the level of scaling is irrelevant. If one's approach is primarily analytic in that features of the curve have theoretical roots and implications, although simple fitting functions may be put to such use, the chances are that some mathematically defined nonlinear transformation into a single new variable (log v, or $\sqrt{v}$) will be mandated, or at least suggested, by the theory. But in this circumstance, it is likely that v is measured on a ratio scale. Normalization, which may be employed either for linearizing, for theoretical reasons, or both, can also be applied at any level of scaling.

6.6.1 Ratio Scales

When, as in ratio scales, there is a true zero point as well as equality of intervals (and almost always only positive values are defined), any method of representation is admissible, but some are likely to be more attractive than others. There is something about this highest form of measurement that invites elegance, for example, regular nonlinear transformations (logs, reciprocals), rather than the gross pulling and hauling of normalization, or the crude representation of nominalization. Still, what finally governs is the purpose of the analysis. If a sociologist wants to represent number of children among other indicators of socioeconomic status, his goals may be better served by polynomials or nominalization than by a log or square root transformation, despite the fact that number of children is unquestionably a ratio scale. An experimental psychologist working with ratio scales in learning, perception, or cognition might well make the other choice.

Proportions as measures are clearly ratio in nature, and the special tail-stretching transformations deserve first consideration, unless a specific mathematical model, which dictates some other nonlinear transformation, is involved. For example, in some mathematical models in learning theory, the logarithm of p is used. This transformation stretches the lower tail, but contracts the upper one.

6.6.2 Interval Scales

Scales whose units are more-or-less equal (see "Scaling," Section 6.2.4), but which have an arbitrary zero point, are the mainstay of the soft behavioral sciences, where MRC is likely to be a particularly useful data-analytic method. The equality of the units, of course, does not assure that interval scales will be linearly related to other quantitative scales. Again, depending on the purpose of the analysis, for independent variables the descriptive–representational polynomial and nominalization procedures are available, as well as normalization. When the variable in question is a dependent variable, the multiple coding of polynomials and nominalization is not available, but normalization is and may accomplish the linearization which is sought. The one-to-one nonlinear transformations (Section 6.5) are not generally attractive because of the arbitrariness of the zero in interval scales. The core idea of this level of scaling is that if an interval scale u is arbitrarily linearly transformed to $u' = cu + d$ (where c and d are any constants), u' contains exactly the same information for correlation as u does (i.e., $r_{uu'} = 1$), and it is a matter of indifference whether one uses u or one of the infinite number of alternatives provided by u'. But if one *non*linearly transforms u by a logarithmic, reciprocal, or geometric transformation, the result will not be perfectly correlated with those obtained by the same transformation applied to u'. There is thus an arbitrary flavor that attaches to such variables as $\log u$, $1/u$, and $\sqrt{u}$ when u is not a ratio scale. Of course, such transformations may "work" empirically, but then the strictures of empirical fitting described in Section 6.5.4 must be kept in mind.

6.6.3 Ordinal Scales

At this, the lowest, level of quantitative scales, neither a true zero nor equality of units exists—we merely take the rank order of the observations as the quantities to be used. As quantities, they do not represent in any sense how much of the attribute is possessed by the object being assessed, but rather how many of the objects fall above or below this object. Ordinal measures will not in general relate linearly to others (except other ordinal measures) because of the likely falsification they represent of the distribution of the attributes if they were actually measured. Ordinal measures (ranks, centiles, deciles, etc.) are necessarily distributed rectangularly, (for example, each rank occurs the same number of times, once). Since few constructs studied by behavioral scientists are conceived as rectangularly distributed, regressions involving ordinal data are not likely to be linear. Note that this is true whether the true distribution is normal, Poisson, or otherwise. If normal, the regression is likely to have two bends, one near each end, reflecting the need to stretch the tails.

With ordinal data, one may, of course, resort to the multiple IV descriptive–representational methods, polynomials or nominalization. These are serviceable procedures that avoid strong assumptions about the shape of the regression, and yet represent the ordinal variable adequately. Note that the real possibility of a double-bended regression line requires the use of at least three polynomial terms and at least four intervals for nominalization.

If there is reason to believe that the ranked attribute is quasi-normal in its distribution, or even if there is no reason to believe otherwise, normalization commends itself in that it will stretch the tails of the ordinal distribution appropriately. For use as a dependent variable, there exists no alternative to normalization (or transformation to some other explicitly hypothesized distribution form).

A word of caution to overeager normalizers. The rating scales which are so often used in soft behavioral science and technology (for example, "evaluation" research), frequently appear to the data analyst to be merely ordinal in character. We have in mind, for example, a one to four coding of the response alternatives "none, mild, moderate, severe" or "never, sometimes, often, always" on a psychiatric symptom rating scale. To be sure, there is no way to prove directly that subjectively equal intervals bounded by adjectival or adverbial modifiers are in fact equal, although the ordinal status of such scales is acceptable. Both theoretical statistical analysis and many empirical results, however, suggest that such scales have "equal enough" intervals for most purposes. Subjectively approximately equal intervals have been found to behave more like equal than like unequal intervals. For example, their regression lines with other variables are usually reasonably straight, and when not, the fault is likely to be with the other variable. We are not advocating that rating scales never be normalized or otherwise treated to allow for curvilinearity, but rather that they not be automatically suspected of requiring such treatment and routinely transformed. Such treatment should be undertaken only on the basis of a priori knowledge or theory, or possibly because of a grossly skewed distribution of the observations over the scale points.

6.7 SUMMARY

Quantitative variables of the ordinal, interval, or ratio type may be treated with provision for representing or studying the shape of their regression as independent or dependent variables in MRC analysis. Several methods of treating curvilinearity are considered:

1. *Power polynomials.* The multiple representation of a research factor V by $X_1 = v$, $X_2 = v^2$, $X_3 = v^3$, etc. makes possible the fitting of regression functions of Y on V of any shape. Hierarchical MRC makes possible the assessment of the size and significance of linear, quadratic, cubic (etc.) aspects of the regression function, and the multiple regression equation may be used for plotting nonlinear regression of Y on V. (Section 6.2)

2. *Orthogonal polynomials.* For some purposes (for example, laboratory experiments, where the number of observed values of V is not large), it is advantageous to code V so that the X_i not only carry information about the different curve components (linear, quadratic, etc.), but are orthogonal to each other as well. Some interpretive and computational advantages and alternative error models are discussed. (Section 6.3)

3. *Nominalization.* By reducing V to a relatively small number g of class intervals (if necessary), the resulting classes can be treated as if V were a nominal scale and represented by dummy variable coding. The MRC output yields the necessary information for plotting the regression function, that is, the Y means for each interval of V, and in R^2, the proportion of variance accounted for by V, whatever the regression shape. (Section 6.4)

4. *Nonlinear transformations.* Unlike the preceding methods, these are one-to-one mathematical transformations, and hence suitable for use as dependent variables, as well as independent variables. The circumstances in which logarithmic, square root, and reciprocal transformations are likely to be effective for linearization are described. Such transformations arise frequently in conjunction with mathematical models which are expressed in nonlinear equations; the nonlinear transformation of variables is resorted to in order to linearize the relationships and thus capitalize on the analytic power of MRC (Sections 6.5.1 through 6.5.4). Three tail-stretching transformations of proportions are described, the arcsine, probit, and logit transformations (Section 6.5.5).

5. *Normalization.* Any quantitative variable can be transformed into one whose frequency distribution is normal by changing its original units ad hoc. The method is described, and its advantages in certain circumstances (e.g., ordinal data) discussed (Section 6.5.6). When a variable is made up of Pearsonian rs (as in Q-sort analysis), the Fisher z' transformation is recommended (Section 6.5.7).

The chapter concludes with a discussion of the methods of representing quantitative scales from the perspective of Stevens' distinctions among the ordinal, interval, and ratio levels of scaling (Section 6.6).

7
Missing Data

7.1 INTRODUCTION

An all too frequent characteristic of research data is that some of the values called for by the structure of the research are missing for some subjects. This can occur for many reasons, to varying degrees, and in various patterns. The purpose of this chapter is to describe the various kinds of missing data, to consider some currently used alternatives for coping with the problem and to describe a simple but powerful method for handling it.

7.1.1 Types of Missing Data

There are a host of circumstances which result in missing data. In survey research, for example, whether carried out by face to face or telephone interviewing, or by questionnaire, some subjects may refuse or simply fail to respond to some items while responding to others. In laboratory experiments, equipment failure, animal mortality, a dropped tray of test tubes, or dropped-out subjects may create some blanks in some columns of the data sheets. In research in school settings, absences or transfer of pupils or teachers may result in incomplete data. And so on. It is only a slight exaggeration to paraphrase Murphy's law for behavior science research to read, "If there are any ways in which data can be lost, they will be."

To come to grips with the problem, we make several useful distinctions about missing data:

1. *Dependent versus independent variables.* Almost all of what follows concerns missing values for IVs. Since Y represents the "outcome" or "effect" of the IVs, when the Y value for a subject is not known, there is little which can be done in MRC but drop that subject. One might regret the apparent loss of information in the IVs when the subject is dropped, but in the "fixed model" situation (see Section 5.3.1), there *is* no information lost, since the investigator

has selected the combination of IV values to be studied. In factorial design analysis of variance (AV), where the investigator plans to have equal (or proportional) numbers of cases in the "cells," the loss of data on the dependent variable results in nonorthogonality (correlation) among the main effects and interactions. This, in turn, frustrates the computational simplicity of the AV, which can be viewed as that special case of MRC where the orthogonality of the effects simplifies the computation. Some treatments of factorial design give complicated exact or simple approximate methods which create values for "missing" dependent variables in order to restore the lost orthogonality and get on with the simplified AV calculations. As we have seen (Section 5.5.4), this circuitous route is quite unnecessary if an MRC analysis is performed; the "missing" Y values are simply nonexistent, and the analysis proceeds without special acrobatics.

When sampling is of the "random model" kind, where n subjects are randomly selected from some population and their X_i values observed rather than selected in advance (Section 5.3.1), one may indeed be losing information when subjects without Y values are dropped. But since the research is focally concerned with Y, we find unattractive, in general, attempts to make up Y values so as to avoid the loss of information in the IVs and the reduction in n. In dropping the subjects without Y values, however, we risk the possibility of the residual sample no longer being representative of the population we originally sampled. (A canonical-analysis approach to this problem is given in Chapter 11.) We turn to this issue next.

2. *Random versus selective loss of data.*   A critical question which an investigator must ask himself is *why* data are missing. In some situations, the presence or absence of information on some IV is randomly related to Y and the other IVs. For example, in an experiment in which recording equipment yielding information on an IV fails intermittently (while subjects are being run in a random or balanced order) the missing data is *randomly missing,* that is, the presence–absence of values on the IV is presumably unrelated to Y or other IVs. Note that it is the presence or absence of values and not their magnitude when present, which is at issue.

In other situations, most frequently occurring in survey research, the presence–absence of data on an IV may well be related to Y and/or other IVs. Consider a study of factors associated with the successful rehabilitation of drug addicts in which reported weekly wages on last job is one of several IVs. Some subjects claim they do not recall or refuse to respond. Another example is a retrospective study of the school records of mentally retarded adults as related to current vocational success where the school record of IQ is abstracted for use as an IV, and found missing in some cases. In neither of these instances would it be prudent to assume that the cases where the datum is missing are on the average like those for whom it is present with regard to Y and other IVs. It is quite possible that the refusal to respond on the income item is associated with lack of success in rehabilitation (Y) or the length of the addiction, or other IVs, and that absence of IQ in the school record is associated with high (or low) vocational status (Y), or ethnicity, or other IVs. The question for the investigator is whether the causal factors which produce absence of information overlap with those operating among the variables under study. An affirmative answer makes critical the decision of how to handle the missing data.

It is under this rubric that one can consider the question, "What if, for some subjects, *all* data are missing?" This is the familar problem of nonresponse in survey (particularly mail questionnaire) research. A carefully drawn sample of (for example) consumers is approached for inquiry, and some fraction does not respond at all—nonreturn of mail questionnaire, refusal to be interviewed, not at home when the interviewer comes, etc. In another context, a research on the relationship between tests of cognitive style (IVs) and IQ (Y) in hospitalized schizophrenics can, by definition, not obtain any data from untestable patients. The conclusions resulting from such studies must perforce be restricted to that subpopulation of the total population from which data are available. This is awkward, since it is usually the total population (consumers, hospitalized schizophrenics) that is of interest, and not the possibly different subpopulation (cooperative compliant consumers, testable hospitalized schizophrenics). The problem is mitigated to the extent to which the nonresponse fraction is small, and to the degree to which it is possible to assume, argue, or just hope that the basis for nonresponse is unrelated to the relationships under study. Given *no* data for the nonresponders, no opportunity for empirically examining this latter assumption is afforded. By contrast, as we will see in detail below, when Y values are available for cases with missing IV data, one can address the question of the implication of the missing data to the research issues.

3. *Many versus few missing data and their pattern.* Of less conceptual weight but nevertheless of practical importance is the question of how large a proportion of the data, either per IV or for all IVs, is missing. Obviously, the problem is of different magnitude if 1% as compared to 40% of the data are missing, and the choice of means of coping with the problem less critical in the first instance. Whether many or few, situations where all the missing data in multiple IVs are contributed by the same subjects require different consideration than those where they are scattered over most subjects. Similarly, we can distinguish circumstances where most missing data occur on a small proportion of the IVs from those where they are scattered over most IVs. We return to this topic in Section 7.4.

7.1.2 Some Alternatives for Handling Missing Data

The missing data problem is frequently handled by one of three methods, none of which is generally to be recommended:

1. *Dropping variables.* When, for one or a few variables, a substantial proportion of cases lacks data, the analyst may simply opt to drop the variables. This is obviously no loss when the variable(s) in question do not contribute materially

to accounting for Y variance. Indeed, it may yield a net gain in the statistical significance and precision of R^2 and the partial statistics of the remaining IVs due to the reduction in the number of IVs. If, however, the variable(s) in question provide criterion-relevant information (a reasonable presumption, else why were they included?), the loss of this information through dropping variables is hardly a satisfactory solution to the missing data problem.

2. *Dropping subjects.* When the pattern of missing IV data is such that they occur exclusively for a proportion of subjects (p_a) the analyst may opt to drop these subjects and perform the analysis on the remainder of the sample. The likelihood of selecting this "solution" increases if p_a is small and/or if n is large. Now, if p_a is small enough and n large enough, there can hardly be a great difference between the results obtained with the n_a subjects dropped and those which would have been obtained from all n. It is difficult to specify "small enough" and "large enough," but no serious objection can be raised to dropping 10 or 15 cases with missing data out of several hundred on practical grounds. But once p_a is more than a few percent and n is not comfortably large, some theoretical objections loom large. Are the data missing randomly? If so, then the results are not biased relative to those of the hypothetical complete data, but the loss of error df will result in the loss of statistical power to reject null hypotheses and larger standard errors (less precision) in estimating population parameters (correlation and regression coefficients). If the data are not missing randomly, this fact will go undiscovered, and the results will be unrepresentative of the population sampled. The statistical power and precision will also be reduced due to loss of error df, but since the conclusions hold properly for the unrepresentative subpopulation with all data present, this is an issue of less concern. Clearly, dropping cases is not a generally adequate solution to the missing data problem.

3. *The "missing-data correlation matrix."* Such a matrix is computed by using for each pair of variables (X_i, X_j) as many cases as have values for both variables. Thus, when data are missing for either (or both) variables for a subject, he is excluded from the computation of r_{ij}. In general, then, different correlation coefficients are not necessarily based on the same subjects or the same number of subjects. This correlation matrix yields R^2 and the other correlation results, and when combined with the m_i and sd_i computed from all available observations for each X_i, produces the B_i and A of the raw score regression equation.

This procedure is sensible if (and only if) the data are *randomly* missing. In this case, each r_{ij}, m_i, and sd_i^2 is an unbiased estimate of its population parameter for full data, and therefore the MRC results are similarly unbiased. This method leaves some awkwardness in statistical inference due to the varying n on which the MRC results are based. They are clearly not as sturdy as if the maximum n had obtained throughout, nor as frail as the minimum n would suggest. Still, working from the missing-data matrix is a tolerable procedure provided that the absence of data can be safely assumed to be random. On this assumption, there are available other methods which are beyond the scope of our presentation. The interested reader is referred to Timm (1970).

TABLE 7.1.1
Illustrative Missing-Data Score Matrix

Subject	Y	X_1	X_2	X_3
1	72	38	6	92
2	84	52	12	114
3	63	47		108
4	81	63	8	
5	47			
6	62		7	110
7	39	31	9	
8	61	56		93
9	71			130
10	46	44	10	86

Note: For example, for X_3, the *m* and *sd* based on all available values and entering r_{Y3} are 104.7 and 14.2, but those entering r_{13} are 98.6 and 10.6, and those entering r_{23} are 100.5 and 11.8.

In most MRC applications with missing data, however, the assumption of randomness is imprudent, if not patently false. When it fails, r_{ij} is not, in general, referrable to the same population as is r_{ig} or r_{jg} (to say nothing of r_{gh}); nor, in general, are the means and standard deviations of the different variables referrable to the same population. Table 7.1.1 presents a small data set with missing values which may clarify the complexity of the makeup of the statistics which go into the MRC when a missing data correlation matrix is used. The criterion correlations (r_{Yi}) are based on nonidentical subsets of subjects ($n_{Y1} = 7, n_{Y2} = 6, n_{Y3} = 7$). The IV correlations ($r_{ij}$), too, are based on nonidentical subsets of subjects ($n_{12} = 5, n_{13} = 5$, and $n_{23} = 4$). The means and standard deviations of X_i entering the r_{ij} differ with the X_j variable being correlated, since whether a given subject's X_i value is included depends on whether he has a value for the given X_j under consideration. Thus, Subject 4's X_1 value of 63 is included in the *m* and *sd* of X_1 which enters r_{12}, but not in the *m* and *sd* of X_1 which enters r_{13}. Finally, in r_{Yi} and in the raw score regression equation, all available values on each IV are used for the *m* and *sd*. Thus, Subject 9's X_3 value of 130 enters into this "official" *m* and *sd* of X_3, even though it was not included in the computation of the *m* and *sd* entering either r_{13} or r_{23}. A veritable mishmash!

Now, if data are missing for reasons related to Y and/or the IVs, the pieces put together for the MRC refer to systematically different subsets of the population. Clearly, the MRC results cannot be coherently interpreted with regard to the entire population, or to any specifiable subpopulation. They are a hodgepodge that offers the analyst the disagreeable choice between hoping that the degree and nature of the nonrandomness will not greatly invalidate the conclusions and simply throwing them away.

Indeed, under the circumstances of nonrandomly missing data, one can obtain a missing-data correlation matrix whose values are mutually inconsistent, that is, one which would be mathematically impossible to obtain were the data complete. For complete data, the correlation between any two variables is constrained by their correlations with a third. Thus, r_{12} may take on any value within the full limits of -1 to $+1$ if and only if X_1 and X_2 correlate 0 with all other variables under the sun. If this condition is not met, then the range of possible values for r_{12} is constrained. Specifically, for complete data, the mathematically possible upper and lower limits for r_{12} are given by

(7.1.1) $$r_{13} r_{23} \pm \sqrt{(1 - r_{13}^2)(1 - r_{23}^2)} .$$

Values for r_{12} outside these limits are inconsistent with the other two rs.

To illustrate, if $r_{13} = .8$ and $r_{23} = .4$, substitution gives $.32 \pm .55$: r_{12} can not take on a value less than $-.23$ or more than $+.87$. If $r_{13} = .8$ and $r_{23} = .6$, r_{12} is constrained to fall within the limits $.48 \pm .48$, that is, $.00$ and $.96$—a negative value for r_{12} is inconsistent, that is, mathematically impossible. Only if one substitutes $r_{13} = r_{23} = 0$ in Eq. (7.1.1), does one get the full ± 1 limits. If one tries to compute the R^2 of any one variable with the other two using inconsistent rs, one obtains mathematically impossible values for R^2, either greater than one or negative. Similarly, partial correlation values outside the ± 1 limits result from inconsistent rs. The constraints on rs and the consequences of using inconsistent values were illustrated for three variables for the sake of simplicity; they obviously obtain for more than three variables.

Now, although inconsistent rs may occur in a missing data correlation matrix where the data are *randomly* missing, this possibility is small, at least for reasonably large n. But when the absence of data is nonrandom, that is, related to the other variables in the system, the possibility of inconsistent rs is obviously larger, since the sample values are no longer estimating necessarily consistent population rs. For the conglomeration of subpopulations, the rs need of course not be consistent at all.

This is not to say that nonrandomly missing data will probably result in inconsistent rs and an error message from the computer (or an absurd result like an R^2 of $-.3$ or pr of 1.17, depending on the fastidiousness of the computer program). Unfortunately, nonrandomly missing data will more often than not yield consistent missing data correlation matrices, but that does not make them right, given the absence of a coherent population about which conclusions may be drawn. One is fortunate to obtain an inconsistent matrix as an absolute preventive of blundering to "conclusions."

Thus, the three methods frequently used to cope with missing IV data are not generally satisfactory. Dropping variables with missing data potentially loses information. Dropping subjects results in lower statistical power due to reduced n and, when missing data is nonrandom, gives results that are nonrepresentative of the original population. Working from a missing data correlation matrix with nonrandomly missing data gives nonrepresentative results (or none at all if the rs are not consistent), and with randomly missing data an ambiguous n for statistical inference. The problems of these latter two methods are of course mitigated when only a small proportion of the subjects have any missing data.

A serious flaw of all the above methods is their failure to use as potential information the *fact* that data are absent for certain subjects on certain research factors. It is desirable to represent as positive information the *absence* of data, and ascertain the degree and statistical significance of the criterion relevance of this information. In fact, what we mean by nonrandomly (or selectively) and randomly missing data is whether or not the absence of data accounts for criterion variance beyond the information resident in the values of the IVs which are present. In the balance of this chapter, we offer methods of coping with missing data which not only avoid the problems detailed above, but positively capitalize on whatever information is contained in the absence of some of the values from the data matrix.

7.2 MISSING DATA IN NOMINAL SCALES

Consider an f-level nominal scale to which we can assign n_p of our n subjects, but we do not know to which of the f levels the remaining n_a $(= n - n_p)$ subjects belong. Then, in fact we have a total of $g = f + 1$ groups, or an $f + 1$-level nominal scale, the additional level representing no information or missing data in regard to membership in one of the other f groups. This $f + 1$-way scale contains all the information we originally had about which among the n_p subjects was in which of the f groups, plus the information about which subjects have missing data.

Concretely, imagine a survey about attitude towards abortion (ATA) in which religious affiliation (G) is to be used as one of several IVs, with provision for G_1 = Protestant, G_2 = Catholic, G_3 = Jewish, and G_4 = Other or no religious affiliation (our running example in Chapter 5). Assume that a subset of the sample declines to respond to this item. We need not drop the item, nor drop these nonresponding subjects completely, nor drop them from correlations involving the set of IVs carrying religious affiliation (missing data matrix method). We merely categorize them on this nominal scale as $(G_a =) G_5$ = No response. To an objection that G_5 is not a religious affiliation and is of a different character from G_1 through G_4, we offer the counterargument that there is no structure in a nominal scale other than what we impose by our coding. *All* information about the subjects' responses to this item is included in our categorization into five groups. A sufficient condition for a nominal scale is mutually exclusive and exhaustive assignment of all cases to the g categories, which we have achieved. The "meaning" of the categorization is expressed in the coding and in the interpretation of the results of the analysis.

Employing this simple device, one proceeds to code the $g = f + 1$ groups of this nominal scale into a set of $g - 1$ $(= f)$ IVs, $X_1, X_2, \ldots, X_f$, each an aspect or function of group membership. Any of the coding methods described in Chapter 5 may be used, since all we are doing is treating G_a as just another group:

1. *Dummy variable coding.* All the formulas and interpretations of Section 5.3 obtain. G_a (the missing-data group) may be used as the reference group if its comparison with each of the others is of particular interest. Or, some other group may serve as reference group, in which case one of the IVs (call it X_a) distinguishes missing-data group membership (1) from membership in one of the original f data-present groups (0). Thus coded, r_{Ya}^2 is the proportion of criterion variance accounted for by whether data on G are in G_a or not, for example, how much variance in ATA (= Y) is accounted for by declining to respond to the religious affiliation item. The t test of its significance is a test of the null hypothesis that nonresponse (as opposed to any response) is random in relationship to (or is uncorrelated with) Y. Even if r_{Ya} is nonsignificant, it may still be the case that membership in G_a yields significant Y discrimination from the reference group or another of the groups. (B_a gives the difference between $\bar{Y}_a$ and the mean of the reference group.[1]) Finally, even if neither of these is true, it is possible that X_a relates in such a way to other research factors in the MRC (for example, age, education, ethnicity) as to significantly account for an increment in Y variance (that is, to serve as a suppressor variable). The important point is that there are several ways in which the positive information of absence of data can bear on the criterion which are assessable by dummy variable coding.

2. *Effects coding.* If the substance of our research suggests treating all g groups, including the missing data G_a, on the same footing, the methods and interpretation of Section 5.4 on effects coding can be used. Whether G_a or another group is selected as the group which is coded by a string of -1s does not matter greatly. If another group, then one of the IVs (call it X_a) will, in the partialled results, carry the distinction in Y between G_a and the unweighted aggregate of all $g = f + 1$ groups. Thus coded, in our running example, B_a is the mean ATA of the nonresponders minus the unweighted mean of the means of all five groups, including the nonresponders (see Footnote 1). sr_a^2 is the proportion of total Y variance accounted for by the effect or "eccentricity" of the missing data group, and its t provides a test of randomness of missing data with regard to Y. This test differs from that described for r_{Ya} with a dummy variable in that in the latter, G_a is compared on Y with all responders pooled (hence each group weighted by its n_i) while in the former, the comparison is with an unweighted aggregate of the groups.

3. *Contrast coding.* To whatever contrast IVs in which one is interested to test hypotheses among the m groups whose membership is known, one can readily add a missing data contrast (X_a) whose coefficients are orthogonal to

---

[1] We assume that only this research factor is being used, as we did in the examples of Chapter 5, for the sake of simplicity. If other research factors (age, education, etc.) are in the MRC, the above holds for means adjusted for these other research factors, in the same sense as in the analysis of covariance. See Section 7.4.3 and Chapter 9.

TABLE 7.2.1
Two Contrast-Coded Sets with Missing Data

G_i	Set I				Set II			
	X_1	X_2	X_3	X_a	X_1	X_2	X_3	X_a
G_1	$\frac{1}{2}$	1	0	-1	$\frac{1}{2}$	$\frac{1}{2}$	$\frac{1}{4}$	-1
G_2	$\frac{1}{2}$	-1	0	-1	$\frac{1}{2}$	$-\frac{1}{2}$	$-\frac{1}{4}$	-1
G_3	$-\frac{1}{2}$	0	1	-1	$-\frac{1}{2}$	$\frac{1}{2}$	$-\frac{1}{4}$	-1
G_4	$-\frac{1}{2}$	0	-1	-1	$-\frac{1}{2}$	$-\frac{1}{2}$	$\frac{1}{4}$	-1
G_a	0	0	0	4	0	0	0	4

those of the other contrasts: each of the f groups is coded -1, and G_a is coded $+f$. On the other contrasts among the f groups, G_a is coded 0. In Table 7.2.1, we illustrate the addition of such a missing data contrast to the contrasts coded as Sets I and II in Section 5.5 (Table 5.5.1). The effect is to leave undisturbed the representation of the contrasts among the f known groups in the partialled results while contrasting Y_a with the unweighted mean of the Y means of the f groups. This method conforms fully to the principles and procedures presented in Section 5.5 to which the reader is referred for details and interpretation. The significance test of B_a here (or equivalently, of the other partial results for X_a) is the same randomness test as described above in effects coding, while that for r_{Ya} here is the same as for (the identical) r_{Ya} of dummy coding.

4. *Nonsense coding.* This curiosity is, interestingly enough, quite relevant to missing data. Missing nominal data occur frequently in survey research, where the responses to an item have been originally coded by arbitrary numbers, as described in Section 5.6. Now, one of these arbitrary codes will be for the "no answer" category, so it is represented. With a total of g ($= f + 1$) levels for an item, each represented by a different number, Section 5.6 describes how its treatment by nonsense coding may proceed by squaring, cubing, etc. up to the $g - 1$ power to fully represent the nominal scale, now containing the missing data information. Of course, none of the results for single nonsense-coded X_i is meaningful, and in the present application this means that no analysis of the missing data is readily deducible from the results. However, this lack may be compensated by saving clerical and/or programming effort.

We remind the reader that whichever coding alternative is selected, the R^2, $\widetilde{R}^2$, F, and Y means for each of the groups will be the same, and, in particular, the information residing in the missing data is positively and fully utilized.

A note of caution: With more than one research factor having missing data, if it is the *same* subjects whose data are missing on two or more variables, the representation by any method of the *same* missing data in the IVs will result in complete redundancy, that is, an R of 1 among IVs coding the nominal scales involved (see Sections 5.3 and 7.4.2). In such circumstances, one must drop an X_i (any one will do) from each of all but one of the research factors carrying the same cases with missing data. For example, consider two nominal scales where f_1 = 4 and f_2 = 3 with data missing for the *same* n_a subjects. If only one were to be used, there would be g_1 = 5 or g_2 = 4 levels, and therefore $g_1 - 1 = 4$ and $g_2 - 1$ = 3 IVs, respectively. But if both were to be used in the same MRC analysis, missing data redundancy could be avoided as follows. Assuming dummy variable coding with the (common) G_a as the reference group, the first scale could be coded normally, but from the second set of X_5, X_6, and X_7, one must be dropped. Were there a third scale with the *same* subjects missing, one of the X_i normally used to code it would also have to be dropped, in order to avoid the redundancy of the implicit missing data information.

Another method of coding nominal scales with missing data is to code a missing data dichotomy and substitute the means of the coded aspects of group membership for the cases with missing data. The details are discussed in Section 7.4.1.

7.3   MISSING DATA IN QUANTITATIVE SCALES

In handling missing values in quantitative scales, we proceed in accordance with the principles described above. We use *all* the information available, the values on the variable which are present *plus,* as positive information, the absence–presence of values on the variable. In so doing, we (*a*) avoid the risk of nonrepresentativeness in dropping subjects if data are missing nonrandomly, (*b*) avoid loss of statistical power due to reduced *n* even if data are missing randomly, (*c*) capitalize on the information inherent in the absence-presence of values on the variable in question, and (*d*) capitalize on the information present on other variables although missing for some subjects on the variable in question. The method is applicable to all quantitative scales as defined in Chapter 6— ordinal, interval, and ratio scales—and to dichotomies.

7.3.1   Missing Data and Linear Aspects of V

Consider a research factor construct V, which we measure quantitatively by v, for which, however, some subjects have no values. In the same sense as we have used the term before, one aspect of V is the absence versus presence of v values. This may or may not be a random phenomenon, but is, in either case, a descriptive feature of them. We represent this aspect of V explicitly by means of a missing data dichotomy, scoring subjects missing v values one, and those with v values zero.[2] We designate this variable d_v for notational convenience, but it is

---

[2] One can, of course, do the reverse: score zero for missing data and one for present data. The signs of correlation and regression coefficients are simply reversed. We slightly prefer one for missing data because we call the variable a *missing* data dichotomy, and follow the convenient convention of naming variables by their upper ends.

generically the same X_a as before (see Table 7.3.1 below). We emphasize the fact that d_v is an aspect of V, and may be directly or indirectly relevant to Y.

Now, what about v, the value (or linear aspect—see Section 6.2) of V? Our data matrix contains blanks for those subjects with missing v values. To correlate V with other variables on all n (= 72) cases requires that values be entered in those blank spaces. We can, in fact, supply *any* constant value c for these blanks, a proposition as contraintuitive as (and indeed related to) that of nonsense coding for nominal scales (Section 5.6). We have now represented a second aspect of V, made up of the v values for those subjects who have them, and a constant c for subjects whose v values are missing. We shall call this "plugged" variable v_c. We submit that these two aspects of V, namely, d_v and v_c, make up a set that *fully represents all the information available on V*.

If we now let $X_1 = d_v$, $X_2 = v_c$ and regress Y on them, the resulting $R^2_{Y \cdot 12}$ is the proportion of Y variance accounted for by these two aspects of V. Let us consider each of these aspects separately:

1. *The missing-data dichotomy:* d_v (= X_1). If the cases with missing data on v have a different Y mean $(\bar{Y}_a)$ from that of the cases with v values present $(\bar{Y}_p)$,

TABLE 7.3.1
Illustrative Data for Y and V, a Variable with Missing Data

S no.	Y	d_v	v	S no.	Y	d_v	v	S no.	Y	d_v	v
1	62	0	6	25	49	0	2	49	104	0	9
2	46	0	1	26	66	0	5	50	78	1	
3	104	0	9	27	35	0	0	51	128	0	9
4	48	0	2	28	53	0	3	52	68	1	
5	72	0	5	29	33	0	0	53	40	0	0
6	56	0	3	30	94	0	7	54	62	0	6
7	78	1		31	108	0	8	55	118	1	
8	100	0	8	32	37	0	1	56	90	1	
9	82	0	6	33	33	0	1	57	44	0	3
10	68	1		34	50	0	3	58	76	0	4
11	98	0	7	35	67	1		59	33	0	0
12	90	1		36	42	0	0	60	37	0	1
13	128	0	9	37	46	0	1	61	44	0	1
14	40	0	0	38	98	0	7	62	112	0	7
15	76	0	4	39	48	0	2	63	86	1	
16	58	0	4	40	123	0	8	64	66	0	5
17	44	0	1	41	72	0	5	65	100	0	8
18	112	0	7	42	56	0	3	66	33	0	1
19	86	1		43	35	0	0	67	82	0	6
20	57	0	4	44	58	0	4	68	57	0	4
21	44	0	3	45	53	0	3	69	50	0	3
22	123	0	8	46	52	0	3	70	67	1	
23	52	0	3	47	94	0	7	71	42	0	0
24	118	1		48	108	0	8	72	49	0	2

then r^2_{Y1} will be nonzero, and if large enough, statistically significant. In the latter case, the null hypothesis of randomness of missing data in relation to Y is rejected, since it indicates that in the population, cases missing v values have on the average larger (r_{Y1} positive) or smaller (r_{Y1} negative) Y values than cases with v values present. From the simple regression equation using the dichotomy $X_1 = 0, 1$ as predictor, $\hat{Y} = B_1 X_1 + A$, one can readily determine the means of the two kinds of cases and their difference:

(7.3.1) $$\bar{Y}_p = A,$$

(7.3.2) $$\bar{Y}_a = B_1 + A,$$

(7.3.3) $$\bar{Y}_a - \bar{Y}_p = B_1 .$$

Thus, use of d_v yields a full analysis of the implication to Y of missing v values.

2. *v plugged with c: v_c ($= X_2$).* We have made the surprising statement that our method permits the use of an arbitrary constant c for the missing v values. Obviously, differing constants will yield differing correlations of v_c with Y, and with any other variable, including d_v. Also, obviously, an arbitrary constant can not be a serious estimate of what v would be if it were present. We say this to reassure those readers made nervous by the idea of "making up" data—c is not a guess nor an estimate, but merely a convenience which permits the use of all n cases. What is by no means obvious is the reason that we can use such an arbitrary constant: $R^2_{Y.12}$ (and its associated F test) will remain constant *no matter what value of c* is used to plug the blanks. The set of two variables *fully and invariantly* represent v in the sense that their R with Y or with any other IVs which may be added to the equation will not change with changes in c. Thus, the same contribution to accounting for variance in Y, with or without other IVs, will be made, no matter what value we assign to c. This means that whatever the value of c, the pr_2, sr_2, and B_2 (but not β_2) will remain invariant, as does their common t test. Also, the Y intercept A remains invariant. But in a simultaneous regression of both these aspects of V, *none* of the partial statistics for d_v, nor their t test, remains invariant.

What the above implies is that the general missing-data model is *hierarchical*, with $d_v = X_1$ the first aspect of v in the equation, and $v_c = X_2$ the second. In the hierarchical model, we are interested in the results of the regression of Y on d_v unaffected by v_c, but when v_c is additionally entered, its results have d_v partialled from them. For differing values of c, differences will occur in r_{Y2} and r_{12}, but, as noted, not in pr_2, sr_2, and B_2; it follows that the increment in Y variance due to v_c, I_2 (which is sr^2_2) is also invariant over changes in c. Thus, what is invariant with changes in c is $X_{2.1}$, but not $X_{1.2}$.

Table 7.3.1 presents data for $n = 72$ cases of which only 60 have values on v, and we have coded the absence–presence of v values by the dummy variable dichotomy d_v. These are the raw research data. Assume that they are repre-

TABLE 7.3.2

Analysis of Missing-Data and Linear Aspects of V

	r			Hierarchical model						
	Y	X_1	X_2	IV	R^2	F	df	I	F	df
$= X_1$	.241	1.000	−.494	X_1	.0580	4.31*	1, 70	.0580	4.31*	1, 70
$= X_2$	.649	−.494	1.000	X_1, X_2	.8373	177.55**	2, 69	.7793	330.49**	1, 69
m	69.69	.1667	3.333			$n = 72$				
sd	27.49	.3727	3.018							

$$pr_2 = .910 \quad pr_2^2 = .8273$$
$$sr_2 = .883 \quad sr_2^2 = .7793 \ (= I_2)$$
$$B_2 = 9.25 \quad t_2 = 18.179**$$

Equation for X_1:

$$\hat{Y} = 17.77 \, X_1 + 66.73 \qquad\qquad R^2_{Y \cdot 12} = .8373$$

$t_{B_1} \quad 2.076* \qquad (df = 70) \qquad\qquad \tilde{R}^2_{Y \cdot 12} = .8258$

Equation for X_1 and X_2:

$$\hat{Y} = 54.75 \, X_1 + 9.25 \, X_2 + 29.75$$

$t_{B_i} \quad (13.29**) \quad 18.18** \qquad (df = 69)$

*$P < .05.$ **$P < .01.$

sented in exactly this way in the computer, blanks and all, and that we are using a computer program which automatically treats blanks as equaling zero. Thus, by leaving v unplugged, we have actually plugged it with $c = 0$, quite arbitrarily. It happens that zero falls at one end of the range of v values, but it does not matter whether c falls within this range or not; it can literally be *any* value, indeed, it is a nonsense value in exactly the same sense as in nonsense coding in nominal scales (Section 5.6).

Table 7.3.2 presents the results of a hierarchical regression analysis of the data in Table 7.3.1. The absence of v data is correlated .241 with Y, and thus accounts for .0580 ($= r^2_{Y1} = R^2_{Y \cdot 1} = I_1$) of the Y variance. This may be tested for significance by the (error Model I) $F = 4.310$ for $df = 1, 70$, or equivalently by the t for B_1 in the equation for Y on X_1 ($t = 2.076$) for 70 df, given the usual relationship $F = t^2$.[3] Since this is significant ($P < .05$), we can reject the hypothesis that the data are missing randomly in regard to Y. Invoking Eqs. (7.3.1), (7.3.2), and (7.3.3), we find that the difference in Y means between

---

[3] A test of r^2_{Y1} using Model II error (which is here $1 - R^2_{Y \cdot 12}$; see Section 4.4.2) will be more powerful whenever there is any material degree of regression of Y on v, which is the usual case. Here, where I_2 is very large, the Model II test is much more powerful, yielding $F = 24.60$ ($df = 1, 69$). As a general rule, Model II error is preferred for testing d_v.

subjects lacking v values and those with values present, $\bar{Y}_a - \bar{Y}_p$, is $17.77 = B_1$, with $\bar{Y}_p = A = 66.73$, and $\bar{Y}_a = 17.77 + 66.73 = 84.50$. Thus, absence of v makes for a 17.77 point higher Y mean than its presence, a difference which we have already noted is significant.

Turning now to X_2, we note first that its r_{Y2}, .649, can hardly be a meaningful quantity, since we have arbitrarily plugged the blanks with zeros. (Had we plugged it with nines, for example, it would be .859.) Combined with X_1, we note that $R^2_{Y \cdot 12} = .8373$, a meaningful quantity that does not change with c, and is here highly significant ($F = 177.55; df = 2, 69; P < .01$), indicating that about 84% of the Y variance is accounted for jointly by the missing data and linear aspects of V. The increase in R^2 due to the addition of X_2 is .7793 ($= I_2$), which is very large and highly significant by Eq. (4.4.2) ($F = 330.49, df = 1, 69, P < .01$). Whatever the value used for c, I_2 always represents the proportion of Y variance due to the linear aspect of V, since the arbitrary c has been partialled by d_v. If one refers this variance not to the total Y variance as base, but to that part of it not associated with d_v, one has $pr_2^2 = .8273$.

Turning to the regression equation in two IVs, we first note that the $B_{1 \cdot 2}$ value 54.75, which depends on the arbitrary c, is not usefully interpretable, nor perforce is its t. It is of course generally the case that the reason the hierarchical model is employed is because we do not wish to partial later-entering IVs from those entering earlier. $B_{1 \cdot 2}$ and its t provide no basis for drawing conclusions about missing data.

On the other hand, $B_{2 \cdot 1}$ ($= 9.25$) provides a most useful result, as one might suspect from the fact that it does not depend on c. Indeed, it also does not depend on the Y values for the 12 subjects who were missing v values. It is thus the slope of the least-square, best-fitting Y on v regression line *for the 60 cases for whom we have paired Y, v observations,* and *not* the slope of the line of Y on v_c, with its plugged c values, for the 72 cases. The latter obviously depends on c, and is as arbitrary as is the choice of c: it is $r_{Y2}(sd_Y/sd_2) = 5.91$ here, an absolutely meaningless quantity.[4]

Turning finally to the Y intercept of the second equation ($A = 29.75$), it too is independent of c, and is the intercept of the Y on v regression line for the 60 cases where data on both are present, not that of the Y on plugged v line for 72 cases.[5]

Any residual uneasiness about plugging the blanks in the data matrix should now be dispelled: we are in no way fudging the regression for the available pairs of observations. For these, a unit increase in v is associated with an increase of 9.25 units in Y, and when v is *observed* to be zero (not plugged, mind you), Y is estimated as 29.75.

In summary, the chief features of this method are:

[4] For example, when $c = 9$, it is .859 (27.49/3.22) = 7.34, equally meaningless, of course.
[5] That value is 49.99 here, and would be 34.23 when $c = 9$, both in principle meaningless.

1. It uses all the sample information, including the fact of the data missing on v, carried by d_v. The importance of this aspect of V, both in terms of r^2_{Y1} and B_1, and its statistical significance is determined. The latter constitutes a test of the randomness (in relation to Y) of the absence of v data. It is thus not necessary to assume randomness—the method works in any case.

2. It is clear that c, the constant we use to fill the blanks in v, being arbitrary, is in no sense an estimate of the missing value, but merely a convenience. All the following statistics are invariant with c, once d_v is in the equation: $R_{Y\cdot 12}$, $\widetilde{R}_{Y\cdot 12}$, pr_2, sr_2 (hence, I_2 which equals sr^2_2), B_2, A, and their respective significance test values. Moreover, B_2 and A give the slope and intercept of the regression line of Y on v only for cases with data present—the missing data pairs do not influence these values.

7.3.2 Missing Data and Nonlinear Aspects of V

Once it is grasped that the absence of data is an aspect of V to be represented by d_v, one can incorporate other aspects of V by any of the methods detailed in Chapter 6: power or orthogonal polynomials, nominalization, and monotonic nonlinear transformation. We illustrate this with power polynomials (Section 6.2), and briefly discuss the use of the other methods for providing for nonlinear relationships in combination with missing data.

Power Polynomials

For the data in Table 7.3.1, in addition to the missing data aspect of V, represented as $X_1 = d_v$, we investigated its linear aspect, $X_2 = v_c$, with $c = 0$. Following the method of the representation of nonlinear aspects of a quantitative variable by a polynomial in integer powers of the variable (Section 6.2), we can continue the hierarchical MRC analysis of these data by providing additional aspects of V as IVs, for example, $X_3 = v^2_c$, and $X_4 = v^3_c$. The results of this continuation are given in Table 7.3.3. We present the entire matrix of rs among Y and the four X_is, and repeat the hierarchical R^2 and I for X_1 and X_2 as given in Table 7.3.2 for convenience and clarity, but do not repeat the two regression equations in the latter.

We note first the typical high positive correlations among the powered v_cs (Section 6.2). The latter show negative correlations with d_v ($-.494$, $-.358$, $-.313$), but, of course, these values are dependent on the purely arbitrary choice of c. (Had we plugged the blanks with $c = 9$, they would have been respectively .579, .668, .653.) But in missing-data polynomials, as in the linear case, the results of interest do *not* depend on c. These are, primarily, the increment in R^2 due to each successive polynomial term and its statistical significance, and the regression coefficients for the polynomial terms and Y intercept for any given order of the polynomial.

Under the heading "Hierarchical model" in Table 7.3.3, the first two lines repeat the results of the analysis for d_v and v_c in Table 7.3.2,—a modest but

TABLE 7.3.3

Power Polynomial Analysis of Regression of Y on V with Missing Data

			r					Hierarchical model			
	Y	X_1	X_2	X_3	X_4	IV^a	R^2	F	df	I	F
$d_v = X_1$	.241	1.000	−.494	−.358	−.313	X_1	.0580	4.31*	1, 70	.0580	4.31* 1
$v_c = X_2$	.649	−.494	1.000	.957	.862	$+X_2$	.8373	177.55**	2, 69	.7793	330.49** 1
$v_c^2 = X_3$	.747	−.358	.957	1.000	.896	$+X_3$	.8631	142.89**	3, 68	.0258	12.81** 1
$v_c^3 = X_4$	.667	−.313	.862	.896	1.000	$+X_4$	.8636	106.01**	4, 67	.0005	.23 1
m	69.69	.1667	3.333	20.22	168.1						
sd	27.49	.3727	3.018	25.24	240.1						

For Equations for X_1 and for X_1 and X_2, see Table 7.3.2.

Equation for missing-data quadratic:

$$\hat{Y} = 47.81\, X_1 + 3.41\, X_2 + .677\, X_3 + 36.69$$

t_{B_i} (11.19**) (2.005*)   3.579** $(df = 68)$

Equation for missing-data cubic:

$$\hat{Y} = 47.91\, X_1 + 3.45\, X_2 + .720\, X_3 − .00558\, X_4 + 36.59$$

t_{B_i} (11.14**) (2.016*) (3.420**)   0.481 $(df = 67)$

*$P < .05.$ **$P < .01.$

^{a}The $+X_i$ notation indicates that all prior variables (orders lower than i) remain as IVs and X_i added.

significant result from d_v, very strongly and highly significantly augmented by v_c (see Footnote 3). Continuing, the addition of the quadratic aspect of V, v_c^2 (= X_3) produces a small but nevertheless highly significant increase in Y variance accounted for: $I_3 = .0258$, and by Eq. (4.4.2), $F = 12.81$, $df = 1, 68, P < .01$. Thus, although a straight line pulls a prodigious amount of Y variance ($I_2 = .7793$), there is nevertheless a small but real bend in the regression of Y on v. Finally, with the addition of the cubic aspect of V, virtually no increase in R^2 occurs: $I_4 = .0005$ ($F = .23$). A necessary concomitant of this is the small value for B_4 in the cubic equation (−.00558), and the very slight changes in the other regression coefficients and A from the quadratic to the cubic equation.

The independence from c of the regression coefficients for the power polynomial terms is due to exactly the same reason as in the linear case—they are determined solely by the 60 cases for whom v values are present, and not on the remaining 12 for whom c was arbitrarily plugged into the blanks in the v column. Having found that the quadratic equation is appropriate, if a graphic portrayal such as Figures 6.2.1 or 6.2.2 is desired, one would omit the 12 missing data cases and the $X_1 = d_v$ part of the quadratic equation; then, the substitution of numerical values for v and v^2 in the remainder of the equation (as described in Section 6.2) will provide a best-fitting regression curve for the 60 cases with data present. The missing data aspect of V having been taken into account, the linear and nonlinear aspects necessarily pertain only to data which are present.

It is not necessary to graphically plot the regression equation in order to obtain a general idea of the shape of the regression curve. In the example, as v increases, Y increases, since the linear B_2 is positive (9.25–Table 7.3.2) but the line is somewhat bent so as to be concave upward, since the quadratic B_3 is positive (see Section 6.2.4).

After one takes note of the special features associated with d_v, and particularly of the independence from c of the hierarchical regression results of other aspects of V, all of the discussion about the polynomial method in Chapter 6 holds for the regression of Y on those values of v which are present.

Orthogonal Polynomials

If any of the conditions which make it advantageous to code V with orthogonal polynomial coefficients obtain (Section 6.3.4), the fact of missing v values is no deterrent to the use of this method. One proceeds hierarchically, as with power polynomials, letting $X_1 = d_v$, and then coding X_2, X_3, etc. by the linear, quadratic, etc. orthogonal polynomial coefficients (Table 6.3.1) where data are present. Where absent, one supplies an arbitrary constant c wherever there is a blank. In power polynomials, c was used for the linear aspect and then was squared, cubed, etc. for higher-order aspects, but in fact it does not matter what constant is used in any X_i, since, whatever its value, it is partialled out by $X_1 = d_v$. The resulting regression coefficients on the coded polynomial terms are again based only on cases with data present, and all the other material in Section 6.3 obtains.

Nominalization

This method of representing V as if it were a nominal scale can be used with missing data by simply applying the method described above in Section 7.2, with the rationale provided in Section 6.4. Using dummy variable coding, one may either explicitly code missing data with d_v ($= X_a$) or use the group with missing data as the reference group. Note that the applicable model here is simultaneous, not hierarchical.

Nonlinear Transformations

Again, no new principles arise when, for a variable which is to be subjected to a one-for-one monotonic nonlinear transformation (Section 6.5), there are missing data. Provided only that d_v is first entered, any constant can be plugged into the blanks before (or after) the transformation is made, and the resulting regression coefficient for the transformed value and the Y intercept will be based only on the present data.

7.4 SOME FURTHER CONSIDERATIONS

7.4.1 Plugging with Means

Quantitative Scales

Since the results of interest do not depend on the value chosen for c, it should be chosen for convenience. In the illustrative example, we assumed the circumstance of a computer program that reads blanks as zeros, so the convenience of c = 0 lies in obviating the necessity for any additional steps in the data preparation.

It is frequently the case that the most convenient value to take for c is $\bar{v}$, the mean of the v values which are present. The primary advantage lies in the fact that when $c = \bar{v}$, the mean-plugged v_c (= X_2) correlates *zero* with d_v (= X_1), that is, these two aspects of V, which are conceptually independent of each other, are also uncorrelated. With r_{12} = 0, $R^2_{Y \cdot 12}$ is simply the sum of the Y variance accounted for by d_v (r^2_{Y1}) and that accounted for by v_c (r^2_{Y2}). Since I_2 in the hierarchical model is sr^2_2, and since when r_{12} = 0, $sr^2_2 = r^2_{Y2}$, and $B_1 = B_{1 \cdot 2}$ (there being nothing to partial), it becomes unnecessary to use the hierarchical model. One simply runs a simultaneous MRC analysis on these two IVs. The hypothesis of randomness of the missing data aspect of V is tested by the t test on r_{Y1}, and of the linear aspect of V by the t test on r_{Y2} (but see Footnote 3). B_1 gives the missing-data effect, $\bar{Y}_a - \bar{Y}_p$, and B_2 and A the slope and Y intercept for the pairs of observations where v is present. If these latter values are of no interest, no computation is necessary beyond that of r_{Y1} and r_{Y2}.

Note that when one plugs with $\bar{v}$ in power polynomials as described above, although d_v is uncorrelated with v_c, it is correlated with v_c^i (where $i > 1$), and, in any case, in order to appraise the nature, amount and significance of the contribution of higher order polynomials, the hierarchical model must be used. However, if one wishes to keep the powered terms uncorrelated with d_v, this can be accomplished by using the means of the present v^is for plugging the blanks, rather than the c^2, c^3, etc. values that result from using v_c^is.

With multiple dependent variables, the saving of the additional step required in the hierarchical linear model when $c = \bar{v}$ may be worthwhile. Further, with additional research factors, d_v and mean-plugged v_c may be used as a set with the resulting B for d_v a meaningful quantity. These advantages also obtain for the polynomial, when one plugs with $\bar{v}^i$ instead of v_c^i.

Nominal Scales

The principle of plugging a missing value with the mean of values which are present in order to render the IV so treated uncorrelated with a missing data dichotomy d_v can be usefully extended to nominal scales, using any of the coding methods discussed in Chapter 5. Earlier in this chapter (Section 7.2) we described how to incorporate a missing group for each of the four methods of nominal scale coding, each method requiring a different procedure consistent with it. Here we describe how plugging with means is applicable to all four methods, and retains the property of making the missing data information orthogonal to, and hence uninvolved in, the contrasts among the groups of known membership.

We use the same notation as in Section 7.2: we have n_p subjects classified into f known groups $(G_1, G_2, \ldots, G_f)$ and an additional n_a of unknown group membership (G_a), a total of n subjects in a total of $f + 1 = g$ groups. For *any* method of nominal scale coding, set up a coding table as follows:

1. Using dummy-variable, effects, contrast, or nonsense coding, code the known groups $(G_1, G_2, \ldots, G_f)$ into the IVs $(X_1, X_2, \ldots, X_{f-1})$ exactly as you would if there were no G_a.
2. Code the missing data dichotomy X_a, using 1 for G_a and 0 for the known groups.
3. Find the means of X_1 through X_{f-1}, taking into account the ns of the groups, and plug the blank in each X_i for G_a with its $\bar{X}_i$.

Table 7.4.1 has been prepared to illustrate this procedure. We revert to our running example of a 4 $(=f)$-level nominal scale of known groups plus a missing data group $(= G_a)$, and show how coding diagrams are constructed for the examples of Chapter 5. Note that there are $g = f + 1$ $(= 5)$ groups and these are represented, as always, by $g - 1 = f (= 4)$ coded X_i. For each diagram, the first 4 $(= f)$ rows and the first 3 $(= f - 1)$ columns are identical with the analagous Tables in Chapter 5 (Table 5.3.1 for dummy, Table 5.4.1 for effects, and Table 5.5.1 for contrast sets I and II). In each, the fth (fourth) column is X_a. For cosmetic reasons, the plugged means are represented by letters in each coding diagram, but their explicit formulas, which are functions of the coding coefficients and the known group of n_is and their total n_p are given below.

It follows from the fact that plugging with $\bar{X}_i$ results in orthogonality with X_a (i.e., $r_{ia} = 0$) that for *all* nominal scale coding methods, using the simultaneous model (all f IVs regressed together on Y), the raw-score regression coefficient for the missing-data dichotomy,

(7.4.1) $$B_a = \bar{Y}_a - \bar{Y}_p ,$$

provides a contrast of the criterion mean for G_a and the weighted mean of the cases of known group membership, that is,

(7.4.2) $$\bar{Y}_p = \frac{\sum Y_p}{n_p} = \frac{n_1 \bar{Y}_1 + n_2 \bar{Y}_2 + \cdots + n_f \bar{Y}_f}{n_p} .$$

Note that $\bar{Y}_p$ is the *weighted* criterion mean of all cases of known group

TABLE 7.4.1

Illustrative Nominal Scale Coding with Plugged Means

G	Dummy				Effects				Contrast I				Contrast II			
	X_1	X_2	X_3	X_a	X_1	X_2	X_3	X_a	X_1	X_2	X_3	X_a	X_1	X_2	X_3	X_a
G_1	1	0	0	0	1	0	0	0	$\frac{1}{2}$	1	0	0	$\frac{1}{2}$	$\frac{1}{2}$	$\frac{1}{4}$	0
G_2	0	1	0	0	0	1	0	0	$\frac{1}{2}$	-1	0	0	$\frac{1}{2}$	$-\frac{1}{2}$	$-\frac{1}{4}$	0
G_3	0	0	1	0	0	0	1	0	$-\frac{1}{2}$	0	1	0	$-\frac{1}{2}$	$\frac{1}{2}$	$-\frac{1}{4}$	0
G_4	0	0	0	0	-1	-1	-1	0	$-\frac{1}{2}$	0	-1	0	$-\frac{1}{2}$	$-\frac{1}{2}$	$\frac{1}{4}$	0
G_a	e	f	g	1	h	i	j	1	q	r	s	1	t	u	v	1

$e = \bar{X}_1 = n_1/n_p$

$f = \bar{X}_2 = n_2/n_p$

$g = \bar{X}_3 = n_3/n_p$

$h = \bar{X}_1 = (n_1 - n_4)/n_p$

$i = \bar{X}_2 = (n_2 - n_4)/n_p$

$j = \bar{X}_3 = (n_3 - n_4)/n_p$

$q = \bar{X}_1 = (n_1 + n_2 - n_3 - n_4)/2n_p$

$r = \bar{X}_2 = (n_1 - n_2)/n_p$

$s = \bar{X}_3 = (n_3 - n_4)/n_p$

$t = X_1 = (n_1 + n_2 - n_3 - n_4)/2n_p$

$u = X_2 = (n_1 - n_2 + n_3 - n_4)/2n_p$

$v = X_3 = (n_1 - n_2 - n_3 + n_4)/4n_p$

membership, and is influenced proportionately by the various groups directly as a function of their n_is.

A further consequence of the fact that the r_{ia}s are all zero is that the other B values carry *exactly* the same meaning as they would were there no missing data; their values depend only on the known groups, and are those which would result if the cases in G_a and hence X_a were dropped and the analysis performed on only the n_p cases. Note that this is exactly what occurs with plugged means in quantitative variables—the regression constants other than B_a provide the best-fitting regression for the values which are present.

Finally, with the r_{ia}s all zero, for all the coding methods, it will be true that

$$(7.4.3) \qquad R_{Y \cdot 12 \ldots f}^2 = r_{Ya}^2 + R_{Y \cdot 12 \ldots (f-1)}^2 \,,$$

which effects a ready distinction between the proportion of Y variance accounted for by missing versus present data, and that which obtains among the f groups of known membership.

The specific implications of the X_a plus plugged mean procedure to the four methods of nominal scale coding are:

1. *Dummy coding.* As presented in Table 7.4.1, G_4 is the reference group, so that the B_i (for $i = 1, 2, 3$) give $\bar{Y}_i - \bar{Y}_4$, and $A = \bar{Y}_4$, just as if there were no missing data. This would also be true if we used the method described in Section 7.2 with G_4 as reference group. The difference lies in the fact that in that method, the B for the X_a dichotomy would give $\bar{Y}_a - \bar{Y}_4$, while with plugged means, the analogous B gives $\bar{Y}_a - \bar{Y}_p$.

2. *Effects coding.* The method described in Section 7.2 is appropriate when G_a is to be put on an equal footing with the known groups, since A is the unweighted mean of the Y means of *all* $f + 1$ groups including G_a. In the present method, A is the unweighted mean of the Y means for *only* the f groups of known membership. The B_is for known groups thus contrasts the $\bar{Y}_i$ with the mean of means of known groups, with $\bar{Y}_a$ excluded. The latter is contrasted with $\bar{Y}_p$ in B_a (Eq. 7.4.1).

3. *Contrast coding.* Again, one distinction between plugging means and the method of Section 7.2 (compare the coding of Table 7.4.1 with that of Table 7.2.1) lies in the fact that the latter yields an A that is the unweighted mean of all Y means of $f + 1$ groups, and the present method excludes $\bar{Y}_a$ from the mean of means. Another difference lies in the regression coefficient for the missing data contrast: the method of Section 7.2 gives the standard comparison of $\bar{Y}_a$ with the unweighted mean of the means of the known groups, whereas the present method uses the weighted mean, $\bar{Y}_p$ (see Eqs. 7.4.1 and 7.4.2).

4. *Nonsense coding.* This hardly needs illustration—there are $f - 1$ nonsense X_is, with G_a containing the respective means, plus the usual X_a. The nonsense B_is and A are individually uninterpretable, but Eqs. (7.4.1), (7.4.2), and (7.4.3) hold.

Summarizing, the X_a plus plugged mean method of coding nominal scales with missing data neatly segregates the missing data effect from those of known groups, leaving their regression coefficients (and srs) a function only of present data, while capitalizing on the total n for statistical power in significance tests. It is relatively simple to program a "plug mean and create X_a" subroutine, which then can be used on option with either quantitative or nominal data. This is a worthwhile addition to a MRC computer program because of the general handiness of the method.

7.4.2 When Not to Use X_a

Having emphasized thus far the desirability of using a missing data IV, we must now consider circumstances where it is optimal *not* to do so. Three come to mind:

1. *When p_a is very small (or very large).* When only a small proportion (.05 or .10) of the cases have missing data on a research factor, the advantage of incorporating that information will likely be outweighed by the probability that r^2_{Ya} will be small and, particularly when n is not large, nonsignificant. This is true because under these circumstances, the variance of X_a is relatively small, and its ability to correlate with other variables constrained. Another way of saying this is that with p_a small, n_a will be small; hence $\bar{Y}_a$ will contribute much unreliability to whatever it enters. This consideration should be set aside when it is clear that $\bar{Y}_a$ is quite extreme compared with $\bar{Y}$.

At the other extreme, if p_a is very high, not only will the missing data variate have low variance, but the other aspect(s) of the research factor will have results based on few observed values, and they will inevitably be very unreliable.

2. *Multiple redundant X_{aj}.* With many IVs with nonrandomly missing data, such as occurs in survey research, it would be dubious practice to have each accompanied by its own missing data aspect, since the resulting set of missing data variables are likely to be substantially (if not perfectly) correlated. Since it is unlikely that each carries uniquely relevant variance, their combined use is likely only to produce generally unstable MRC results owing to the loss of error df and high correlation among IVs (see Section 4.6.2). In such circumstances, the use of a single IV of "tendency to have missing data," scored as the number of items on which data are missing, is likely to be much superior to either all or none of them being included.[6] Means are used for plugging.

3. *Randomly missing data.* When it is *known*, or can be safely assumed, that absence of data on a research factor is random, then the inclusion of X_a simply weakens the statistical stability and power of the MRC analysis. However, the assumption of randomness should not be made lightly, particularly when n is very large and the number of candidates for missing data variates very few, since then the power loss is small relative to the risk of being wrong. It is when n is

[6] An elegant solution worthy of consideration is to factor analyze the matrix of missing-data dichotomies, and to score and include the missing data factors that emerge. The recommendation of summation into a single score presumes that such an analysis would yield one dominant factor.

small or moderate that a sound judgment that X_as can be dispensed with, particularly if there would be more than one, may make the difference between a successful and unsuccessful analysis.

It is most important to note that if, for whatever reason, the missing-data aspect of a quantitative or nominal research factor is not included in the analysis, one *must* plug the blanks with means. Without X_a being partialled, the results for a plugged X_i are no longer invariant over c, and the choice of c cannot be arbitrary. We can only use that unique value for c which renders the plugged X_i uncorrelated with X_a, so that when the latter is omitted, B_i is unaffected—it is exactly what it would have been had the cases with missing data been dropped, that is, it is based only on cases with known values on X_i. This is easily seen for a simple bivariate regression coefficient for cases with data present, whose formula can be written in terms of deviations of X and Y from their respective means as

(7.4.4)
$$B_{YX} = \frac{\sum xy}{\sum x^2}.$$

When we add the missing-data cases, we are adding cases whose $x = \bar{X} - \bar{X} = 0$; both their additional xy products in the numerator and x^2 in the denominator are zero, leaving the B_{YX} for plugged X the same as it was with missing-data cases omitted.[7]

Thus, when missing-data information is not to be included as an IV, for any aspect X_i of a quantitative or nominally scaled variable, plugging the blanks with $\bar{X}_i$ assures that B_i will not be affected and hence not distorted by the plugging process.

7.4.3 Missing Data and Multiple Independent Variables

Thus far, for the most part, we have considered a single research factor with missing data, either a quantitative V, or a nominally scaled G. For concreteness, assume that it is made up of a missing-data dichotomy (X_a) and one or more X_i that have been plugged with means. Now, if one or more other variables (X_j) are added and regressed simultaneously on Y, the following consequences occur:

1. Even if X_a is uncorrelated with Y, the possibility arises that X_a relates materially to an X_j variable. Now, if r_{Yj} is material, X_a will have substantial partial correlation and regression coefficients by acting as a suppressor variable (Section 3.4.4). Thus, the question of the randomness of missing data transcends

[7] This is not true for A. Perhaps more important, it also does not hold for r_{Yi}^2, which is always smaller (strictly, nonlarger) for the mean-plugged X_i than it would be with missing-data cases dropped. This can be seen from the fact that, with added cases, the $\sum y^2$ in the denominator of Eq. (2.3.4) increases (strictly, nondecreases).

that of whether it relates directly to Y; with other variables present, it may be indirectly related.

2. With X_j variables in the equation, they are partialled from X_a, so that B_a is no longer the difference $\bar{Y}_a - \bar{Y}_p$, but the difference between these means *adjusted* for the X_j, in exactly the sense as in the analysis of covariance, with the X_j as covariates. (See Chapter 9.)

3. Similarly, the X_js are partialled from the X_is, so that, for example, if the X_is are aspects of a nominal scale, the B_is are functions of X_j-adjusted Y means.

4. Finally, the B_js are similarly adjusted for both X_a and the X_is. Thus, missing-data information can even serve as a covariate when X_j is of special interest (Section 9.3.5).

7.4.4 General Outlook on Missing Data

Consistent with our approach throughout this book, we look upon missing data in a research factor as a aspect of that factor just as we would a logarithmic aspect or a contrast function of G, or an orthogonally coded polynomial aspect, or any other, and code it as an IV. This aspect X_a of either a quantitative or nominally scaled factor, may be directly or indirectly criterion relevant, or it may not, as is any other aspect of a research factor under consideration. The spirit is empiric, and the inclusion of X_a precludes the necessity, in general, of assuming the randomness of missing data—that issue is resolved as a finding from the analysis. On the other hand, the investigator may choose a priori to omit X_a, in exactly the same way as he may omit v^4, or a distinction between two groups, or any other of the many possible aspects of a research factor.

We thus view missing data as a pragmatic fact that may be investigated, rather than as a disaster to be mitigated. Indeed, implicit in this philosophy is the idea that like all other aspects of the sample data, missing data are a property of the population to which we seek to generalize. It is perfectly clear that in this framework, the arbitrary constant c, be it zero or the mean or whatever, is in *no* sense an estimate of the missing value. Since some outlandish value for c far outside the range of observed values will work as well as any other, we are not estimating with c, but merely neutrally plugging the gap so that we can get on with the analysis. As we have seen, the regression coefficients for plugged variables and A are based only on the cases for which information is present. It could not be otherwise—how can one fit a regression function to missing data? We have seen that, indeed, we do not—the c values simply "disappear" in the determination of the B_is, having been neatly extracted by the partialling of X_a, or, when c is the mean, they "disappear" even with no X_a provided. Yet the cases with missing data contribute to the statistical power and reliability of all the parameters, since they contribute to the df for error.

Returning to the other methods for coping with missing data discussed in the introduction, it should be clear why we prefer the method of X_a with c-plugged X_i. It runs no risk of a mistaken randomness assumption, nor of producing an inconsistent correlation matrix. It uses all the X_i and all the n. It hews realistically to the population actually sampled, missing data and all.

For completeness, we note that when the randomness assumption can be made, in contrast to the methods proposed here, there are available methods which estimate missing values and/or "true" correlations. Timm (1970) describes and compares these.

7.5 SUMMARY

In the introduction, the nature and circumstances of occurrence of missing data are discussed, and three frequently used methods for coping with missing data in research factors are described: dropping subjects, dropping variables, and working from a missing-data correlation matrix. These methods are criticized variously on grounds of loss of information, loss of statistical power and precision of estimates, and dependence on the assumption of the randomness of occurrence of missing data, which, when false, results in biased or nonrepresentative results. (Section 7.1)

Methods that are not subject to these criticisms are presented which are applicable to both quantitatively and nominally scaled research factors. They centrally involve the treatment of the absence of data as one aspect of the research factor which is treated together with other aspects.

For nominal scales, this means treating the cases whose group membership is unknown as simply another group. One then may use any of the methods of Chapter 5 to code the total sample, fully utilizing the information of missing data group membership with that of membership in known groups. (Section 7.2)

For quantitative scales, one constructs a missing data dichotomy which directly represents that aspect of the factor and then, as a second aspect, plugs the blanks in the factor with any constant, selecting one which is convenient. When the hierarchical model is used, all the results of interest are independent of the constant chosen, making it clear that the constant used for plugging is in no sense an estimate, but merely a data-analytic convenience. If nonlinear aspects of the factor are of interest, any of the methods of Chapter 6 may be applied together with the missing data dichotomy. The B coefficients for aspects other than that of missing data and A are determined only by cases with values present, that is, not distorted by the plugging of the constant. (Section 7.3)

A useful special case, applicable both to quantitatively and nominally scaled research factors, occurs when one uses for the plugging constant of each aspect the mean of the values present for that aspect. This results in these aspects correlating zero with the missing data dichotomy, which in turn means that the missing-data aspect is excluded from any influence on the other aspects in a simultaneous analysis. (Section 7.4.1)

Circumstances when one would omit the missing-data dichotomy are discussed, those being chiefly when missing values are few or are clearly random (Section 7.4.2). The consequences to the meaning of the missing-data dichotomy and other aspects of the research factor of the addition of other IVs into the MRC are briefly presented (Section 7.4.3).

The chapter concludes with a summary of the theoretical model implied in the "missing-data plus plugged-blanks" method, and its superiority over the methods described in the introduction is argued (Section 7.4.4).

8
Interactions

8.1 INTRODUCTION

The idea of an interaction being carried by an IV in MRC analysis was introduced in Chapter 5, in the discussion of contrast coding for a 2 X 2 design (Section 5.5.4). We saw there that the coding of the interaction contrast required the multiplication of the coding coefficients for the two main effects (X_1 and X_2), and that the resulting X_3 could be interpreted as any other IV in an MRC analysis. (A review of Section 5.5.4 at this point probably would be helpful.) It was also stated there that "throughout MRC, whether we are dealing with nominal or quantitative IVs or combinations of these, *interactions are carried by products of variables* (or variable sets)," and a footnote claimed that the implications of this statement for data analysis are far-reaching. In this chapter we seek to draw these implications out fully, and deliver on that promise.

In this section we lay a foundation for the understanding of the nature of interaction effects on which the later illustration and discussion of specific instances can build a sound structure.

Interaction effects usually come into the ken of behavioral scientists through the analysis of variance. In AV, which works perforce with nominally scaled research factors, interactions are functions of cell means; usually differences between differences of means, differences between other linear contrasts among means as in "trend analysis" (Edwards, 1972), or among unweighted means of means, as illustrated in Section 5.5.4, particularly Table 5.5.5. But, as this chapter will demonstrate, the idea is far more general than as represented in AV, which we have seen is itself a special case of MRC oriented toward nominal scales. Interactions are defined and may be studied among quantitative scales, or among combinations of quantitative and nominal scales. In fact, one can form a single IV to carry the effect of the interaction of any aspect of any research

TABLE 8.1.1
2 X 2 Tables of Means Illustrating No Interaction,
Crossed Interaction, and Uncrossed Interaction

	No interaction			Interaction (crossed)			Interaction (uncrossed)		
	F	C	$\bar{m}$	F	C	$\bar{m}$	F	C	$\bar{m}$
D	6	4	5	14	4	9	8	2	5
P	10	8	9	10	12	11	12	10	11
$\bar{m}$	8	6	7	12	8	10	10	6	8

	No interaction	Interaction (crossed)	Interaction (uncrossed)
$\bar{m}_D - \bar{m}_P = B_1$	$5 - 9 = -4$	$9 - 11 = -2$	$5 - 11 = -6$
$\bar{m}_F - \bar{m}_C = B_2$	$8 - 6 = 2$	$12 - 8 = 4$	$10 - 6 = 4$
$m_{DF} - m_{DC} = B_D$	$6 - 4 = 2$	$14 - 4 = 10$	$8 - 2 = 6$
$m_{PF} - m_{PC} = B_P$	$10 - 8 = 2$	$10 - 12 = -2$	$12 - 10 = 2$
$B_D - B_P = B_3$	$2 - 2 = 0$	$10 - (-2) = 12$	$6 - 2 = 4$
A	7	10	8

factor (nominal or quantitative) with any aspect of another research factor (nominal or quantitative), and so on for interactions of higher order.

But what *is* an interaction? Two variables, u and v, are said to interact in their accounting for variance in Y, when *over and above* any additive combination of their separate effects, they have a joint effect. We encounter frequent confusion of a simple linear combination of two main effects with their *joint* effect, which is something quite different. Table 8.1.1 gives three 2 X 2 tables of means designed to illustrate this distinction. For concreteness, we revert to the content of Section 5.5.4; a study of errors in retention (Y) in rats as a function of Drug versus Placebo (X_1) and Frontal versus Control Lesion (X_2). For the equations for regression constants in Table 8.1.1 to hold, we assume the contrast coding of Table 5.5.1, Set II. Each 2 X 2 table is bordered by $\bar{m}$, the unweighted mean of the Y means in its row or column, and the lower right hand corner value is the unweighted mean of all four Y means. Neither the size nor equality of sample sizes nor the amount of error (within-cell) variance is relevant to the *values* of main or interaction effect contrasts (although they are relevant to the proportion of variance accounted for and to statistical significance), so the reader may assume whatever he pleases about these.

In all three 2 X 2 tables, we posit the existence (nonzero value) of both main effects, Drug and Frontal. Note that a main effect is an *average* effect, that is, it is the effect of a factor averaged over the other factor (or factors in multifactorial designs). In the first table, for example, Drug has an average effect of 4 fewer errors than Placebo $(5 - 9 = -4)$, and Frontal Lesion has an average effect of 2 more errors than Control Lesion $(8 - 6 = 2)$. But each of these average effects is *constant* over the levels of the other, for example, the Frontal Lesion's effect under the Drug condition, $6 - 4 = 2$, is exactly the same as its effect under the Placebo condition, $10 - 8 = 2$. Similarly, and necessarily, since there is only one *df* for the interaction in a 2 X 2 table, the average (main) effect for Drug versus Placebo of -4, holds for both Frontal Lesions $(6 - 10 = -4)$ and Control Lesions $(4 - 8 = -4)$. No interaction exists—each effect, whatever it may be, operates quite independently of the other, or equivalently, quite uniformly for each level of the other.

Consider, by way of contrast, the second 2 X 2 table, "Interaction (crossed)." Here, the average (main) Frontal effect is 4 more errors than Control, $12 - 8 = 4$. But this is *not* uniform: for the drugged animals, the Frontal cases average 10 *more* errors than Control $(14 - 4 = 10)$, whereas for those on Placebo, the Frontal cases average 2 *fewer* errors $(10 - 12 = -2)$. These two separate effects average out to the Frontal main effect $(10 - 2)/2 = 4$, but are obviously quite different and the fact of their difference constitutes the interaction. The interaction is called "crossed" because the effects are of opposite sign $(+10$ and $-2)$. The result is the same if one takes the separate Drug effects for Frontal $(14 - 10 = 4)$ and for Control $(4 - 12 = -8)$ Lesions.

The oppositeness of the effects is what makes the interaction *crossed,* but what makes for the interaction is the fact that the effects are *different.* The third 2 X 2 table of means illustrates an uncrossed interaction: The Frontal effect for Drug animals is $8 - 2 = 6$, and for Placebo $12 - 10 = 2$; they are of the same sign, hence uncrossed. Although for both Drug and Placebo groups, the Frontal Lesion results in more errors, this occurs to different degrees, hence an interaction is present. Again, of course, the results are the same if one takes differences vertically instead of horizontally.

Table 8.1.1 should make clear what is meant by a joint effect. Only in the first 2 X 2 table can one account for the means by positing only a constant influence for Drug versus Placebo and another for Frontal versus Control. In the second and third tables the two factors in addition operate jointly—the combination Drug–Frontal and the combination Placebo–Control have more errors, and the Drug–Control and Frontal–Placebo combinations fewer errors, than their main effects account for.

Now, assume that these 2 X 2 tables were analyzed by MRC using the coding coefficients in Table 5.5.1, including the interaction contrast constructed as the product of the main effect contrasts, X_3 equals X_1 times X_2. B_1, B_2, B_3, and A of the regression equations, which produce (and thus account for) the four cell means, are given below each 2 X 2 table. (See below for the meaning of B_D and B_P.) In full accord with what was said above, the interaction B_3 is zero for the first 2 X 2 table, but takes on nonzero values for the two others, reflecting the requirement that over and above whatever main or average effects the two research factors have a third source of Y variation, namely, their joint or interaction effect, D X F, is operating in the latter two tables. We stress again the fact that this effect is not a vague derivative second-class IV, but stands on equal footing with the others as a distinct IV in its own right. (Note also that with contrast coding A is the mean of all the cell means.)

We have belabored this very special narrow case of an interaction of two single dichotomous variables because its complete understanding makes possible a progression up the ladder of increasing generalization to its highest rung—interaction among multiple sets of IVs. But before we begin the ascent, there is an important implication of joint or interactive effects that the humble 2 × 2 table may clarify which facilitates an understanding of the generalization we seek. Another way of saying that u and v operate jointly in relating to Y is to say that they operate *conditionally:* the relationship between u and Y depends on the value of v, being greater for some values of v than for others. This conditional relationship is also symmetrical, the relationship between v and Y depending on the value of u. To state the matter more exactly, a nonzero $u \times v$ interaction effect means that the *regression* of Y on u varies with changes in v (and that the regression of Y on v varies with changes in u).

Let us consider the 2 × 2 table from this perspective. The regression coefficient for Y on any single dichotomous variable coded $\frac{1}{2}, -\frac{1}{2}$ (or 1, 0) is simply

$$(8.1.1) \qquad B = \bar{Y}_h - \bar{Y}_l,$$

that is, the Y mean of whichever group was assigned the algebraically higher coding coefficient (Frontal) minus the Y mean of the other group (Control). Now, we can consider each row of the 2 × 2 table separately. For the D animals taken alone, the regression of retention error on the Frontal-Control dichotomy is $m_{DF} - m_{DC} = B_D$, and for the P animals, it is $m_{PF} - m_{PC} = B_P$. Thus the difference between these differences in the two halves of the experiment is

$$(8.1.2) \qquad B_D - B_P = B_3,$$

the partial regression coefficient for X_3, the interaction IV of the entire experiment. Because of the already noted symmetry, we could proceed with the separate regressions of Y on the Drug–Placebo dichotomy for the Frontal and Control Lesion portions of the data to the same result.

Now Eq. (8.1.2) is immediately generalizable. If instead of the frontal-control dichotomy, we had any quantitative variable, v (= X_2), age or number of prior exposures or size of litter of origin, it would still be true that the regression coefficient B_3 for an IV constructed by multiplying the X_1 (contrast or dummy) code by the X_2 value ($X_3 = uv = X_1 X_2$) would equal the *difference* between the regression coefficient of v (or slope of the Y on v line) for the Drug group and that for the Placebo group. If the interaction is present, we could say that the regression of Y on v is conditional, that is, it depends on (varies with) whether we are considering the D or P condition. (See Section 8.3.)

Nor is it necessary for one of the variables to be a dichotomy. With u (= X_1) and v (= X_2) both quantitative IVs (ordinal, interval, or ratio), it remains true that the interaction information is carried by literally u times v (i.e., $uv = X_3$), and the conditional interpretation holds: if B_3 is significant, we can conclude that the Y on u regression line slope changes with or depends on the value of v, and equivalently that the Y on v slope changes with or depends on u. (See Section 8.4.)

We pause to note that the term "moderator" variable has come into use in psychometric psychology to describe a variable v which interacts with another so as to enhance predictability of a criterion. In such usage, v taken alone usually shows no consequential relationship with the criterion. Mathematically, however, symmetry obtains, and u moderates the v, Y relationship as much as the reverse. Also, the typical psychometric moderator which is uncorrelated with Y invites confusion of interaction with suppression (see Section 3.4), two completely distinct MRC phenomena: in an interaction, the uv term, a third IV, is invoked, whereas suppression depends only on the relationships involving the two variables u and v.

Whatever the nature of u and v then, the $u \times v$ interaction is carried in the uv product and can be understood in terms of joint or conditional or nonuniform relationships with Y, as described. But we say the $u \times v$ interaction *"is carried by,"* not "is" the uv product. This is because, in general, uv will be linearly correlated with both u and v, often quite substantially so. Only when u and v have been linearly partialled from uv does it, in general, become the interaction IV we seek, thus,

$$(8.1.3) \qquad\qquad u \times v = uv \cdot u,v.$$

To help keep this important distinction in mind, we generally use the "$u \times v$" notation to represent the *interaction* and uv to represent the product, which only becomes the interaction when its constituent elements are partialled. As far as u and v are concerned, for the purpose of accounting for Y variance, we do not in general wish to partial uv from either of them (but see Section 8.2.1). The linear relationship between uv and u and between uv and v are understood to be wholly due to u and v, respectively; partialling uv from u or v would be stealing their rightful variance, exactly as partialling v^2 and v^3 from v in a powered polynomial would be stealing variance from v (see Section 6.2). We use the same solution as for power polynomials, the hierarchical model—uv enters the regression after u and v, and it is thus the partialled uv results which we interpret as the $u \times v$ interaction.

The partialling of u and v from uv is a signally important step. Since the scaling of IVs in behavioral science is generally arbitrary, some methodologists have been misled by the discovery that linear transformations of u and/or v will change the r of uv with Y, as well as with u, v, and anything else (Althauser, 1971). But uv is not the interaction, $uv \cdot u,v$ is. It is demonstrable that the correlation of $uv \cdot u,v$ with Y (or anything else) does *not* change with linear transformations of u and v. Thus, the sr and pr with Y for uv with u and v partialled, and their t value, are invariant over the arbitrary scaling of u and v. Also, if the rescaling involves only a mean shift (that is, no sd change), B for $uv \cdot u,v$ is invariant (Cohen, 1973a). The same principle holds for sets U and V, and their analogous multiple semipartial and multiple partial correlations (see below).

The next order of generalization we make is to more than two variables. Whatever their nature, the IVs u, v, and w may form a three-way interaction in their relationship to Y, that is, they may operate jointly in accounting for Y variance *beyond* what is accounted for by u, v, w, $u \times v$, $u \times w$, and $v \times w$. This would mean, for example, that the Y on u regression varies with differing $v \times w$ *joint* values, or is conditional on the specific v,w combination, being greater for some than others. The symmetry makes possible the valid interchange of u, v, and w in the above statement. The $u \times v \times w$ interaction is carried by the uvw product, which requires refining by partialling of constituent variables and two-way products of variables, that is,

(8.1.4) $u \times v \times w = uvw \cdot u,v,w,uv,uw,vw.$

Again, this is accomplished by using hierarchial MRC, and interpreting the partialled results at the appropriate point in the hierarchial cumulation. Also, the invariance of correlations with this term over linear transformation of u, v, and w holds as discussed above.

Higher order interactions follow the same pattern, both in interpretation and representation: a d-way interaction is represented by the d-way product from which the constituent main effect variables, the two-way, three-way, etc. up to $d - 1$-way products have been partialled, most readily accomplished by hierarchical MRC.

The fact that the mathematics can rigorously support the analysis of interactions of high order, however, does not mean that they should necessarily be constructed and used: interactions greater than three-way are most difficult to conceptualize, not likely to exist, and are costly in statistical inference (see Section 8.8).

Now for a vaulting generalization: the statements made above about interactions among single variables u, v, w also hold for *sets* of variables U, V, W. Specifically, if U is a set of k_U IVs (u_i; $i = 1, 2, \ldots, k_U$), and V a set of k_V IVs (v_j; $j = 1, 2, \ldots, k_V$), then we can form a UV product set of $k_U k_V$ IVs by multiplying each u_i by each v_j. The $U \times V$ interaction is found in the same way, as

(8.1.5) $U \times V = UV \cdot U, V$

by hierarchical MRC, and the same three-way Eq. (8.1.4) and generally d-way interaction relationships hold for sets as for single variables. (See Sections 8.5–8.8)

Interpretively, an exactly analogous joint or conditional meaning obtains for an interaction of sets, $U \times V$: the regression coefficients that relate Y to the u_i variables of the U set are not all constant, but vary with changes in the v_j values of the V set (and, too, when U and V are interchanged in this statement). Stated in less abstract terms, this means that the nature and degree of relationship between Y and U varies, depending on V. Note again that whatever is found to be true about the relationship of Y with U alone is true for *average* v_j values in the V set, but when a U $\times$ V interaction is present, this relationship changes with changes in the v_j values of V. (Again, symmetry permits interchanging U and u_i with V and v_j.)

The importance of this analytic strategy lies in the fact that some of the most interesting findings and research problems in behavioral and social science lie in conditional relationships. For example, the relationship between performance on a learning task (Y) and anxiety (U) may vary as a function of psychiatric diagnostic group (V). As another example, the relationship between income (Y) and education (U) may vary as a function of race (V). As yet another example, in aggregate data where the units of analysis are urban neighborhoods, the relationship between incidence of prematurity at birth (Y) and the female age distribution (U) may depend on the distribution of female marital status (V). The reader can easily supply other examples of possible interactions. The reason we represent these research factors as sets is that it may take more than one variable to represent each, or in the language of our system, each research factor may have more than one aspect of interest to us (Section 4.1). Thus, the construct anxiety may be represented in terms of two polynomial aspects ($u_1 =$ a, $u_2 = a^2$), age distributions (say, of census tracts) may be represented in terms of their first three moments (u_1 = mean, u_2 = sd, u_3 = skewness), and any nominal scale of more than two levels requires for complete representation two or more aspects.

One final and most important feature of the above procedure is the interpretability of each of the single product variables $u_i v_j$. As noted, the multiplication of the k_U variables of the U set by the k_V variables of the V set results in a product set which contains $k_U k_V$ IVs. Since each u_i is a specifiable and interpretable aspect of U, and each v_j is a specifiable and interpretable aspect of V, then when partialled, each of these $k_U k_V$ IVs, $u_i \times v_j$, represents an interpretable aspect of U by aspect of V interaction, a *discrete* conditional or joint relationship, and like any other IV, its B, sr, pr, and their common t test are meaningful statements of regression, correlation and significance status. Thus, if Y is a measure of learning, u_1 is the linear aspect of an anxiety measure (U), and v_3 is an effects-coded aspect of psychiatric diagnosis (V) with obsessive neurotics coded 1, then the $u_1 \times v_3$ interaction IV represents the source of variance due to the difference in slope in the regression of learning score on anxiety between obsessive compulsives and the unweighted mean of the slopes of all diagnostic groups in V. One could similarly interpret all the rest of the single interaction IVs, for example, quadratic aspect (degree of curvature) by hysterical versus all groups.

As discussed in Chapter 4, the concept of set is not constrained to represent aspects of a single research factor like age, or psychiatric diagnostic group, or marital status distribution. Sets may be formed that represent a functional class of research factors, for example, a set of variables collectively representing socioeconomic status, or a set made up of the subscales of a personality questionnaire or intelligence scale, or, as a quite different example, one made up of covariates, that is, variables which one wishes to statistically control while studying the effects of others (see Section 8.3.3 and Chapter 9).[1] However defined, the global $U \times V$ interactions and their constituent $u_i \times v_j$ single interaction IVs are analyzed and interpreted as described above.

Having presented the nature of the interaction and generalized verbally from those of 2×2 tables to those which make up multifactorial sets, the next several sections are devoted to concrete illustrative examples that provide specific procedures and interpretive details.

8.2 THE 2 × 2 REVISITED

To avoid undue repetition we do not present another worked example here; the reader is referred to the example and detailed discussion of Section 5.5.4, as well as in the introduction to the present chapter. We will briefly summarize that material and go on to other considerations about the 2×2 factorial.

8.2.1 The Special Case of Orthogonal Coding

When the two dichotomous main effects of the 2×2 (X_1, X_2) are coded by the method of contrast coding set forth in Section 5.4, that is, with coding coefficients of $\frac{1}{2}$ and $-\frac{1}{2}$, and their product constructed to represent the interaction IV $(X_3 = X_1 X_2)$, the coding table of Set II in Table 5.5.1 results. We saw that a *simultaneous* MRC using the three resulting IVs yielded the usual (coding-invariant) $R^2_{Y \cdot 123}$ and its F ratio, but also meaningful B_i, sr_i, pr_i, the t test they share, and A.

This may seem to contradict the general principle described above for interactions involving any u and v, which was that a hierarchical model was appropriate in order that the uv product would not be partialled from, and hence steal the rightful variance of, u and v. What makes the simultaneous model nevertheless usable for contrast-coded main effects and their products is the orthogonality of the a coding coefficients, as described in Section 5.4. It is *generally* the case that

---

[1] Although U and V must be partialled out of UV to create $U \times V$, and hence the hierarchical model is invoked at this point, the question of the model to be employed in re U and V remains open. If the content of the problem dictates that U is causally prior (an "antecedent condition") and hence should be partialled from V (as when U is intended as a covariate set), then the model is fully hierarchical, and U is entered prior to V; one then has in three steps $U, V \cdot U$, and the interaction $UV \cdot U, V$. If no such priority obtains between U and V, they are entered simultaneously, followed by UV; the result is $U \cdot V$ and $V \cdot U$ in the first step, and then the interaction $UV \cdot U, V$ in the second.

when research factors are coded using orthogonal coefficients (that is, contrast or orthogonal polynomial), whether dichotomous or multileveled, with products representing interactions, a simultaneous solution will correctly represent all the main effect aspects as well as the interactions. When the two research factors are each dummy coded, and their product used for interactions, the MRC analysis may be performed, but would need to be done hierarchically, since a simultaneous solution would result in distorted main-effect *sr* and *pr* values as described for the general case. (We shall see below that the *B* values in these circumstances are nevertheless interpretable.) This would also be the case (for factors with more than two levels) for effects-coded variables, since their coefficients are also not orthogonal. Finally, all the above generalizes to three-way and higher-order interactions and to interactions of sets.

8.2.2 When *Not* to Use the 2 X 2 Factorial Design

Consider Professor Doe. He is investigating the effects of authoritarianism, as measured by the California F scale (f) and general ability (g) on a cognitive style variable (Y), using high school students as subjects. He is interested not only in the main effects of f and g, but also in the $f \times g$ interaction. He gives the three tests to his sample, and because he is interested in the interaction and finds the 2 X 2 factorial design as comfortable as an old shoe, he proceeds to dichotomize the f and g distributions as near their medians as possible into two high–low groups, yielding the 2 X 2 structure with cells containing the Y scores. Now he almost certainly finds that the n_is of the four cells are not proportional—there are rather more high–low and low–high cases than high–high and low–low, reflecting the fact that f and g are (negatively) correlated. Now, let us charitably assume that he will not approach this problem of disproportionality with a meat axe—that he will neither drop cases (dutifully randomly) to achieve proportionality (zero r between f and g) with its inevitable cost in statistical power, nor use some approximation such as "unweighted means" or "weighted means," such as his colleague Professor Roe does in such a fix (and which is acceptable in respectable journals!). Professor Doe knows how to accomplish the exact solution of "fitting constants" (Winer, 1971, pp. 404–416), which is identical with the MRC for the 2 X 2 factorial we have described, and so proceeds. So, what's wrong with this picture?

What's wrong is what happened at square one, when he dichotomized the graduated f and g distributions to create the 2 X 2 cells. It is intuitively obvious that when one reduces a graduated many-valued scale to a two-point scale, one is willfully throwing away information. This has immediate negative consequences to the amount of variance such a crippled variable can account for and concomitantly to the power of the statistical test of its contribution. Concretely, if we assume a normal distribution for the median-dichotomized variable (and Y), its (point biserial) r^2 with Y (r_{pb}^2) will be only .64 as large as would the r^2 with Y of the original graduated variable (r_g^2) (and even less as the cut departs from the median, as it often must).[2,3] As an example of the effect of dichotomization on the power of statistical tests, the t value (when $r_g \neq 0$) for the r_{pb} will be only

$$(8.2.1) \qquad q = .8 \sqrt{\frac{1 - r_g^2}{1 - .64\, r_g^2}},$$

as large as the t for r_g. The value of q is .69 when $r_g = .70$, it is .76 when $r_g = .50$, and it is never as large as .80. Few behavioral scientists can afford the luxury (loss of statistical power) of "working on the qt"(!)—obtaining t values some two-thirds or three-quarters (q) of what they should be, say 1.5–1.9, instead of more than 2.0. With $n = 50$, for example, an r of .28 is significant at the two-tailed .05 level. But for r_{pb} to equal .28, the r_g must be .35. If the latter is less, the former will not be significant.

We remind the reader that when we describe drops in r_{pb}^2 relative to r_g^2, and concomitant drops in t, we are perforce describing decreases in the sums of squares, mean squares, and F ratios of the AV, given the identity between MRC and AV (Section 5.4). Also, this drop occurs in the r_{pb}^2 with Y of the product of the two dichotomies, X_3 (Section 5.4.4) and that analogous drops occur in all the *partialled* effects and *their* detectability (power) and significance test results, which is the very heart of the analysis.

Now, the MRC gives identically the same results as the (exact) AV, *once* the IVs have been dichotomized, and there is the rub! It would be dismal enough a prospect if it were necessary to dichotomize in order to analyze the data, but of course no such necessity exists. As we state above (Section 8.1), and illustrate below (Section 8.4), the general model for interaction holds that for quantitative (graduated) variables, as indeed for variables or variable sets of any kind, the interaction $u \times v$ is represented by $uv \cdot u, v$, and its effects readily determined by the MRC model. Dr. Doe need not dichotomize anything: he can take for X_1 and X_2 the graduated scores on f and g and for X_3 the "score" which results when each subject's f score is multiplied by his g score. This X_3 carries the full graduated information available for the interaction of f with g, and not the impoverished information of the dichotomous interaction. There is no need whatever for the abuse of the data required to cast it into 2 × 2 factorial design form which then results in reduced variance accounting, power, and significance. The graduated data can be appropriately and exhaustively analyzed by MRC, as we demonstrate in Section 8.4.

The circumstance in which one should not use the 2 × 2 design (and by obvious extension, the 2^d design), then, is when either (any) of the research factors are graduated variables which must be dichotomized to that end. The 2 × 2 is admissable when both variables are natural dichotomies, for example,

---

[2] Also, median cuts on two normally distributed variables will result in an r^2 between them (r_ϕ^2) which is only 40% as large as r^2 would be on the original variables.

[3] With other than normally distributed variables of the kind encountered in real data, those values will be approximated.

male–female, schizophrenic–nonschizophrenic, experimental–control, and the example of Sections 5.4 and 8.1, that is, truly nominal scales of two levels. This structure is a special case of a more general principle, made possible by the system of MRC analysis to which this book is devoted. For optimal data exploitation, do not represent quantitative research factors by a few class intervals, particularly if interactions are to be studied. The sole merit of the method of nominalization to represent quantitative research factors (Section 6.4) is simplicity when the goals are descriptive rather than analytic. It is not generally recommended when interactions are studied.

8.3 A DICHOTOMY AND A QUANTITATIVE RESEARCH FACTOR

8.3.1 Introduction

Going beyond the interaction of two dichotomies, we next encounter the case where the two main effect IVs are a dichotomy (u) and a quantitative variable (v) and the third IV, their product (uv) which contains the $u \times v$ interaction, as in Eq. (8.1.3). Since a dichotomy necessarily partitions the cases into two groups, this section describes the MRC analysis of data structures that involve two groups, a quantitative independent variable (interval, ratio or possibly ordinal) and the usual dependent variable (Y). We shall see that this common data structure lends itself to alternative analytic and interpretive strategies, depending on the logical function of the IVs, the purpose of the research, and the importance to these of the $u \times v$ interaction. We shall also see that the principles adduced in this case are capable of useful generalization.

For concreteness, we consider a sample of $n = 72$ cases, each with a dependent variable score Y, an IV that is quantitative ($v = X_2$), and membership in one of two groups which we score as a dummy variable ($u = X_1$): one for the $n_1 = 32$ members of G_1 and zero for the $n_0 = 40$ members of G_0. In addition, we create a third IV, which is the product for each case of the scores on X_1 and X_2, e.g., $uv = X_1X_2 = X_3$. Cases in G_1 will thus have for X_3 the same score as on X_2, since $X_3 = 1 \times X_2 = X_2$, whereas those in G_0 will all have zero scores on X_3, since $X_3 = 0 \times X_2 = 0$. This strange looking column X_3 on the data sheet, made up of 32 real-looking scores and 40 zeros, is uv, the product term that contains the $u \times v$ interaction.

Table 8.3.1 presents the results of simple correlation and regression analysis, which we will interpret as a review and for the sake of continuity before we attack the MRC results.

Taking the two main effect variables, we note from Table 8.3.1 first that group membership ($u = X_1$) accounts for some 11% of the Y variance, a statistically significant amount by the standard test ($t = 2.965$, $df = n - 2 = 70$, $P < .01$). Moreover, the constants of the regression equation for Y on X_1 indicate that the Y mean of G_0 is 67.55 and that the Y mean of G_1 is 19.33 points higher, that is,

TABLE 8.3.1

Simple Correlation and Regression for a Dichotomy (u),
a Quantitative Variable (v), and their Product (uv)

	r				r^2_{Yi}	t_i (df = 70)	Regression equation
	Y	X_1	X_2	X_3			
$u = X_1$	.334	1.000	.115	.896	.1116	2.965**	$\hat{Y} = 19.33\,X_1 + 67.55$
$v = X_2$	.593	.115	1.000	.405	.3511	6.154**	$\hat{Y} = \quad .79\,X_2 + 31.33$
$uv = X_3$	.535	.896	.405	1.000	(.2859)	(5.294)**	
m	76.14	.444	57.03	26.58		$n = 72$	
sd	28.75	.497	21.68	31.17			

**$P < .01$.

67.55 + 19.33 = 86.88 (Section 5.3). Since r_{Y1} is significant, so is identically the 19.33 mean difference.

The quantitative variable $(v = X_2)$ accounts for 35% of the Y variance, ($t = 6.154$, $df = 70$, $p < .01$). Just as in assessing the simple relationship of the group membership dichotomy to Y we ignore scores on X_2, so here the fact of group membership is ignored, an issue we will shortly return to. The Y on X_2 regression equation indicates that a unit change in score on X_2 is associated with a .79 change in Y on the average in the total sample of 72 cases, and that when $X_2 = 0$, Y is estimated at 31.33.[4]

Thus in this illustrative example, X_1 and X_2 each account for a nontrivial and significant portion of the Y variance. Also, note that these two IVs are not quite mutually uncorrelated, since $r_{12} = .115$.

Table 8.3.1 gives the correlations with X_3, the product of the two main effect variables, but its r^2_{Y3} and t are placed in parentheses and its regression equation is omitted. It is thus being slighted for good reason—although it contains the $u \times v$ interaction, it is an arbitrary hodgepodge that also necessarily contains chunks of u and v variance, as has been stressed (Section 8.1). This can readily be seen in its substantial correlations with u (.896) and v (.405). In general, simple correlations yielded by products with other variables are functions of the arbitrary scaling of the variables in the product and hence not interpretable; not until the constituents of the product have been linearly partialled from it does it become interpretable as an interaction.

In the following two sections, we present two substantive alternatives of the nature of the variables and research purposes of the illustrative example of Table

---

[4] As is often the case in behavioral science data, the Y intercept of 31.33 is not necessarily a meaningful quantity, since $X_2 = 0$ may not be an interpretable score, for example, when X_2 is IQ.

8.3.1, interpreting the MRC analysis in each case and particularly the role and meaning of the interaction.

8.3.2 A Problem in Differential Validity of a Personnel Selection Test

As the first example, we posit an inquiry into whether a test used as a basis for hiring is of equal validity in groups of minority and white job applicants. This issue is hardly academic—a test that is less valid for a minority group is patently unfair (Jones, 1973). Moreover, it violates the Equal Opportunity guidelines and has been held by the Supreme Court to be unconstitutional.

Thus, we interpret Y as an acceptable criterion of job adequacy, the dichotomy u (= X_1) as the distinction between white (1) and minority (0) group membership, and v (= X_2) as the score on the selection test. The presumption is that the test was not used as a basis for selection for the 72 cases under consideration, but its results merely filed until the Y data were available.

So conceived, Table 8.3.1 indicates that the white group has a significantly higher mean criterion score (86.88) than the minority group (67.55), the dichotomy accounting for some 11% of the criterion variance. Also, for the two groups combined as the total sample, the r_{Y2} of .593 means a highly significant 35% of Y variance accounted for by the selection test, with, on the average, a one-point increase in the test associated with an increase of .79 in the criterion.

Note that each of the above statements about the IV ignores the other. Now, consider a hierarchical MRC with the dichotomy entered first, then the test, and finally the uv (= X_3) product. Table 8.3.2 shows that R^2 as each IV is added successively, the increments (I) in R^2, and the relevant F tests.[5] We have already considered the simple correlation and regression of Y on the white-minority dichotomy ($u = X_1$). When the selection test ($v = X_2$) is added to the dichotomy, R^2 increases by .3112, a highly significant amount. It is worth pausing to consider the partial regression and correlation coefficients for this level of the hierarchy. Partialling out the dichotomy means taking out the difference between the group means of Y. The pr_2 at this stage, literally $r_{Y2 \cdot 1}$, equals .591; this is a correlation of deviations of the subjects' Y and X_2 scores from the

---

[5] We follow the practice here, as in Chapter 6, of using Model I error: for each I_i, it is $1 - R^2_{Y \cdot 12 \dots i}$ with $df = n - i - 1$. As noted in the discussion of the special case of polynomials (Section 6.3.2), but true quite generally, this practice results in F being negatively biased when some later-entering X_j accounts for substantial variance, since the latter is a part of the Model I error of variables entering earlier. Despite this, we have a slight preference in general for Model I error for the sake of simplicity and consistency and particularly so when n is not large, or the research is exploratory, or when IVs are in decreasing order of likely importance. We would not quarrel with a purist who, taking seriously all k IVs to be studied, prefers Model II error $(1 - R^2_{Y \cdot 12 \dots k}, df = n - k - 1)$, and tests the I_is with Eq. (4.4.4). We would certainly not quarrel with him in the present instance, where n is not small, k is only three, and all effects are quite credible. The successive Fs for Model II error are 14.163**, 39.495**, and (necessarily not changing since the error is the same) 5.264*, all with $df = 1, 68$.

means *of their respective groups.* Thus, Subject j in Group g would contribute the deviations $Y_j - \bar{Y}_g$ and $X_{2j} - \bar{X}_{2g}$; the 32 paired deviations of the white subjects from their means and the 40 paired deviations of the minority subjects from *their* means are combined to make up this correlation. This is the "pooled within group" correlation, a simple correlation between own-mean deviated scores, but easily found by partialling the X_1 dichotomy out of the r_{Y2} correlation. We stress that this is the "*averaged out*" validity of the selection test. The zero-order r_{Y2} is a correlation of deviations from the combined group (total sample) means, and is the validity ignoring the fact that we deal here with groups having different Y and v means. Neither $r_{Y2 \cdot 1}$ nor r_{Y2} bears on the possibility that the groups have *different* validity.

Now, the fact that $r_{Y2 \cdot 1}$ and r_{Y2} are here so close in value (.591 and .593) is a coincidence characterizing this particular set of data, and should not mislead the reader into believing that they are necessarily similar. The discussion of partial r (Section 3.5.5) makes clear that it can be of greatly different value than its unpartialled counterpart, larger or smaller, even of opposite sign. Nor should the similarity in value lead one to believe that the degree of relationship between Y and X_2, that is, the validity of the selection test, is similar in the two groups; quite the contrary turns out to be the case here, as we shall see.

The regression equation for u and v is found to be

$$\hat{Y} = 15.59\,X_1 + .74\,X_2 + 26.74$$

$$t \quad 2.928** \quad 6.099** \qquad (df = 69, **P < .01).$$

We focus on the slope of Y on v for the pooled within group residuals, $B_{2 \cdot 1} = .74$, the partial regression coefficient analogous to the partial correlation coefficient discussed above. It is a kind of *average* of the separate validity slopes of the white and minority groups (and not much different in this example from the total slope of v, given in Table 8.3.1 as .79). Like all averages, it gives no information on the dispersion of its units, and nothing in the analysis thus far tells us whether the separate slopes (and hence validities) of the two groups are similar or different, which is the point of this research.

We also note, in passing, that $B_{1 \cdot 2} = 15.59$ $(P < .01)$ is the difference in the group's criterion means *adjusted* by means of the averaged within group slope for the selection test. Although not a particularly meaningful quantity in the present interpretive context, we will see in Section 8.3.3 (and Chapter 9) that it is identically the quantity sought in the analysis of covariance.

The Interaction and the Issue of Differential Validity

We come finally to the major point of the analysis, the question of the differential validity of the personnel selection test in the white and minority groups. This issue is identical with the $u \times v$ or group by selection test interaction.

When we add the uv term as X_3 in the hierarchical MRC, since u and v (X_1 and X_2) are already in the IV set, the partialling process removes the latter from uv

TABLE 8.3.2

Hierarchical MRC Analysis of Illustrative Data Conceived
as a Problem in Differential Test Validity

IVs	cum R^2	F	df	IVs added	I	F_I	df
X_1	.1116	8.791**	1, 70	$u = X_1$	.1116	8.791**	1, 70
X_1, X_2	.4228	25.266**	2, 69	$v = X_2$	.3112	37.195**	1, 69
X_1, X_2, X_3	.4642	19.639**	3, 68	$uv = X_3$	.0415	5.264*	1, 68

Regression equations:

(8.3.1) Complete: $\hat{Y} = -15.72\ X_1 + .49\ X_2 + .54\ X_3 + 40.71$

  t -1.077 $3.016**$ $2.294*$ $(df = 68)$.

(8.3.2) When $u = 1$: $\hat{Y}_1 = (.49 + .54)v - 15.72 + 40.71 = 1.03\ v + 25.0$.

(8.3.3) When $u = 0$: $\hat{Y}_0 = -15.72\ (0) + .49\ v + .54\ (0) + 40.71 = .49\ v + 40.7$.

*$P < .05$.   **$P < .01$.

so that the operative partialled $X_{3 \cdot 12}$ *is* now the $u \times v$ interaction. Table 8.3.2 indicates that it accounts for a significant additional .0415 of the Y variance, that is, none of the increase in R^2 from .4228 to .4642 is attributable to either u or v (since *their* contribution is the $R^2_{Y \cdot 12} = .4228$)—all of it is $u \times v$ interaction variance.[6] The implication of a significant $u \times v$ interaction is that the regression lines determined separately in the white and minority groups have significantly different slopes, hence the selection test has significantly different validity in the two groups. Note that this interpretation is fully consistent with our most general understanding of interaction as conditional relationship: the degree of relationship (as measured by slope) between Y and the selection test (v) is conditional or dependent upon the value of u, that is, whether it is 1 (white group) or 0 (minority group).

More detailed information is available from the constants of the regression equation at this final stage,

$$\begin{array}{cccc} B_1 & B_2 & B_3 & A \end{array}$$
(8.3.1) $\hat{Y} = -15.72\ X_1 + .49\ X_2 + .54\ X_3 + 40.71$.

(See Table 8.3.2 for t values).

---

[6] Note the parallel with the principle discussed in polynomial MRC, which is also hierarchical (Section 6.2). There, the unpartialled v^2 carries (much) more than simply quadratic information because of its (high) linear correlation with v. It is the effect of $v^2 \cdot v$, accomplished in the hierarchical procedure, that is relevant. This is more than a verbal analogy—polynomials are mathematically a special case of interactions: if we let $u = v$, then $v^2 \cdot v$ is exactly interpretable as a $v \times v$ interaction, meaning that the slope of Y on v is conditional on or varies with v, a roundabout but valid way of describing a curved regression line.

This regression equation in three independent variables defines the best-fitting three-dimensional hyperplane to the Y observations for the variables u, v, and uv, and is impossible to visualize. However, it contains the information for the equations for each of the Y on v regression lines for the two groups. Recall that $X_1 = u$ is coded 0 for minority and 1 for white, so that $X_3 = uv$ is 0 for all minority group members, while the whites each have their v score as X_3 and also as X_2. Thus, for the whites, the regression equation is

$$\hat{Y}_1 = B_1 \,(1) + B_2 v + B_3 v + A,$$

or, rearranging terms,

(8.3.2) $$\hat{Y}_1 = (B_2 + B_3)v + B_1 + A.$$

Thus, the slope of the line for the white group is $B_2 + B_3$, and its Y intercept is $B_1 + A$. For the illustrative example, the regression equation for the white group is therefore $\hat{Y}_1 = 1.03\,v + 25.0$ (Table 8.3.2).

In the minority group with both u and uv always 0, the equation is

$$\hat{Y}_0 = B_1\,(0) + B_2 v + B_3\,(0) + A,$$

or

(8.3.3) $$Y_0 = B_2 v + A.$$

The slope of the line for the minority group is B_2 and its Y intercept is A, which for the example gives this group's regression equation as $\hat{Y}_0 = .49\,v + 40.7$ (Table 8.3.2). In both Eqs. (8.3.2) and (8.3.3), with this 0–1 coding, u and uv drop out, leaving Y as a simple linear function of v.

Our earlier statement that the significant increment in R^2 due to X_3 denotes a significant interaction effect and different slopes can now be further detailed. Comparing Eqs. (8.3.2) and (8.3.3), we see that the line for $u = 1$ has slope which is by the amount B_3 larger than the slope for $u = 0$. Recall that all the partial coefficients for an X_i share the same significance test, so our finding that the slope difference B_3 is significant ($t = 2.294$, $df = 68$) is exactly the same as the finding that I_3 ($= sr_3^2$) is significant ($F = 5.264 = t^2$, $df = 1, 68$).

We thus find that the slope and hence the validity of v for the white group is significantly greater than that for the minority group. An increase of one point in the selection test is associated with an average increase of 1.03 points in Y for the white sample; this is twice as great as that for the minority group sample, .49. The implications of the difference in the regression lines for the two samples are best understood by reference to Figure 8.3.1. The greater slope of the white sample denotes a greater degree of predictability of the criterion and hence of validity than for the minority group members.

Since the lines are not parallel, they must cross (thanks to Euclid) although not necessarily within the range of observed values for v and Y. The value of v

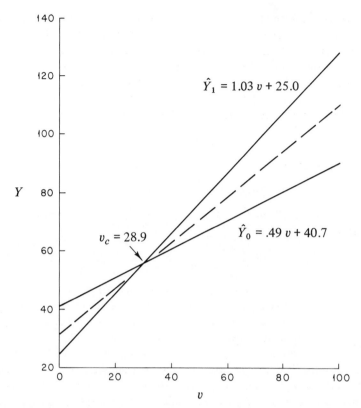

FIGURE 8.3.1 The regression lines for whites ($u = 1$) and minority group members ($u = 0$). (The dashed line is for the total sample.)

where they intersect, formed by simultaneously solving Eqs. (8.3.2) and (8.3.3) is

(8.3.4)
$$v_c = \frac{-B_1}{B_3},$$

which, for these data, gives $-(-15.72)/.54 = 28.9$. Now, *this* value of v is well within the observed range of v. The difference in validity is inherently unfair, but the crossing of the regression lines within the range of v reveals another anomaly: persons with v scores above 28.9 are expected to have higher criterion scores if they are white, while those with v scores below 28.9 are expected to have higher criterion scores if they are minority group members. (At $v = 0$, the difference is $B_1 = -15.72$, which is not significant, but also not likely to be particularly meaningful in this kind of application, where v is a psychological test whose zero is arbitrary and not even necessarily a possible value.)

Figure 8.3.1 shows as a dashed line the regression line for Y on v of the total sample of 72 cases, that is, the line which ignores the distinction between groups (Table 8.3.1). As would be expected, it is a compromise between the significantly different lines of the two groups and therefore necessarily not an adequate description of either group's regression.

Thus, to summarize the main point of this section, we have seen that an interaction, $u \times v = uv \cdot u,v$, where u is a dichotomy and v a quantitative variate, is interpretable as a slope difference between the two groups in their Y on v regression lines. In the interpretive context, it is understood as indicative of differential test validity.

8.3.3 A Covariance Analysis in a Learning Experiment

We now take the same data structure and values as before: a dichotomy (u), a quantitative variable (v), and their product (uv) related to a dependent variable (Y), and analyze it as a simple two-group analyses of covariance (ACV), using the same illustrative data. (A full-dress treatment of ACV is reserved for Chapter 9.)

The ACV involves the analysis of (the residuals of) Y when one or more other variables (the covariates) have been partialled out, according to the same principles and using the same experimental design structures as used in the analysis of variance (AV). Thus, if we designate a set of covariates as set A, and use them to estimate Y by multiple regression, then we can obtain estimated values for each subject, $\hat{Y}_A$. In ACV, the residual that is analyzed is $Y - \hat{Y}_A$ for each subject, that is, Y "adjusted for set A," or "with the effects of A removed," or "with A partialled out," the terms being used interchangeably, all in exactly the same way as Y itself is analyzed in AV. In ACV as in AV, the details of the experimental design partition the observations into groups ("cells," "plots") defined by research factors or their combinations. When we estimate $\hat{Y}_A$ by multiple regression for an ACV, the coefficients of the equation are not those of the "total" regression, since to do so would make the effects of their research factors on Y part of the regression adjustment and we would thus be removing from Y, in part, exactly what we mean to study. To avoid this stepping on one's own toes, one must take out of each group's Y observations its own mean, $\overline{Y}_j$, which reflects the effects of the research factors; so it is in fact the deviations from own mean for Y and for the variables in set A from which are computed the coefficients which produce the estimate $\hat{Y}_A$ in the residual $Y - \hat{Y}_A$. Thus, it is the *within-group* regression coefficients obtained from pooling these deviations from all the groups which are used to adjust or residualize Y in covariance analysis.

But what now if it should be the case that a covariate relates differently to Y in one population than in another, specifically in that its regression coefficient (slope) differs? This is a violation of a key assumption in ACV, indeed, a violation of its logic. A B_i derived from the pooled results used in the adjustment is a compromise between two truly different values and is thus not appropriate for either group, overadjusting in one and underadjusting in the other. Analysis of covariance presumes populations of parallel regression surfaces of Y on the A set and the absence of this condition precludes its meaningful application.

Now let us return to the data and its analysis in an ACV framework. Consider that it was produced in a learning experiment, in which Y represents a postscore, v a prescore and the dichotomy, u, distinguishes the E (1) from the C (0) group. Returning to Table 8.3.1, the simple correlation results for these redefined variables indicate that there is a significant difference in postscore means between the E and C groups, that is, $r_{Y1} = .334, B_1 = 19.33 \ (= \bar{Y}_E - \bar{Y}_C)$, each with $t = 2.965**, df = 70$. However, there is some difference between the groups in prescore, that is, $r_{12} = .115$, and the pre–post correlation, as usual in such problems, is substantial (.593).

The role of the v prescore is to serve as a covariate in the analysis of the Y postscore, so that the effective dependent variable is $Y - \hat{Y}_v$, or $Y{\cdot}v$. Logically, then, in the hierarchical model, $v = X_2$ is the first IV entered. For the total sample, r_{Y2}^2 is .3511 and the slope of the regression of post on pre is .79 (Table 8.3.1). Now this is the total sample regression described above and not appropriate for adjusting the group means, since it is partly a function of them. However, when we next add the dichotomy ($u = X_1$) into the regression system, the partialling process takes the "between group" variance out of both Y and v, and the resulting regression coefficient, $B_2 = .74$, is the pooled within groups slope. $R_{Y{\cdot}21}^2$ increases to .4228 (Table 8.3.2) from .3511. (The order in which X_1 and X_2 enter has, of course, no effect on the R^2, that is, $R_{Y{\cdot}21}^2 = R_{Y{\cdot}12}^2$.) Now, the significance of the $I_{1{\cdot}2} = .0717$ is the same as that for $B_{1{\cdot}2} = 15.59$ in the regression equation in X_1 and X_2, namely $t = 2.928 \ (P < .01)$. We repeat this equation for convenience here:

$$\hat{Y} = 15.59 \, X_1 + .74 \, X_2 + 26.74.$$

If we now substitute for X_1 the value 1 for E and 0 for C of the u dichotomy, we obtain for each group a separate equation. For the E group,

$$\hat{Y}_1 = 15.59 \,(1) + .74 \, v + 26.74,$$

(8.3.5) $$\hat{Y}_1 = B_2 v + B_1 + A,$$

which yields $\hat{Y}_1 = .74 \, v + 42.33$, and for the C group,

$$\hat{Y}_0 = 15.59 \,(0) + .74 \, v + 26.74,$$

(8.3.6) $$\hat{Y}_0 = B_2 v + A,$$

which yields $\hat{Y}_0 = .74 \, v + 26.74$.

Equations (8.3.5) and (8.3.6) define two distinct regression lines, which, because their slope ($B_2 = .74$) is the same, are parallel, and separated by 15.59 (= B_1) Y units. B_1 is the difference in Y (post) means between the E and C groups "adjusted for" the v (pre) covariate, which we have seen is statistically significant. The two adjusted Y means are readily obtained by substituting $\bar{v}$, the total sample mean of v (= 57.03, see Table 8.3.1) for v in Eqs. (8.3.5) and (8.3.6); the results are 86.88 and 67.55, respectively, necessarily differing by (B_1 =) 15.59, since the addition of B_1 in (8.3.5) is the only difference between the Eqs. (8.3.5) and (8.3.6). Thus, we obtain *exactly* the results of a conventional ACV performed on these data (but see below).

The Interaction as a Violation of an Analysis of Covariance Assumption

But there is something obviously wrong here, which is highlighted by comparing these two equations with their counterparts in the previous section, Eqs. (8.3.2) and (8.3.3), derived from Eq. (8.3.1). The latter included the $uv = X_3$ term, which resulted in the slope for the E group ($u = 1$) being 1.03 and for the C group .49 (Table 8.3.2). Since in Eq. (8.3.1) B_3 was significant, the crucial ACV assumption of equal covariate regression coefficients or parallel slopes in the two populations is patently unsupportable, as is made graphic in Figure 8.3.1. Thus, *the ACV described in the preceding paragraph is invalid.* It presumes a constant difference between adjusted E and C when the difference is seen to vary, and in the illustrative example, even to change direction as v changes.[7]

Here, then, is another role for interactions. We can use MRC analysis to readily accomplish ACV, even for designs of far greater complexity than are illustrated in textbook treatments of the latter (see Chapter 9). However, in addition, the hierarchical MRC model readily supplies through the inclusion of the *UV* product set of IVs (where *V* is a set of IVs of interest and *U* is conceived as a covariate set), a test of the ACV assumption of parallel slopes or regression surfaces. The statistical significance of the interaction set rules out this parallelism and invalidates the ACV method, although it does not preclude a fully meaningful MRC analysis.

8.4   TWO QUANTITATIVE VARIABLES

Recall the hapless Professor Doe of Section 8.2.2, who sought to analyze the relationships of a cognitive style dependent variable (Y) to authoritarianism (f), general ability (g) and the $f \times g$ interaction, and cast his data on the Procrustean bed of a 2 $\times$ 2 factorial design AV. We now fulfill the promise of showing how such a set of data can be analyzed by MRC without the loss of information and statistical power entailed in reducing graduated variables such as f and g to dichotomies. Table 8.4.1 gives, for a sample of $n = 108$ highschool students, the simple correlation and regression results for f (= X_1), g (= X_2) and fg, the product of these scores, ($X_3 = X_1 X_2$). Although a "score" of (24) (86) = 2064

---

[7] With nonparallel regression lines (surfaces), one may ask the question "For what range of values of the covariate(s) do the adjusted Y values of the groups differ (and in what direction) significantly?" The Johnson-Neyman technique, designed to answer this question, is given in detail by Walker and Lev (1953, pp. 398–404).

TABLE 8.4.1

Simple Correlation and Regression for Two Quantitative Variables
(f and g) and Their Product (fg)

	Y	X_1	X_2	X_3	r^2_{Yi}	t ($df = 106$)	
			r				
$f = X_1$	.115	1.000	−.274	−.759	.0132	1.271	$\hat{Y} = .047 X_1 + 28.6$
$g = X_2$	−.483	−.274	1.000	.740	.2330	−5.675**	$\hat{Y} = −.210 X_2 + 39.9$
$fg = X_3$	−.455	−.759	.740	1.000	(.2069)	(−5.259)**	
m	30.44	39.33	45.33	1987.8			
					$n = 108$		
sd	11.50	28.25	26.47	2261.9			

**$P < .01$.

"authoritarianism–general ability" points seems intuitively to be a nonsensical mixture of apples and oranges (or even horses and galaxies), it nevertheless carries the interaction information, as do all products of variables.

Table 8.4.1 indicates that f taken alone accounts for a nonsignificant 1.3% of the Y variance, while g alone accounts for a highly significant 23.3% (negatively). We note that a unit increase in g is associated with a change in Y of −.210 (= B_2) units, on the average. We place in parentheses r^2_{Y3} and its t, since $X_3 = fg$ contains, in addition to $f \times g$ variance, variance linearly associated with f and g, as can be seen from the typically high values for r_{13} and r_{23} .

In considering the MRC analysis of these data, we must first decide whether we wish to appraise each main effect IV with the other partialled, or to order them so that the first is unpartialled and the second has the first partialled from it. The latter procedure was used in the previous section (and generally in the polynomial regression of Chapter 6). Assume that we have no basis for assigning priority here, and simply enter f and g simultaneously as a set of main effects (see "Intermediate Models," Section 4.4.1). The MRC results are given in Table 8.4.2a. We find that the two variables together account for .2333 of the Y variance, with g making by far the major contribution, since if it were omitted, R^2 would drop by .2201 (= sr^2_2), a highly significant amount materially as well as statistically ($t = −5.491$, $P < .01$). The unique contribution of f in accounting for variance in Y is virtually nonexistent ($sr^2_1 = .0003$), even less than its small nonpartialled contribution ($r^2_{Y1} = .0132$, Table 8.4.1).[8]

With the addition of the product $fg = X_3$ as a third IV, some dramatic changes occur in the regression, given in Table 8.4.2b. Recall first that with f and g

[8] It is barely worth noting that f is a suppressor variable in this equation, since the sign of its partial coefficients is negative, while that of its r with Y positive. We say "barely" because of the miniscule and emphatically nonsignificant degree to which this occurs.

already in the equation and hence partialled from fg, the latter is $fg \cdot f,g$ and hence the $f \times g$ interaction, as in Eq. (8.1.3). Its entry increases R^2 from .2333 to .3324, that is, by .0991 ($= R^2_{Y \cdot 123} - R^2_{Y \cdot 12} = I_3 = sr_3^2$), a substantial and highly significant amount ($t = -3.932$, $df = 104$, $P < .01$). We thus find that an important conditional relationship exists in these data: the regression of this cognitive style (Y) on authoritarianism (f) depends on (or varies with) the level of general ability (g), or, equivalently, the regression of Y on general ability depends on the level of authoritarianism. (We remind the reader that linear transformations of f and g, although affecting simple correlations of the resulting $f'g'$, will *not* effect the value of I_3 or its significance; see Section 8.1.)

TABLE 8.4.2

MRC Results for Two Variables and for Two Variables
plus Interaction

---

a. Results for f and g:

$$R^2_{Y \cdot 12} = .2333, F = 15.978^{**} \qquad (df = 2, 105)$$

	pr_i	sr_i^2	B_i	$t_i \, (df = 105)^a$
$f = X_1$	−.020	.0003	−.008	.209
$g = X_2$	−.472	.2201	−.212	−5.491**

$$\hat{Y} = -.008\,f - .212\,g + 40.3$$

b. Results for f, g, and fg:

$$R^2_{Y \cdot 123} = .3324, F = 17.264^{**} \qquad (df = 3, 104)$$

	pr_i	sr_i^2	B_i	$t_i \, (df = 104)^a$
$f = X_1$	−.324	.0783	−.232	−3.493**
$g = X_2$	.025	.0004	.0172	.251
$fg = X_3$	−.360	.0991	−.00466	−3.932**

$$\hat{Y} = -.232\,f + .0172\,g - .00466\,fg + 48.1$$

or

$$\hat{Y} = (.0172 - .00466\,f)g + (-.232\,f + 48.1)$$

Letting $f = 11.1$, 39.3, and 67.6 results in the equations:

"Low" ($f = 11.1$): $\hat{Y}_L = -.035\,g + 45.5$;

"Average" ($f = 39.3$): $\hat{Y}_A = -.166\,g + 39.0$;

"High" ($f = 67.6$): $\hat{Y}_H = -.294\,g + 32.4$;

**$P < .01$.
^{a}See Footnote 5.

Some simple tinkering with the regression equation will help make clear how this conditionality works. When $X_3 = X_1 X_2$, the relevant regression equation can be written

(8.4.1) $$\hat{Y} = B_1 X_1 + B_2 X_2 + B_3 X_1 X_2 + A,$$

which can also be expressed in either of the following ways:

(8.4.2) $$\hat{Y} = (B_1 + B_3 X_2) X_1 + (B_2 X_2 + A),$$

(8.4.3) $$\hat{Y} = (B_2 + B_3 X_1) X_2 + (B_1 X_1 + A).$$

Taking Eq. (8.4.3) for concreteness, it can be viewed as a family of equations of regression lines whose slope is $B_2 + B_3 X_1$ and whose Y intercept is $B_1 X_1 + A$. Each different value we assign to X_1 defines a different line, thus justifying the conditional formulation: the regression of Y on X_2 depends on the value of X_1. Interchanging X_1 and X_2 provides the symmetrical interpretation of Eq. (8.4.2).

For the illustrative example, substituting the literal notation and coefficients found in Table 8.4.2b, the equation

$$\hat{Y} = -.232\, f + .0172\, g - .00466\, fg + 48.1$$

becomes, when the terms are organized as in Eq. (8.4.3),

$$\hat{Y} = (.0172 - .00466\, f)\, g + (-.232\, f + 48.1),$$

as shown in Table 8.4.2b. The table provides equations for three representative members of this family of lines, one for a "low" f score (1 sd below the mean), one for an "average" f score (at the mean), and one for a "high" f score (1 sd above the mean). Note that these regression lines differ both in slope and intercept. For these data, low f cases show little or no linear regression of Y on g, while high f cases show substantial (negative) linear regression; as f increases, the slope of Y on g becomes steeper.

The varying regression can be subjected to further interesting analysis. The slopes of these lines are given by $B_2 + B_3 f$ and the Y intercepts by $B_1 f + A$, as noted. For these data, $B_2 = .0172$ is not significant ($t = .251$), leading to the conclusion that the differential slope of Y on g is not materially dependent on g as such, but rather on fg (i.e., $B_3 = -.00466$, $t = -3.932**$), which is why there is little or no regression of Y on g when f is small. The Y intercept does, however, vary significantly with f (i.e., $B_1 = -.232$, $t = -3.493**$).

One is impressed by the difference in the model with and without the interaction. When the equation in f and g alone (Table 8.4.2a) is similarly specified for "low," "average," and "high" values of f, one obtains three necessarily *parallel* lines, since they share the same regression coefficient for Y on g, $-.212$. (In these data, because $B_1 = -.008$ is so small, the three lines are quite close together.) The equation in two IVs can only carry a common slope for all levels of f, thus (as we see from the model including interaction) overestimating the slope for low f, and underestimating it for high f. Tailoring the Y on g regression to the level of f results in increasing the Y variance accounted for by .0991.

It was, of course, quite arbitrary to carry f as the "moderator" variable and look at the Y on g regression which it "moderates."[9] By substituting in Eq. (8.4.2) instead of Eq. (8.4.3), and interchanging B_1 and B_2 and X_1 $(=f)$ and X_2 $(= g)$ in the ensuing discussion, we could show the details of how the regression of Y on f is conditional on the level of g. This operation is left to the reader as an instructive exercise.

This, then, is the analysis which Professor Doe should have carried out. No information has been lost through degrading multivalued IVs to dichotomies. Not only has the kind of information normally provided by the AV of these data been obtained, but it has more metric precision and hence, more statistical power. Also, the contrast of the model with and without interaction is instructive. Finally, and as always, the various correlation and regression coefficients which are normal MRC by-products allow for a more deeply etched portrait of the phenomena under study.

8.5 A NOMINAL SCALE AND A QUANTITATIVE VARIABLE

In this section, we extend our consideration of interactions from a dichotomy (two groups) and a quantitative variable (Section 8.3) to the more general nominal scale (g groups) and a quantitative variable.

For concreteness, we consider an example with $g = 3$ groups. For coding the nominal scale (G) in multifactor problems we have available all the coding methods for nominal scales described in Chapter 5: dummy, effect, contrast, and even nonsense coding. Our choice will be dictated by the aspects of group membership toward which our hypotheses direct our attention, as was the case where G was the only research factor.

We flesh the bones of the illustrative data of this section with the following substantive interpretation: for a sample of $n = 72$ subjects, Y is some performance measure, the quantitative IV (w, which we represent as X_1) is some measure of level of motivation, and the three groups are given different instructions: G_1 is given control instructions, and G_2 and G_3 are given different experimental instructions. For the sake of generality, the n_i are unequal: $n_1 = 22$, $n_2 = 24$, and $n_3 = 26$. Consistent with the structure of the experiment, assume that the researcher's interest is in comparing G_1 with the two experimental groups combined, and the latter with each other. This is directly accom-

---

[9] In the psychometric literature, the term "moderator" variable refers to an IV which potentially enters into interaction with "predictor" variables, while having a negligible correlation with the criterion itself. In such usage, f in the example would be the "moderator" variable. But, as we have noted, the relationship of two interacting variables is symmetric—if f "moderates" the regression of Y on g, then g "moderates" the regression of Y on f.

plished by using contrast coding: following the methods described in Section 5.5, the 2 (= $g - 1$) contrast IVs have orthogonal coding coefficients assigned, in order, as follows: X_2 = 1, $-\frac{1}{2}$, $-\frac{1}{2}$, and X_3 = 0, 1, −1 (see "coding diagram" in Table 8.5.1). Thus, X_2 and X_3 constitute a set of IVs representing G.

With one IV representing motivation ($w = X_1$), and two IVs representing, as contrasts, group membership ($G = X_2$, X_3), there are (1) (2) = 2 product variables, together comprising a set carrying the $w \times G$ interaction, that is, X_4 = $X_1 X_2$ (= wX_2), and X_5 = $X_1 X_3$ (= wX_3). For example, a subject in G_2 whose w score is 52 has the following values for the five IVs: X_1 = 52, X_2 = $-\frac{1}{2}$, X_3 = 1, X_4 = (52) $(-\frac{1}{2})$ = −26, and X_5 = (52) (1) = 52; with the same w score, the IVs for a subject in G_1 are: X_1 = 52, X_2 = 1, X_3 = 0, X_4 = 52, X_5 = 0.

We will dispense with the usual tabular presentation of all the correlations among the 6 variables, but present in Table 8.5.1 for each variable its m and sd, and for the main effect IVs, r_{Yi} and its t value (df = 70). We note that r_{Y1} = .254 ($P < .05$), and r_{Y2} = .491 ($P < .01$). They indicate respectively that for the total sample there is a significant modest relationship of Y with the motivation measure, and that there is a substantial difference between the Y mean of the control group (G_1) and that of the two experimental groups (G_2, G_3) combined and treated as a single sample.

As is customary, we will use the hierarchical MRC model in analyzing the data. Since the motivation measure w assesses an enduring trait and is antecedent to the experimental instructions, it is entered first as X_1. Table 8.5.1 indicates that for the total sample, it accounts linearly for .0645 of the variance in performance, a statistically significant amount (F = 4.825, df = 1, 70, redundant since this $F = t^2$ for r_{Y1}). The regression line of Y on X_1 has a slope of .25 (= B_1). Note that these results can be looked upon as averaged over the entire 72 cases which are at this stage not differentiated as to group membership.

At the second stage of the analysis, we add the set of two variables which carry the information of group membership, contrast coded as described above, G = X_2, X_3. This brings the Y variance accounted for from .0645 (= $R^2_{Y \cdot 1}$) to .2757 (= $R^2_{Y \cdot 123}$), an increase of .2112 (= $I_{23 \cdot 1}$, or $I_{G \cdot w}$). The F test of this I based on 2 df can be accomplished by means of the general formula for the significance of the increase in R^2 of an added set B containing k_B variables to that of a set A containing k_A variables using Model I error (but see Footnote 5). The formula Eq. (4.4.2), is used throughout, although up to this point it has been specialized for k_B = 1. We restate it here for convenience:

(4.4.2) $F = \dfrac{R^2_{Y \cdot AB} - R^2_{Y \cdot A}}{1 - R^2_{Y \cdot AB}} \times \dfrac{n - k_A - k_B - 1}{k_B}$ ($df = k_B, n - k_A - k_B - 1$).

$$= \frac{.2757 - .0645}{1 - .2757} \times \frac{72 - 1 - 2 - 1}{2}$$

$$= 9.914 \quad (df = 2, 68),$$

which is significant ($P < .01$). This indicates that when adjusted for w by the

TABLE 8.5.1

Hierarchical MRC Analysis for a Quantitative Variable (w), a Set
Representing G by Contrast Coding, and an Interaction Set

Coding diagram	Set	X_i	m	sd	r_{Yi}	t (df = 70)
	w	X_1	55.00	26.41	.254	2.196*
		X_2	−.042	.691	.491	4.710**
	G	X_3	−.028	.833	−.016	−.134
	wG { $wX_2 = X_4$		.444	44.46		
	$wX_3 = X_5$		.222	48.28		
		Y	45.33	26.47		

Coding diagram:

	X_2	X_3
G_1	1	0
G_2	$-\frac{1}{2}$	1
G_3	$-\frac{1}{2}$	-1

Hierarchical MRC

Set	IVs added[a]	cum R^2	F	df	I	F	df
w	X_1	.0645	4.825*	1, 70			
+ G	X_2, X_3	.2757	8.626**	3, 68	.2112	9.914**	2, 68
+ $w \times G$	X_4, X_5	.3670	7.652**	5, 66	.0913	4.760*	2, 66

Regression equations

$$\text{for } X_1: \quad \hat{Y} = .25\,X_1 + 31.30$$
$$t \quad 2.196* \quad (df = 70)$$

$$\text{for } X_1, X_2, X_3: \quad \hat{Y} = .19\,X_1 + 17.75\,X_2 - 1.30\,X_3 + 35.70$$
$$(\text{ACV}) \quad t \quad 1.793 \quad 4.439** \quad -.397 \quad (df = 68)$$

$$\text{for } X_1 \text{ through } X_5: \quad \hat{Y} = .22\,X_1 + 41.91\,X_2 - .36\,X_3 - .42\,X_4 + .03\,X_5 + 35.40$$
$$t \quad 2.147* \quad 4.768** \quad -.056 \quad -3.057** \quad .228 \quad (df = 66)$$

*P < .05.   **P < .01.
[a]We list only the IVs added, and not the prior IVs.

within group regression, as in ACV, the Y means of the three groups are not all
equal. Indeed, this *is* an ACV, exactly as was performed in Section 8.3.3 where
only two groups were involved.[10]

[10] As was the case there, the covariance assumption of homogeneous (parallel) regression
proves to be invalid when the interaction set (to be added next) turns out to be significant.
We nevertheless continue the ACV to show how it would proceed when the interaction set is
not significant and the regression homogeneity assumption is tenable.

Observe that at this point we know that the three (w-adjusted) means differ, since the G set carries the group membership information, but we do not yet know the status of the separate null hypotheses carried by each of the two contrast aspects X_2 and X_3. This would be answered by the partial coefficients for X_2 and X_3 in the analysis for the first three IVs. Table 8.5.1 gives the regression equation, and hence B_2 and B_3, together with their t ratios; B_2 is found to be highly significant ($t = 4.439**$), and B_3 not at all ($t = -.397$). Thus, on the ACV assumption of equality of the three groups' Y on w slopes, we can say more specifically that the motivation-adjusted performance (Y) mean of G_1 (control instructions) is significantly higher than the unweighted mean of the adjusted means of G_2 and G_3 (which were coded $1, -\frac{1}{2}, -\frac{1}{2}$ for X_2). We can also state the value of this contrast on adjusted means; in Section 5.5.3, we find (Eq. 5.5.3) that it is $C = 17.75 (1 + 2)/(1) (2) = 26.62$. (For X_3, the reader should verify that the sample contrast value is -2.60.) For a correlational statement of importance of these two contrasts, one can determine that $pr_2 = .474$, $pr_3 = -.048$. (See Section 9.3.2.)

As Footnote 10 warned, the above analysis conceived as ACV fails to meet the assumption of regression homogeneity, as the next step in the hierarchical MRC reveals: when the $w \times G$ interaction set made up of X_4 and X_5 is next added to the main effect IVs, R^2 increases to .3670, and $I_{45 \cdot 123} = .0913$, a significant increment ($F = 4.760*$). There *are* slope differences among the three groups, so that the Y means were "adjusted" on the basis of a common slope which we now know misrepresents the true circumstances.

Although the significance of the interaction set renders the ACV analysis invalid, it would be a serious error to conclude from this that the research is uninterpretable. Quite the contrary! The MRC analysis we have performed provides a rich yield of information from the data which may materially increase the investigator's insight into the phenomena under study. To illustrate, let us rewrite the complete regression equation using literal notation instead of subscripted Xs for the sake of clarity in the manipulations. We will symbolize the first contrast (X_2) as c (for control vs. experimental) and the second (X_3) as e (for between experimental), and use asterisks to indicate significance:

$$(8.5.1) \quad \hat{Y} = B_w w + B_c c + B_e e + B_{wc} wc + B_{we} we + A,$$

$$\hat{Y} = .22* w + 41.91* c - .36 e - .42* wc + .03 we + 35.40.$$

As we saw in the previous section, we can reorder the terms to provide an equation for a set of Y on w lines with a composite variable slope and composite variable intercept:

$$(8.5.2) \quad \hat{Y} \equiv (B_w + B_{wc} c + B_{we} e)w + (B_c c + B_e e + A),$$

$$\hat{Y} = (.22* - .42* c + .03 e)w + (41.91* c - .36 e + 35.40).$$

This equation, of course, still has five IVs in it and thus defines a five-dimensional hyperplane. But it becomes a set of simple regression line equations, one equation for each group, when the group's coded values are entered. Thus, G_1 is coded 1 on $c (= X_2)$ and 0 on $e (= X_3)$, so, substituting in Eq. (8.5.2),

(8.5.3) $\hat{Y}_1 = [.22 - .42(1) + .03(0)] w + [41.91(1) - .36(0) + 35.40]$,

$\qquad \hat{Y}_1 = -.20\, w + 77.31.$

G_2 is coded $-.5$ on c and 1 on e, so

(8.5.4) $\hat{Y}_2 = [.22 - .42(-.5) + .03(1)] w + [41.91(-.5) - .36(1) + 35.40]$,

$\qquad \hat{Y}_2 = .46\, w + 14.08.$

G_3 is coded $-.5$ on c and -1 on e, so

(8.5.5) $\hat{Y}_3 = [.22 - .42(-.5) + .03(-1)] w + [41.91(-.5) - .36(-1) + 35.40]$,

$\qquad \hat{Y}_3 = .40\, w + 14.80.$

The three regression lines are clearly quite different, the reasons for which are laid bare in the analytic equations (8.5.2) to (8.5.5). Thus, the slope for G_1 is $-.20$ and quite different from those for G_2 and G_3, .46 and .40, the latter hardly differing. Inspection of the equations shows that the highly significant B_{wc} (= B_4) of $-.42$ is an overriding part of the slope for G_1, whereas $-\frac{1}{2}$ of it (+.21) contributes to the slopes for G_2 and G_3. Indeed, the contrast between the slopes provided by the $w \times c$ interaction term (X_4) is *exactly* the same as that between means, that is, the *slope* for G_1 minus the unweighted mean of the *slopes* for G_2 and G_3, and its value is *exactly* given by the same formula which provides it for means, Eq. (5.5.3), that is, $-.42(1 + 2)/(1) (2) = -.63$, which is easily checked: $-.20 - (.46 + .40)/2 = -.63$. Similarly, the small slope difference between G_2 and G_3 is a consequence of the fact that it depends solely on B_{we} (= .03, not significant), which is added in the slope composite for G_2 and subtracted in that for G_3. The significance test for B_{we} is thus a test of the contrast hypothesis that the population slopes for G_2 and G_3 are equal, exactly as represented in the $w \times e$ interaction term (X_5), and the sample contrast value is again given by Eq. (5.5.3), $.03(1 + 1)/(1) (1) = .06$ (i.e., $.46 - .40 = .06$).

Thus, each interaction B is a contributor to the separate slopes and its t tests an hypothesis about the slopes defined by the coding of the groups. This is quite general, holding for any number of quantitative variables and any number of groups. Moreover, it holds not only for contrast coding but for any kind of nominal scale coding—the slope comparison is the same as the comparison of means provided by the coding of the nominal scale. For example, the B for the interaction of a dummy variable D with another variable w is the difference in slopes between the group coded 1 on D and the reference group (Section 5.3).

The source and nature of the intercept differences for the three groups is similarly apparent from a study of the analytic equations (8.5.2) to 8.5.5), the intercept composite being identical in structure with the slope composite. Again, the respective Y intercepts of 77.31, 14.08, and 14.80 reflect the two coefficients B_c = 41.91 (t = 4.768**) and B_e = −.36 (t = −.056). The first is a function of the difference between the G_1 intercept and the unweighted mean of the intercepts of G_2 and G_3, whose sample value is, from Eq. (5.5.3), 41.91 (1 + 2)/(1) (2) = 62.86, and the latter is a function of the trivial and nonsignificant intercept difference between G_2 and G_3, whose sample value is −.36 (1 + 1)/(1) (1) = −.72.

In summary, the analysis indicates markedly different (linear) relationships between level of motivation and performance as a function of instructions. The two experimental groups show virtually identical relationships, a rising performance level with increasing motivation scores, markedly (and highly significantly) different from the control group, which shows a decline. Further, at low motivation levels (that is, where w is near zero), the experimental groups both show low performance, whereas the control group is high. The failure of the ACV assumption evidenced by the significant interaction means only that it is not meaningful to compare performance scores adjusted for motivation by a common adjustment which represents the three regression lines as parallel. But its very failure is illuminating and highly interpretable. In this case, if the three lines are plotted, it can be seen that the G_1 line is above those of G_2 and G_3 over virtually the entire range of w from 0 to 100, thus the experimental instructions make for poorer performance almost no matter what the motivation score. Had the lines crossed near the middle of the w distribution, these results would indicate the level of motivation for which a type of instruction leads to better or poorer performance (see Footnote 7). A valid (interaction not significant) ACV would make it possible to infer a *constant* effect of instructions over the range of the covariate; significant interaction, when fully analyzed, makes it possible to detail the nature of the variability of the effect as a function of the covariate.

8.6 SET INTERACTIONS: QUANTITATIVE X NOMINAL

We come now to the first case of interactions between sets of two or more IVs. In this section, we represent a quantitative research factor W as a set using a second order power polynomial (Section 6.2), explicitly $X_1 = w$, $X_2 = w^2$, and another set representing a three-level nominal scale G (X_3, X_4), and focus on the resulting $W \times G$ interaction set.

To save space, we will expand the problem of the last section, using the same fictitious data, by adding the quadratic aspect of the motivation measure to form the quantitative set. Although not necessary to our central purpose, we will enrich the problem by recoding the nominal scale from contrast coding to dummy-variable coding (Section 5.2), designating the two aspects of G as D_1, D_2. We will set G_1, the control group, as the reference group, for example, with D_1, D_2 (X_3, X_4) coded as 0, 0; G_2 is coded 1, 0; and G_3 is coded 0, 1.

As described above in Section 8.1, we form the interaction set by multiplying each of the two IVs of W by each of the two dummy variable IVs of G, resulting in the following four IVs:

$$X_5 = X_1 X_3 = wD_1$$
$$X_6 = X_1 X_4 = wD_2$$
$$X_7 = X_2 X_3 = w^2 D_1$$
$$X_8 = X_2 X_4 = w^2 D_2$$

Thus, the problem is as before, except for the representation of motivation (W) not only by its linear aspect (w), but also by its quadratic aspect (w^2), which makes possible the detection and description of simple (parabolic) curvature in the regression of Y on W.

Table 8.6.1 presents the chief results of the hierarchical MRC analysis. In the interests of focusing attention and saving space, we omit the less important results: the ms, sds, rs of the simple correlation analysis, the F ratio for each cumulative R^2, and most of the partialled results per IV. We focus on the I for each *added* set or subset of IVs, their statistical significance using Model I error, and the relevant regression equations with the t values for their B_i. It is instructive to compare these results with those of Table 8.5.1.

The entry of w provides a significant .0645 of the Y variance (as before). The addition of w^2 increases R^2 by only .0039, a small and quite nonsignificant amount, indicating that *for the entire undifferentiated sample,* the data do not warrant providing a bend in the straight line regression of Y on W (Section 6.2).

At the next level of the hierarchy we introduce the group membership information G (X_3, X_4, expressed as dummy variables instead of contrasts) and note a large (.2079) and highly significant increment in Y variance accounted for.[11] We see then that the three groups do differ substantially in Y, with W not only linearly but also quadratically partialled out ("controlled," "adjusted for"). In these data the quadratic aspect is trivial; nevertheless, the method is a valid one for making allowance for simple *non*linear regression. What we have done, then, is a "between-group" ACV on Y adjusted for the possibility of a simple (that is, second order polynomial) nonlinear relationship with the covariate, represented by Eq. (I) in Table 8.6.1.

Unfortunately, as we have already seen in the previous section, the ACV is invalidated by the evidence of significant $w \times G$ interaction, that is, group differences in the slopes of the Y on w regression lines. Eq. (I) in Table 8.6.1 is presented, nevertheless, to demonstrate the regression equation form of a quadratically adjusted ACV. In the absence of interactions, this equation implies

[11] Note the similarity to its analogous I in Table 8.5.1 (.2112). They *should* be similar, since the analyses differ only in the presence here of w^2 which we have seen makes a trivial contribution. (The use of dummy instead of contrast coding of course has no effect on the proportion of variance accounted for by G.)

TABLE 8.6.1
Results of a Hierarchical MRC Analysis for a Quantitative
Set ($W = w, w^2$), a Set Representing G by
Dummy Variable Coding (D_1, D_2)
and an Interaction Set

coding for G	Set	IVs added[a]	Cum. R^2	I	F	df
D_1 D_2	W	$w = X_1$	.0645	.0645	4.825*	1, 70
X_3 X_4		$w^2 = X_2$	.0684	.0039	.291	1, 69
G_1 0 0	$+ G$	$D_1, D_2 = X_3, X_4$	.2763	.2079	9.623**	2, 67
G_2 1 0	$+ W \times G$	$wD_1, wD_2 = X_5, X_6$	.3671	.0908	4.660**	2, 65
G_3 0 1		$w^2 D_1, w^2 D_2 = X_7, X_8$	.4311	.0640	3.544*	2, 63

Regression equations:

For w, w^2, D_1, and D_2 (W and G)

(I) $\hat{Y} = .31\,w - .0011\,w^2 - 28.67\,D_1 - 25.54\,D_2 + 56.02$

t .612   −.244   −4.007**   −3.690**   ($df = 67$)

For complete model: $w, w^2, D_1, D_2, wD_1, wD_2, w^2 D_1, w^2 D_2$ ($W, G, W \times G$)

(II) $\hat{Y} = .60\,w - .0070\,w^2 - 5.57\,D_1 - 61.11\,D_2 - 2.02\,wD_1 + .60\,wD_2 + .0235\,w^2 D_1 + .0001\,w^2 D_2 + 60.49$

t .751   −1.026   −.194   −2.279*   −1.761   .538   2.349*   .005   ($df = 63$)

(IIa) $\hat{Y} = (.60 - 2.02\,D_1 + .60\,D_2)w + (-.0070 + .0235\,D_1 + .0001\,D_2)w^2 + (60.49 - 5.57\,D_1 - 61.11\,D_2)$

     B_w B_{wD_1} B_{wD_2} B_{w^2} $B_{w^2 D_1}$ $B_{w^2 D_2}$ A B_{D_1} B_{D_2}

For $G_1, D_1 = 0, D_2 = 0$, so $\hat{Y}_1 =$ .60 w − .0070 w^2 + 60.49;

For $G_2, D_1 = 1, D_2 = 0$, so $\hat{Y}_2 =$ −1.42 w + .0165 w^2 + 54.92;

For $G_3, D_1 = 0, D_2 = 1$, so $\hat{Y}_3 =$ 1.20 w − .0069 w^2 −   .62.

*$P < .05$.   **$P < .01$.
[a] We omit listing the prior X_is in each row.

three *parallel* curved regression lines (when one enters coded values for D_1 and D_2) for the three groups, the shape determined by averaged out within-group values. On this (here unsatisfactory) basis, G_2 and G_3 are seen to average, respectively, 28.67 and 25.54 adjusted Y units below G_1 at all values of W, and highly significantly so. (See Section 9.3.4 for a similar data structure analyzed successfully as an ACV.)

In the next stage, we add the product of G (D_1, D_2) with w alone, reserving the product with w^2 for the following stage. (This is done for the same reason that w^2 was added after and separately from w—the "variance stealing" that occurs with power polynomials, as extensively discussed in Sec. 6.2). We remind the reader that wD_1 and wD_2, with w, D_1 and D_2 already in the equation and hence partialled, become respectively the interaction IVs $w \times D_1$ and $w \times D_2$, according to Eq. (8.1.3). Again we find, as in Table 8.5.1, a substantial (.09) and significant increment, indicating the heterogeneity of the slopes of best-fitting lines between groups, and hence the invalidation of the covariance model.

Finally, we add the subset $w^2 D_1$ and $w^2 D_2$. Given the partialling of their elements (w^2, D_1, D_2) this set allows for the possibility that the degree of quadratic *curvature* is not equal for the three groups, that is, these terms represent collectively the quadratic $\times$ groups interaction. Table 8.6.1 shows that I for this interaction is .0640, which is significant (Eq. 4.4.2, $F = 3.544$, $df = 2$, 63, $P < .05$). Given the dummy variable coding with G_1 as reference group, the regression coefficient for $w^2 D_1$ is literally the difference in coefficients of quadratic curvature: G_2 minus G_1. Table 8.6.1, Eq. (II), gives it as .0235, which is significant. The analysis of this equation into separate equations for each group (see below) shows that this quadratic difference operates in exactly the same manner as the linear (slope) difference does: it is a component of the quadratic coefficient for G_2 and (with this coding) not for G_1. Similarly, the B for the other term, $w^2 D_2$ carries the difference in quadratic coefficients, G_3 minus G_1, found here to be essentially zero (.0001).

Equation (II) of Table 8.6.1 is the regression equation of the "full model"—all IVs entered simultaneously. It is expressed in literal (rather than X_i notation) to facilitate interpretation. As was done in the previous section, we can reorder the terms so that $\hat{Y}$ is written as a function of W, with composite coefficients which are themselves a function of group membership; this is done in Eq. (IIa). We see that *each* composite coefficient is made up of three coefficients from the full equation (as many as there are groups): the first is a constant, the second a function of D_1, and the third a function of D_2. Taking the composite coefficient for w^2, for example, it is the sum of B_{w^2}, $B_{w^2 D_1}$, and $B_{w^2 D_2}$. The other two composite terms are similarly constituted. The terms involving G (D_1 and D_2 in this example) are of course not constants in the equation as a whole, but when the coding coefficients which define each group are substituted, a separate equation for each group results, as shown at the bottom of Table 8.6.1. With 1, 0 dummy variable coding, this amounts to either adding (1) or not adding (0) each of the variable terms in the composite. From this, it can be seen how each regression coefficient and the Y intercept A of the full equation can be interpreted in terms of the properties of the regressions of, and differences in regressions between, the separate groups, according to the interpretation specific to dummy variable coded nominal scales:

1. B_w, B_{w^2}, and A are the constants defining the regression for the reference group (G_1 here). Since neither B_w nor B_{w^2} is significant, the reference group's Y on W regression does not differ significantly from that of a horizontal straight line.

2. $B_{w D_1}$ is the amount by which the linear coefficient (slope) for G_2 is larger than that for G_1, $B_{w^2 D_1}$ is the amount by which the quadratic coefficient for

G_2 is larger than that for G_1, and B_{D_1} is the amount by which the Y intercept of G_2 is larger than that for G_1. Since $B_{w^2 D_1}$ is significant and positive, the curvature for G_2 is significantly more concave upward (Section 6.2.4) than that for G_1 (see Figure 8.6.1 below). Neither its slope nor Y intercept differs significantly from those for G_1.

3. Similarly, $B_{w D_2}$, $B_{w^2 D_2}$, and B_{D_2} carry differences of the reference G_1 from G_3 in regard to slope, curvature and Y intercept. Only B_{D_2} (= −61.11) is significant, indicating no evidence that the *shape* of G_3's population regression line differs from that of G_1, but that the former falls considerably lower than the latter when $w = 0$.

The above is analytic, and not easy to follow in detail. A graphic plot of the best fitting quadratic regression equation for each of the three groups would undoubtedly help and can readily be prepared by substituting a few values for W (w and w^2) in each of the three equations at the bottom of Table 8.6.1, and drawing smooth lines through the resulting $\hat{Y}$ (Section 6.2.2). The results of this procedure are presented in Figure 8.6.1. The statements made above about the best fitting curve of Y on W for G_1 (control group), and differences between the latter and each of the other two groups can be visualized directly from the curves in the figure.

Had we elected to code G differently, for example, using the contrast coding posited in the previous section, the B coefficients would take on different meaning consistent with the new coding. For example, $B_{w^2 C_2}$ is a function of the quadratic contrast between G_2 and G_3; it equals (see Eq. 5.5.3) one-half of the difference in the quadratic coefficients of G_2 and G_3, i.e., .0117. An analogous change in meaning would occur were we to use effects coding (Section

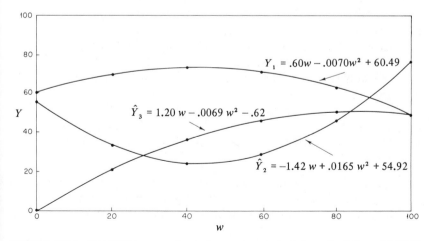

FIGURE 8.6.1 Best-fitting curves for each group as derived from the full model (Table 8.6.1, Eq. IIa).

5.3) for G. However, *whatever* coding coefficients are employed, when those of each group are substituted in the resulting regression equation, the three equations at the bottom of Table 8.6.1 are produced. This is because of the principle that, however coded, group membership is completely represented whether as a main effect or in interactions with other variables.

As long as we follow the rule that constituents of products precede in the hierarchy the products in which they enter, other sequences of entry into a hierarchical MRC model can be validly employed. For example, if the investigator judged, a priori, that the probability of quadratic nonlinearity was not great but wished merely to check it, this feature could be given lesser prominence by using the sequence w, G, wG, w^2, and $w^2 G$. Given the use of Model I error, having w^2 follow rather than precede G and wG may increase the statistical power of the test of the latter two sets and decrease that for w^2.

Most behavioral scientists would find a quadratic of w by G_1 versus the unweighted mean of those for G_2 and G_3 a rather exotic issue to put to their data. We present it primarily to illustrate the general applicability of the concept and methods of interactions to *sets* of IVs. The material of this section would apply if one set were made up of a nominal scale at g levels and the other of socioeconomic indicators or of a collection of psychiatric symptom scales, or, indeed, if both sets were nominal scales. We turn to an explicit consideration of the latter case.

8.7 SET INTERACTIONS: NOMINAL × NOMINAL; FACTORIAL DESIGN ANALYSIS OF VARIANCE

What is called "factorial design" in AV is a data structure and model which is the special case in MRC of two (or more) IV sets each representing a nominal scale and a set (or sets) of their products which contain the interaction(s). The general principles previously described for sets of IVs hold in this special case. We shall see that with MRC analysis we can not only duplicate exactly the results of an AV of such data structures, but that their analysis via MRC may provide some advantages.

Consider a research study comparing the efficacy (as measured by Y) of three different treatment procedures, two experimental (T_1, T_2) and one control (T_c). These three different procedures have each been employed on randomly assigned samples of suitably selected patients at each of four different hospitals $(H_1, H_2, H_3,$ and $H_4)$, thus making possible an appraisal of the uniformity of the treatment effects across hospitals. This would be identified as a (fixed effects[12]) 4 × 3 factorial design, with the interaction $(H \times T)$ directly addressing the issue of uniformity of effects. Each of the twelve "cells" contains

---

[12] Fixed, since our conclusions are intended to apply only to these three treatments and these four hospitals; we have used this model throughout.

efficacy scores for the patients treated by one of the three procedures at one of the four hospitals, and in the interest of generality, we posit differing and disproportionate numbers of cases in the 12 subgroups. The total n for the entire study is 108.

Each of the research factors H and T is a nominal scale and could be represented by IVs using any of the four coding methods of Chapter 5. The choice should be dictated by convenience in making the inferences required by the substantive content of the study. H involves comparisons among the four hospitals all on an equal footing, so that any comparison conveniently proceeds as being between the Y mean of any given H_i and the unweighted mean of means of all four hospitals. We saw in Section 5.4 that effects coding provides just that comparison in its B weights. The four hospitals are thus represented by three IVs as shown in Table 8.7.1a. (The choice of category to which to assign the string of -1s being arbitrary, on esthetic grounds, we use the last group.) For the T factor, on the other hand, the presence of a control group to which each of the experimental groups is to be compared makes most convenient the use of dummy variable coding (Section 5.3). The three treatments are thus represented by two IVs, with the control condition assigned the string (here only two) of zeros, as shown in Table 8.7.lb. Thus, each patient is characterized for hospital of origin by three IVs and for treatment group by two IVs, which together produce the coding of X_1 through X_5 in Table 8.7.1c. Note in Table 8.7.1c that all H_1 patients (whatever their T condition) are coded $1, 0, 0$ on X_1, X_2, X_3; all H_2 patients $0, 1, 0$; all H_3 patients $0, 0, 1$; and all H_4 patients $-1, -1, -1$ (that is, effects coded). Also, all T_1 patients (whatever their hospital of origin) are coded $1, 0$ on X_4, X_5, all T_2 patients $0, 1$; and all T_c patients $0, 0$ (that is, dummy coded).

X_1 through X_5 assign each patient to an H_i and, separately, to a T_j, so that these five variables carry the H and T main effects as two sets of IVs. However, the *joint* status H_iT_j, which represents the hospital-treatment *combination*, is not carried in X_1 through X_5. The latter information, of course, must be separately represented in IVs, that is, in the interaction set $H \times T$ which is represented by X_6 through X_{11}. Exactly as in previous sections and following the same principle, these six variables are constructed by multiplying each of the three aspects (X_1, X_2, X_3) of H by each of the two aspects (X_4, X_5) of T. Thus, $X_6 = X_1X_4, X_7 = X_1X_5, X_8 = X_2X_4, \ldots, X_{11} = X_3X_5$. The 11 IVs *together* exactly identify the H_iT_j cell of each patient.

Note that this is to be expected. We emphasized in Chapter 5 that to fully represent G, made up of g groups, $g - 1$ IVs are necessary, whatever the form of coding. This 4×3 design results in 12 groups, hence full representation requires 11 IVs. There are an infinite number of ways in which one can code membership in one of 12 groups (remember nonsense coding); the coding given in Table 8.7.1c represents these 12 groups as three effects-coded IVs for H, two dummy-coded IVs for T, and the six interaction-bearing IVs that result from their set by set multiplication. Note the identity in numbers of IVs for H, T, and $H \times T$ with

TABLE 8.7.1
Coding for the $H \times T$ (4 × 3) Factorial Design

	a. H Coding (Effects)				b. T Coding (Dummy)	
	X_1	X_2	X_3		X_4	X_5
H_1	1	0	0	T_1	1	0
H_2	0	1	0	T_2	0	1
H_3	0	0	0	T_c	0	0
H_4	−1	−1	−1			

c. Joint Coding of the 12 Groups (Cells) of the $H \times T$ Factorial

| | | H | | | T | | $H T$ | | | | | |
| | | | | | | | $X_1 X_4$ | $X_1 X_5$ | $X_2 X_4$ | $X_2 X_5$ | $X_3 X_4$ | $X_3 X_5$ |
Cell	X_1	X_2	X_3	X_4	X_5	X_6	X_7	X_8	X_9	X_{10}	X_{11}
$H_1 T_1$	1	0	0	1	0	1	0	0	0	0	0
$H_2 T_1$	0	1	0	1	0	0	0	1	0	0	0
$H_3 T_1$	0	0	1	1	0	0	0	0	0	1	0
$H_4 T_1$	−1	−1	−1	1	0	−1	0	−1	0	−1	0
$H_1 T_2$	1	0	0	0	1	0	1	0	0	0	0
$H_2 T_2$	0	1	0	0	1	0	0	0	1	0	0
$H_3 T_2$	0	0	1	0	1	0	0	0	0	0	1
$H_4 T_2$	−1	−1	−1	0	1	0	−1	0	−1	0	−1
$H_1 T_c$	1	0	0	0	0	0	0	0	0	0	0
$H_2 T_c$	0	1	0	0	0	0	0	0	0	0	0
$H_3 T_c$	0	0	1	0	0	0	0	0	0	0	0
$H_4 T_c$	−1	−1	−1	0	0	0	0	0	0	0	0

the numbers of df associated with each source when perceived as an AV 4 × 3 factorial design, that is, 3, 2, and 6, respectively. The sample sizes of these 12 groups, which are neither equal nor proportional, are given later in this section in Table 8.7.3.

Given the efficacy scores (Y) for the 108 patients, together with the representation of each patient as to hospital and treatment group membership using the $k = 11$ IVs of Table 8.7.1, the data matrix is completely defined. In order to proceed to a hierarchical MRC analysis, a decision must first be made as to the ordering of the research factors. The disproportionality of the cell n_{ij} means that

H and T are not "balanced," for example, a given treatment mean might be high because more patients from a generally high-scoring hospital are represented in that treatment. Another way to say this is that the sets H and T are non-orthogonal or correlated with each other; this fact is demonstrable in various ways, for example, a 4 X 3 table of the cell frequencies (see Table 8.7.3 below) would have a nonzero χ^2 contingency value.[13] Just as with two single IVs, when two sets of IVs, A and B, are correlated, $R^2_{Y \cdot AB} \neq R^2_{Y \cdot A} + R^2_{Y \cdot B}$. By way of review of the discussion in Chapter 4, particularly as illustrated in the "ballantine" of Figure 4.3.1, this correlation produces an area of overlap between A and B in Y, which we called c. This area c cannot, on mathematical grounds, be assigned unambiguously to either A or B, or divided between them except arbitrarily. In hierarchical MRC, we *posit* on grounds of a priori causal hypotheses, or the dictates of the substantive issues of the experiment in its logical structure, an ordering of the research factors, that is, a specific hierarchy in which the Y variance of every prior set is entirely removed before considering the Y variance contribution of every subsequent set. Thus, if A is prior to B, the overlapped c area is wholly assigned to A, and we consider only the b area, which is $R^2_{Y \cdot AB} - R^2_{Y \cdot A}$ or $I_{B \cdot A}$, that is, the proportion of Y variance which B accounts for over and above A. (See Section 9.7.)

This issue is critically relevant to our ordering of H and T in the present analysis. The overlap in Y represents an indeterminate mix of antecedent hospital and treatment method effects. Since our experimental interest lies in T, we wish a "pure" expression of T, untrammelled by any H effects. We therefore first take out all of the Y variance associated with H, including the overlap, and then determine what T additionally accounts for. True, in denying T any of the overlap, we may be assigning to H and withholding from T variance in efficacy for which T is responsible. But in the absence of random assignment of patients *to hospitals* (a practical impossibility and a dangerous assumption to make for reasons of convenience), we accept this conservative assessment since it affords us "pure" T.

The traditional ACV also partials out the covariate one seeks to control, but as applied to experiments, makes the additional assumption as usually stated that the covariate (which would be H here) "has no effect on" the research factor of interest (T). In experiments in which there is random assignment of subjects to groups, the covariate(s) and research factor(s) are not (except by chance) correlated, as, for example, H is correlated with T here.[14] In analyzing data using hierarchical MRC, we can avoid this frequently untenable assumption by

---

[13] Another way is via canonical correlation, which is a generalization of multiple correlation to the case where a set of variables is related to another *set* of variables, rather than to a single variable. (See Chapter 11.) The (first) canonical R for X_1, X_2, X_3 versus X_4, X_5 is not zero.

[14] This is of course quite separate from the assumption of parallelism of regression lines (surfaces), another assumption of ACV that need not be made in the general MRC analysis of such data structures, as we have already seen, and will see again shortly.

TABLE 8.7.2
Results of Hierarchical MRC Analysis for the 4 × 3 Factorial Design

Set	IVs added	cum. R^2	I	Model I (Eq. 4.4.2)		Model II (Eq. 4.4.4)	
				F_I	df_I	F_{II}	df_{II}
H	X_1, X_2, X_3	.0270	.0270	.961	3, 104	1.417	3, 96
$+ T$	X_4, X_5	.1823	.1553	9.686**	2, 102	12.231**	2, 96
$+ H \times T$	$X_6, X_7, \ldots, X_{11}$	.3907	.2084	5.473**	6, 96	5.473**	6, 96

$n = 108, k = 11$ $1 - R^2_{Y \cdot H, T, H \times T} = .609325,$ $df = 96$

Regression equations
(I): Effects-coded H:

$$\hat{Y}_H = 4.64\ X_1 + 4.49\ X_2 + 3.16\ X_3 + 81.74$$
$$t \quad .615 \quad\quad .619 \quad\quad .452 \quad\quad\quad\quad (df = 104)$$

(II): Effects-coded H and dummy-coded T:

$$\hat{Y}_{H,T} = 3.04\ X_1 + 3.40\ X_2 + 3.96\ X_3 - 14.78\ X_4 + 26.91\ X_5 + 78.79$$
$$t \quad .433 \quad\quad .506 \quad\quad .634 \quad\quad -1.581 \quad\quad 2.744** \quad\quad (df = 102)$$

(III): H, T, and the $H \times T$ interaction:

$$\hat{Y}_{H,T,H \times T} = 5.17\ X_1 - 19.50\ X_2 + 23.17\ X_3 - 13.90\ X_4 + 22.75\ X_5 - 20.77\ X_6$$
$$t \quad .430 \quad -1.869 \quad\quad 2.220* \quad -1.648 \quad\quad 2.580* \quad -1.317$$

$$+ 18.25\ X_7 + 23.90\ X_8 + 47.92\ X_9 - 31.35\ X_{10} - 19.42\ X_{11} + 78.83$$
$$t \quad 1.142 \quad\quad 1.636 \quad\quad 3.231** \quad -2.242* \quad -1.309 \quad\quad\quad (df = 96)$$

accepting the possible conservatism of using only Y variance which is uncontroversially attributable to the research factor, what we have called "pure" T. (See the more detailed discussion of this issue in Section 9.7.)

We proceed to a hierarchical MRC analysis of this data structure, using the IV sets in the order H, T, and finally the HT product set, the chief results of which are given in Table 8.7.2.

When Y is regressed on the set of IVs representing H (X_1, X_2, X_3), $R^2_{Y \cdot 123}$ is found to equal .0270; thus, less than 3% of the variance in efficacy scores is associated with gross hospital group membership. We say "gross" because the imbalance of H with regard to T means that some "confounding" T influence may be entering here (either positively or negatively). Note, for example, that in H_1 more than a third of the patients are in T_1 and less than a third are T_c, whereas in H_3 the reverse is true (see Table 8.7.3 below).

For significance testing of the I, we offer both the results of using error Model I, the residual from the cumulative R^2, as well as those of error Model II the residual from the R^2 for all the $k = 11$ IVs, i.e., .6093 (see Section 4.4). Since Model II would be preferred here, and is, in any case, the standard model for fixed factorial design AV, we present its results together with those of Model I. As is so often the case when $n - k - 1$ is reasonably large, the results in terms of statistical significance do not differ for any of the I values. Thus, neither F_I nor F_{II} are significant for the .0270 I value for H.

Although the H effect is neither significant nor of material substantive interest, for completeness and review, Table 8.7.2 gives the regression equation for its three effects-coded IVs as Eq. (I), and the t values, which (it comes as no surprise) are small. Recall that the B_is give the effect or "eccentricity" of the group coded 1 on X_i, and A is the unweighted mean of the Y means of all the groups (Section 5.4.3). Thus, the latter is 81.74 and, for example, the Y mean for H_1 is 4.64 units higher, hence 86.38, etc. (see Table 8.7.3 below).

At the next level of the hierarchy, X_4 and X_5, the dummy-coded IVs representing T are added to those of H, and $R^2_{Y \cdot H, T}$ is found to equal .1823,[15] an increment of .1553 over $R^2_{Y \cdot H}$ which is highly significant using either error model ($F_I = 9.686$, $F_{II} = 12.231$). The differences among the three treatments, after allowing for (partialling out) the (small and nonsignificant) differences between hospitals, are given in Eq. (II), Table 8.7.2, by the regression coefficients for the dummy variables: the adjusted Y mean of T_1 is 14.78 units below that of T_c, and that of T_2 is 26.91 units above it; the former is nonsignificant ($t = -1.581$), while the latter is highly significant ($t = 2.744**$). The adjusted mean of T_c is given by $A = 78.79$.[16]

As we have shown in previous sections in this chapter, in the absence of interaction-bearing product IVs, the adjustment for hospital effects in the treatment means of Y presumes that these effects are uniform from treatment to treatment, that is, the standard ACV model, here realized with a nominal scale covariate, H. This is implicit in Eq. (II) of Table 8.7.2, which is an ACV equation; we can make it explicit by substituting in (II) the dummy variable-coded values for X_4 and X_5 of Table 8.7.1 and collecting terms, thus generating for each treatment group an equation yielding its means in the four hospitals:

$$\text{For } T_1: \quad \bar{Y}_i = 3.04\, X_1 + 3.40\, X_2 + 3.96\, X_3 + 64.01;$$

(8.7.1) $\quad\quad$ $$\text{For } T_2: \quad \bar{Y}_i = 3.04\, X_1 + 3.40\, X_2 + 3.96\, X_3 + 105.70;$$

$$\text{For } T_c: \quad \bar{Y}_i = 3.04\, X_1 + 3.40\, X_2 + 3.96\, X_3 + 78.79.$$

The Y intercepts of these equations are the treatment means adjusted for hospital effects (see also Table 8.7.3 below), obviously *uniformly,* for example,

---

[15] Note that we do not bother to present significance tests on the cumulative R^2 in Table 8.7.2 in order to conserve space, and because they are not usually of much interest.

[16] For completeness, we note that, symmetrically, the B_is for the effects-coded H variables in Eq. (II) carry differences among the Y means of the hospitals, similarly adjusted for treatment differences. Since they are effects-coded, each is a difference between a treatment-adjusted hospital mean and the unweighted mean of all four treatment-adjusted hospital means.

TABLE 8.7.3

Cell Y Means and Various Main Effect Means for the 4 × 3 Design
(Parenthetical Values are Sample Sizes)

	T_1	T_2	T_c	Eq.(I)[a] $\bar{Y}_H$
H_1	49.33	125.00	84.00	86.38
	(9)	(9)	(6)	(24)
H_2	69.33	130.00	59.33	86.23
	(9)	(9)	(9)	(27)
H_3	56.75	105.33	102.00	84.90
	(12)	(9)	(9)	(30)
H_4	84.33	46.00	70.00,	69.45
	(9)	(6)	(12)	(27)
	(39)	(33)	(36)	(108)

$\bar{Y}_T$ adjusted for H:	64.01	105.70	78.79	Eq.(II)[a]
$\bar{Y}_T$ adjusted for H, $H \times T$:	64.93	101.58	78.83	Eq.(III)[a]

[a]See Table 8.7.2.

each is increased by 3.04 for H_1, by 3.40 for H_2, by 3.96 for H_3, and reduced by 10.40 (= 3.04 + 3.40 + 3.96) for H_4. (See Section 9.3.6 for an explicit ACV analysis of a revised version of this problem.)

The presumption of uniformity of T effects over H (and, by symmetry, necessarily H effects over T) is found to be untenable when the 6 interaction-bearing IVs, $X_6, X_7, \ldots, X_{11}$, are added to $X_1, X_2, \ldots, X_5$ at the next and final level of the hierarchy. The increment in R^2 is .2084, which is highly significant (since this is the last entry in the hierarchy, $F_I = F_{II} = 5.473$, $df = 6$, 96; see Table 8.7.2). Apparently, the treatments do *not* have uniform effects in the four hospitals, that is, the effects are conditional on the other factor—there is a substantial $H \times T$ interaction. Were I for $H \times T$ not significant, we would be content to let stand the treatment means adjusted for H and draw the conclusion without further qualification (from the significant $B_5 = 26.91$) that T_2 is significantly more effective than T_c.

In the regression equation for all 11 IVs, (III) in Table 8.7.2, each of the last 6 terms represents a specific facet of the non-uniformity of effects. Each of these IVs represents a specific effect aspect of H by dummy aspect of T, its B gives the size of this discrete interaction effect, and the accompanying t provides its significance test. Two of these six are significant. The first is X_9, which is X_2 × X_5, and B_9 = 47.92, which indicates that the difference between the means of T_2 and T_c for H_2, 130.00 − 59.33 = 70.67, is 47.92 units higher than the analogous difference between the means of all four hospital means for T_2 and T_c, 101.58 − 78.83 = 22.75 (see Table 8.7.3 below). Thus, one significant departure from uniformity of effect is that experimental treatment 2 is more effective in H_2 than it is in general (on the average over all hospitals). The second of these significant, specific interaction effects is X_{10}, which is constructed as $X_3 X_4$; its B_{10} of −31.35 indicates that the difference between the means of T_1 and T_c for H_3, 56.75 −102.00 = −45.25, is 31.35 units lower than the analogous difference between the means of all four hospital means for T_1 and T_c, 64.94 − 78.83 = −13.89 (see Table 8.7.3). Thus, although T_1 is generally ineffective relative to T_c, it is *particularly* and hence nonuniformly ineffective in H_3.

We have referred above to individual $H_i T_j$ cell Y means. These can be found from (III) in Table 8.7.2 by substituting the 11 values on the IVs which define membership in the 12 groups given in Table 8.7.1 (or, of course, rather more directly). These are given in Table 8.7.3, together with the sample sizes in parentheses, and certain marginal Y means.

The regression coefficients for the first six IVs in the full Eq. (III) of Table 8.7.2 are also meaningful, since B_4 and B_5 are functions of the Y means for the T_j adjusted for both H and also H × T, and B_1, B_2, and B_3 are functions of the Y means for the H_i adjusted for both T and H × T. Two of these are significant. First, B_3 = 23.17 gives the difference for T_c (since it is coded 0 on X_4 through X_{11}) between the Y mean of H_3 and the mean of all four hospital means (since H is effects-coded) adjusted for T and H × T, i.e., 102.00 − 78.83. Of substantively greater importance here, B_5 = 22.75 gives the difference between the means of the four hospital means (since H is effects coded) of T_2 and T_c (since T is dummy coded) adjusted for H and H × T, i.e., 101.58 −78.83.

In terms of the purpose of the research, then, we may conclude not only that the treatment effects are not uniform over the hospitals (particularly T_2 in H_2 and T_1 in H_3), but also that when we allow both for H and this significant nonuniformity (H × T), T_2 is *generally* more effective than T_c.

By "generally," we do not necessarily mean in all four hospitals, but, as we have seen, rather on the average, that is, in terms of unweighted means of means. It can be seen from the cell means in Table 8.7.3 that although T_2 is much larger than T_c in H_1 and even more so in H_2, it is only slightly so in H_3 and is substantially *smaller* in H_4. The significance of the difference between any pair of cell means of a fixed factorial design can be determined by the following formula, which is then illustrated for those of $H_4 T_2$ and $H_4 T_c$:

(8.7.2) $$t = \frac{\bar{Y}_{ij} - \bar{Y}_{ef}}{\sqrt{ \widetilde{sd}^2_{Y-\hat{Y}} \left(\frac{1}{n_{ij}} + \frac{1}{n_{ef}} \right)}} \qquad (df = n - k - 1),$$

where $\tilde{sd}^2_{Y-\hat{Y}}$ is the usual estimate of the error variance (residuals from regression), given by Eq. (5.3.11) (the latter value for this problem, not given in the tables, is 426.5633):

$$t = \frac{46.00 - 70.00}{\sqrt{426.5633\left(\dfrac{1}{6}+\dfrac{1}{9}\right)}} = \frac{-24.00}{10.89} = -2.205 \qquad (df = 96);$$

therefore, in H_4, T_2 is significantly $(P < .05)$ *less* effective than T_c.

It is of interest to substitute in the full equation which includes the interaction IVs, Eq. (III) of Table 8.7.2, the dummy variable-coded values for X_4 and X_5 in order to generate separate equations for each treatment group, as was done in (II), which omitted the interaction IVs. Keep in mind that X_4 and X_5 appear not only by themselves, but as part of product terms, that is, $X_6 = X_1 X_4$, $X_7 = X_2 X_4, \ldots, X_{11} = X_3 X_5$ (Table 8.7.1c). When simplified, these equations are

$$\text{For } T_1: \quad \bar{Y}_i = -15.60\,X_1 + 4.40\,X_2 - 8.18\,X_3 + 64.93;$$

(8.7.3) $\quad \text{For } T_2: \quad \bar{Y}_i = 23.42\,X_1 + 28.42\,X_2 + 4.25\,X_3 + 101.58;$

$$\text{For } T_C: \quad \bar{Y}_i = 5.17\,X_1 - 19.50\,X_2 + 23.17\,X_3 + 78.83.$$

In contrast with the equations derived from (II), these have varying regression coefficients between the treatments, due to the combination of main effect and interaction regression coefficients; for example, for T_1, the term $-15.60\ X_1$ comes from

(8.7.4) $\quad B_1 X_1 + B_6 X_6 = B_1 X_1 + B_6 X_1 X_4 = B_1 X_1 + B_6 X_1\ (1) = (B_1 + B_6)X_1$
$\qquad = (5.17 - 20.77)X_1 = -15.60\ X_1.$

Note also that each Y intercept in Eqs. (8.7.3) is similarly a composite; for example, for T_1,

(8.7.5) $\quad B_4 X_4 + A$ [from (III)] $= B_4\ (1) + A = -13.90 + 78.83 = 64.93.$

The Y intercepts of Eqs. (8.7.3) are the respective treatment means of Y, adjusted for H and $H \times T$, which appear in Table 8.7.3; they are incidentally, simply the unweighted means of the four hospital means for each treatment.

It is important to note that none of the above analysis requires that the cell n_{ij} be either equal or proportional as in conventional orthogonal AV. Whatever disproportionality exists is reflected in the correlations between the IVs of the respective research factors, and is automatically taken care of in MRC which is designed to take into account correlation among IVs, as we have abundantly seen.

The exact "fitting constants" or "least-squares" procedure in factorial design AV for disproportional cell n_{ij} is, in fact, a simultaneous multiple-regression model, and would yield exactly the same adjusted means, as well as the same significance test results when error Model II is used. (See Winer, 1971, pp. 498–503.)

An additional important advantage of approaching problems like these via MRC rather than through AV lies in the modularity of the former. Assume, for example, that in the above research there was an additional set of variables Q. It might be a nominal scale, a set representing a quantitative scale (for example, q, q^2), or even possibly a group of sets. Assume further that they are antecedent variables which the investigator wishes to control (for example, diagnostic category, age, rating of severity of illness). This set could precede the others and be followed by its interaction with the other sets in a hierarchical MRC analysis, viz. (in order): $Q, H, T, H \times T, Q \times H, Q \times T, Q \times H \times T$ (see Sections 8.8 and 9.4.4). Indeed, since H was itself conceived as a covariate, it could be incorporated with Q into a single covariate set A, and the analysis proceed $A, T, A \times T$, which, then, would not include in the model $Q \times H$ or $Q \times H \times T$. Either of these would proceed essentially as the analysis described above.

As an alternative assumption, imagine that set Q represents patient characteristics (demographic, personality) of interest to a social psychologist. Since these will be more or less (and not necessarily significantly) correlated with H, T, and $H \times T$, a conservative assessment of the relationship of Q to Y could be obtained from the hierarchy $H, T, H \times T, Q$. Whatever overlap in Y exists between Q and the prior sets would be withheld (as noted, conservatively) from Q. Further, one could then go on to add $Q \times T$ (and indeed if of interest, $Q \times H$ and $Q \times H \times T$) to describe and assess the possibility that the effectiveness of the treatments was differential, that is, conditional on the patients' characteristics, both taken as a set and also in terms of the regression coefficients of single interaction IVs.

In order to avoid undue further complexity, the only partial coefficients presented and discussed in this section, as in most of this chapter, are B coefficients. For each single IV, however, one can compute meaningful semi-partial and partial correlation coefficients, and their squares as proportions of variance. Thus, any t_i for a B_i in this chapter can be converted into an sr_i or pr_i (or their squares) by equations given in Chapters 3, Eq. (3.7.7) and 9, Eq. (9.2.5); see also Appendix 3. Thus, for example, the product variable X_9, which when partialled is an interaction IV, with $t = 3.231$ [(III), Table 8.7.2)], has a $sr^2 = .0662$ from Eq. (3.7.7); hence, some 6.6% out of the 39.0% of the variance accounted for by H, T, and $H \times T$ is due to the nonuniformity of $T_2 - T_c$ for H_2 relative to the average of all four hospitals.

In summary, then, we have seen that a factorial design data-structure can be treated as a special case of MRC for two (nominally scaled) sets and their product set. Unlike the standard AV, there is no need for orthogonality of main effects (proportionality of cell frequencies). By selecting a coding method for each nominal scale suited to the nature of the inquiry, not only are parameter estimates of interest produced by means of the partial regression coefficients (or correlations), but relevant tests of statistical significance are automatically provided. Each AV "source of variance," be it a main effect or interaction, is represented not only as a set, but also in terms of discrete IVs, the importance of which in regard to size and statistical significance is separately determinable. The MRC method is thus truly *analytic.*

8.8 INTERACTIONS AMONG MORE THAN TWO SETS

In the introduction to this chapter, the principle was stated that an interaction is a product from which its constituent elements have been linearly partialled. To make this latter formulation explicit, in the simplest case of two single IVs, u and v, if one were to "estimate" their product by linear regression, we would obtain

(8.8.1) $\widehat{uv} = B_u u + B_v v + A,$

which would necessarily not produce uv exactly. The $u \times v$ interaction is literally this error of estimate, or residual, that is,

(8.8.2) $uv - \widehat{uv} = u \times v,$

for which we have been using the notation $uv \cdot u, v$ [Eq. (8.1.3)]. We have shown that it is not necessary to literally generate such residuals for all n cases, since the later entry of the product than its constituents in the hierarchical model automatically accomplishes the necessary partialling.

We noted that this can be generalized to three single variables,

(8.1.4) $u \times v \times w = uvw \cdot u, v, w, uv, uw, vw,$

and beyond that to d single variables: the d-way interaction is the d-way product from which all d single variables and all possible lower-order products are partialled.

With regard to sets of variables, we stated that the same partialling principle applied, and gave

(8.1.5) $U \times V = UV \cdot U, V.$

There exists a direct matrix-algebraic analog to Eqs. (8.8.1) and (8.8.2), and again, the hierarchical model assures the operation of such partialled sets by entering UV after U and V.

We now take the obvious final steps in generalization. A three-way interaction among sets can be accomplished as

(8.8.3) $U \times V \times W = UVW \cdot U, V, W, UV, UW, VW,$

and beyond that to d sets: the d-way interaction among d sets of variables is the d-way product from which all d sets and all lower order two-way, three-way, $\ldots$, $d-1$-way product sets have been partialled. Here again, the partialling is automatically accomplished by the hierarchical model. One simply goes up the hierarchy, computing the I for each added level of interaction (or particular interaction at a given level) and tests it for significance.

The principles adduced for interpreting single $u_i \times v_j$ components of a $U \times V$ two-way set interaction generalize to the $u_i \times v_j \times w_p$ components of a $U \times V \times W$ three-way set interaction, and beyond that to the individual d-way interaction components of a d-way set interaction. A $u_i \times v_u \times w_p$ interaction component, carried by a single IV, can be interpreted either as a $u_i \times (v_j \times w_p)$, or a $v_j \times (u_i \times w_p)$, or a $w_p \times (u_i \times v_j)$ interaction. This means that the effect of u_i on Y is not uniform over all combinations of v_j and w_p, and that the effect of v_j on Y is not uniform over all combinations of u_i and w_p, etc. This also generalizes, for example, a $u_i \times v_j \times w_p \times z_q$ interaction component means that the effect of u_i varies with the joint *triple* combination of v_i, w_p, and z_q values, etc., and also that the $u_i \times v_j$ interaction (joint) effect varies with the $w_p \times z_q$ combination, etc. The specific interpretation can be as varied as the combination of possible methods which have been used for coding U, V, etc.

As for Y means of the g levels of a nominal research factor U, adjusted for other research factors and interactions among them, we have seen in previous sections how one can derive from the regression equation computed at the highest level of the hierarchy a set of g equations, one for each level. One simply substitutes in this equation the coding coefficients used for each level of U, and simplifies. In a $U \times V \times W$ design, with g levels of U and h levels of another nominal scale V, one can proceed to obtain adjusted Y means by generating from the final equation a set of gh equations by substituting in it the coding coefficients which carry the U, V, and $U \times V$ information. Each of the resulting gh equations gives $\hat{Y}$ as a different linear function of W. Were there a fourth research factor Z, these gh equations would each give Y as a different linear function of W, Z, and WZ terms. And so on.

Although the preceding paragraph considered U as a set representing a nominal scale G with g levels, its central idea holds when it is instead a set representing a quantitative variable, for example, via power polynomial representation. Instead of g levels of U, g equations, and g adjusted Y means, we can obtain as many "levels," equations and similarly adjusted $\hat{Y}$ estimates as there are admissible values of u, although we will ordinarily only be interested in a few representative ones of these (for example, "low," "average," "high," as in Section 8.4).

Although the above is quite abstract and does not easily yield its meaning, we will not illustrate it with worked examples (but see Chapter 9). We do not deem it to be worth the space in an already long chapter, or a conscientious reader's effort, to pursue the analytic details of interactions of high order. Our reason is simple: over the span of behavioral science research, three-way and higher-order interactions are only rarely of central interest. Most of the theories are not of a degree of complexity such as to warrant positing relationships of that order, and not many variables are measured with sufficient precision to demonstrate such relationships even when they are posited.

Yet another reason for not pursuing interactions of high order resides in the increased risk of errors in statistical inference. Section 4.6 was devoted to a consideration of the general issue, and the principles discussed there are fully applicable here. We saw that as the number of hypotheses posed in an investigation increases, the risks of both the spurious occurrence of significance (Type I errors) and of failure to detect real effects (Type II errors) mount rapidly. With d research factors, it is a simple matter, given enough clerical help or a transgeneration computer routine, to generate all the interaction IVs of a d-way design. But there are, in addition to d main effect sets, $2^d - d - 1$ different interaction *sets*. For $d = 5$ (a relatively modest number for survey research), that comes to 26 interaction sets: 10 two-way, 10 three-way, 5 four-way, and 1 five-way. Now if the design were a $2 \times 3 \times 4 \times 5 \times 5$ factorial, these interaction sets would require a total of $k = 585$ IVs! Not only does this exceed the capacity of any statistical program package of which we know, it is likely to exceed n, or even if not, play havoc with the stability of the parameter estimates and result in low power. The latter is particularly enfeebling if Model II error, which has $n - k - 1$ df for all tests, is used (see Section 4.6.3). Further, considerations of experimentwise rates of Type I errors should give one pause in the face of the prospect of performing $2^d - 1$ significance tests on main effect and interaction sets when $d = 5$ (or even 4).[17] When the 31 tests for $d = 5$ are performed on the I values of the sets on the basis of the $\alpha = .05$ criterion per set, the probability of one or more being "significant" when all null hypotheses are true is in the vicinity of $1 - .95^{31} = .80$. Even if all tests were performed at the $\alpha = .01$ criterion, with its inevitable cost in power, the experimentwise rate would be in the vicinity of $1 - .99^{31} = .27$. Protecting these many F tests by the requirement that the overall F be significant would hardly help, and the mind boggles in contemplating the experimentwise Type I error rate for the (rather poorly) "protected" ts of the 599 single IVs (Section 4.6.4)!

Short of setting unrealistic requirements in research that sample sizes be in the thousands and that true effect sizes be very large so that very small α levels (.01 or .001) per test may be used, the only solution to the problem is *not* to represent all the possible interactions at all possible orders up to d. No interaction set should be included in the IVs unless it is seriously entertained on substantive grounds (except for regression homogeneity checks when the ACV model is employed—see Chapter 9). This requires as a minimum condition that it be understood by the investigator, and on practical grounds that it can be clearly explicable to his audience. This counsel makes it unlikely that interactions beyond three-way will be pursued.[18,19] But by no means should the investiga-

[17] The issue of per comparison versus experimentwise Type I error rates is debated by Ryan (1959) and Wilson (1962).

[18] This argument is closely related to that in Section 6.2.4 on the order of polynomial to use. Recall (Footnote 6 above) that the powered variate, e.g., v^3, is a special case of an interaction, e.g., $v \times v \times v$. High-order interactions, which are thus generalizations of high-order polynomials, are equally problematic to valid statistical inference and scientific generalization.

[19] It should be kept in mind that if a triple product UVW *is* included, for it to represent $U \times V \times W$ requires that UV, UW, and VW have also been included, as well as U, V, and W.

tor feel obligated to include all three-way or even all two-way interactions, even if they are somehow rationalizable on theoretical grounds. For example, with 10 research factors (a not unusually large number outside the experimental laboratory), there are 45 two-way and 120 three-way interaction sets, and heaven knows how many individual IVs. The frequent admonition in research courses to clearly state a limited number of hypotheses in a given investigation, that is, the "less is more" principle of Section 4.6.2, is well taken, and particularly so in regard to interactions.

By omitting possible interaction sets carrying chance variance from the equation, as many df as there are omitted IVs become available to the error term, thereby increasing the accuracy of parameter estimates (proportions of variance, regression coefficients) and the statistical power of the tests which are performed. This can be understood in homey terms as "not asking foolish questions." The questions are doubly foolish: not only are many of the interaction sets not seriously entertained as hypotheses, but their inclusion harms the accuracy of estimation and the power of the tests of the effects that are, particularly when the AV error Model II is used.

The hierarchical model provides some protection of the overzealous shotgun researcher when error Model I is used, as discussed in Section 4.6.3 ("least is last"). Since interaction sets enter in progressively higher order with the entire 1 $- R^2$ residual at each point used for error, the loss of error df from generally unlikely higher order interaction sets does not affect the power or accuracy in testing and estimating main effects and lower order interactions. Even when one proceeds in this manner, it is prudent to give limited credence to the apparent significance of higher-order interaction sets, treating them as exploratory.

In summary, the fact that we are able to represent all interactions of any order as sets of IVs does not mean that we should. Interaction IVs, like any other kind, should only be included if there is serious reason to believe that they are real. Otherwise, the value of the conclusions from the research investigation as a whole, both positive and negative, is jeopardized.

8.9 SUMMARY

Interactions are interpreted and illustrated as conditional relationships between Y and two or more variables or variable sets, for example, a $U \times V$ interaction is interpreted as meaning that the regression (relationship) of Y on U is conditional on (depends on, varies with, is not uniform over) the status of V. An interaction between two variable sets U and V is represented by multiplication of their respective IVs and then linearly partialling out the U and V sets from the product set. The hierarchical model accomplishes this partialling procedure automatically, for example, entering U, V, and then UV. The increment to R^2 due to UV in such a sequence is thus attributable to the $U \times V$ interaction and may be tested for significance by the standard F test. The partial regression coefficients for the interaction IVs are also meaningful, for example, as slope differences, and testable for significance by the usual t test. These basic ideas are illustrated for the simplest form of an interaction, that of two dichotomous variables, and then generalized in discussion to multiple sets of variables, nominal and quantitative (Section 8.1). In successive sections, progressively more complex interactions among research factors are illustrated with worked examples and discussed.

The 2 X 2 fixed factorial using contrast coefficients is reviewed (Section 8.2.1). The misuse of this design for two graduated variables which have been dichotomized ad hoc is criticized for its unnecessary sacrifice of measurement information (Section 8.2.2).

The role of interaction between a dichotomy and a single quantitative research factor is illustrated with a data set, and given two different interpretations: as a problem in differential test validity in white and minority groups (Section 8.3.2), and as a covariance analysis in a learning experiment (Section 8.3.3). The regression coefficient for the interaction IV is demonstrated to be literally the between group difference in slope, and hence in test validity, and is found in the example to be significant. When, for the same data set, the hierarchical order is changed for interpretation as a covariance analysis of learning data, with the quantitative variable playing the role of the covariate, it is shown that the significance of the interaction invalidates the ACV assumption of parallel Y-on-covariate regression lines for the two groups (see Chapter 9).

The case of two quantitative variables and their interaction is then illustrated, demonstrating the lack of necessity for reduction to a 2 X 2 factorial design. It is shown how, when interaction is present, each variable "moderates" the regression of Y on the other, for example, how the regression of Y on f depends on (varies with) the value of g. One may generate a family of Y on f regression equations (or lines), one for each different value of g which is posited (Section 8.4).

We next move to a data structure of a quantitative variable w and a nominal scale G at g levels. The illustration is for $g = 3$, contrast coded. It is shown how the regression coefficients of the IVs in the interaction set represent contrasts in slopes directly analogous to the contrasts in means provided by the IVs for G. The final regression equation containing the interaction IVs can, by substitution of the G coding coefficients, be reduced to a set of simple regression equations for $\hat{Y}$ as a function of w, one for each of the three groups (Section 8.5).

Next, the interaction of two sets, one nominal ($g = 3$, dummy coded) and one a polynomial (w, w^2) is illustrated. It is shown how the regression coefficients for the four interaction IVs relate to differences in slope and curvature between each of two groups and the reference group, and again how separate (now quadratic) equations can be generated for the three groups from the equation which includes interaction terms (Section 8.6).

When the two sets being studied both represent nominal scales, the data structure is that made familiar by AV as a factorial design. This is illustrated by the usual hierarchical MRC analysis of four hospitals (H, effects coded) by three treatments (T, dummy coded), with an efficacy measure as Y. It is shown that the MRC does not require equal or proportional cell frequencies, and yields (in this case) six interaction IVs, each of whose B values is meaningfully interpretable as a specific hospital effect by treatment versus control component of the total interaction. The adjustment of T effects for H, and for both H and the $H \times T$ interaction, is accomplished by substituting the T coding coefficients in the appropriate regression equation, which also yields the equations, one for each treatment, as a function of the IVs for H (Section 8.7).

Finally, the generalization to more than two sets of variables is discussed as a simple extension from two sets. We stress that it is generally *un*desirable to include higher-order interaction sets or even routinely all possible lower-order sets in the interests of controlling the rates of spuriously significant results and failures to detect real effects (Section 8.8).

At many points during this chapter, we discussed the interpretability of the data structures illustrated as covariance designs, since, in principle, the first set in a hierarchical MRC can be interpreted as a covariate set to be controlled, with the second set the one of research interest. However, since this chapter is devoted to interactions, for pedagogical reasons, all interactions were constructed to be significant, thus making each illustrative example one in which the ACV model is invalid. We have seen that this is no bar to a completely meaningful analysis via MRC. However, with similar data structures, the interactions in the examples of the next chapter are not significant, which makes possible their analysis in terms of the conventional ACV, to which we now turn.

PART III

APPLICATIONS

APPLICATIONS

9

The Analysis of Covariance and Its Multiple Regression/Correlation Generalization

9.1 INTRODUCTION

At several points in this book and particularly in Chapters 4 and 8, the ACV has received some incidental attention. Since its function, the statistical control of irrelevant sources of variation, is frequently necessary in research inference in the behavioral and social sciences, it warrants being treated in a separate chapter. This is particularly true since the conventional ACV, as it is presented in textbooks in "experimental design" and AV (cf. Edwards, 1972; Hays, 1973; Winer, 1971) is an unnecessarily limited special case of what we will show to be a highly general form of ACV as accomplished by the MRC system as it has been developed in this book.

Our previous encounters with ACV occurred whenever we discussed two or more sets of IVs, at least one of which, call it set B, represented group membership, that is, a nominal scale, coded by whichever method of Chapter 5 was preferred. Let set A designate the remaining IVs in the analysis. Then, an MRC analysis in which Y is regressed on sets A and B *is* an ACV, since the IVs of set A, which may be deemed "covariates," are partialled from the IVs of set B, the research factor of interest; thus, it is $B{\cdot}A$ on which we focus our attention in ACV (Chapter 4).[1] The partial coefficients for the IVs of set B provide us with all the interpretive material which we have seen arises from the nominal scale coding method used, but now further *partialled* by the IVs of set A. Hence, we

---

[1] Indeed, from the more general perspective of MRC, since set B is also partialled from set A, we can view this symmetrically as an ACV of set A conceived as the research factor(s) of interest, with set B serving as the covariate set. But this would transcend the limited perspective of conventional ACV, in which the research factor is always a nominal scale (that is, a set of groups), and covariates are always quantitative variables. From our point of view, however, the analysis of $A \cdot B$ is not formally different from that of $B \cdot A$, however sets A and B are constituted. We generalize to this broader perspective in Section 9.4.

have in addition to the usual yield of the ACV, the whole array of correlational and proportion of variance results yielded by MRC. We defer discussion of the logic and limitations of partialling to Section 9.7.

When Y is regressed on set A alone, the partial regression coefficients for its IVs and Y intercept are not the appropriate coefficients for "adjusting" (that is, partialling or regressing out) the set A covariates from the group (or cell) Y means, since set B, which carries the group-membership information, is not in the equation and thus is ignored. These are the "total" regression coefficients for the set A IVs, which, because they ignore the effects of group membership, are contaminated by them. But when sets A and B are simultaneously regressed on Y, the partial coefficients of the IVs of set A are the proper "pooled, within-group" values needed for adjustment, since the group membership information of set B (that is, the group or cell means) has also been partialled from the covariate set A. The difference (as discussed at some length in Section 8.3.3) is that the "total" regression takes deviations of each observation from the *grand* means of Y and the covariates, whereas the "within-group" regression takes deviations of each observation from its *own* group or cell mean and is thus not affected by group membership (for example, "treatment" effects).

To reiterate: A conventional ACV may be performed using MRC with two sets of variables, one of which (set B) represents a nominal scale of group membership, and the other (set A) contains the covariates which are to be controlled or adjusted for in the analysis of Y. The sets may contain as few as one IV each (as in Section 8.3.3). But the ACV model presumes that the relationship (regression) of Y on the covariates (set A) is the same across the populations from which the groups are drawn, since no provision is made by the ACV model for the possibility that these regressions differ among groups, that is, that regression is heterogeneous. The latter possibility is identically that of *interaction* between covariates and group membership, since interaction is most generally defined as a nonuniform or conditional relationship, that is, the regression of Y on the IVs comprising set A varies with changes in the values of set B, hence with group membership.

With heterogeneous regression, the use of a single "averaged-out" regression of the covariates would be over- or underadjusting for some groups relative to their appropriate population values. Another way of understanding this is that the ACV asks the question, "What is *the* difference in $\hat{Y}$ between groups for any given set of values for the covariates?" This presumption of a constant difference between groups is not met when their regression lines (more generally, surfaces) are not parallel, which renders the question unanswerable (see Sections 8.3.3, 8.3.5, and 8.3.6). Thus, in the presence of nontrivial $A \times B$ interaction, the ACV is invalid.

This was abundantly demonstrated in Chapter 8. Wherever a set B represented a nominal scale, the ACV was invoked, with set A serving as a covariate set. But because Chapter 8 was concerned with interactions, and our examples were contrived so that the $A \times B$ interaction was in each case significant, each attempt to interpret an example in ACV terms was aborted by the evidence of heterogeneity of regression demonstrated by the significant $A \times B$ interaction. As we saw, such an outcome, although precluding a valid ACV, is nevertheless amenable to a detailed MRC analysis of the data.

In overview, then, a conventional ACV is accomplished by hierarchical MRC as follows. Regress Y on the covariate set A, then add the group membership information as set B, and then add the AB product set. If the increment of the latter is adjudged nonexistent or trivial (for example, by being nonsignificant), the MRC results of the second stage (of $B \cdot A$) are readily interpreted in conventional ACV terms and also yield partial coefficients as useful measures of effect size.

We discuss below the general MRC approach to conventional ACV in detail (Section 9.2), its application to less conventional problems (nonlinearly related, nominal, and missing-data covariates—Section 9.3), the use of a quantitative set B via a generalization to the Analysis of Partial Variance (Section 9.4), the problem of unreliable covariates (Section 9.5), applications to the study of change (Section 9.6), and finally, the role and limitations of these methods (Section 9.7).

9.2 ANALYSIS OF COVARIANCE VIA MULTIPLE REGRESSION/CORRELATION: GENERAL

For fashioning an MRC approach to ACV, we have the following elements:

1. In the preceding chapters, we have seen that we can represent virtually any information using sets of IVs, that is:

a. nominal scales with alternative coding methods to facilitate the expression and testing of the aspects of group membership that are of interest (Chapter 5);
b. quantitative scales, coded so as to carry linear or nonlinear aspects of the regression by various methods (Chapter 6);
c. missing data information in nominal and quantitative research factors (Chapter 7); and
d. interactions among sets representing research factors (Chapter 8).

2. In Chapter 4, we developed the idea of using sets of IVs, either structural or functional, as units of analysis (Section 4.1), and

a. showed that sets could be related to Y by measures of correlation and proportions of variance analagous to those of single IVs, that is, multiple, multiple semipartial, and multiple partial correlations (Sections 4.2, 4.3);
b. developed the notion of partialled sets, for example, $B \cdot A$, as sources of Y variance, and offered methods of significance testing (Section 4.4) and power analysis (Section 4.5) for these.

3. In the introduction to this chapter, we described the ACV as generally proceeding by hierarchical MRC: a covariate set A enters first, followed by a set B carrying group membership information, and the resulting regression on $B \cdot A$ constitutes the ACV. Additionally, as a check on the validity of the assumption of regression homogeneity, the AB product set is entered to appraise the $A \times B$ interaction.

Since any kind of information can be expressed as a set A, *it follows that there is virtually no constraint on what can serve as covariates in the ACV performed via MRC.* The entire gamut of paragraphs 1a through 1d above, and combinations thereof, are available.[2] This means that not only can one use as covariates linear aspects of one or more quantitative variables as described in standard treatments of ACV, but also nonlinear aspects of quantitative variables, nominal scales, research factors containing missing data, and interactions among research factors of any kind. Set B is constrained by the conventional definition of ACV to carry group membership information (nominal scales and/or their combinations as in "factorial design"), and we will so treat them for the present. (In Section 9.4, we shall liberate set B from this constraint to achieve an even higher order of generality called the Analysis of Partial Variance.)

In conventional ACV, there are two major foci of interest. The first is on whether for the covariate-adjusted Y values, the F ratio of the between-groups mean square to the within-groups mean square is large enough to be significant. This is equivalent to an appraisal of the proportion of Y variance which set B accounts for over and above the proportion accounted for by the covariate set A, that is, an appraisal of $R^2_{Y \cdot AB} - R^2_{Y \cdot A}$ in a hierarchical MRC. As extensively discussed in Chapter 4, this quantity is the squared multiple semipartial correlation of Y with $B \cdot A$, which we have variously symbolized as sR^2_B, I_B, or more explicitly as $R^2_{Y \cdot (B \cdot A)}$; it is represented by area b in the ballantine (Figure 4.3.1). The overall ACV F test is identically a test of the null hypothesis that, in the population, set B accounts for no Y variance beyond that accounted for by set A, that is, $R^2_{Y \cdot AB} - R^2_{Y \cdot A} = 0$. This null hypothesis is tested by the very general equation given in Chapter 4 for the F test,

(4.4.2) $$F = \frac{R^2_{Y \cdot AB} - R^2_{Y \cdot A}}{1 - R^2_{Y \cdot AB}} \times \frac{n - k_A - k_B - 1}{k_B},$$

with $df = k_B$, $n - k_A - k_B - 1$ (k_A, k_B are the number of IVs in each set). Equation (4.4.2) gives identically the same results as the ACV F test (which is a special case of it) yet has an advantage in that it focussed upon a proportion of

[2] The statistical purist will object, pointing out that our framework is "fixed-model" throughout and that the model assumes no measurement error in the IVs. We have already addressed ourselves to the first objection (Section 5.3.1), and will consider the second in Section 9.5. Although the italicized proposition can be argued, it is certainly more true than it is false.

variance, $R^2_{Y \cdot AB} - R^2_{Y \cdot A} = sR^2_B$, which is an interpretable measure of effect size, rather than an adjusted between groups mean square expressed in (often arbitrary) raw units of Y. Furthermore, it is a short step away from an even more meaningful effect size measure for ACV, the squared multiple *partial* correlation, pR^2_B (or $R^2_{YB \cdot A}$), which removes variance that is due to the covariate set A from both B *and* Y, as discussed in Section 4.3.3, that is,

$$(4.3.11) \qquad pR^2_B = \frac{R^2_{Y \cdot AB} - R^2_{Y \cdot A}}{1 - R^2_{Y \cdot A}} = \frac{I_B}{1 - R^2_{Y \cdot A}},$$

represented as b/(b + e) in the ballantine (Figure 4.3.1). This takes as a base for the proportion of $B \cdot A$ variance not the total Y variance as does sR^2_B, but that portion of it not associated with set A. This is a natural base for ACV, whose focus is in fact not on Y, but on $Y \cdot A$ (Section 4.3.3). For even greater notational simplicity, we will sometimes refer to this "residual" $Y \cdot A$ variable as Y'.

Yet another advantage of the MRC formulation of the overall F test in ACV is the simplicity of the procedures available for statistical power analysis. The power of the F test of Eq. (4.4.2) for given k_B, n, and f^2, and the sample size necessary for given desired power, k_B, and f^2 are readily determined by the methods of Section 4.5; that section also discusses statistical power analysis in detail, and its application to the special case of ACV is straightforward.

One way to conceive of the overall F test of conventional ACV is as a test of the equality of the covariance (set A)-adjusted Y means of the g groups in the population. A second focus of interest in ACV is on the adjusted sample Y means, which we will symbolize as $\bar{Y}'_h$ ($h = 1, 2, \ldots, g$). These are readily determined from the constants of the regression equation containing the IVs of sets A and B by means of the equation

$$(9.2.1) \qquad \bar{Y}'_h = \sum B_j X_j + A',$$

where j designates an IV from set B (coded by any of the methods of Chapter 5), the summation being over all k_B of these, and A' is the adjusted Y intercept. The latter may be found from

$$(9.2.2) \qquad A' = A + \sum B_i m_i,$$

where A is the Y intercept of the complete ACV equation and i designates an IV from set A, the summation being over all k_A of these. Since

$$(9.2.3) \qquad A = \bar{Y} - \sum B_j m_j - \sum B_i m_i,$$

one can alternatively compute A' from

$$(9.2.4) \qquad A' = \bar{Y} - \sum B_j m_j.$$

We will illustrate the determination of the adjusted Y means throughout this chapter.

Beyond this, the MRC route to accomplish the ACV offers its usual advantages. For one thing, by selecting the most relevant method of coding group membership as the X_js, the B_js carry the same comparisons or contrasts of means described in Chapter 5, but now they are functions of covariance-adjusted means, and the accompanying t_j values provide significance tests of these comparisons. Similarly, A' bears the same relationship to the adjusted means as A does to the means. Finally, there is available a natural measure of effect size, the proportion of the variance of covariance-adjusted Y (that is, of $Y - \hat{Y}_A = Y \cdot A = Y'$) accounted for by each of these aspects of group membership in the form of pr_j^2 values. If not given in the computer output they may be readily determined from

(9.2.5)
$$ pr_j^2 = \frac{t_j^2}{t_j^2 + n - k_A - k_B - 1}. $$

With the illustrative material which follows, we will show that the MRC way of doing ACV is much more versatile yet simpler than standard textbook methods.

9.3 MULTIPLE, NONLINEAR, MISSING-DATA, AND NOMINAL SCALE COVARIATES

9.3.1 Introduction

In order to illustrate the diversity of application and analytic power of the MRC approach to ACV and still avoid undue repetition, the examples in this section were so constructed that:

1. all have covariate sets with 2 ($= k_A$) IVs in set A;
2. all have 4 ($= g$) groups, hence 3 ($= g - 1 = k_B$) IVs in set B, with the same unequal sample sizes per group ($n_1 = 26, n_2 = 18, n_3 = 12, n_4 = 16$);
3. all will share the same presumed correlation results for the sets involved, and, given the same k_A, k_B, and total n ($= 72$), the same results of F tests on R^2s and Is.

However, the examples differ in the nature and results of both the individual covariate IVs (the X_i of set A) and group membership IVs (the X_j of set B). This will provide the opportunity to exemplify the variety of types of covariates together with alternatives in coding methods for the nominal scale defined by set B. Although the examples all have two covariate IVs and four groups, it will be found to be very easy to generalize to more (or fewer) covariates and groups. The inequality of the group sizes, as in Chapter 5, is in the interests of generality.

9.3.2 The Hierarchical R^2 Analysis by Sets

Each of the three illustrative examples in the following sections are characterized by the results of the setwise hierarchical R^2 presented in Table 9.3.1, so that these results are to be understood as integral to each example. Whatever the specific interpretations to be given later to the variables, in this section they will be abstractly designated as sets constituted as follows:

Set A (covariates): X_1, X_2
Set B (group membership): X_3, X_4, X_5
Set $A \times B$ (interaction): $X_6 (= X_1 X_3), X_7 (= X_1 X_4), \dots, X_{11} (= X_2 X_5),$

that is, $k_{A \times B}$ = 6 IVs constructed by multiplying the value of each of the set A IVs by the value of each of the set B IVs ($k_A k_B$ = 6).

Following the procedure described in the previous section, the regression of Y on set A results in $R^2_{Y \cdot A}$ = .2413; thus, some 24% of the total Y variance is accounted for by the covariates, which by the standard F test for an R^2, Eq. (3.7.1), yields $F = 10.973**$ for $df = k_A$, $n - k_A - 1 = 2, 69$ (Table 9.3.1). Thus, our covariate set is "working"—it is pulling out a large and highly significant amount of Y variance. But it is the remaining $1 - R^2_{Y \cdot A}$ = .7587 of the Y variance, the variance of Y' (the adjusted or residual Y variance) which is of interest to us, and particularly how much of it is accounted for by group membership.

Accordingly, we now add the set B variables to those of set A and regress Y on all 5 (= $k_A + k_B$) IVs. We find the (cumulative) R^2 to be .3965, an increase of .1552 (= $I_B = sR^2_B$) of total Y variance, which is tested by Eq. (4.4.2):

$$F = \frac{.3965 - .2413}{1 - .3965} \times \frac{72 - 2 - 3 - 1}{3} = 5.656$$

for $df = k_B$, $n - k_A - k_B - 1 = 3, 66$, which is highly significant. This is the overall F test of ACV, but we do not yet know whether we have a valid ACV insofar as meeting the assumption of homogeneity of regression of Y on set A among groups is concerned.

This issue is assessed by adding the ($k_{A \times B}$ = 6) IVs carrying the $A \times B$ interaction. The R^2 for all 11 IVs is found to be .4627, increasing by only $I_{A \times B}$ = .0662. Substituting in Eq. (4.4.2), letting our $A \times B$ set be set B of the

TABLE 9.3.1
A Hierarchical R^2 Analysis by Sets for Analysis of Covariance

Sets added	IVs added		cum. R^2	I	F_I	df	
A	X_1, X_2	$R^2_{Y \cdot A}$ = .2413		.2413	10.973**	2, 69	
+ B	+ X_3, X_4, X_5	$R^2_{Y \cdot A,B}$ = .3965		.1552	5.656**	3, 66	pR^2_B = .2046
$A \times B$	+ $X_6, X_7, \dots, X_{11}$	$R^2_{Y \cdot A,B,A \times B}$ = .4627		.0662	1.232	6, 60	

$n = 72$ $k_A = 2$ $k_B = 3$ $k_{A \times B} = 6$

$(n_1 = 26,$ $n_2 = 18,$ $n_3 = 12,$ $n_4 = 16)$

**P < .01.

formula and our sets A and B collectively be the formula's set A, we find

$$F = \frac{.4627 - .3965}{1 - .4627} \times \frac{72 - 5 - 6 - 1}{6} = 1.232,$$

for $df = k_{A \times B}$, $n - k_A - k_B - 1 = 6, 60$, which is close to the chance-expected value of 1 and far below conventional criteria for statistical significance (Appendix Table D.2: at $\alpha = .05$, $F = 2.25$ for $df = 6, 60$). We accept these results as quite consistent with the null hypothesis of parallelism of the four groups' separately determined best-fitting regression planes (Y on X_1 and X_2) in the population, that is, of homogeneity of regression.

It is worth pausing a moment to consider the nature of this decision. A statistically significant $I_{A \times B}$ invalidates the ACV, in which case we turn to the methods of Chapter 8 to analyze the data. But to reach a positive conclusion that the regression is homogeneous between groups requires the logically impossible feat of proving a null hypothesis. We must therefore settle for results *consistent with* this null hypothesis, that is, we posit homogeneity in the absence of evidence to the contrary. A nonsignificant F ratio, particularly one well below the value at the conventional $\alpha = .05$ criterion, and ideally one that is close to the chance-expected value of 1, constitutes such evidence.

Having found the assumption of regression homogeneity acceptable, we may interpret the first two stages of the hierarchical analysis as an ACV. As noted, the $I_B = .1552$ is large and highly significant; $F = 5.656$, $df = 3, 66$. The latter is the overall "between-groups" F test of ACV, and the exact meaning of the proportion of variance of .1552 deserves review in this context. Since $I_B = sR_B^2 = R_{Y(B \cdot A)}^2$, it represents the portion of the total Y variance accounted for by $B \cdot A$, that is, group membership *after* the overlap (redundancy) in the relationship to Y between covariates and group membership has been removed from the latter. (Such redundancy exists to the extent that groups differ in their covariate means.) Recall that I_B is the b area in the ballantine (see Figure 4.3.1), the portion of Y variance *uniquely* accounted for by group membership, that is, freed of the ambiguity of joint overlap with the covariates (area c). Its statistical significance thus indicates that variance in Y associated with group membership (the Y, B overlap) can*not* be entirely accounted for by group differences in covariates (the A, B overlap). Furthermore, of necessity, this area b is entirely in the covariate-adjusted portion of the Y variance, $1 - R_{Y \cdot A}^2$, which is the portion to which the ACV is directed, that is, the variance of Y'. In this example, the latter is $1 - .2413 = .7587$. Thus, a more suitable measure than sR_B^2 for the effect size of group membership in ACV is pR_B^2 (Eq. 4.3.10), which has the same $B \cdot A$ variance in Y in its numerator, but then uses the relevant $1 - R_{Y \cdot A}^2$ as its base, that is,

$$pR_B^2 = \frac{.1552}{.7587} = .2046.$$

Group membership freed of covariate differences thus accounts for some 20% of ACV-adjusted Y variance (with the same statistical significance as for sR_B^2; see

Section 4.4.1). We will see that, analogously, we use pr_j^2 as an effect size measure for the single X_j of the B set.

In this problem we found both $R_{Y \cdot A}^2$ and I_B to be statistically significant. What happens when this is not so?

1. *Nonsignificant set A.* In order to find I_B, we first determine $R_{Y \cdot A}^2$, and its F test is routine. But the regression of Y on set A is the *total* regression, while the efficacy of the covariates depends strictly on the magnitude of the *within-groups* regression, whose variance proportion is $R_{Y \cdot (A \cdot B)}^2$. This can be determined and F tested in the usual way. However, the ACV proceeds in exactly the same way whether $R_{Y \cdot (A \cdot B)}^2$ is significant or not. It would not be correct, if $R_{Y \cdot (A \cdot B)}^2$ (or $R_{Y \cdot A}^2$) is nonsignificant, to then drop the covariates and simply regress Y on set B. An ACV is performed when, for substantive and logical reasons, our interest is in Y', which is conceptually a quite different variable from Y. An arbitrary significance criterion should not deflect us from the variable relevant to our interests. Further, a policy of dropping nonsignificant covariate sets (or subsets) results in capitalization on chance. True, it is generally disappointing to find $R_{Y \cdot (A \cdot B)}^2$ nonsignificant or small, since other things equal, the smaller it is the larger are the denominators of both the F test (Eq. 4.4.2) and pr_B^2 (Eq. 4.3.10), but the solution does not lie in dropping set A; it may lie in a reconceptualization of the research issue under study, likely followed by gathering new data.

2. *Nonsignificant B·A.* When I_B fails to be significant, the question arises whether one can scrutinize the partialled results for the k_B individual IVs in set B, and interpret those with "significant" t values, or perform t tests on other contrasts among the adjusted group means. Following the general protected t procedure of Section 4.6.4 that we have used throughout, our answer is no—ignoring the failure of I_B to be significant strips the t tests of their protection from high experimentwise Type I error rates. Other multiple comparison procedures may be used in ACV as in AV, depending on the type of hypothesis and treatment of Type I error, which do not require that the F for set B be significant. These were reviewed briefly and referenced in Section 4.6.1. We prefer the general protected t test for its combination of simplicity and effectiveness in balancing Type I and Type II errors in inference, and would therefore, as a rule, not pursue the t tests of set B's IVs or test other contrast functions of the adjusted group means when the F for I_B is not significant.

Getting back on the track of the illustrative example, we will now clothe it in three different costumes by varying the meaning of the covariates and the coding of group membership, and present and interpret the partialled results for each IV in each analysis.

9.3.3 Multiple Linear Covariates

The simplest realization of a set of $k_A = 2$ covariates is for each of them to be a linear aspect of a quantitative variable. Imagine that Table 9.3.1 describes the results of a survey of attitudes toward minority group hiring (AMH = Y) of men classified in 4 (= g) different ethnic groups (G_h; h = 1, 2, 3, 4). Our interest lies not in AMH as such, but rather in AMH *after* it is freed of the influence of a covariate set A made up of age (= X_1) and years of education (= X_2), both linearly, that is, in Y'. Since our purpose is to compare each ethnic group with the aggregate of groups, the nominal scale coding method employed for the X_j of set B is effects coding (Section 5.4). The coding for X_3, X_4, X_5 is exactly as given in Table 5.4.1 (where they are designated X_1, X_2, X_3).

We learned from Table 9.3.1 that age and education do indeed account for much Y variance, and that, freed of their influence, a material and significant portion of the remaining variance (pR_B^2 = .2046 of it) is accounted for by ethnic group membership. (We also found, as a necessary validity check, that the relationship of AMH with age and education is sufficiently similar among the four groups so as to yield a near-chance F ratio for $I_{A \times B}$.)

Table 9.3.2 gives the results of the ACV for the individual variables. The usual raw score regression equation with t values for the coefficients is interpreted as an ACV for these data. As a matter of incidental interest, the B_i for age and education (−1.41 and 3.54) are the "pooled" or "within-group" coefficients appropriate to ACV (see Sections 9.1 and 8.3.3) and are both significant. They may be interpreted in the usual way: within groups, an increase of one year in age is associated on the average with a drop of 1.41 AMH units, and an increase of one year of education is associated on the average with an increase of 3.54 AMH units, each with the other held constant. Since $I_{A \times B}$ was small, these values obtain quite uniformly across the groups.

Of primary interest are the B_j and their t_j. Since the X_j are effects coded, each gives the "effect" or "eccentricity" of a group. But since the covariates are in the equation and hence partialled, the constants are functions of *adjusted* means. Thus, we can rewrite Eqs. (5.4.2), (5.4.3), and (5.4.4) for effects-coded coefficients with covariates present in terms of adjusted values:

(9.3.1) $$B_j = \bar{Y}_h' - \bar{\bar{Y}}',$$

where $\bar{\bar{Y}}'$ is the unweighted mean of all the g adjusted Y means, that is,

(9.3.2) $$\bar{\bar{Y}}' = \frac{\sum \bar{Y}_h'}{g}.$$

This last quantity turns out to be the adjusted Y intercept of Eqs. (9.2.2) and (9.2.4), that is,

(9.3.3) $$A' = \bar{\bar{Y}}'.$$

Thus, since X_3 codes G_1 as 1, B_3 = 8.28 indicates that the adjusted mean of G_1, $\bar{Y}_1'$, is 8.28 units higher than the mean of the adjusted means $\bar{\bar{Y}}'$, a nonsignificant departure (t_3 = 1.722, $P >$ .05). B_4 and B_5 indicate that $\bar{Y}_2'$ and $\bar{Y}_3'$ are respectively 14.73 units below and 14.28 units above $\bar{\bar{Y}}'$, both significant

TABLE 9.3.2
Age and Education as Covariates, Effects-Coded Ethnic Groups

		Set A		Set B		
	Y	Age X_1	Ed. X_2	X_3	X_4	X_5
m	76.14	39.72	10.23	.139	.028	−.056
sd	28.75	8.16	2.36	.751	.687	.621

ACV regression equation:

$$\hat{Y} = -1.41\,X_1 + 3.54\,X_2 + 8.28\,X_3 - 14.73\,X_4 + 14.28\,X_5 + \;\;95.98$$

t	−3.736**	2.633*	1.722	−2.874**	2.540*	($df = 66$)
pr^2	.1745	.0951	.0430	.1112	.0891	
pr	−.418	.308	.207	−.333	.298	

Adjusted Y intercept:

(9.2.2) $A' = A + B_1 m_1 + B_2 m_2$
$\qquad = 95.98 \;\; -1.41(39.72) + 3.54(10.23) = 76.19 = \bar{\bar{Y}}'$

Adjusted means:

(9.2.1) $\bar{Y}'_h = B_3 X_3 + B_4 X_4 + B_5 X_5 + A'$
$\qquad = 8.28\,X_3 - 14.73\,X_4 + 14.28\,X_5 + 76.19$
$\bar{Y}'_1 = 8.28(1) - 14.73(0) + 14.28(0) + 76.19 = 84.17$
$\bar{Y}'_2 = 8.28(0) - 14.73(1) + 14.28(0) + 76.19 = 61.46$
$\bar{Y}'_3 = 8.28(0) - 14.73(0) + 14.28(1) + 76.19 = 90.47$
$\bar{Y}'_4 = 8.28(-1) - 14.73(-1) + 14.28(-1) + 76.19 = 68.36$

*$P < .05$.   **$P < .01$.

departures. The departure of $\bar{Y}'_g$ ($= Y'_4$ here) is given by

(9.3.4) $$\bar{Y}'_g - \bar{\bar{Y}}' = -\sum B_j$$

in this example, $-(8.28 - 14.73 + 14.28) = -7.83$. The t test for the significance of this departure (effect) is given by applying Eq. (5.4.8).[3] The resulting $t_4 = 1.561$, which is not significant. One can also perform pairwise comparisons among the $\bar{Y}'_h$ by applying Eqs. (5.3.10) and (5.4.8).[4] Recall, however, that

[3] Slight modification is required. The R^2 in the variance error of estimate, Eq. (5.3.11), becomes $R^2_{Y \cdot AB}$ and $k = k_A + k_B$. Also, the subscripts obviously need to conform to the usage of this chapter: B_i becomes B_j and n_i becomes n_h.
[4] The same change in variance error of measurement and similar changes to make subscripts conform as described in Footnote 3 are required.

none of the above t tests would be permitted by our "protected t" rule were I_B not significant (Section 9.3.2).

The actual values of the $\bar{Y}_h'$ implied in Eqs. (9.3.1) through (9.3.4) may be produced mechanically by the application of Eqs. (9.2.1) and (9.2.2), as demonstrated in Table 9.3.2. These may, of course, differ quite substantially from the unadjusted $\bar{Y}_h$, but as we have argued, the latter are not relevant: the variable of interest is *not* Y, but Y'. If desired, they are readily found by regressing Y on set B alone (or, more simply, by computing them directly).

Finally, the pr^2 (or pr) for the IVs, computable from Eq. (9.2.5), provide unit-free measures of effect size. Each pr_j^2 is a proportion whose numerator is variance in Y' due to a given group's "effect" and whose denominator is Y' variance *not* accounted for by the remaining $g - 1$ groups' effects. In other words, it is identically as described in Section 5.4.3, except for adjusted (Y') variance instead of unadjusted (Y) variance. Thus, of the age- and education-adjusted variance in AMH not accounted for by the effects of the other three groups, the effect of G_2 accounts for .1112 ($= pr_4^2$).[5] As is always the case, such proportion of variance measures have the advantage of being free of the unit in which Y is measured, but not free of their dependence of the relative sizes of the groups, a fact which needs to be taken into account in interpretation. (See the discussion in Section 5.4.3, all of which holds here when Y is replaced by Y'.)

Thus, the results of Table 9.3.2 in combination with those of Table 9.3.1, provide all the usual yield of an ACV, and more: proportion of Y' variance due to group membership, the departures of each group's $\bar{Y}'$ from their $\bar{\bar{Y}}'$ expressed both in units of Y (the B_j) and as proportions of variance (pr_j^2) or correlations (pr_j). There is also the proportion of Y variance due to the covariate set A, and the within-group partial regression and correlation coefficients of each covariate. Finally, all the above quantities are accompanied by their appropriate F or t values.

We selected $k_A = 2$ quantitative (linear) covariates in our example for economy. Clearly, the use of more than two (or of only one) offers no difficulty in conceptualization and computation, nor does a change in the number of groups.

Since there is no relationship between the type of covariate set and the method of representing G, we could have used any other nominal scale coding method, with the result that the B_js, pr_js, and their t_js would change in value and interpretation. The use of other coding methods for set B will be illustrated in succeeding sections.

9.3.4 A Nonlinear Covariate Set

Chapter 6 described a group of methods for representing nonlinear (curvilinear) aspects of research factors by means of sets of IVs. When an investigator has

---

[5] To find the pr^2 for G_g ($= G_4$ here), enter the t value found from Eq. (5.4.8) as modified in Footnote 3 above in Eq. (9.2.5). The result here is $pr^2 = .0356$.

reason to believe that a quantitative covariate is nonlinearly related to Y, whichever of these methods is deemed suitable may be used to represent the research factor as the covariate set A. (A previous effort along these lines [Section 8.6] aborted when the regression homogeneity assumption was violated by a highly significant interaction.)

We will illustrate the use of a quadratic power polynomial (Section 6.2), tied in with the (same) setwise results of Table 9.3.1. Conceive of an investigation in psychodiagnosis involving the performance of three psychiatric diagnostic groups (G_1, G_2, G_3) and one normal control group (G_4) on a memory for designs (MD) test (Y). Since age is related to memory (and may be related to G), we elect to use the ACV with age as a covariate, but we have reason to believe that over the age range studied the relationship of age to Y is not linear, and, further, that it is adequately represented by a quadratic polynomial. Accordingly, we let X_1 be age and X_2 be age squared (hence, $k_A = 2$). Since our interest is primarily in comparison of each diagnostic group with the control (G_4), we use dummy variable coding (Section 5.3) for set B, coding the 4 $(= g)$ groups into 3 $(= g - 1 = k_B)$ IVs, exactly as shown in Table 5.3.1 (but designated X_3, X_4, X_5). The six products of each of the X_i with each of the X_j provide us with the set carrying the $A \times B$ interaction (for example, X_{11} has age squared for the $n_3 = 12$ cases in G_3 and zero for all other cases). Table 9.3.1 is now interpreted as indicating that age (quadratically) accounts for 24% of the MD variance, that group membership accounts for an additional 15.5% of the MD variance (and 20.5% of the adjusted MD $= Y \cdot A = Y'$ variance), a statistically significant amount. The small and nonsignificant $I_{A \times B}$ indicates that there is no reason to believe that the (population) Y on age quadratic regression curves for the groups are not parallel, that is, we accept the regression homogeneity assumption necessary to the ACV model.

Table 9.3.3 presents the per-variable results for this interpretation. The ACV regression equation, with the t values for the regression coefficients, contains the key results. Following the general principle previously expressed, each B_j for a dummy variable coded X_j now gives the difference between the adjusted Y means for the group coded 1 on X_j and the reference group, G_g:

(9.3.5) $$B_j = \overline{Y}_h' - \overline{Y}_g'.$$

Thus, the adjusted mean of G_1 is 13.32 points below that of the control group G_4 (with a significant $t = -2.562$), $\overline{Y}_2'$ is a significant 11.68 points below $\overline{Y}_4'$, and $\overline{Y}_3'$ is 8.96 above $\overline{Y}_4'$, a nonsignificant difference. The adjusted mean of the reference group is the adjusted Y intercept A' from Eq. (9.2.2) above, exactly parallel to Eq. (5.3.6):

(9.3.6) $$\overline{Y}_g' = A',$$

that is, 53.30 in this example (see Table 9.3.3). All of this is implied by the general equation for $\overline{Y}_h'$ (Eq. 9.2.1), which for dummy coded X_j simplifies to the exact analogue of Eq. (5.3.7):

(9.3.7) $Y_h' = B_j + A'$

                                      $= B_j + \bar{Y}_g'$,

where B_j is for the X_j on which G_h is coded 1, as illustrated in Table 9.3.3.

If t tests between adjusted means not involving the reference group are desired, they can be found by applying Eq. (5.3.10), using the changed variance error of estimate (see Footnote 3), and making the conforming changes in subscripts. For example, the t value between $\bar{Y}_1'$ and $\bar{Y}_3'$ is

$$\frac{-13.32 - (+8.96)}{\sqrt{(19.18)^2(1 - .3965)\left(\frac{72}{66}\right)\left(\frac{1}{26} + \frac{1}{12}\right)}} = -4.102,$$

a highly significant result for $df = 66$.

TABLE 9.3.3

Quadratic Polynomial of Age as Covariates, Dummy Coded Diagnostic Groups

	Y	Set A Age X_1	Set A Age² X_2	Set B X_3	Set B X_4	Set B X_5
m	47.06	32.68	1147.56	.361	.250	.167
sd	19.18	9.04	182.32	.480	.433	.373

ACV regression equation:

$\hat{Y} = 2.01\, X_1 - .0608\, X_2 - 13.32\, X_3 - 11.68\, X_4 + 8.96\, X_5 + 57.38$

t (3.137**)   −2.092*   −2.562*   −2.117*   1.479 $(df = 66)$

pr^2 (.1298)   .0622   .0905   .0636   .0321

pr (.360)   −.249   −.301   −.252   .179

Adjusted Y intercept:

(9.2.2) $A' = A + B_1 m_1 + B_2 m_2$

          $= 57.38 + 2.01(32.68) - .0608(1147.56)$

(9.3.6) $A' = 53.30 (= \bar{Y}_4')$

(9.3.7) $\bar{Y}_h' = B_j + A'$ (for $h = 1, 2, \ldots, g-1$)

          $\bar{Y}_1' = -13.32 + 53.30 = 39.98$

          $\bar{Y}_2' = -11.68 + 53.30 = 41.62$

          $\bar{Y}_3' = 8.96 + 53.30 = 62.26$

*P < .05.   **P < .01.

The pr_j^2 (Eq. 9.2.5) express the difference between G_h and the reference group as a proportion of the variance of Y' not otherwise accounted for. Thus, $pr_3^2 =$ .0905 indicates that 9% of the quadratically-adjusted-for-age MD variance from which the other two comparisons have been partialled is accounted for by the G_1 versus G_4 distinction; $pr_3 = -.301$ states the relationship as a (point-biserial) correlation with Y', the negative sign indicating that $\overline{Y}_1' < \overline{Y}_4'$. In general, the discussion in Section 5.3.3 holds here with Y' substituted for Y.

The above represent the major results of interest in an ACV. Some comment on the partial coefficients for the X_i of a set A is, however, in order, particularly in regard to those for age, X_1. As noted in detail in Section 6.2, the simultaneous regression of powers of a variable results in "variance stealing" and renders ambiguous the interpretation of the magnitude, sign and t values for the partial coefficients of lower order powers. This affects X_1 here (r_{Y1}, not given, is *negative,* but $r_{Y1 \cdot 2}$ is positive) but not X_2, and the evidence for curvilinear within-group regression is unambiguously given by $t_2 = -2.092*$, with the negative sign of B_2 indicating that the curvature is concave downward.[6]

Again we see that we have gotten the usual yield of an ACV, and more, this time for a nonlinearly (quadratically) related covariate. Higher degrees of complexity in the relationship could be accomodated by more polynomial terms, or a different method of representation of the covariate research factor, or both, but this does not change in any essential way the nature of the analysis.

9.3.5 A Covariate with Missing Data

Even a variable with missing data may be used as a covariate, since all its information can be incorporated in set A using the methods of Chapter 7. We now reinterpret the results in Table 9.3.1 as having come from the following investigation:

In a ten-year follow-up, the vocational adjustment of expupils of special classes for the retarded in a Southwestern city is assessed using a weighted scoring of interview responses (Y). The research factors of interest arise from the cross classification of the 72 cases by sex (M–F) and membership in one of two cultural groups, one defined by a Spanish surname and the other made up of white "Anglos" (S–A). The four groups thus defined (with their unequal sizes as given in Table 9.3.1) can be appraised as a 2 X 2 factorial design, in accordance with our interest in the effects of sex, cultural group, and their interaction. We code set B (X_3, X_4, X_5) for these three contrasts following the principles of

---

[6] For a detailed analysis of the shape of the regression of Y on age for the total sample, one would need to enter X_1 and X_2 hierarchically, as described in Section 6.2. Also, the separate regression equations for Y on age for each group are available from the ACV equation by substituting the coded values for X_3, X_4, and X_5 that describe each group, as was illustrated in Chapter 8. Since there are no interaction terms, the separate equations share the same $B_1 = 2.01$, $B_2 = -.0608$, but have different intercepts, $B_j + A$, that is, the curves are necessarily parallel. Over the age range 20–50, the Y values *decrease,* and since the curve is concave downwards, do so at an accelerating rate.

Section 5.5 (particularly Section 5.5.4). The coding coefficients are exactly those of Table 5.5.1, Set II, with the groups redefined as G_1 = MS, G_2 = MA, G_3 = FS, and G_4 = FA.

We wish to free the vocational adjustment score insofar as is possible from the effect of variability in IQ scores from the subjects' records while in school. A search of these records reveals that for 17 of the 72 cases, no IQ appears.[7] For reasons extensively discussed in Chapter 7, we define IQ as a set made up of two variables: X_1 is a dummy variable dichotomy (d) coded 1 for cases missing IQs and 0 where present, and X_2 contains the IQ values where available and their mean where values are missing (see Sections 7.3.1 and 7.4.1). These two variables, which contain all the information available for IQ, constitute the covariate set A.

Reinterpreting the results in Table 9.3.1 in this context, we see that this set A, linear IQ plus missing data, accounts for a substantial and significant chunk of the vocational adjustment score variance; that, beyond this, group membership (set B) adds a material and significant amount, and that there is no reason to suppose that the ACV is invalidated by different regressions among the groups of Y on set A. Since our interest lies in fact on $Y'(= Y \cdot A)$, we note that group membership uniquely accounts for 20.5% of IQ-plus-missing-data-adjusted (partialled) vocational adjustment score variance.

The details of this version of the problem are given in Table 9.3.4. As we have noted, the ACV is an analysis of the relationship between set $B \cdot A$ and $Y \cdot A$, and the relationship of set A to Y is generally of secondary interest. However, we note here incidentally that both covariates are substantially and highly significantly related to Y within groups. Now, although this is not surprising in regard to (mean-plugged) IQ (X_2), we note that those subjects who are missing IQ information on the average show a 6.01 (= B_1) point higher adjusted Y mean within groups than those with IQs present (amounting to a pr = .320). This suggests that a breakdown occurred in the process of obtaining and recording IQs of these "mentally retarded" pupils. Specifically, the suspicion arises that more able pupils (as evidenced by superior later vocational adjustment) were more likely not to be tested (or their IQs not recorded).[8] It is conceivable that these findings might surpass in importance with regard to social policy those of the ACV proper.

[7] It is not unlikely that the distribution of missing IQs (X_1) is quite disproportionate in the four groups, that is, $R^2_{1 \cdot 345}$ may be substantial (see Footnote 8). Whether or not this is the case does not, however, effect the utility of this distinction as a covariate, since r^2_{Y1} may well be of consequential size. But even if both of these quantities are small, the method remains appropriate, as we have argued, although power may be slightly reduced.

[8] Some light on how this may have occurred might be shed by study of the matrix of simple rs. For example, a large positive r_{14}, indicating that Spanish-surname pupils were far more likely to be missing IQ data, might suggest that nonretarded pupils in that group were assigned to retarded classes because of language handicap. Similarly, a high positive r_{13} would suggest that the above practice occurred more with boys than with girls.

TABLE 9.3.4
IQ with Missing Data Dichotomy as Covariates, 2 X 2 Factorial
Contrast Coded Groups

		Set A		Set B		
	Y	d X_1	IQ X_2	M–F X_3	S–A X_4	M–F X S–A X_5
m	20.79	.236	63.86	.111	.028	.042
sd	8.69	.425	6.02	.488	.499	.246

ACV regression equation:

$$\hat{Y} = 6.01\,X_1 + .436\,X_2 + 2.45\,X_3 + 5.22\,X_4 + 7.14\,X_5 - 9.19$$

t	2.747**	2.994**	1.363	2.625*	2.065*	$(df = 66)$
pr^2	.1026	.1196	.0274	.0945	.0607	
pr	.320	.346	.165	.307	.246	

Adjusted Y intercept:

(9.2.2) $A' = 20.07 = \bar{\bar{Y}}'$

Adjusted means:

(9.2.1) $\bar{Y}'_h = 2.45\,X_3 + 5.22\,X_4 + 7.14\,X_5 + 20.07$

Adjusted means in 2 X 2 form:

	Spanish	Anglo	Mean of $\bar{Y}'$	Diff.
Male	25.69	16.90	21.30	
				$2.45 = B_3$
Female	19.67	18.02	18.85	
Mean of $\bar{Y}'$	22.68	17.46	$\bar{\bar{Y}}' = 20.07 = A'$	
Diff.		$5.22 = B_4$		

Interaction effect = $(\bar{Y}'_1 - \bar{Y}'_2) - (\bar{Y}'_3 - \bar{Y}'_4) = 7.14 = B_5$

*$P < .05$.   **$P < .01$.

Given the contrast-coding of the four groups as a 2 X 2 factorial, we can conclude that the M–F distinction, averaged over the S and A groups, is not significantly related to Y', IQ-plus-missing-data-adjusted vocational adjustment scores ($t_3 = 1.363$). The Spanish are significantly higher than the Anglos when averaged over the sexes ($t = 2.625*$), and the Spanish–Anglo difference for men is significantly greater than that for women ($t = 2.065*$). The $\bar{Y}'_h$ are computed, as usual, from Eq. (9.2.1), substituting the contrast coding coefficients for each group as X_3, X_4, and X_5 (as demonstrated in Table 5.5.5). The regression coefficients for set B are interpreted, in general, *exactly* as described in Sec. 5.5 on contrast coding, with adjusted means replacing means. All the formulas of that section apply, and Eq. (5.5.3) results, for the 2 $\times$ 2 table, in the B_j giving the contrasts as functions of adjusted means, as shown at the bottom of Table 9.3.4. Again we see the operation of the principle that the B_j and A' in an ACV have exactly the same meaning in terms of the $\bar{Y}_h'$, for any given method of nominal scale coding, as the B_i and A do for the $\bar{Y}_i$ in the absence of covariates.

This same principle governs the interpretation of the pr_j^2 for a contrast-coded X_j; it gives the proportion of that part of the adjusted Y variance not accounted for by the other contrasts uniquely accounted for by the X_j contrast.

To perform a t test between the $\bar{Y}_h'$s of a pair of groups (cells), substitute the adjusted means for the Bs in the numerator of Eq. (5.3.10) and modify the denominator for ACV as described in Footnote 3.

The above illustrates the use in ACV of only the linear aspect of a variable with missing data. The representation of missing data can readily be combined with that of curvilinear regression of Y on a quantitative covariate by any of the means described in Section 7.3.2, for example, by power polynomials, as illustrated there.

Analysis of Covariance for More Complex Factorial and Other Experimental Designs

Although the primary focus of this section is on various types of covariates in set A, we take the opportunity provided by the last example of a 2 $\times$ 2 layout to describe the simple extension to general factorial design in the representation of group membership in set B. As was illustrated in Section 8.7, a $U \times V$ factorial design (U has u levels, V has v levels) results in $uv = g$ cells, or groups. What constitutes the factorial design is that the $k_B = g - 1$ ($= uv - 1$) IVs representing these groups are made up as follows:

Effect	df; number of IVs
U	$k_U = u - 1$
V	$k_V = v - 1$
$U \times V$	$k_U k_V = (u - 1)(v - 1) = uv - u - v + 1$

$$k_B = uv - 1 \ (= g - 1)$$

Recall that the form of coding of U and V, and hence of their interaction set product is at the disposal of the data analyst. (Table 8.7.1 illustrates the coding of a particular 4 $\times$ 3 design.)

For concreteness, consider the illustration in Section 8.7 in an ACV context. Set A could be any type of covariate set (for example, age and amount of prior hospitalization; $k_A = 2$), and precedes the factorially organized group membership information for hospitals (H) and treatments (T). The hierarchical analysis then proceeds

Set	Subset		df	
A			2	$= k_A$
$+ B$	$\left\{\begin{array}{l} H\,(= U) \\ T\,(= V) \\ H \times T = (U \times V) \end{array}\right.$	$\left.\begin{array}{l} u - 1 = 3 \\ v - 1 = 2 \\ (u-1)(v-1) = 6 \end{array}\right\}$	$11 = g - 1 = k_B$	
$+ A \times B$	$(A \times H, A \times T, A \times H \times T)$		22	$= k_A\,k_B$

Following the principle of requiring homogeneity of regression of Y on covariates between groups, the increment in R^2 when the 22 IVs of $A \times B$ are entered is appraised by an F test. Note that this $I_{A \times B}$ is an agglomerate of interactions of set A with the subset constituents of set B, and the ACV requires that this aggregate yield a suitably low F.[9] Assuming this test is met, one can then interpret the factorial design exactly as described in Section 8.7, except that the dependent variable is $Y' = Y \cdot A$, and the set effects are, hierarchically, of $H \cdot A$, $T \cdot A, H$, and $HT \cdot A, H, T = (H \times T) \cdot A$. Further, the partial coefficients of the X_js of set B are also interpretable exactly as described in Section 8.7 with the same proviso that the effective dependent variable is not Y, but Y'; for example, the pr_j^2s refer to variance in Y' and the B_j and A' refer to the adjusted $\bar{Y}$'s.

Obviously, the above generalizes to ACV factorial designs of higher order, as we have seen for AV factorial design in Sections 8.8 and 8.9. Further, more exotic fixed factor ACV experimental designs, e.g., incomplete factorials, Latin squares, etc. (Winer, 1971), can be accommodated in this scheme by coding the research factors appropriately to make up set B.[10]

9.3.6 Nominal Scales as Covariates

Nominal scales are no exception to the principle that anything which can be expressed as a set of IVs can serve as a covariate in ACV. One simply represents

---

[9] A more detailed hierarchical analysis of the constituent interactions may be of interest, since it is possible that one of them yields an unacceptably large F while the F for overall $A \times B$ is acceptably small. Should that one be $A \times H$, it would hardly matter, since it is T which is of central interest—H is a variable to be controlled, in effect a covariate itself (see Section 9.3.6), and interactions among covariates do not invalidate the ACV.

[10] For example, a Latin Square of order p will have p^2 cells. If one codes (by any appropriate means) a subset for rows, another for columns, and a third for "Latin letters," each will be represented by $p - 1$ IVs for a total of $3(p - 1)$ IVs in set B. The remaining $p^2 - 3(p - 1) - 1$ which would be needed to represent all p^2 cells are the "Latin square error" (LSE) and need not be coded, in which case they become pooled with the within cell error to make up the $1 - R_{Y \cdot AB}^2$ error term, based on $df = n - k_A - k_B - 1 = n - k_A - 3(p - 1) - 1$. If one wishes the LSE segregated in order to have "pure" (Model II) error, the LSE may be represented by any coding of the $p^2 - 3(p - 1) - 1$ IVs necessary to fully account for the p^2 cells, and place this subset of IVs in the hierarchy after the row, column, and "Latin letter" effects.

362 9 ACV AND ITS MRC GENERALIZATION

the group membership information which is to serve as a covariate in set A, coded as desired, and proceeds as described above in Section 9.2. Specifically, all the equations and relationships given there apply: Eq. (4.4.2) gives the ACV overall F test for the significance of $B \cdot A$; Eq. (4.3.10) gives the squared multiple partial correlation pR_B^2 (= $R_{YB \cdot A}^2$); Eqs. (9.2.1) with (9.2.2) provide the adjusted means of Y for the levels of the nominal scale research factor of interest, that is, that represented in set B, and the pr^2 for the IVs of the ACV equation may be found from Eq. (9.2.5).

Indeed, in Section 8.7 of the preceding chapter we had a data structure of exactly this type presented as a 4×3 factorial design: the efficacy (Y) of three different treatment procedures (T), each applied in four different hospitals (H). Just above (Section 9.3.5), we augmented this problem by a covariate set to illustrate an ACV factorial design, the 12 groups defining set B as a 4×3 factorial. We now return to the problem as originally formulated, with only H and T. Let membership in one of the 3 T groups be defined as set B, and let H constitute the covariate set A. An attempt at analysis along these lines in Section 8.7 (see Table 8.7.2) was aborted when it was found that $H \times T$ produced a quite large (.2084) and highly significant ($F = 5.473^{**}, df = 6, 96$) increment in Y variance. Now imagine exactly the same results as in Table 8.7.2 for H and T, but with an $H \times T$ increment that is small (say, .0411) and emphatically nonsignificant ($F = .847$). Thus revised, this *is* a valid ACV, with H serving as a nominal scale covariate. Table 8.7.2 provides the ingredients for the ACV results. The hierarchical R^2 analysis for H as set A and T as set B gives the overall F ratio of ACV (= 6.835) via Eq. (4.4.2) and provides the ingredients for pR_B^2 [= .1553/(1 − .1823) = .1889] from Eq. (4.3.10). Equation (III) in Table 8.7.2 is the ACV regression equation, since it is an equation for Y as a function of sets A and B (but omits $A \times B$). Thus, its regression coefficients for the X_j of set B are interpreted as functions of covariate-adjusted Y means; its pr_j^2 values (computable from the t_j by Eq. 9.2.5) give proportions of covariate-adjusted Y' variance; and the equation's constants together with the means of the IVs (not given in Table 8.7.2) can be substituted in Eq. (9.2.2) to give A', and the latter in Eq. (9.2.1) to give the formula which generates the adjusted Y means for the T groups. (Since T was dummy-coded, the somewhat more specialized Eqs. 9.3.5 to 9.3.7 apply here.) In short, the fact that set A defines a nominal scale in no way inhibits the use of the formulas and interpretations for an ACV analysis via MRC developed in this chapter.

The reader may wonder by what legerdemain we have converted a 4×3 factorial design to a one-way ACV using the other factor as a nominal covariate. Indeed, he may ask, "*Is* there any difference between the two? Can all AV factorial designs be so converted?" There *is* a difference, but a subtle one. Recall the point emphasized earlier that the dependent variable of an ACV is $Y \cdot A$, its set of IVs of research interest is $B \cdot A$, and its $A \times B$ is only of negative interest, to be gotten out of the way so the ACV can proceed validly. We prefer to use the AV factorial design model when the factors as well as their interactions are (more or less) of direct research interest. They are then all, including the interaction(s), part of the model, and strictly speaking, Model II error should be used (see Sections 4.2.2 and 8.7). In ACV, the asymmetry of our interest is expressed in both the fact that it is $Y \cdot A$ and not $Y \cdot B$ that constitutes the dependent variable, and that it is the effect of $B \cdot A$ and not $A \cdot B$, which we are studying. Of course, as we have seen in Section 8.7, when a valid ACV is precluded by the discovery of a large $A \times B$ interaction, we must switch from the ACV model to one which includes the interaction, and although more complex, its yield may be all the more enriching of our understanding.

9.3.7 Mixtures of Covariates

Although the reader may well have inferred this by now, for the sake of completeness we point out that nothing prevents the analyst from combining subsets of covariates of different (or, of course, the same) types into set A. Thus, a consumer psychologist comparing groups exposed to different advertising copy may wish to control for age nonlinearly, product usage linearly, and type of household (a nominal scale) by combining appropriate representations of these (and possibly their interactions) into a covariate set A. Similarly, an experimental social psychologist comparing learning in animals as a function of different social conditions may wish to control for litter size nonlinearly and for which of three research assistants actually ran the animal; these two subsets could be combined as set A. All the formulas and interpretations given in this chapter apply to mixed covariates.

9.4 THE ANALYSIS OF PARTIAL VARIANCE: A GENERALIZATION OF ANALYSIS OF COVARIANCE

9.4.1 Introduction

The use of nominal scales and missing data variables as set A covariates represents a clear departure from the ACV as presented in the literature, but we now offer an extension of ACV which is an even more radical departure: the use of quantitative research factors in set B rather than restricting it to those representing group membership.

In the clarity of retrospective vision, we can see that the term "ACV" is not very descriptive of the actual process, if not a downright misnomer. After all, it is not the covariance of Y with set A which is analyzed; rather, *that* is thrown away as irrelevant and its residual variance, that is, the variance of $Y \cdot A$, which is analyzed. Further, as we have seen, it is not set B which is the source of variance of interest, but $B \cdot A$ (or, more exactly $\hat{Y}_B \cdot \hat{Y}_A$ —see Section 4.3.1). A more descriptive name than "ACV" would be "Analysis of Partial (or residual) Variance," and thus a special case of hierarchical MRC. But it is an unnecessarily narrow special case to restrict set B to the representation of group membership.

In the free-wheeling data-analytic spirit of our fixed-model MRC system, once a research factor is coded as a set of IVs representing its aspects of interest, it may be used as a unit of analysis, whole or partialled, whatever its structural or functional characteristics (see Chapter 4). We saw in the previous section that any kind of research factor could be used as (or in) the covariate set A, and we now assert the same freedom, in principle, for the research factor(s) of direct interest in set B. Linear, nonlinear, and/or missing-data aspects of one or more quantitative variates, and not only aspects of nominal scales, may constitute set B.

9.4.2 The Analysis of Partial Variance

Let us call this generalization of the (fixed model) ACV the (fixed model) Analysis of Partial Variance (APV), and formulate it explicitly (largely by recapitulation):

1. A set (A) of IVs, deemed "covariates," *of any formal type,* is believed to carry irrelevant and thus potentially distorting variance in Y; this irrelevant variance is to be removed (partialled) from Y. If we were to estimate Y from set A by the usual means, we could obtain $\hat{Y}_A$, and for each subject determine $Y - \hat{Y}_A$. This residual or adjusted Y, which we denote as $Y \cdot A$ or more simply Y', contains no set A variance; its variance is represented by the areas a + e in the ballantine (Figure 4.3.1). We thus have created from Y and set A, a *new* dependent variable, Y'.

2. Another set (B) is made up of IVs representing aspects of one or more research factors, *of any formal type,* whose bearing on Y' constitutes the research focus. Indeed, the effective source of variance is $B \cdot A$ ($= \hat{Y}_B \cdot \hat{Y}_A$), that is, the overlap in Y of sets A and B (area c of Figure 4.3.1) is removed from set B with the result that our conclusions can be framed with regard to set B with set A statistically controlled (partialled, held constant). This requires the assumption of regression homogeneity (tested via the $A \times B$ interaction), which in APV is the assumption that for any given set of observed values for the X_j in set B, the regression coefficients for Y on the X_i of set A in the population are the same as those for any other set of X_j, that is, the regression surfaces for Y on set A do not change shape for varying set B values.[11] Note that the within group regression homogeneity assumption of ACV is a special case of this more general APV assumption.

Thus, once one drops the presumption that set B is a nominal scale, *all of Sec. 9.2* (ACV via MRC: General) *holds for APV:* The setwise hierarchical MRC proceeds with A, B, and the $A \times B$ test of regression homogeneity. The F test of $B \cdot A$ via Eq. (4.4.2) appraises the overall null hypothesis of APV, and pR_B^2 of Eq.

---

[11] Again, the purist will point out that in the fixed model which we assume, this holds only for the specific sets of X_j values present in the data. This is strictly so, but as we have already argued, as a practical matter, this is likely to hold for any sets of X_j values falling within the observed range, not only for the specific sets of X_j values observed in the data.

(4.3.10) is the appropriate measure of effect size, i.e., the proportion of Y' variance accounted for by $B \cdot A$.

The APV equation is the usual regression equation estimating Y from the X_is of set A and the X_js of set B simultaneously, which we can write as

(9.4.1) $$\hat{Y} = \sum B_i X_i + \sum B_j X_j + A.$$

Note that is not a new equation, but an ordinary least-squares multiple regression equation in which we have segregated the terms of the two sets; the first summation is over the k_A IVs of set A and the second is over the k_B IVs of set B. Now, substituting the m_i for the X_i of the covariate set, the first and third terms on the right are absorbed into a new constant, the adjusted Y intercept, A', as given in Eq. (9.2.2) of Section 9.2:

$$A' = A + \sum B_i m_i.$$

Now, Eq. (9.2.1) merely substitutes Eq. (9.2.2) in Eq. (9.4.1) to give the formula for adjusted Y means $(\bar{Y}_h')$ for the ACV special case where set B defines a nominal scale (groups). In APV we have exactly the same equation, but now written generally for a set B of any formal type, which estimates not Y, but Y':

(9.4.2) $$\hat{Y}' = \sum B_j X_j + A'.$$

That Eq. (9.2.1) is a special case of Eq. (9.4.2), which replaces $\hat{Y}'$ by $\bar{Y}_h'$, is apparent from a moment's reflection: when the X_js define group membership in G_h, the best (least-squares) estimate of Y' is the mean of the Y' for G_h, since the mean of any collection of values is that value from which the squared departures of the values in the collection sum to a minimum.

In Eq. (9.4.2) we have what is, in fact, a regression equation in which Y' (rather than Y) is estimated. Its regression coefficients (B_j) and intercept (A') are interpreted *exactly* in accordance with the nature of the X_j, as discussed throughout the book, except that the dependent variable is the covariate-adjusted Y', rather than Y. Similarly, the pr_j^2, computable from Eq. (9.2.5), refer the partialled X_j to the *partial* variance of Y, that is, the variance of Y'.

9.4.3 Analysis of Partial Variance with Set B Quantitative

For the sake of concreteness, we briefly illustrate an APV with set B made up of aspects of quantitative variables. The setwise hierarchical R^2 analysis proceeds exactly as described in Section 9.3.2 and we will take its results to be those of Table 9.3.1 (yet again). Since it yields a small $I_{A \times B}$, the APV proceeds, and since I_B is significant, set B accounting for .2047 $(= pR_B^2)$ of the Y' variance, we can proceed to an interpretation of the individual partial results for the X_j variables. The regression equation (Eq. 9.4.1) for $\hat{Y}$ is produced, and the $\sum B_i m_i + A$ are absorbed into a new constant, A' of Eq. (9.2.2), the adjusted Y intercept (or Y' intercept). We need not specify the nature of the covariate set

A; it may be of any of the types illustrated in Section 9.3 (or mixtures thereof). Whatever its nature, the APV proceeds within a framework in which set A has been partialled both out of Y and out of set B. We can now put our results in the form of Eq. (9.4.2), which, for these fictitious data let us assume is

$$Y' = B_3 X_3 \; + \; B_4 X_4 \; + \; B_5 X_5 \; + \; A'$$
$$= 9.72 \, X_3 + 12.23 \, X_4 - 1.46 \, X_5 + 21.56 \quad (df = 66)$$

t_j	2.486*	3.093**	1.03
pr_j^2	.1011	.1309	.0397

The interpretation of this regression equation is as always: a unit increase in X_j is associated with B_j units increase in Y' (and when all X_js are at zero, $\hat{Y}' = A'$). More specific interpretation of the B_js depends on exactly what the X_js are. In ACV they are aspects (dummy variable, effects, contrasts) of one or more nominal scales. With set B quantitative, they are aspects of one or more quantitative scales as described in Chapters 6 through 8: linear, variously coded nonlinear functions, interaction-bearing products. Thus, for example, X_3, X_4, and X_5 may be respectively linear aspects of three different quantitative research factors u, v, and w. In interpreting the B_js one must keep in mind that each is a *partial* coefficient—not only are the covariates partialled, but also all the other variables in set B. Thus, B_4 is actually $B_{Y'4\cdot35}$ (or $B_{Y4\cdot1235}$) and indicates that a unit change in v is associated with a (statistically significant) 12.23-unit increase in Y' with u and w held constant. Or X_3, X_4, X_5 may be respectively a missing-data dichotomy d_v, a mean-plugged missing-data variable v_c, and some other variable w. In this case, B_3 indicates that, holding w and v_c constant,[12] the $\bar{Y}'$ of those missing v values is (a statistically significant) 9.72 units higher than that of those with v values present (see Section 7.3.1). Or X_3, X_4, X_5 may be the first three orthogonal polynomial terms of a quantitative variable v, in which case the equation indicates that the regression of Y' on v is a generally rising (B_3 is positive and significant), concave upward (B_4 is positive and significant) function, adequately fitted without a cubic component (B_5 nonsignificant). Similarly, the interpretation of other partial functions of X_j, for example, pr_j^2, proceeds exactly as described in previous chapters, with the dependent variable now being Y'.

When the set B variables are of a kind that make it desirable that their analysis be hierarchical, for example, for power polynomials or interactions, one simply adds them cumulatively after set A. The result is a series of APV equations, which are each translatable into equations for $\hat{Y}'$ of increasing order, whose constants (A' and the B_j) change, exactly as was the case in the general hierarchical model as applied in Chapters 6 through 8. The interpretation is again the same, but for Y' instead of Y.

---

[12] It isn't really necessary to include v_c in this phrase, since the mean-plugged v_c is uncorrelated with d_v, and partialling an uncorrelated variable has no effect on the value of a partial coefficient. We do so in the interest of generality.

To recapitulate: The APV is a very general method of data analysis applicable whenever the variance associated with a covariate set A (of any type) is to be removed from Y, in order to appraise the effects of research factors in set B (of any type). The appropriate regression equation for Y on sets A and B is readily translated into an equation in which the partialled or covariate-adjusted Y' (= $Y \cdot A$) is written as a function of adjusted B (= $B \cdot A$). Such equations may be simultaneous or a hierarchical series in the X_js of set B, and their constants are interpreted exactly as in general MRC analysis, but with Y' as the dependent variable. Indeed, *all* partial correlation and regression coefficients are so interpreted.

9.4.4 Sequential Analysis of Partial Variance and Setwise Hierarchical Multiple Regression/Correlation

The following circumstances occur with some frequency in research in the behavioral and social sciences, particularly (but not exclusively) in nonexperimental investigations: With regard to some dependent variable Y, there are IV data for h presumed or potential causal entities, each expressed as a set. The investigator can organize these in a hypothesized order of causal priority on temporal, commonsense, or theoretical grounds (see Section 9.7). He thus has h sets $T, U, V, \ldots, Z$, in that hierarchical order of assumed decreasing causal proximity to Y.

What we have called the hierarchical MRC model for sets simply adds each set cumulatively, in this a priori order, determining at each stage the increment in R^2 due to that set, for example, $I_V = R^2_{Y \cdot TUV} - R^2_{Y \cdot TU}$, and testing it for significance. In Chapter 4, we illustrated this for a sample of psychiatric hospital admissions with Y as length of stay and sets of demographic variables (D), MMPI scores (M), and hospitals (H) as presumed causal entities (in that order). In the discussion there, the APV was briefly alluded to but not pursued. The simple version of the hierarchical model presented called only for the determination of each increment and its significance testing.

From our present enlarged conception of APV, however, a fuller analytic exploitation of such a hierarchical series becomes possible. Since at each stage, all prior sets are covariates, one can perform a series of complete APVs beginning with the second set. For each stage, the newly entering set is set B, and all prior sets are aggregated into set A. For the APV of each stage to be valid, the check on regression homogeneity via the F test of that stage's $A \times B$ interaction should be passed successfully, in which case the interaction set is dropped. Then, the full analytic power of the APV, including the regression equation in $\hat{Y}'$ and the interpretation of the partial relationships of that stage's Y' with its X_js, is available. If the interaction is substantial, one reverts to the analysis of the full model including interaction, as described in Chapter 8, for that stage. When this occurs, subsequent stages continue to carry the significant interaction set as part of their set A aggregates. Because of the increasing number of IVs accompanied by probably decreasing *I* values, we recommend that in Sequential APV, Model I error be used throughout for optimal power (see Sections 4.4 and 4.5).

Thus, in the Chapter 4 example of *Y* on the hierarchy *D, M, H,* one would perform a Sequential APV as follows: the first stage would proceed exactly as described in Sections 9.2 and 9.4.3, with *D* as set *A* and *M* as (a quantitative) set *B*. The second stage would aggregate *D* and *M* as a (mixed) covariate set *A,* and use *H* as set *B*. Exactly the same procedure would be used. Since *H* is nominal, this is the ACV special case of APV. The interaction check at this stage is performed by adding to *D, M* (= set *A*) and *H* (= set *B*), the (*D, M*) $\times$ *H* = *D* $\times$ *H,* *M* $\times$ *H* (= *A* $\times$ *B*) aggregated interaction set, and assessing the increment for significance. If the null hypothesis is acceptable, a full-dress interpretation as an APV (here, an ACV) may be made. If in the first stage, the *D* $\times$ *M* interaction is significant (indicating that the relationship between personality characteristics and length of stay varies with the patients' demographic characteristics), no valid *APV* for this stage is possible, but application of the methods of Chapter 8 provide a full analysis down to the level of single IVs. In this event, at stage two set *A* would include the *D* $\times$ *M* set in addition to sets *D* and *M,* and the interaction check for this stage would include in the interaction aggregate *D* $\times$ *M* $\times$ *H* as well as *D* $\times$ *H* and *M* $\times$ *H.* With more sets, the process continues in the same way.

We do not penetrate the surface of the conception underlying the assignment of "presumed causal priority" to the entities under consideration. This failure is not because the issues are unimportant, quite the contrary. We have stressed at several points the dependence of the results per set (or single IV) on its placement in the hierarchy. A responsible discussion of these matters is simply beyond the scope of this presentation (but see Section 9.7). There is a rapidly emerging social science methodology called "causal models" which is addressed to the analysis of nonexperimental data, and which is heavily dependent on regression and correlation analysis (Blalock, 1964, 1971; reviewed by Cohen, 1973a). Some of its development (for example, "structural equations," due to econometricians) transcends ordinary least-squares fixed-model regression analysis, the model of this book. Sociologists and psychologists have exploited another aspect of this field, "path analysis" (invented by a geneticist!), but have thus far restricted themselves to linear aspects of quantitative variables. Some exciting methodological possibilities, too new to have been exploited, lie at the intersection of these ideas with the more versatile set-oriented system of MRC presented in this book. We leave these as a challenge to the reader.

9.5 THE PROBLEM OF PARTIALLING UNRELIABLE VARIABLES

9.5.1 Introduction

Early in this book we presented the idea that simple correlations between variables were "attenuated," that is, reduced in absolute size, by the operation of random measurement error in the variables, in accordance with classical psychometric theory (see Section 2.11.2). One result presented there was that the correlation between the "true" scores (X^* and Y^*) was a simple function of the correlation between the observed scores (X and Y) and their reliability coefficients (r_{XX} and r_{YY}),

(2.11.5)
$$r_{X^*Y^*} = \frac{r_{XY}}{\sqrt{r_{XX}r_{YY}}}.$$

(We presume population values throughout this section; with sample values, these are estimates.) This is the classical formula for "correction for attenuation in both variables." We can also correct for attenuation in either one of the variables, and not the other, for example,

(9.5.1)
$$r_{X^*Y} = \frac{r_{XY}}{\sqrt{r_{XX}}}.$$

Thus, correction for attenuation due to the fallibility (lack of perfect reliability) of any given variable involves dividing its correlation with another variable by the square root of its reliability coefficient. If the fallibility of both variables is to be corrected for, this operation on r_{XY} is, in effect, performed twice, as in Eq. (2.11.5).

Whether performed once or twice, however, the important fact to note is that random measurement error (unreliability, fallibility—we use the terms interchangeably) *attenuates,* that is, diminishes the absolute size of rs, hence the size of r^2s. In both Eqs. (2.11.5) and (9.5.1), given less than perfect (but not zero) reliability, corrected rs (if not zero) are thus further from zero and retain their original sign. Conversely, then, as observed with fallible measures, simple correlations are numerically smaller than would be the case if X and/or Y could be measured without error.

Part of the reason for the weak relationship which characterize so many of the areas which comprise behavioral and social science lies in this attenuation phenomenon. It is not surprising that R^2 values as large as .50 are rare, when perhaps as much as half of the Y variance is random error (that is, $1 - r_{YY} = $ .50), and, by definition, inaccessible to correlation with IVs. In these research areas there exists, as a chronic state, a normative expectation of weak correlations and analogous small mean differences. One unfortunate historical consequence of this is an undue emphasis on null hypothesis testing, with the incorrect use of statistical significance as an arbiter of substantive or practical significance (Cohen, 1965, pp. 101–111), against which the MRC's emphases on correlational and proportion of variance measures of effect size serve as a counterforce. But we digress. The point we wish to stress here is that most behavioral/social scientists take for granted that, because of measurement error, the relationships they observe are universally weaker than they should be. The point we are about to make is that this is *not* necessarily the case for *partial* relationships, and hence those involving covariates.

9.5.2 The Effect of a Fallible Partialled Variable

That measurement error may decrease *or* increase, or even change the sign, of a *partial* relationship holds for all our means of expressing partial relationships: B, β, *sr*, *pr*. Since they all share the same numerator component (and significance test), it is sufficient for our purposes to present the case for the three-variable partial r. As presented in Chapter 3, the partial r for Y with X_2, partialling X_1, is given by Eq. (3.3.10). We rewrite it here (in slightly different form) as

$$(9.5.2) \qquad r_{Y2 \cdot 1} = \frac{r_{Y2} - r_{Y1} r_{12}}{\sqrt{(1 - r_{Y1}^2)(1 - r_{12}^2)}}.$$

It turns out, after some algebraic manipulation, that the numerator does *not* change when one corrects for attenuation the correlations for either Y or X_2 (or both) which enter this equation, and that the terms in the denominator remain positive values between 0 and 1. For example, if applying Eq. (9.5.1), r_{Y2} is replaced by $r_{Y*2} = r_{Y2}/\sqrt{r_{YY}}$ in Eq. (9.5.2), we obtain after simplification

$$(9.5.3) \qquad r_{Y*2 \cdot 1} = \frac{r_{Y2} - r_{Y1} r_{12}}{\sqrt{(r_{YY} - r_{Y1}^2)(1 - r_{12}^2)}}.$$

If we correct for X_2 and not Y, the results are perfectly symmetrical, leaving the numerator unchanged. If we correct *both* Y and X_2, that is, apply Eq. (2.11.5) to obtain r_{Y*2*} and Eq. (9.5.1) to obtain r_{Y*1} and r_{12*}, and then substitute in Eq. (9.5.2) and simplify, we obtain

$$(9.5.4) \qquad r_{Y*2* \cdot 1} = \frac{r_{Y2} - r_{Y1} r_{12}}{\sqrt{(r_{YY} - r_{Y2}^2)(r_{22} - r_{12}^2)}}.$$

Again, no change occurs in the crucial numerator. In both Eq. (9.5.3) and Eq. (9.5.4) the denominator is smaller, and so the effect of correcting for either or both Y and X_2 is invariably to increase the absolute value of the partial r between them, that is, the usual effect of correcting for attenuation discussed in the previous section. But such is *not* the case when one corrects for measurement error in X_1, *the partialled variable*. Whether or not one corrects for measurement error in Y and X_2, correction for X_1 results in a change in the numerator to $r_{11} r_{Y2} - r_{Y1} r_{12}$. For example, correcting for X_1 and also for Y and X_2 by applying Eq. (2.11.5) to obtain r_{Y*2*}, r_{Y*1*}, and r_{1*2*}, substituting again in Eq. (9.5.2) and simplifying, we find

$$(9.5.5) \qquad r_{Y*2* \cdot 1*} = \frac{r_{11} r_{Y2} - r_{Y1} r_{12}}{\sqrt{(r_{11} r_{YY} - r_{Y1}^2)(r_{11} r_{22} - r_{12}^2)}}.$$

Finally, if one corrects for *only* X_1, one finds

$$(9.5.6) \qquad r_{Y2 \cdot 1*} = \frac{r_{11} r_{Y2} - r_{Y1} r_{12}}{\sqrt{(r_{11} - r_{Y1}^2)(r_{11} - r_{12}^2)}},$$

that is, the same numerator, but with a larger denominator.

TABLE 9.5.1
Effects of the Fallibility of a Partialled Variable

Example	r_{Y2}	r_{Y1}	r_{12}	r_{11}	$r_{Y2 \cdot 1}$, Eq. (9.5.2)	$r_{Y2 \cdot 1*}$, Eq. (9.5.6)
1	.3	.5	.6	.7	.00	−.23
2	.5	.7	.5	.7	.24	.00
3	.5	.7	.6	.7	.14	−.26
4	.5	.3	.8	.7	.45	.57
5	.5	.3	.6	.7	.42	.37

When the partialled variable's fallibility is corrected for, the intrusion of its reliability coefficient (r_{11}) into the numerator of $r_{Y2 \cdot 1*}$ (= pr_2, and thus also into the numerators of B_2, β_2, and sr_2) can produce effects which are counterintuitive. It serves our purpose to concentrate on Eq. (9.5.6), the simplest case where only the effect of measurement errors in X_1 is taken into account. Table 9.5.1 gives several illustrative examples comparing uncorrected $r_{Y2 \cdot 1}$ of Eq. (9.5.2) with $r_{Y2 \cdot 1*}$ of Eq. (9.5.6). All these examples use r_{11} = .7, a representative value.

In Example 1, failure to allow for the fallibility of X_1 results in an observed partial of zero, when the true-partialled (X_1^*) relationship is −.23. Thus, a real partial relationship is wiped out by the measurement error of the covariate. In Example 2, the converse occurs: an observed partial of .24 is actually zero when the imperfect reliability of the partialled variable is taken into account. Example 3 is most unexpected, and quite dangerous in its implications. Here, an apparently positive partial turns out to be negative (and numerically greater) when corrected for the fallibility of X_1. That this possibility is not merely academic is evidenced by a convincing argument that mistaken conclusions were drawn about the compensatory education program Head Start because a circumstance like Example 3 obtained (Campbell & Erlebacher, 1970). Example 4 is the only one in which there has been simple *attenuation* (reduction) because of measurement error, that is, where a nonzero correlation retains the same sign and is absolutely increased when unreliability is taken into account. Although this may occur, the more frequent outcome is that displayed in Example 5: correction for the imperfect reliability of X_1 results in a reduction in absolute value.

The cases in Table 9.5.1 are not intended to be representative of empirical results. The discrepancy between $r_{Y2 \cdot 1}$ and $r_{Y2 \cdot 1*}$ is a complex function of their constituent four values (which are mutually constraining[13]) and almost anything can occur with correction. An extensive investigation covering all possible combinations of selected values over the positive range for the four parameters suggests the following:

[13] Neither r_{Y1} nor r_{12} (nor any other population correlation involving X_1) can exceed $\sqrt{r_{11}}$, in accordance with the assumptions of classic psychometric theory. Also, the consistency constraints of Eq. (7.1.1) must obtain among the three observed rs.

1. Because the effect of r_{11} in the numerator in reducing the first term generally is greater than its effect in the denominator, the most frequent result of the correction is an algebraic reduction, that is, $r_{Y2 \cdot 1*} < r_{Y2 \cdot 1}$. Reductions from positive to negative values, though infrequent, can and do occur, as we have seen.

2. The lower r_{11}, the more extreme the results are likely to be. This is particularly true as r_{Y1}^2 and/or r_{12}^2 approach r_{11} in magnitude, since this makes the denominator become small and thus exaggerates the effect of r_{11} in the numerator.

A useful perspective on the relationship between Eqs. (9.5.2) and (9.5.6) is to note that when $r_{11} = 1.00$, $r_{Y2 \cdot 1*} = r_{Y2 \cdot 1}$. Thus, when we fail to correct for the fallibility of the partialled variable, we are acting as if its reliability is perfect, and the further this is from the truth (that is, the lower r_{11}), the greater the probability of being misled about the size and even the sign of the partial relationship between Y and X_2. Failing to correct when reliability is .8 or more will not matter much, but the risk is very great when it is as low as .5 or .6.

Details aside, the central fact we wish to emphasize is that unreliability in a partialled variable may yield grossly inaccurate results when it is ignored. This fact ramifies in several directions:

1. It holds whether or not either or both of the variables being correlated are corrected. As noted, the numerator remains the same. Study of Eq. (9.5.5) for $r_{Y*2* \cdot 1*}$ reveals that such further corrections can only effect an absolute increase over $r_{Y2 \cdot 1*}$, that is, a true correction for the latter's attenuation due to measurement error in Y and X_2.

2. As noted, it holds for all measures of partial relationship including partial regression coefficients. When X_2 is a dummy-coded binary variable, $B_{Y2 \cdot 1*}$ is an X_1^*-adjusted difference between Y means, and may differ in size and sign from $B_{Y2 \cdot 1}$, which is X_1-adjusted.

3. It holds not only for single partialled variables, but for sets of such variables.

4. This, in turn, makes clear why this section appears in a chapter on the ACV and its generalization to APV. In previous sections we presented these models as devices for studying $Y \cdot A$, where set A is a set of covariates, variables to be partialled from Y. Patently, from what we have seen, the issue of the reliabilities of the X_i of set A must be attended to, and corrective action taken, particularly so when they are not high. The most effective corrective action is to increase the reliability of the X_i by appropriate psychometric procedures, but, failing that, we must allow for their fallibility.

9.5.3 A General Method for Analysis of Partial Variance with Fallible Covariates: AP*V

Although the nature of the problem of fallible covariates is widely understood among methodologists, the solution to the problem is still in the realm of debate, the pros and cons of which are well beyond the scope of this textbook. Rather than ignore the problem we offer a method for coping with it which is coherent with the rest of the MRC system and not unduly complicated. We must warn the reader, however, that this method has not been proved mathematically nor even tested by extensive computer trials on data of known characteristics ("Monte Carlo" methods). It rests on no more than the judgment of the present authors and some of our colleagues. We debated its inclusion in this textbook, but finally decided to include it because of our belief that, whatever its deficiencies, its use is preferable to treating unreliable covariates as if they were perfectly reliable, and that its results are more likely to be right than wrong. Given this rather qualified endorsement, it will not surprise the reader that we far prefer to use highly reliable covariates and thus obviate the need for this method (as we argue in Section 9.5.4). This method is offered for only those occasions when unreliable covariates must be used.

The Analysis of estimated true-Partialled Variance (AP*V) is a general method (of which AC*V is a special case) whose results are exactly parallel to those of APV (of which ACV is a special case), but for "true" rather than observed covariate measures. It allows for measurement error *only* in the covariates, the X_i of set A, which, as we have seen, is where the problem lies. It does not correct for any unreliability in Y or in the research factor(s) in set B. Doing so would only correct for the attenuation which we have learned to live with, and for which we have not been correcting throughout the book. Thus, the AP*V examines the relationship between $Y \cdot A^*$ and set $B \cdot A^*$, exactly as APV examines the relationship of $Y \cdot A$ with set $B \cdot A$; all the varied measures of effect size found and interpreted in APV have their counterparts with the same meaning in AP*V, save that they are for estimated true rather than observed covariates. An important deficiency of AP*V resides in the absence of exact statistical significance tests and power analysis, and we will discuss how this difficulty may be coped with in the spirit of data analysis.

Reliability Coefficients

We require the r_{ii}s, the reliability coefficients of the covariate X_is of set A, that is, the partialled variables. These should ideally be exact values of parallel-form reliabilities for the population under study.[14] Population values are of course never available, and odd–even or Cronbach α values obtained on standardization samples,[15] or even on the sample at hand, may have to do, provided

---

[14] The reader who is unfamiliar with the elements of psychometric theory will need to refer to a treatment of this area (e.g., Cronbach, 1970; Nunnally, 1967).

[15] Remember that r_{ii} is the proportion of observed variance which is true variance and thus depends on the population. If we have a sample whose observed sd_i^2 is a times that of the standardization population for which we have an r_{ii} estimate (where a is either greater than or less than one), we would need to revise that estimate to $r'_{ii} = 1 - (1 - r_{ii})/a$ for use in our analysis. Thus, a very homogeneous sample whose variance of X_i is only half (i.e., $a = .5$) of that of a reference population whose $r_{ii} = .8$ would have an estimated r'_{ii} of $1 - (1 - .8)/.5$ = .6.

they are not spurious overestimates. All too frequently, no empirically based coefficients are available. Under these circumstances we believe it preferable to use informed "guesstimates," particularly when they are low, than to make no correction for unreliability, which we have seen is equivalent to assuming $r_{ii} = 1.00$. When uncertain, one might analyze the data making low and high estimates and only draw conclusions consistent with both.

The AP*V Procedure

As a prelude, first perform the usual complete APV (or ACV) on observed data, following the procedures described in Sections 9.2, 9.3, and 9.4. (If the setwise interaction $A \times B$ is large, the AP*V is aborted; see below.)

The AP*V (or AC*V) as such then proceeds by correcting all the rs in the basic correlation matrix which involve the covariates (the X_is) for attenuation due to *their* unreliability, and all the sd_i to sd_{i*} (as explained below). The result is our best estimate of the correlation matrix and standard deviations which would have resulted if we could have measured the covariates without random measurement error, that is, the true X_is instead of the observed X_is.[16] In detail:

1. Correct the correlations among the k_A covariates (r_{ih}) for attenuation due to both variables by dividing each by $\sqrt{r_{ii}r_{hh}}$ to produce r_{i*h*}; in other words, apply Eq. (2.11.5) $k_A(k_A - 1)/2$ times.

2. Correct for attenuation due to the X_i the correlations with Y of the k_A covariates (r_{Yi}) by dividing each by its $\sqrt{r_{ii}}$ to produce r_{Yi*}. Similarly, correct the $k_A k_B$ correlations of covariates with the X_j variables representing the research factor(s) of set B (r_{ij}) for attenuation due to the X_i by dividing each by its $\sqrt{r_{ii}}$ to produce r_{i*j}. In other words, apply Eq. (9.5.1) k_A times to obtain the r_{Yi*} and $k_A k_B$ times to obtain the r_{i*j}.

3. Correct each of the covariate sd_i to the estimated sd of true scores,

$(9.5.7)$[17]
$$sd_{i*} = sd_i \sqrt{r_{ii}}.$$

Note that since the mean of a true score equals that of an observed score, no change is made in the m_i. Note again that *all* correlations involving an X_i and only those are attenuation corrected; the r_{Yj}s, sd_Y, and sd_js remain exactly as computed from the raw data. The exact mechanics of the above corrections

---

[16] If any attenuation-corrected correlations exceed (or even closely approach) 1.00, the procedure must be abandoned. This anomaly must be due to underestimating reliability and/or an unrepresentatively high sample correlation. Also, some applications founder on the rocks of correction for attenuation: covariates which include powered values and variables with missing data are not suitable for AP*V.

Note that this matrix does not include the $A \times B$ interaction set, which is not needed.

[17] This formula is a simple derivation from the definition of the reliability coefficient in Eq. (2.11.3). Indeed, the covariance of two true scores equals that of their respective observed scores; the reason for the attenuation of an observed r is the inflation, by error, of the sds of the observed scores in the denominator. Thus, when simulating the behavior of true scores, one must apply Eq. (9.5.7) to correctly shrink the sd.

depend on the method of computation, and if done by computer, on the details of the computer program (see Appendix 3).

Once the above changes have been made to simulate the results from the X_i^*s of set A^*, one proceeds exactly as in APV. With some qualifications to be discussed below, all the applications and interpretive procedures described in Sections 9.2 and 9.3 on ACV, and their generalization in Section 9.4 to APV, can be extended to the results. For example, the pR_B^2 and pr_i^2 have the same meaning, except that it is the estimated true set A^* instead of the observed set A which is being partialled. The regression constants (the B_js and Y' intercept, A') which were functions (depending on coding) of covariate-adjusted Y means in ACV, are now the same functions of estimated *true* covariate-adjusted Y means in A*CV. The ACV Eq. (9.2.1), which yields the covariate-adjusted Y means, is now applied to yield true covariate-adjusted Y means as AC*V. Similarly for the generalization of this equation in APV to Eq. (9.4.2): its application here yields true covariate adjustments and constitutes AP*V. This exact parallelism for all measures of effect size makes new illustrative examples for AP*V unnecessary: simply precede all references to covariate adjustment by "estimated true" in Sections 9.2, 9.3, and 9.4, and AP*V (AC*V) is effectively illustrated.

However, none of the standard significance tests applied to these AP*V results remains valid. The process of correction for attenuation using reliability coefficients (which themselves may be subject to error) changes the sampling distributions of the correlations (and sds) of the X_is. Then, in turn, the complex functions of these which are distributed as F and t under the relevant null hypothesis in the *observed* data, can no longer be expected to be so distributed. Nor is it known how they *are* distributed, as we have no way of adjusting the Fs and ts to properly compensate for our attenuation adjustments. (Note, in this connection, that AP*V is not least squares for the data in hand as they were observed, but rather for data which imply estimated values for the X_i^*.)

We attempt to cope with this dilemma by giving the significance test results of AP*V (henceforth in quotes, for example, "significant") a limited degree of credence and using them in conjunction with the significance test results of the preliminary APV on the observed data. Specifically, we distinguish between the assessment of homogeneity of regression and the tests of null hypothesis about the results of the AP*V.

Homogeneity of regression.   Recall that the APV assumes as a precondition that for all combinations of values of the X_js of set B (in ACV, for all groups), the regression of Y on the X_is of set A is the same in the population. Sections 9.3 and 9.4 describe the F test of the $A \times B$ interaction set used to assess the tenability of this assumption. The same assumption must be made in AP*V, but the decision process does *not* use an $A^* \times B$ "significance" test (there are, in fact, no $A^* \times B$ IVs in the analysis), but rather the F test of the observed $A \times B$ interaction of the preliminary APV analysis. Thus, if we can conclude that the assumption of regression homogeneity for observed set A is tenable, then that assumption extends automatically to the homogeneity of regression on Y of the

X_i^*s of set A^*.[18]   If, on the other hand, the decision about regression homogeneity in APV is negative, the AP*V, like the APV, is aborted.

AP*V results.   We will refer to the standard significance tests when applied to AP*V results as F^* and t^*, and characterize them as "significant" or "nonsignificant." Although not strictly valid, they can nevertheless be used as rule-of-thumb "ballpark" judgments, which, when used in conjunction with the APV significance test results, provide a basis for tentative significance decisions.

Each coefficient found in an AP*V (e.g., pR_B^*, B_j^*), together with its F^* or t^* ratio has a counterpart uncorrected for the X_i measurement error in the APV (e.g., pR_B, B_j) with its F or t. For any given result, we look for evidence of a substantial change in magnitude from APV to AP*V. For concreteness, consider a B_j, t_j shifting to B_j^*, t_j^*:

1. If t_j is not significant and t_j^* is not "significant," one simply draws the usual conclusion associated with nonrejection of the null hypothesis. Assuming reasonable power (*n* large relative to $k_A + k_B$), B_j and B_j^* are both small and the relationship in the population between Y and X_j, when partialled for either observed or estimated true covariates, is no more than trivial.

2. If t_j is significant and t_j^* is not "significant," one tentatively concludes that the large B_j is a spurious effect of unreliability in set *A,* and that the small B_j^* more accurately describes the partial relationship. When this occurs, one's disappointment as a working researcher should yield to gratitude as a scientist that it is likely that a fallacious positive conclusion due to covariate unreliability has been avoided.

3. If t_j is not significant, while t_j^* is "significant," one has presumably uncovered a partial relationship which lay buried in the unreliability of the covariates as observed. Here, the jubilance of the working researcher should be tempered by the caution of the scientist, and due respect paid to the uncertainty attached to the AP*V method and the "significance" of t^*. One's confidence in the positive conclusion about B_j^* would be bolstered by a more stringent significance criterion (for example, .01 rather than .05). In any case, prudence dictates that the positive conclusion be leavened with a certain amount of diffidence.

4. When t is significant and t^* is "significant," substantial confidence is justified in asserting the positive conclusion about the true-partialled relationship. This assumes that B_j and B_j^* are of the same sign. In the unlikely (but possible, particularly when power is very large for even small effects) event that they are of opposite sign and both significant, the caution advised in the preceding paragraph applies.

---

[18] The rationale for this extension is that in going from the X_i to the X_i^*, the regressions on Y will change similarly, so if they are homogeneous for the X_i they will remain homogeneous for the X_i^*.

9.5.4 Some Perspectives on Covariates in APV and AP*V

In this section, we consider some of the issues involved in the choice of covariates and their correction for unreliability.

Our detailed consideration of APV and AP*V procedures and the ease which they can be mechanically applied to data should not mislead the reader into a feeling of unconcern about either the quantity or quality of the covariates. The flexibility of MRC analysis in handling partial variances may easily tempt an investigator to use a large number of covariates to blanket (perhaps even smother) the research factors he wishes to control. We have in mind, for example, the employment in set A of a dozen or two income, rental, welfare status, occupational, educational, etc. variables to cover the concept of "socio-economic status," or a similar number of questionnaire scale scores to represent "personality."

Early in this book (Section 4.6) we raised a red flag against such practices in general MRC analysis. The "less is more" principle (Section 4.6.2) holds with even greater force in AP*V. First, large numbers of IVs in covariate sets, even when of perfect reliability, serve to depress power and precision of parameter estimation in set B. Second, the larger k_A is made in a conscientious effort to cover an area, the greater the likelihood of redundancy and suppression among the IVs in set A, thus reducing the power and precision of their estimation and increasing difficulty in their interpretation. Then, the more covariates the greater the risk of including one or more of inappropriate causal relationship to Y and/or set B, resulting in variance stealing (see Section 9.7). Now, if these problems are compounded further by those posed by unreliability, the prospects of useful results become dim indeed.

The fact that AP*V affords a method for correcting for the fallibility of covariates should not lull the investigator into complacency about measurement error. The method requires reliability coefficients for the covariates appropriate to the population under study. Although small errors in estimating these may not seriously distort the results, even modest errors are likely to have increasingly serious cumulative effects as k_A increases. Quite apart from this, there remains the ambiguity which attaches to significance testing and power analysis in AP*V.

The investigator should go to great lengths to achieve few, highly reliable covariates and analyze by APV. The means to this goal include the factor-analytic reduction of the candidate measures to factor scores, or to key representative variables of high reliability, and conventional test-construction procedures for increasing reliability (Nunnally, 1967). Patently, ten questionnaire or interview items which share a common factor beyond which their variance is largely due to measurement error are far better added up to a single score than dealt with as ten variables with whose poor (and likely poorly estimated) reliabilities we must deal.

Thus, recourse to AP*V is not done lightly. It is best applied when one must *unavoidably* use covariates of low reliability (say, less than .80), and the fewer of these the better. When avoidable, it should be avoided, and a straight APV used.

Finally, our emphasis on the reliability of covariates should not have the impression that unreliability in Y and the X_js of set B is of little importance. Remember that unreliability in the variables being correlated systematically underestimates the magnitude of all measures of relationship. This in turn reduces the statistical power with which we determine their existence. The AP*V does not attempt to correct for these sources of attenuation. We have noted that the problem of measurement unreliability is chronic in the "softer" part of the spectrum of behavioral science, but this is not to condone it. Many of the subtleties of theoretical formulation are swamped in the measurement error (and invalidity) of the variables we use to operationalize the concepts. There is bound to be research payoff in improving the quality of all the measures we use in research, not only those which serve as covariates.

9.6   THE STUDY OF CHANGE AND APV–AP*V

9.6.1   Introduction

The importance of the study of change in the behavioral sciences as a methodological tool is self-evident. Whether the context is one of manipulative experiment or the nonexperimental observation of events in natural flux, the very heart of the study of causal mechanisms and systems is intimately bound up with the assessment of change and its correlates.

Its importance acknowledged, it must immediately be noted that the study of change is not the simple, straightforward proposition it appears to be. Indeed, this is a methodological area fraught with booby traps, where intuitive "doing what comes naturally" is almost certain to lead one astray.

Not only are the issues here of a complexity not immediately apparent to the naked eye, but they have not yet been settled by the methodologists. A symposium wholly devoted to this topic (Harris, 1963) provides some excellent guidelines, yet the fact that it left much to be resolved is evidenced by the continuing debates in the journal literature.

A detailed discussion of the issues in assessing change is obviously beyond the scope of this presentation, and definitive solutions are beyond our modest purview. All we hope to do is present the major ideas in their simplest form, and show how the methods of this chapter may be applied to yield useful results. Specifically, after demonstrating the inadequacy of simple change scores, we will show how the use of the postscore as the dependent variable and the prescore as a covariate makes of the analysis of change a special case of general APV, with covariate fallibility handled by AP*V.

9.6.2 Change Scores

The elements of the situation we are considering are as follows: We have scores on some variable of interest (v) at two points in time. We designate those at the first point, the prescores, as b (for *before*) and those at the second point, the postscores, as a (for *after*). Intervening between these two points in time are one or more research factors which we intend to relate to change in the variable v. These may be contrived experimental conditions or observed natural variations (or both).

Now what could be more obvious than that the way to proceed is to relate the research factor to

$$(9.6.1) \qquad\qquad a - b = c,$$

the simple change in v, the algebraic difference as one goes from pre to post? This is how physical scientists measure change, and, indeed, how we measure it in everyday life: a weight change (loss) of $120 - 135 = -15$ pounds, a temperature change (increase) of $47° - 41° = +6°$, an IQ change (gain) of $90 - 82 = +8$ points, etc. This operation is virtually dictated by the concept "change" and is so compelling that it is hard to see why "doing what comes naturally" does not serve our purpose.

Very early in this book, we noted some problems with difference scores, of which $c = a - b$ is the special case where a and b are measures of the same variable, v, obtained on different occasions. In Section 2.11.2, in discussing the effect of unreliability on correlation, we pointed out that the reliability of a difference score is likely to be materially lower than that of the variables being differenced, and that low reliability sharply attenuates correlation with other variables. If we let r_{vv} represent the reliability of v for both occasions, and r_{ab} the pre–post correlation within groups,[19] Eq. (2.11.6) specializes to the reliability of simple change scores as

$$(9.6.2) \qquad\qquad r_{cc} = \frac{r_{vv} - r_{ab}}{1 - r_{ab}}.$$

Concretely, if we take $r_{vv} = .81$ for both occasions, and $r_{ab} = .64$, then the reliability of the change score works out to .47. This result is not atypical—since a and b are both measures of v, they will usually be fairly highly correlated, to the inevitable detriment of r_{cc}, unless r_{vv} is close to 1.00. But although low reliability is a frequent characteristic of c, this is not the fundamental problem.

---

[19] Do not confuse the reliability of v with the pre–post correlation. r_{vv} should ideally be an estimate of the parallel-form reliability at a given point in time, and may be a valid internal consistency estimate. r_{ab} is not a reliability coefficient, not even of the dubious test–retest variety, since a contains inter alia the effect of individual difference in change, whose variance is not part of b. (See Nunnally, 1967.)

We came closer to the heart of the problem when, in Chapter 2, we noted that correlations of change scores with prescores and postscores are rather like part-whole correlations, and may be spurious (Section 2.11.4). This point requires further development here, and particularly in regard to r_{bc}, the correlation of the prescore with simple change. Since $c = a - b$, this correlation is between b and $a - b$ and may be conceived as a (weighted) combination of r_{ab}, the correlation between b and a, and $r_{b(-b)}$, the correlation of b with $-b$, which necessarily equals -1. Now, the combination of some positive midrange value of r_{ab} with -1 is likely to produce a negative value,[20] although, as we shall see, the critical point is not that the resulting $r_{bc} = r_{b(a-b)}$ is likely to be negative, but rather that it will almost certainly not equal zero.

It is possible to write the equation of the correlation of the prescore with change as a function of r_{ab} and the two sds as

(9.6.3) $$r_{bc} = r_{b(a-b)} = \frac{r_{ab}sd_a - sd_b}{\sqrt{sd_a^2 + sd_b^2 - 2r_{ab}sd_a sd_b}}.$$

Note that the numerator subtracts from a fraction (r_{ab}) of sd_a all of sd_b. For some typical values: when $r_{ab} = .64$, $sd_a = 12$, and $sd_b = 10$, Eq. (9.6.3) yields $r_{bc} = -.24$. If r_{ab} were smaller (say, .50), r_{bc} would be even further from zero ($-.36$). Note that this formula is an identity, that is, an algebraic necessity which makes no special assumptions. In particular, note that it contains no reference whatever to the reliability of v—variables at all levels of reliability, including 1.00, are subject to it. Thus, the dependence of change on initial level operates irrespective of unreliability, although, all things equal, unreliability will enhance it.[21]

The heart of the problem with simple change scores lies squarely in their necessary dependence on prescores. The details of Eq. (9.6.3) are not of great importance, nor is the fact that it is likely to produce a negative value, but rather that, in general, $r_{bc} \neq 0$. Consider that we begin with b and a scores and seek to derive from them a measure of change per se by "doing what comes naturally."

---

[20] This depends, of course, on the weights, which we shall shortly see are the standard deviations of a and b. As a practical matter, however, it is very unlikely that the weighting will offset the negative bias.

[21] There is a delicious irony here. Since r_{ab} is constrained by measurement error so that its effective upper limit is r_{vv}, it must follow that the lower the reliability, the lower the r_{ab} and hence the *larger* (absolutely) the negative r_{bc}. Thus, an r_{ab} of .20 that is due to low reliability will have, for the same sds, $r_{bc} = -.54$, a value large enough to be highly "significant" for even a relatively small sample. There are, unfortunately, many studies in the behavioral science literature triumphantly reporting and interpreting such correlations as evidence that some procedure is maximally effective for cases with the lowest initial level, unaware that the "effect" found is wholly artifactual. However, note that the high negative r_{bc} is not due to unreliability per se, but rather to the low r_{ab}, which may be due to large *true* individual differences in change, rather than apparent ones due to measurement error. We thus reiterate that the phenomenon described by Eq. (9.6.3) is not inherently one of unreliability.

That is, by subtracting the prescore from the postscore, it is our intention to remove the effect of initial level b, which has nothing to do with change, from the final level a, in order to produce a measure which *only reflects change* in v. However, this intuitively obvious means of accomplishing this goal fails, as Eq. (9.6.3) testifies: r_{bc} is *not* in general equal to zero, thus our c measure contains some variance wholly due to b. The relationship of c to other variables is therefore distorted. It may be enhanced or reduced by their relationship to the b which c contains. The correlation of c with some other variable u is given by

$$(9.6.4) \qquad r_{uc} = r_{u(a-b)} = \frac{r_{ua}sd_a - r_{ub}sd_b}{\sqrt{sd_a^2 + sd_b^2 - 2r_{ab}sd_a sd_b}}.$$

The presence of r_{ub} in this equation makes explicit the contamination of the correlation of u with change in v by its relationship with b, the initial level of v.

Once we have identified the problem as being that c contains unwanted variance due to b, it is readily solved. We have repeatedly shown how to partial one variable from another, for example, how to produce such partial or residual variables as $f \cdot g\ (= f - \hat{f}_g)$ or correlations of such variables with others as $r_{(f \cdot g)u}$, a semipartial r, and $r_{fu \cdot g}$, a partial r. The partialled variable $f \cdot g$ is *guaranteed* to correlate exactly zero with g by dint of its construction. The APV (including its special case, ACV) provides us with the most general approach to handling such problems.

Thus, we could solve our problem by working with $c \cdot b$. It is, however, unnecessary to create c. After all, $c = a - b$, so $(a - b) \cdot b$ quite unnecessarily partials b from $-b$, as well as from a. It is sufficient to simply partial b from a, that is, to work with $a \cdot b$, which correlates perfectly with $c \cdot b$, and, necessarily, correlates zero with b.

The structure of such "regressed" or "partialled" change scores helps clarify the nature of the defect of simple change scores. The variable $a \cdot b$ is literally $a - B_{ab}b - A$, where B_{ab} is the regression coefficient of a on b, that is, the amount of change in a per unit increase in b. Since A is a constant, it may be ignored when we are considering correlation, so we are effectively dealing with $a - B_{ab}b$. Now, $B_{ab} = r_{ab}sd_a/sd_b$, a quantity which will almost certainly be less than one (and almost certainly positive). For example, for such representative values as $r_{ab} = .50$, $sd_a = 12$, and $sd_b = 10$, $B_{ab} = (.50)(12)/(10) = .60$. This says that for each unit increase in b, we expect .6 of a unit increase in a, and that for an optimal, uncorrelated-with-b index of change, we must use $a - .6\ b$. But the simple change score is $c = a - b$, which we can write $c = a - 1.00\ b$. The trouble with using the simple change score is that it presumes that the regression of a on b has a slope of 1.00 instead of the actual B_{ab}, that is, that there is a unit change in a associated with each unit change in b, instead of what in the example was only a .6-unit change in a. Regression coefficients of postscore on prescore of 1.00 almost never occur in the behavioral sciences; as noted, they are almost always less than one. For B_{ab} to equal 1.00 requires that $r_{ab} = 1.00$ when the sds are equal, and more generally that $r_{ab} = sd_b/sd_a$, a most unlikely occurrence.

Thus the effect of using c scores is typically one of *over*correction of the postscore by the prescore.[22]

This analysis incidentally illuminates why simple change scores work quite well in the physical sciences. While in the behavioral sciences, individual differences in true change and measurement error both operate to reduce B_{ab}, neither of these factors operates significantly in the physical sciences, where B_{ab} *is* generally, to a close approximation, equal to 1.00, so that simple change scores are also automatically regressed change scores. This analysis also alerts us to the possibility of encountering in a behavioral science application the circumstance that B_{ab} approaches 1.00, in which case simple change scores will approximate regressed change scores, and may be used in their place. But only in such circumstances is the intuitive $c = a - b$ appropriate.

9.6.3 The APV in the Study of Change

It is of course not necessary to literally create regressed change scores in order to study their relationship with other variables. We have seen that APV (including its special case ACV) is a general method for studying partialled (regression-adjusted, residualized) variables, in which a covariate set A is regressed from the dependent variable Y, and related to a set B of research factors of interest (from which set A is also partialled). Its application to the study of change in v is thus very simple: let the postscore a of v be Y, and the covariate set A contain a single variable, the prescore b of v; the research factor(s) we want to relate to change then comprise set B. The APV then proceeds exactly as described in Sections 9.2 through 9.4. Thus, the method proposed for studying relationship to change is that special case of APV wherein $k_A = 1$, and the single covariate is a prescore for the same variable v on which the dependent variable Y is the postscore.

A Simple Case

To see how this works most clearly, we will take the simplest possible case: where set B represents membership in one of two groups, so that it is represented by a single variable which we will dummy code $E = 1$, $C = 0$. We find that the interaction-bearing product IV (X_3), when tested in the usual way, can be discarded. Thus, the assumption of parallel regression lines for the two groups necessary for a valid APV is accepted, and the APV proper can proceed. We have in all three variables: Y (post-v), X_1 (the pre-v covariate), and X_2 (the E versus C group-membership dichotomy). We assume a total n of 60 cases, divided unequally (for the sake of generality) with $n_E = 36$, $n_C = 24$. Since we are proceeding with an APV, we are obviously assuming that the reliability of the

---

[22] Another way to put this is that if both a and b scores are standardized, then the measure of regressed change is $(z_a \cdot z_b = z_a - \hat{z}_a =) z_a - r_{ab}z_b$, while the simple change score is $z_a - z_b$. The latter subtracts too much, all of z_b from z_a, instead of the fraction r_{ab} which is just the right amount to render the result uncorrelated with z_b.

pre-v covariate is quite high. (We will make the contrary assumption in the next section and see its consequences.)

Table 9.6.1 presents the details of a complete analysis. Much of the information in the table is redundant, but is presented in the interests of thorough description; for example, the separate group means and sample sizes are implicit in the means, sds, and simple correlations of the three variables. We also, for the sake of continuity, follow the format of the general hierarchical setwise MRC analysis for APV–ACV, although each of our sets contains only one variable.

TABLE 9.6.1
APV (ACV) Analysis of Regressed Change for E and C Groups

Basic data:

		m	sd			m_E	m_C
Post-v	Y	60	12	$r_{Y1} = .52$	Y	64.70	52.95
Pre-v	X_1	50	10	$r_{Y2} = .48$	X_1	53.35	44.98
E vs. C	X_2	.6	.49	$r_{12} = .41$	$n_E = 36, n_C = 24; n = 60$	

Hierarchical MRC:

(Sets)	IVs	cum. R^2	I	F_I	df	
(A)	X_1	.2704	.2704	21.496**	1, 58	
(B)	$+ X_2$	.3560	$.0856 = sr_2^2$	7.573**	1, 57	$(pR_B^2 =) pr_2^2 = .1173$
$A \times B)$	$+ X_3 = X_1 X_2$	.3692	.0132	1.172	1, 56	

APV regression equation:

$$\hat{Y} = .466 X_1 + 7.86 X_2 + 31.98$$
$$t \quad 3.334** \quad 2.752** \qquad (df = 57)$$

Adjusted Y intercept (9.2.2):

$$A' = 31.98 + .466(50) = 55.28.$$

Adjusted Y means (9.2.1):

$$\bar{Y}'_h = 7.86 X_2 + 55.28,$$
$$\bar{Y}'_E = 7.86(1) + 55.28 = 63.14,$$
$$\bar{Y}'_C = 7.86(0) + 55.28 = 55.28 \quad (= A').$$

Adjusted (regressed) change means (9.6.7):

$$\bar{c}'_E = 63.14 - 50 = 13.14,$$
$$\bar{c}'_C = 55.28 - 50 = 5.28,$$
$$\bar{c}'_E - \bar{c}'_C = 7.86 = B_2.$$

**$P < .01$.

The large number of values in Table 9.6.1 should not obscure the fact that the central result is given by pr_2, B_2, and their common t (or F) value.

However, some of the detail is worth noting. The basic data show not only a large correlation between the group dichotomy and post-v (r_{Y2} = .48), but also a sizable correlation between the group dichotomy and pre-v (r_{12} = .41). These are equivalently reflected by the differences in the two pairs of means. The difference in initial status indicates that the groups (almost certainly) did not arise by random assignment; clearly, the two groups preexisted the experiment (see Section 9.7).

In hierarchical MRC, we find that .2704 ($= r_{Y1}^2$) of the post-v variance is linearly accounted for by pre-v. Once this is removed, the remaining .7296 is variance in regressed change ($Y \cdot A = Y \cdot X_1 = a \cdot b$), that is, the variance of the residuals, which necessarily correlate zero with pre-v. When the group dichotomy is entered, it accounts for an additional .0856 of the *total* post-v variance, but this constitutes .0856/.7296 = .1173 of the variance in regressed change. The former value is for sr_2^2 (literally $r_{Y(2 \cdot 1)}^2$), while the latter is for pr_2^2 (literally $r_{(Y \cdot 1)(2 \cdot 1)}^2 = r_{Y2 \cdot 1}^2$). They are highly significant, sharing the same $F = 7.573$** or $t = \sqrt{F} = 2.752$**. We can thus assert that the E–C group distinction, when adjusted for their pre-v mean difference (that is, $X_2 \cdot X_1$), accounts significantly for variance in regressed change ($Y \cdot X_1$).

The regression equation in Table 9.6.1 provides further information. It gives as B_1 = .466 the common (pooled, within-group) slope of the two regression lines. More important, it indicates that difference in adjusted Y (post-v) means is 7.86 ($= B_2$), with E the larger. We have already seen that it is highly significant (since, as always, its significance is the same as that of pr_2 and sr_2). To find the adjusted Y means themselves, we specialize Eq. (9.2.2) to find the adjusted Y intercept

(9.6.5) $A' = A + B_1 m_1,$

and substitute it in a specialized version of Eqs. (9.2.1) or (9.4.2),

(9.6.6) $\overline{Y}_h' = B_2 m_2 + A'.$

As Table 9.6.1 illustrates, this equation produces the adjusted post-v means of $\overline{Y}_E'$ = 63.14, and $\overline{Y}_C'$ = 55.28. (Note that $\overline{Y}_E' - \overline{Y}_C'$ = 7.86 = B_2.) These are post-v means adjusted for the groups' pre-v status; if one wishes the similarly adjusted (regressed) *change* means, they can be found by subtracting the grand pre-v mean from each of the adjusted post-v means, simply

(9.6.7) $c_h' = \overline{Y}_h' - m_1.$

Table 9.6.1 shows these to be 13.14 for E and 5.28 for C. Their difference is necessarily B_2 = 7.86, so that B_2 is not only the difference between adjusted post-v means, but also the difference in means of adjusted (regressed) change.

The solid lines in Figure 9.6.1 in the next section are the regression lines for the E and C groups. (The dashed lines are for later discussion.) Their equations are found by substituting the coded values for X_2 (1, 0) in the regression equation for X_1 and X_2 in Table 9.6.1. The slopes are the same (.466), as presumed in ACV, and the parellel lines are separated by B_2 = 7.86 units on the Y scale. Thus, for any given prescore, the expectation is that an E subject will be 7.86 units higher on post-v than a C subject. This is yet another interpretation of the meaning of B_2.

Summarizing, we have found that .1173 (= pr_2^2) of the variance in regressed change is associated with E versus C group membership adjusted for pre-v mean difference, and that the group difference in regressed change means (or adjusted post-v means) is 7.86 (= B_2), a highly significant result (t = 2.752, df = 57, $P <$ .01).

It is of interest to see what the outcome would be if we were to use raw change scores $c = a - b = Y - X_1$. These could be created for each subject, and a t test performed on the difference in mean change. A more informative and less cumbersome approach would be to determine the correlation between c and the E versus C group membership dichotomy X_2. This does not require literally creating the c values for each subject, since we can determine the correlation directly by substituting in Eq. (9.6.4), letting $u = X_2$:

$$r_{c2} = \frac{.48(12) - .41(10)}{\sqrt{12^2 + 10^2 - 2(.52)(12)(10)}} = .152,$$

and r_{c2}^2 = .0231. The standard t test is nonsignificant (t = 1.712, df = 58). Thus, while we found that a robust and highly significant 11.73% of regressed change variance was accounted for by the (adjusted) dichotomy, we find here a puny and nonsignificant proportion of raw change variance accounted for by the (unadjusted) dichotomy. Two things account for this marked difference. The first is the already emphasized overcorrection of raw change scores (i.e., $c = a -$ 1.00 b). The second is the fact that the raw change approach uses the dichotomy without partialling out its correlation with b, that is, uncorrected for the pre-v mean difference. The regression approach relates $Y \cdot X_1$ to $X_2 \cdot X_1$, whereas the correlation with raw change relates $Y - X_1$ to X_2; the latter is a flawed (post is overcorrected) and *semi*partial approach, while the former is an optimal and fully partial approach.

The second difference, that between X_2 and $X_2 \cdot X_1$, needs some scrutiny. The APV (and conventional ACV) approach, as we have seen, relates $Y \cdot X_1$ to $X_2 \cdot X_1$. Explicity, the formula for pr_2 is

(9.6.8) $$r_{Y2\cdot 1} = \frac{r_{Y2} - r_{Y1}r_{12}}{\sqrt{1 - r_{Y1}^2}\sqrt{1 - r_{12}^2}}.$$

What if, instead, we related $Y \cdot X_1$ to an unpartialled X_2, as the approach using c does? The key correlation would then be the semipartial

(9.6.9) $$r_{(Y\cdot 1)2} = \frac{r_{Y2} - r_{Y1}r_{12}}{\sqrt{1 - r_{Y1}^2}}.$$

Applied to the data of Table 9.6.1, this turns out to equal .312, and $r^2_{(Y \cdot 1)2} =$ .0976. Note that this is smaller than the partial $r^2_{Y2 \cdot 1} = .1173$, but that is hardly news—semipartials are never larger than partials based on the same variables. But note also that it is distinctly larger than the now fully analogous $r^2_{c2} = .0231$.

It is also worth noting that this semipartial is not the same as the $sr^2_2 = I_2$ (= .0856) in Table 9.6.1 (that is, the squared semipartials produced automatically in hierarchical MRC). The latter partials X_1 from X_2 and relates it to the unpartialled Y, that is,

$$(9.6.10) \qquad r_{Y(2 \cdot 1)} = \frac{r_{Y2} - r_{Y1}r_{12}}{\sqrt{1 - r^2_{12}}}.$$

This equals .293, and $r^2_{Y(2 \cdot 1)} = .0856$. This does not relate to regressed *change* $(Y \cdot X_1)$ at all, but to post-v (Y). Nevertheless, the numerators of Eqs. (9.6.10) and (9.6.9) are identical. These two formulas differ, when squared to present proportions, in the variance bases in regard to which they are taken: $r^2_{Y(2 \cdot 1)}$ gives the proportion of total post-v variance accounted for by group membership adjusted for difference in pre-v, whereas $r^2_{(Y \cdot 1)2}$ gives the proportion of regressed change variance $(Y \cdot X_1)$ accounted for by (unadjusted) group membership. The former is not likely to be of any interest to us, since it does not involve change. The latter may be, since it *is* about change, but it does not remove from the group membership variable that part of its variance associated with pre-v. However, since the two equations have the same numerator, which is also the numerator for the partial of Eq. (9.6.8), they must all yield identical significance test results. However these three proportions differ descriptively, then, they must give the same answer in regard to the status of the null hypothesis. This, in turn, implies that the null hypothesis must be logically the same. But this is not surprising, since zero divided by three (or more) different bases yields an identical zero.

But this still poses the problem of the descriptive choice between relating $Y \cdot X_1$ to X_2 as in Eq. (9.6.9), or to $X_2 \cdot X_1$ as in the pr_2 of Eq. (9.6.8). The ACV (and APV generally) automatically does the latter, while some investigators also working with regressed change scores do the former (see Tucker, Damarin, & Messick, 1966). Here, however, we can easily show one facet of the problem. We acknowledged in the example of Table 9.6.1 that the presence of a fairly large (.41) correlation between groups and pre-v indicates that the data could not have arisen from a true experiment, that is, one which features randomization. When cases are assigned randomly to groups, the expected correlation of pre-v (or anything else) to group membership is zero. Substituting $r_{12} = 0$ in the formulas for the semipartial (Eq. 9.6.9) and the partial (Eq. 9.6.8) yields the same result:

$$(9.6.11) \qquad r_{(Y \cdot 1)2} = r_{Y2 \cdot 1} = \frac{r_{Y2}}{\sqrt{1 - r^2_{Y1}}},$$

that is, when pre-v is unrelated to group membership, the result of adjusting group membership for it is nil. The use of randomization as a principle of experimentation obviates some of the ambiguities about causality in **APV** by assuring no relationship (in the population) between the covariate (pre-v) and group membership (see Section 9.7).

Multiple Groups

The generalization of **ACV** in studying change from two groups (and hence one IV in set B) to g groups (and hence $g - 1$ IVs in set B) is quite obvious. Recall that any nominal scale coding method may be used. The further generalization to **APV**, where set B does not necessarily represent group membership is also obvious. With pre-v as a covariate, one can transform the regression equation in Y (post-v) to one in Y' (adjusted post-v or regressed change) using the standard APV–ACV Eqs. (9.2.2) and (9.2.1) or (9.4.2). Whatever the nature of set B, we end up automatically referring to the appropriately adjusted post-v values. The B_js and pr_js of set B are functions of adjusted Y or regressed change; what functions they are depends only on the nature of the X_j, as was liberally illustrated in Sections 9.2 through 9.4.

Nonlinear Change

We need not restrict ourselves to linear change, as do the standard methods. If there is reason to believe that the (common within-group) regression of post-v on pre-v is not linear, we have seen that various methods are available for representing the pre-v information so as to allow for this nonlinearity. One simply uses one of these methods in the representation of pre-v in set A and proceeds. For example, if we revise the example of Section 9.3.4 so that Y is a posttreatment score, and the covariate set is X_1 = prescore and X_2 = squared prescore, we would be allowing for a quadratic nonlinear relationship between initial and final score in our adjustment.

Multiple Time Points; Other Covariates

We can easily extend the analysis beyond that of the two measures of v, pre-v and post-v. With measures of v at more than two points in time, post-v as Y can be adjusted for all or any of the prior measures of v serving as a set of covariates. This transcends the idea of regressed change, which implies a single baseline measure. Here, post-v is adjusted using whatever additional information is supplied by measures of v other than the immediately prior measure.

Similarly, one may adjust Y (= post-v) for covariates additional to pre-v measures. Obviously, a set of multiple covariates may contain, in addition to one (or more) pre-v measures, other non-v covariates of any kind deemed relevant. The analysis may then be interpreted as one of regressed change further adjusted by the additional covariates in set A.

In all such extensions (and others which may occur to the reader), the standard procedures of Sections 9.2 through 9.4 apply. But so do our frequent strictures about the cost in statistical power and bloated experimentwise Type I error rates of increasing the number of IVs in an MRC analysis "just in case" they may contribute. This caveat is particularly appropriate here. Although possible, it is generally not likely that covariates added to pre-v (or the immediately preceding v when there are more than one) will materially or significantly increase the variance in Y accounted for beyond what pre-v supplies, or materially effect the adjustment of set B. The reason for this is that whatever relationship to post-v these other candidates may have is likely to already be represented (and to an even greater degree) in their relationship to pre-v; hence their inclusion provides little or no new information. We do not offer this point to provide the basis for a rule of procedure to exclude other covariates in the presence of a prescore, but merely as an expectation to help guide the investigator.

We end this section by reiterating, for the sake of emphasis, the point made at its beginning: the analysis of change may proceed as a simple special case of APV, where Y is the postscore and the prescore serves as the (or in some instances, a) covariate.

9.6.4 The AP*V in the Study of Change

The simple example used in the preceding section can serve us again here, but now we assume that the reliability of our pre-v covariate falls substantially short of 1.00, and that therefore we must make allowance for this unreliability by using AP*V instead of APV, that is, by correcting correlations involving the covariate and its standard deviation by the methods of Section 9.5. We use the same basic data as in Table 9.6.1, to which we add the estimated covariate reliability of r_{11} = .60. By using a simple example, the consequences of r_{11} = .60 rather than the implicit r_{11} = 1.00 of APV can be made evident, and the use of the same data makes possible direct comparison and further clarification.

Before launching into the example, we call the reader's attention to the fact that although the reliability of post-v (Y) is also presumably low, we do not correct for it in AP*V. We showed in Section 9.5.2 that it is the reliability of partialled variables (hence covariates) which is critical, so it is only the unreliability of the pre-v that we will be correcting for in the application of AP*V.

We proceed then to an AP*V of the same data, with r_{11} = .60. As described in Section 9.5.3, as a prelude, an APV is required, both for the assessment of the homogeneity of regression on Y of the covariate(s), and for its unambiguous significance tests. The APV has already been done (Table 9.6.1); it yielded a near-chance F ratio for the interaction (X_3), so we take the precondition of regression homogeneity as satisfied for the AP*V.

The AP*V proceeds by correcting for unreliability the correlations involving covariates, and their standard deviations (see Table 9.6.2). In this simple example that has only one covariate, there are no correlations between covariates, but only the correlations with this single covariate of the other two variables (Y and X_2) to correct. This involves simply dividing each by $\sqrt{r_{11}}$ = $\sqrt{.60}$ = .7746, that is, applying Eq. (9.5.1). Thus, r_{Y1*} = .52/.7746 = .671, and r_{1*2} = .41/.7746 = .529. These are simple corrections for attenuation due to the covariate's unreliability and produce estimates of what the correlations would be if we had X_1^*, that is, the pre-v measured without random measurement error. Similarly, using Eq. (9.5.7), we find sd_{1*}, the estimated standard deviation of X_1^*, as 10(.7746) = 7.75. Only these values involving X_1^*, appearing as part of the basic data in Table 9.6.2, are changed from those of Table 9.6.1. All means remain unchanged as do the other sds and r_{Y2}.

Again we report the results quite fully in Table 9.6.2, and in the format of setwise hierarchical MRC, despite the modesty of this application. The amount

TABLE 9.6.2
AP*V (AC*V) Analysis of Regressed Change for E and C Groups

Basic data, as revised for X_1^* ($r_{11} = .60$):

		m	sd			m_E	m_C
Post-v	Y	60	12	$r_{Y1*} = .671$	Y	64.70	52.95
Pre-v*	X_1^*	50	7.75	$r_{Y2} = .48$	X_1	53.35	44.98
E vs. C	X_2	.6	.49	$r_{1*2} = .529$	$n_E = 36, n_C = 24; n = 60$	

Hierarchical MRC:

(Sets)	IVs	cum. R^2	I	F_I^*	df	
(A)	X_1^*	.4507	.4507	47.582	1, 58	
(B)	+ X_2	.4723	.0216	2.332	1, 57	$(pR_B^2 =) pr_2^2 = .0393$

AP*V regression equation:

$$\hat{Y}^* = .898\, X_1^* + 4.24\, X_2 + 12.55$$
$$t^* \quad\;\; 6.898 \quad\;\; 1.527 \qquad\qquad (df = 57)$$

Adjusted* Y intercept (9.2.2):

$$A'^* = 12.55 + .898(50) = 57.45$$

Adjusted* Y means (9.2.1):

$$\bar{Y}_h'^* = 4.24\, X_2 + 57.45$$
$$\bar{Y}_E'^* = 4.24(1) + 57.45 = 61.69$$
$$\bar{Y}_C'^* = 4.24(0) + 57.45 = 57.45\ (= A'^*)$$

Adjusted* (regressed*) change means (9.6.7):

$$c_E'^* = 61.69 - 50 = 11.19$$
$$c_C'^* = 57.45 - 50 = \;\; 7.45$$
$$c_E'^* - c_C'^* = \;\; 4.24 = B_2$$

*Estimated true covariate.

of detail must not be allowed to obscure the fact that the results are pretty well summed up by pr_2^2 and its t^*, to which we will turn in a moment.

Note first that attenuation correction of r_{Y1}^2 (= .2704) to r_{Y1*}^2 ($= r_{Y1}^2/r_{11}$ = .4507) results in a substantially larger proportion of post-v variance accounted for by estimated true pre-v, necessarily the case when r_{11} is small. It is also necessarily more "significant" ($F^* = 47.582$), the quotes being a reminder of the dubious validity of standard significance tests in AP*V. When the X_2 dichotomy of group membership is entered, it increases R^2 by only .0216 (= I_2 = sr_2^2), which yields an F^* of 2.332, not "significant" (Appendix Table D.2: at α = .05, F = 4.00 for df = 1, 60). The pr_2^2 = .0216/(1−.4507) = .0393, and B_2, the difference between estimated true pre-v-adjusted post-v means, is 4.24. These are, of course, exactly as "nonsignificant" as I_2 (t = 1.527 = $\sqrt{2.332}$). These three partial coefficients are literally $r_{Y(2 \cdot 1*)}$, $r_{Y2 \cdot 1*}$, and $B_{Y2 \cdot 1*}$, and are different from their APV counterparts $r_{Y(2 \cdot 1)}$, $r_{Y2 \cdot 1}$, and $B_{Y2 \cdot 1}$. The latter were significant (t = 2.752**); these are not. Following the rules for interpreting the "significance" tests of AP*V in Section 9.5.3, we conclude in this instance that when the fallibility of the covariate is taken into account, the adjusted mean difference shrinks (from 7.86 to 4.24), as does the pr_2^2 (from .1173 to .0393), and becomes "nonsignificant." We thus can offer no evidence for an experimental effect.

How this occurred may be clarified by reference to Figure 9.6.1. As noted above, the parallel solid lines are the regression lines for the E and C samples in the APV analysis, whose equations are generated from the APV regression equation of Table 9.6.1 by successively substituting 1 and 0 for X_2. The dashed lines are the counterpart lines of the AP*V analysis, produced in exactly the same way from the AP*V regression equation in Table 9.6.2. Recall the regression lines always go through the point defined by the means of their samples, as they are shown here. Now, it will always be true that the common within sample slope for the AP*V lines ($B_{Y1* \cdot 2}$) will be steeper than that ($B_{Y1 \cdot 2}$) for the APV lines, since it can be shown that

(9.6.12)
$$B_{Y1* \cdot 2} = B_{Y1 \cdot 2} \left(\frac{1 - r_{12}^2}{r_{11} - r_{12}^2} \right),$$

from which one can see that the term in parenthesis must be greater than one when r_{11} is less than one, and that it increases as r_{11} decreases. In the present example, with r_{11} = .60, B_{Y1*2} = .898, which is almost twice as large as $B_{Y1 \cdot 2}$ = .466. The increased slope of the lines together with the necessity of their passing through the means results, *in this example*, in the separation of the lines and hence the adjusted mean difference $B_{Y2 \cdot 1*}$ = 4.24 being much smaller than $B_{Y2 \cdot 1}$ = 7.86.

We stress *in this example*, because it finally depends upon where the mean points are. Holding the Y means constant, if the X_1 means were equal, there would be *no* change in separation with the steeper regression slope, that is, $B_{Y2 \cdot 1*}$ = $B_{Y2 \cdot 1}$. If, on the other hand (again holding the Y means constant),

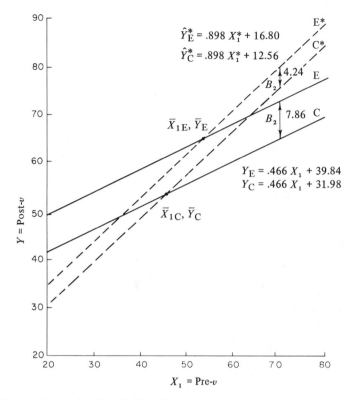

$$\hat{Y}_E^* = .898 \, X_1^* + 16.80$$
$$\hat{Y}_C^* = .898 \, X_1^* + 12.56$$

$$Y_E = .466 \, X_1 + 39.84$$
$$Y_C = .466 \, X_1 + 31.98$$

FIGURE 9.6.1 Regression lines for E and C groups in APV (solid lines) and AP*V (dashed lines).

the C mean for pre-v were greater than that for E, the steeper slope would result in an *increase* in separation, that is, in $B_{Y2 \cdot 1*}$ over $B_{Y2 \cdot 1}$. It is even possible, with a large enough difference between the pre-v means of E and C favoring the former, for $B_{Y2 \cdot 1*}$ to be negative, while $B_{Y2 \cdot 1}$ is positive. But we have already seen all these possibilities illustrated for $r_{Y2 \cdot 1*}$ compared to $r_{Y2 \cdot 1}$ in Table 9.5.1 and its accompanying discussion, in which we argued that the demonstration there held for all true-partialled relative to observed-partialled correlation and regression coefficients, including $B_{Y2 \cdot 1*}$.

The results of the analysis with observed-partialled pre-v in the preceding section and true-partialled pre-v in the present section may be taken as a morality tale. If we take the data of Tables 9.6.1 and 9.6.2 to represent the same study, the investigator in the previous section performed an ordinary APV (more specifically ACV), thus implicitly treating his pre-v measure as if its reliability were perfect. The resulting conclusion of a significant difference in adjusted post-v means (hence, adjusted change) is seen to be in error, when the $r_{11} = .60$ is taken into account in the AP*V (AC*V) analysis. This type of error is by long tradition the most obnoxious to the scientist, since it leads to a false positive claim.[23] In this connection, it may be of interest to answer the following question: Given that an APV has been performed which results in some nonzero observed-adjusted Y mean difference $(B_{Y2 \cdot 1})$, what value of r_{11} must be posited which would result in an AP*V with a *zero* true-adjusted Y mean difference $(B_{Y2 \cdot 1*})$? This implies that the two regression lines for the latter merge into one line which would then pass through the two mean points. (Reference to Figure 9.6.1 will facilitate visualization.) When the formula for the slope of this line is set equal to zero, and solved for r_{11}, one finds

$$(9.6.13) \qquad\qquad r_{11} = \frac{r_{Y1} r_{12}}{r_{Y2}}.$$

This formula may yield a psychometrically impossible result. Since the psychometric theory by which we correct for attenuation holds that r_{11} can not be exceeded by r_{Y1}^2 or r_{12}^2, a solution to this effect calls for an impossibly low r_{11}. Negative and zero values automatically fall in this class. Equally impossible, of course, are solutions for r_{11} which exceed 1.00. Finally, when $r_{Y2} = 0$, Eq. (9.6.13) is indeterminate and for our purposes (assuming $r_{12} \neq 0$) also an impossible result for r_{11}. In any of these instances, we conclude that there is no possible value of r_{11} which can render $B_{Y2 \cdot 1*} = 0$. (This is not necessarily to say that no value of r_{11} can render it so small as to be nonsignificant.) If we have found $B_{Y2 \cdot 1}$ in an APV to be significant and have no estimate of r_{11}, an impossible result from applying Eq. (9.6.13) is reassuring.

Now consider possible results. For example, in the running illustration of Table 9.6.1, we find using Eq. (9.6.13) $r_{11} = (.52)(.41)/.48 = .44$. If we had no estimate of the actual reliability of X_1, we would understand that if it were as low as .44, our true score adjusted Y sample means would not differ. It turned out that for the $r_{11} = .60$ which we posited for the AP*V, $B_{Y2 \cdot 1*} = 4.24$, nonzero but "nonsignificant." Conversely, note that if the actual r_{11} is *smaller* than a possible value derived from Eq. (9.6.13), then $B_{Y2 \cdot 1*}$ is of *opposite* sign from $B_{Y2 \cdot 1}$, that is, the true-adjusted means differ in the opposite direction (although not necessarily significantly so). Again, this last case was already illustrated for the partial $r_{Y2 \cdot 1*}$ (Table 9.5.1, Example 3).

Although the specific details of the preceding hold only for the simple two-group case, the principles involved extend to the g-group case. Imagine g groups to be represented in Figure 9.6.1. In an ACV, we would have a set of g parallel regression lines of slope $B_{Y1 \cdot 2}$ each passing through the point defined by the group's means on X_1 and $\bar{Y}$. The vertical distance between any pair of lines would then be the observed X_1-adjusted difference in Y, and the relative amount of vertical spread of the set of g lines would be carried by the pR_B^2 (and sR_B^2). Correcting for the unreliability of X_1 in an AC*V results in a new set of g (dashed) lines, of common slope $B_{Y2 \cdot 1*}$, *necessarily steeper* than the ACV

---

[23] It is thus analogous to a Type I error in statistical inference which is conventionally guarded against by setting α at some low value like .01 or .05.

lines, and again passing through the mean points. The amount of vertical spread of the dashed lines may be greater or smaller than before, and any pair of these lines may be further separated, closer together, or even reversed, depending on the configuration of the groups' mean points.

Again we note that although the illustration was in terms of an ACV–AC*V comparison with set B carrying group-membership information, the same principles apply generally for APV–AP*V in which set B may carry quantitative information: the $B_{Y1*.2}$ of our example can no longer be interpreted as a within-group slope, but its analog $B_{Y1*.j}$, the regression of Y on X_1^* partialling X_j variables in set B, will be absolutely larger than the APV's $B_{Y1.j}$, that is, in any region of the space defined by the X_j values of set B, the regression line of AP*V will be steeper than that of APV. As before, however, this increase in steepness may result in enhancing, decreasing, or reversing the apparent effect of set B.

We end this section with a bit of irony. We have argued against the use of raw change scores, and also against the use, without adjustment, of only modestly reliable covariates. Recall that the first results in overcorrection and the second in undercorrection. Thus, distortion occurs when one of these errors is made. But when they are made simultaneously, as they are by so many methodologically unsophisticated behavioral and social scientists, they tend to cancel out! The distortion is thus minimized (but not, of course, the attenuating effect of the unreliability of c scores).

In summary, the AP*V (AC*V) provides a method for studying change which makes allowance for the fallibility of the prescore measure. We have seen that failure to do so may produce invalid results in terms of the size and even direction of effects as well as their significance. But recall also the caveats offered in connection with AP*V in Section 9.5.3.

9.7 SOME PERSPECTIVES ON THE ROLE AND LIMITATIONS OF PARTIALLING IN ANALYSIS OF PARTIAL VARIANCE AND ANALYSIS OF COVARIANCE

9.7.1 Introduction

We now come, perhaps belatedly, to a consideration of the nature and implications of the operation we have been invoking throughout this book and most particularly in this chapter—partialling. In what follows, we assume that the precondition of regression homogeneity obtains. Also, although we do not invoke AP*V/AC*V explicity, the ideas are of sufficient generality to extend to partialling estimated true scores.

The symbol $Y \cdot A$ represents the endpoint of a process which, if literally performed, would involve having a variable Y estimated by regression on one or more covariate variables constituting a set A, yielding $\hat{Y}_A$; this estimate is then subtracted from the original Y, thus producing a new variable $Y - \hat{Y}_A = Y \cdot A$

(Section 4.3). We have seen that the literal production of $Y \cdot A$ values by the above operation is generally unnecessary, yet it is important to understand that whether we create a literal $Y \cdot A$ value for each case or not, it is that variable which is invoked by partial and semipartial correlation and regression coefficients. This process is also variously called "residualization" of Y by A ($Y - \hat{Y}_A$ is what is left in Y when $\hat{Y}_A$ is subtracted), "adjusting" Y for A (in the context of ACV), and Y "with A held constant statistically." The latter usage refers to an important property of $Y \cdot A$: it contains no variance associated with set A, or, to put it exactly, $R^2_{(Y \cdot A) \cdot A} = 0$, that is, the MRC of $Y \cdot A$ on set A yields multiple correlation and regression coefficients of zero.[24] It is exactly this property of partialled variables which is the raison d'être of simple partial and semipartial coefficients and of APV/ACV.

A more general form of partialling is implied by the notation $B \cdot A$. As was pointed out in Section 4.3.1 (and particularly in the discussion of the ballantine of Figure 4.3.1), this literally means $\hat{Y}_B \cdot \hat{Y}_A$ (= $\hat{Y}_B - \hat{Y}_A$); hence, $\hat{Y}_B$ is residualized by $\hat{Y}_A$. Thus, $B \cdot A$ is a variable whose relationship to Y is attributable to what is unique to set B, that is, independent of set A. Specifically, $R^2_{(B \cdot A) \cdot A}$ (or, literally $r^2_{(\hat{Y}_B \cdot \hat{Y}_A) \hat{Y}_A}$) = 0—no part of the $B \cdot A$ variance in Y can be accounted for by set A. This is the mechanism whereby, in ACV, the effect of group membership (set B) is adjusted or equated for group differences in the covariates (set A). The same mechanism operates in the more general APV: whatever the nature of set B, its effects on the dependent variable is adjusted for its relationship to set A.

Now, it must be borne in mind that the variables created by this process, $Y \cdot A$ and $B \cdot A$, are statistically manufactured residual variables designed to correlate zero with set A, and in only that literal sense can we talk of "holding set A constant." We shall see that difficulties may arise when causal interpretations are assigned to the results of this process, as when results of APV are expressed in the subjunctive, for example, "*were* the groups equal in set A, *then* they would differ in $Y \cdot A$." Such interpretations *may* be valid, but if so, their validity stems from considerations extraneous to the partialling process itself.

In the interest of convenience and simplicity, we will frequently couch the discussion which follows in ACV terms, that is, with set B as defining group membership or "treatments." The principle will hold for the more general case when set B is quantitative or made up of both nominal and quantitative scales.

Although produced by the same statistical procedure of partialling, the two entities we relate, $Y \cdot A$ and $B \cdot A$, need to be distinguished conceptually:

1. $Y \cdot A$: It is $Y \cdot A$ which is by definition the dependent variable, not Y, as we have argued (Section 9.3.1). Since $Y \cdot A = Y - \hat{Y}_A$, it follows that $Y = Y \cdot A + \hat{Y}_A$. Were we to ignore set A and use Y as our dependent variable, its variance would be composed of both the relevant $Y \cdot A$ variance and the irrelevant $\hat{Y}_A$

---

[24] Indeed, if E is a subset of A, it will also be true that $R^2_{(Y \cdot A) \cdot E} = 0$.

variance. We would then not know whether, how, and to what extent the relationship of set B to Y was referable to the irrelevant $\hat{Y}_A$ component. Moreover, significance tests would use in their error terms $1 - R^2_{Y \cdot B}$, which is larger (and often much larger) than $1 - R^2_{Y \cdot AB}$, thus reducing the power of the test (Section 4.5).

The content of set A is defined by the substantive nature of the research, as indeed, is Y. We make the assumption that the research factor(s) constituting set A will bear causally on Y, and not vice versa. This is assured, for example, when set A describes the subjects at a point in time prior to that at which they are described by Y.

2. $B \cdot A$: The independent variables are the contents of $B \cdot A$, set B "adjusted for" set A, and the same argument, which relies finally on substantive issues, is made for $B \cdot A$ in preference to B when ACV is performed. The creation of $B \cdot A$ is the step in which the groups' inequalities in covariate means are "taken into account" in appraising their differences in the dependent variable, that is, the groups are "statistically equated" for set A. Much "after the fact" use of ACV is motivated by the desire to wipe out these "extraneous effects" of the covariates. As with $Y \cdot A$, we are assuming that, if they are related, the direction of causality is from A to B, not vice versa.

Serious problems may arise in interpretation in regard to this function of ACV, due to ambiguity about the direction of causal flow between covariates and group membership. Such ambiguity may arise in nonexperimental or quasi-experimental research where preexisting groups make up set B. We will consider this case in the next section (Section 9.7.2), and the case where there is no relationship (and hence no causal relationship) between sets B and A in the one that follows (Section 9.7.3).

9.7.2 Preexisting Groups in Analysis of Covariance

Much research in the behavioral and social sciences involves contrasting groups whose existence or definition precedes the initiation of the research. The purpose of the investigation may be the study of the "effects" of the defining (sometimes "organismic") characteristics which differentiate the groups, for example, psychiatric diagnostic categories, men versus women, college youth versus noncollege youth, or rural areas versus small towns versus cities. Or, experimental treatments may be applied to existing groups such as different fourth grade classes in a school, or groups of patients bearing the same diagnosis from different hospitals. The ACV is often undertaken in such research when it is known (or discovered) that the groups differ on one or more variables which are related to Y, but which are extraneous to the research issue; if they are not coped with, they offer an unwanted competing interpretation. These extraneous variables are then set up as covariates, that is, are partialled. Sets A and B are related—overlap in their relationship to Y. (The casual reader may not have the ballantine etched on his cortex—see Figure 4.3.1 and accompanying discussion.)

This state is reflected in covariate mean differences among groups, or, more generally, in the fact that $R^2_{Y \cdot AB}$ does not in general equal $R^2_{Y \cdot A} + R^2_{Y \cdot B}$.

Now, when we find sR^2_B , and the other partialled functions of set B and of its constituent X_j, we are attributing all the Y variance shared by sets A and B (that is, the $A-B$ overlap) to set A. If the substantive meanings of A, B, and Y are such as to make it clear that the flow of causality is from A to B (as it is from A to Y) and not vice versa, this attribution of all the shared Y variance to set A is straightforward. It makes sense to take out the *effects* of set A on the relationship between set B and Y. The causal language is explicit and intended—the covariates are causally prior. Thus, we partial causes from variables containing their effects so as to obtain residuals which are free of the effects of these causes. That is, at least part of the reason that our preexisting groups differ in Y is *because* of their difference in the covariates, for which we are seeking to make due allowance.

But what happens if the causal relationship is in reality reversed: set B causes (in part) the variation in set A? Under these circumstances, the $A-B$ overlap is reflecting the causation of A by B, and when we assign all of this to set A, we are stealing Y variance from its rightful owner, set B. We are partialling effects of set B out of set B. We may however wish to do this on policy grounds in order to see whether or to what extent, despite this partialling of some of set B's causal effects, it nevertheless bears on the dependent variable.

Causation need not be unidirectional. In the complexity of the real world where nonexperimental and quasiexperimental research is used, the covariates and group membership may influence each other, or both be influenced by yet other variables. Thus, in part, the process of natural selection whereby individuals become members of one group rather than another may be *due* to elements in set A and, in part, whatever is entailed in group membership may cause variation in the same or other elements in set A. To the extent to which the latter circumstances obtain, when the ACV partials all A from B, it underestimates[25] the amount of Y variance properly assigned to set B (in pR^2_B) and biases the partial coefficients of the X_js of set B.

By constraining set A to descriptors which characterize the subjects at a point in time prior to that at which they are characterized by Y, one can largely meet the requirement of causal direction from set A to Y. But with preexisting groups, the causal precedence between sets A and B is not usually determinable on temporal grounds. Are subjects members of their groups because of set A differences, or has group membership (or unmeasured correlates thereof) produced the set A differences? Many an important scientific and social controversy hinges on the answer to questions of this form. Clearly, one does not answer

---

[25] The alert reader may object, pointing out that under conditions of suppression (Section 3.4), overestimation may occur. Although this is true, the likelihood of the occurrence of suppression in applications of APV is small enough to be negligible.

such questions with ACV—rather, the valid interpretation of ACV assumes the correctness of the answer, and is therefore conditional on it.

Given this uncertainty, some methodologists reject the use of ACV with preexisting groups, or at least in cases where there are significant group differences in covariates (generally, any significant relationship between sets A and B). They would restrict its use to experiments in which assignment to groups is random (Section 9.7.3). To be sure, this solves the methodologists' problem, but not that of the researcher seeking to cast some light, however dim, on some important issue of behavioral/social science, practice, or policy. He simply does not have the option of randomly assigning infants (let alone foetuses) to racial, sexual, or psychiatric diagnostic "conditions" in order to answer his questions. Nor, as a practical matter, does the experimenter who is permitted to work with existing groups (for example, school classes) which must remain intact for administrative reasons, have the option of randomization.

As throughout this book, we take a "data-analytic" position and ask the question, "Given data of this kind, what kind of analysis yields what kind of meaning?" We use ACV (APV) where sets A and B are related, that is, where we will "adjust" group means for covariate differences, when we are prepared to assume that, if sets A and B are related, set A is causally prior to set B (and Y). To the extent that our assumption is incorrect, our results may be invalid. Or, in an exploratory spirit, we may make alternate assumptions and appraise their consequences to the results.

Another device for addressing the dilemma of ambiguous causal direction is to remain strictly descriptive: Given a significant $B \cdot A$ effect in an ACV, one can assert that for any specified value(s) of the covariate(s), the groups have different expected (mean) Y values in the population. Such a statement can be made with no causal assumptions whatever, but, of course, it begs the question of interpretation: to say what such differences (or their absence) mean returns us to the causal assumptions. Yet the descriptive formulation may itself be helpful in thinking through the causal issues.

We have not illustrated this discussion with examples of straightforward versus ambiguous causal relationships, since such examples are as varied as the fields spanned by behavioral and social science. In the earlier examples of this chapter, we have generally avoided instances of ambiguous causality (but the reader may be prone to disagree—aye, there's the rub!). Only an expert in a given area can make such judgments, and in some areas, his equally expert colleagues may disagree with him. But we will permit ourselves a single illustration, drawn from agronomy. The example is taken from an illuminating article by Lord (1969), although our discussion goes beyond, without disagreeing with, his.

An agronomist, investigating the difference in yield (Y) between "black" and "white" varieties of corn (set B), grows 20 pots of each under the same conditions to maturity. Upon harvesting, he finds a considerably higher yield of marketable grain for the white variety. But he also observes that the white variety averages 7 feet high at flowering time (set A), compared to 6 for the black variety. The yield-on-height slopes being the same, he performs an ACV and finds that the adjusted mean yield difference is near zero and nonsignificant.[26] What does this mean?

First, note that the dependent variable is no longer yield (Y), but yield adjusted for flowering height ($Y \cdot A$). The use of ACV does not change the fact of the yield difference of the two varieties.

The assignment of all the yield variance which is shared by height and variety to height alone can not be justified on the usual causal grounds. Rather, height is presumably a property of the variety. However, the results of the analysis *are consistent with* the proposition: the reason for the yield difference is the flowering height difference (or its correlates). We stress "are consistent with," since, of course, such propositions can never be proved in nonexperimental research where assignment to groups is nonrandom (Blalock, 1964, 1971). Yet another way to put this interpretation is that height is a *sufficient* basis on which to account for the yield difference; but this does not mean that it is the *correct* basis. Finally, one can make the purely descriptive statement: black and white plants of the same flowering height have the same (expected) yield. This is a fact about the data at hand, but again any causal interpretation is an assumption and not a product of the analysis.

What are clearly *not* warranted by the results of such analyses are subjunctive formulations like: "*If* black and white varieties *were* of equal height, *then* they would have equal yields." We can not know what the consequence of equal height would be on yield without specifying the means of producing equality in height. We might add fertilizer or water to the black variety, or stretch it when it is young, or use other means to actually equate the varieties. There is no reason to believe that such alternative methods of equating height would have the same consequences, or that any of them would result in equal yields. More technically, such intervention in the process might well change the groups' yield-on-height regression lines, so that those used for adjustment in the analysis are not necessarily descriptive of "what would happen if. . . ." Only if heights were equated by excluding short black plants or tall white ones could the regression lines be assumed to hold, and the interpretation be admissible.[27] But this is not what we usually wish to know. Only by making the appropriate changes which accomplish height equality in a new experiment and determining the consequent yields can the question be answered. Unfortunately, operations analogous to

---

[26] Although for simplicity of exposition we imagine a near-zero difference, the discussion which follows would only change in incidental detail were we to posit any other value for the adjusted mean difference.

[27] This is effectively the same as the statement, "For plants of the same given height from the two varieties, the yields are the same." Whether one compares yields by making overall height means equal for the two varieties by exclusion ("equating covariate means"), or addresses oneself to any given level of height ("holding the covariate constant"), one is describing the data at hand gathered under the conditions which obtained.

these in behavioral and social science applications are more often than not simply impossible, and such subjunctive questions can not be answered by ACV, or indeed, by any method of analyzing data.

9.7.3 Randomized Groups in Analysis of Covariance

The difficult problems and uncertainties of causal inference are largely avoided when cases from a given source are randomly assigned to groups which are then subjected to differential treatment conditions (set B). The randomization method assures that the expected (population) differences between groups for *all* variables (and therefore the covariates) at the time of randomization are zero. More generally, with randomization, it must be the case that, for the population,

$$(9.7.1) \qquad R^2_{Y \cdot AB} = R^2_{Y \cdot A} + R^2_{Y \cdot B} ,$$

that is, in the ballantine of Figure 4.3.1, visualize pulling the circles for sets A and B apart until there is no longer any overlap between them and hence no area c. Because of chance factors, the above equation will not hold exactly in the sample, but its validity in the population permits us to proceed to do a standard ACV, which assigns the (usually small) chance overlap to set A. Since there can be no true causality between covariates and treatments, the difficulties encountered with preexisting groups cannot arise. As we have noted, some methodologists restrict the use of ACV to these happy circumstances.

For the independence of sets A and B to be assured, the covariates must characterize the subjects prior to the institution of the treatment. Otherwise the treatments may affect the covariates, and we are back with problems of variance stealing. Measuring covariates prior to the institution of differential treatments also assures the desired causal precedence of the covariates to Y.

Clearly, then, the ACV with randomized groups is not for the purpose of adjusting or equating groups for covariates, since randomization accomplishes this goal. Why, then, use it? Primarily because, as we have argued, the substantive issues concern $Y \cdot A$, not Y. Take, for example, the use of ACV with Y as a postscore and set A as a prescore. If change (as a function of set B) is what the research is about, its dependent variable is not Y, post, but $Y \cdot A$, post from which pre has been partialled, that is, change. We think it important to stress the fact that partialling covariates is not some casual touching up of the dependent variable, but a fundamental change in its definition. Secondarily, as some authors stress, the use of $Y \cdot A$ (rather than Y) brings increased statistical power, since the F ratio uses the smaller error component $1 - R^2_{Y \cdot AB}$ (rather than $1 - R^2_{Y \cdot B}$). We consider this admitted benefit to be incidental, a reward, as it were, for studying the relevant dependent variable.

A word about accidental, as contrasted with randomized, groups is in order. Often one must work with preexisting groups about which one is prepared to assume no population relationship with set A, evidenced in part by the approximation to Eq. (9.7.1) in the sample data. When one then proceeds to an ACV unconcerned by variance stealing, one is replacing the randomization mechanism by substantive a priori judgment buttressed by some sample evidence. This practice strikes us as reasonable provided that one makes explicit the conditional nature of the conclusions drawn.

In closing this section, we wish to lay heavy stress on the fact that APV/ACV is not a mechanical data-massaging procedure, but rather a powerful technique which, perhaps more than others, requires intelligent and substantively knowledgable use. It is of the utmost importance that the investigator carefully think through the implications of "Y from which covariates have been partialled," and "equating groups on covariates" to assure himself that they make substantive theoretical sense. Consider the fact that the difference in mean height between the mountains of the Himalayan and Catskill ranges, adjusting for differences in atmospheric pressure, is zero! This is worth pondering. . .

9.8 SUMMARY

The ACV is described as a special case of setwise MRC, which proceeds hierarchically as follows: With Y as the dependent variable, a set of IVs which represents covariates (set A) is entered in the first step, to which is then added another (set B) carrying the group membership information. The requirement of homogeneity of regression between groups is assessed by adding in the third step the AB product set; if the null hypothesis for the $A \times B$ interaction is found acceptable, the analysis in the second step using sets A and B is interpreted as an ACV, which is then understood as being an assessment of the relationship between $Y \cdot A$ and $B \cdot A$.

Some of the advantages of this approach arise from our previous demonstrations (Chapters 5 through 8) that information in virtually any form may be expressed by means of sets of IVs. Thus, there is virtually no constraint on what may serve as covariates in set A: one can employ nonlinear as well as the usual linear aspects of quantitative variables, nominal scales coded by various methods, variables with missing data, interactions, and combinations of these. We adopt the setwise methods of Chapter 4 for the covariate-adjusted group membership set, $B \cdot A$: effect size measures (pR_B^2, sR_B^2), statistical power analysis, and significance testing. The constants of the regression equation in sets A and B yield adjusted group means $\overline{Y}_h'$, and the partial coefficients (pr_j, B_j) of the X_js of set B provide effect size measures for the relationship of single aspects of adjusted group membership with the adjusted $Y \cdot A$ (or Y'). (Section 9.2)

The above is then illustrated in detail, using an example that is interpreted variously for multiple, nonlinear, missing-data, and nominal scale covariate sets. The opportunity is seized to illustrate how alternative methods for coding group membership (effects, dummy, contrast) yield regression constants which are functions of adjusted Y means. The ACV for general factorial design is shown to be readily accomplished by the set B coding, and the use of mixtures of different types of research factors as subsets of the covariate set is discussed. (Section 9.3)

Having liberated the covariates from the constraints of conventional ACV, we turn to set B, and point out that its restriction to the representation of group membership (nominal scales) is unnecessary in MRC analysis. Removing this restriction, we offer a generalization of ACV to the Analysis of Partial Variance (APV), in which either quantitative or nominal research factors (or combinations thereof) may be used as set B. The APV logic and procedure is exactly the same as that for ACV, and it yields an equation for $\hat{Y}'$ ($= \hat{Y} \cdot A$) which is a generalization of the equation which produces the adjusted means of ACV. The APV is thus a highly general method for the study of partial (residualized) variance which may use any type(s) of research factors as covariates and any type(s) of research factors whose covariate-adjusted effects are of interest. The APV may be used sequentially for h sets in a hierarchical MRC, the sets ordered in presumed causal priority, wherein at any given stage, all prior sets function as an aggregated covariate set relative to the set entering at that stage. The resulting analysis, a series of $h - 1$ progressive APVs, provides a useful data-analytic method particularly for complex nonexperimental research. (Section 9.4)

The effects of errors of measurement in APV (ACV) are discussed, initially for the simple three variable case, $r_{Y2 \cdot 1}$. While unreliability in Y and X_2 has the well-known attenuating effect, the effect of unreliability in X_1, the partialled variable, may be to either raise, lower, or even change the sign of the true correlation. Generalizing, unreliability in the covariate set A may yield results of the APV greatly different from what would have been obtained were the covariates perfectly reliable. The problem is best avoided by using only highly reliable covariates. When covariates of modest reliability must be used, a method for estimating true relationships, AP*V (or its special case, AC*V) is proposed. It requires estimates of the reliability coefficients of the covariates, and its exact sampling properties are not known. It is therefore an approximate method from which only tentative conclusions may be drawn. (Section 9.5)

When change in some variable is to be studied as a function of others, the intuitively attractive simple change ("post- minus pre-") scores are shown to have the generally undesirable property of dependence on the prescore. The methods of this chapter can be applied so that regressed change, which is uncorrelated with prescore, is the dependent variable by letting Y be the postscore, set A be (or include) the prescore, and set B the research factor(s) whose effects on change are to be studied. Since APV relates $Y \cdot A$ to $B \cdot A$, the dependent variable is effectively the postscore regression-adjusted for (hence uncorrelated with) the prescore. A simple illustrative example where set B contains only a single dichotomous variable is thoroughly discussed and then generalized to multiple groups or other multiple variables in set B, nonlinear change, and multiple time points. Finally, with the same data, we posit modest reliability for the prescore covariate and present the resulting AP*V (AC*V). From this comparison, we discuss the effect of unreliability in the study of change, a frequently recurring problem, which, if ignored, may lead to serious blunders in interpretation. (Section 9.6)

The chapter is concluded by addressing the role and limitations of partialling which are implied by the logical framework which underlies its use in APV methods. It is argued that a causal model is implied in which the direction of causality is from the partialled variables (the set A covariates) to the variable(s) from which they are partialled. Thus, $Y \cdot A$ is meaningful when the causal flow is from set A to Y, a condition which is usually not difficult to satisfy. On the other hand, in research which uses preexisting groups in set B, the assurance that the causal flow (if any) runs from set A to set B is frequently hard to come by. When the causality is in fact reversed, Y variance which properly belongs to set B is stolen by set A. This ambiguity is eliminated when subjects are randomly assigned to the groups (treatments) represented by set B, since that model assures no relationship between sets A and B in the population and hence no possibility of reverse causality and variance stealing. The option to randomize, when available, should certainly be taken. When preexisting groups must be used, the causal implications of partialling to the substantive issues of the research require careful thought and circumspect interpretation. An example from agronomy indicates why the use of APV (ACV) does not warrant interpretations cast in subjunctive form, for example, "*were* the groups equal with regard to the covariates, *then* the differences in the dependent variable would be thus and such." One cannot know this without creating the conditions which produce such equality (assuming their possibility and meaningfulness) and then studying the associated effects on the dependent variable. Nothing guarantees that the results of such an experiment would be correctly anticipated by the differences among the covariate-adjusted means. (Section 9.7)

10

Repeated Measurement and Matched Subjects Designs

10.1 INTRODUCTION

This book has thus far been concerned only with data structures in which each of the Y observations has come from a different sampling unit or "subject." There is another class of research designs in which each of the n subjects gives rise to a Y observation under each of c (> 1) different conditions (C). The data layout has n rows, c columns, and hence a total of $N = nc$ Y observations. Such "repeated measurement" designs are frequently used in psychological experimentation. The statistical issues are much the same for "matched-subject" designs: instead of a single subject being observed under the c conditions, each row contains a set of c subjects who are homogeneous with regard to variable(s) which are believed to be correlated with Y, but extraneous to the purpose of the experiment as would be the case, for example, for litter mates. Such matching (or "blocking") serves to control this extraneous variance in comparisons involving C. Thus, whether a row represents a single subject or a "c-tuple"[1] of matched subjects, variance in Y due to systematic "between-row" differences can be identified and removed, with salutary effects on precision and statistical power in studying the effects of C.

The above may either constitute the research design or be part of a more complex design where the subjects are members of different groups ("split-plot" designs), which may be variously structured, for example, as in factorial design. Also, C may be variously structured, for example, as points along a quantitative continuum analyzable by orthogonal polynomials ("trend analysis"). In textbooks on AV and experimental design, this general class of designs is a special

---

[1] In what follows, in order to avoid repeated use of this awkward collection of letters, the words "subjects" or "rows" will be generally used to include matched sets of c subjects ("c-tuples").

case of "Model III" or "mixed-model" designs, the latter referring to the combination of "fixed" C with "random" (-ly selected) subjects or sets of matched subjects.

The purpose of this chapter is to outline the MRC analysis of such designs. Here, as elsewhere, the analytic apparatus of the MRC system provides generality and flexibility. In addition to its provision of practical data-analytic procedures, MRC conceptions illuminate, and in turn are illuminated by, this type of application. Also, as design complexity increases, so does the advantage of MRC analysis over AV.

In this brief treatment, we forego detailed consideration of the complex issues of experimental design and of the statistical assumptions which underly repeated measurement and split-plot designs, but not because we deem them unimportant. Quite the contrary. But to do them justice requires an extended treatment, such as is to be found in Edwards (1972), Hays (1973), or Winer (1971), while our main purpose here is to sketch out the MRC analysis of these designs. For example, one assumption which has important consequences is that the $c(c - 1)/2$ population correlations between conditions (or between conditions within groups in the split-plot design) are equal, a circumstance more likely to obtain for matched subjects than for repeated measurements of a single subject. The above references show how to assess the status of this assumption and how to cope with its failure. For our purposes, we will simply assume that the necessary assumptions are met, or that any failure occurs under conditions of robustness such that the significance tests are not materially invalidated.

10.2 THE BASIC DESIGN: SUBJECTS BY CONDITIONS

Visualize the data layout which is basic to all these designs, a two-way $(n \times c)$ table of values of Y in which each row represents a subject or matched set of subjects (S), and each column a condition or treatment (C). We will call this design "subjects by conditions" $(S$ by $C)$. With n rows and c conditions, the total number of observations is $N = nc$. For concreteness, assume $n = 50$ subjects, $c = 4$ conditions, and hence $N = 200$ Y observations. A fundamental characteristic of such data structures is that the total variance of Y for these N observations can be partitioned into two additive portions of quite different character, each of which is then separately analyzed, the "between-S" variance and the "within-S" variance.

10.2.1 Between-Subjects Variance

The between-S Y variance represents the "individual differences" in overall Y among the 50 $(= n)$ subjects: each subject has 4 $(= c)$ Y values, and their aggregate (sum or mean) varies from subject to subject. This portion of the total variance can be determined in various ways. One possibility that occurs immediately to an MRC enthusiast is to treat S as a nominal scale of n $(= 50)$ levels, code it as $n - 1 = k_S$ $(= 49)$ IVs, do an MRC analysis for the N $(= 200)$

observations, and find $R^2_{Y \cdot S}$, the proportion of Y variance accounted for by subjects (S), which is exactly the quantity we seek. There are, however, practical problems in this approach: apart from the clerical task of coding $n - 1$ IVs, with large n one may find that the available computer program does not have sufficient capacity. Even for this example, the number of IVs, $k_S = n - 1 = 49$, exceeds the capacity of at least one popular MRC computer program. With a larger n of a more complex design where more IVs are required to accomodate all the research factors, one can easily exceed for most programs the capacity for IVs which would be required by this approach.

A more practical route to determining $R^2_{Y \cdot S}$ is to literally break the analysis in two, and attend to the between-S variance by itself. Specifically, we begin by analyzing the n subjects' mean Y scores for the c ($= 4$) conditions, that is, for the pth subject, we find

$$(10.2.1) \qquad \bar{Y}_p = \frac{\sum Y_p}{c}$$

(where $p = 1, 2, \ldots, n$), and treat the resulting means as a new dependent variable in an analysis where the number of observations is now n ($= 50$). Note that we have discarded (temporarily) all the information about the effect of C; the current part of the analysis is only about overall differences between subjects.

In more complex designs, we will subject these $\bar{Y}_p$ scores to MRC analysis, but here, for the basic S by C design, we need them only for the purpose of determining $R^2_{Y \cdot S}$. This is easily accomplished: we find *their* variance $(sd^2_{\bar{Y}_p})$ in the usual way (that is, with n as the divisor, not $n - 1$) and find what proportion it constitutes of sd^2_Y, the variance of the total set of N ($= 200$) observations, that is,

$$(10.2.2) \qquad R^2_{Y \cdot S} = \frac{sd^2_{\bar{Y}_p}}{sd^2_Y}.$$

Concretely, if the variance of the 50 subjects' means is 48, and of the 200 individual Y values is 80, then $R^2_{Y \cdot S} = 48/80 = .60$; thus, 60% of the total Y variance is accounted for by subjects, that is, is between-S variance.

Although our present concern with $R^2_{Y \cdot S}$ is to get it out of the way, its magnitude is of some interest. The larger it is, the more Y variance that is irrelevant to our assessment of C effects has been removed. The purpose of these mixed model designs is to get rid of this individual differences variance which, in fixed model designs, makes up much of the error variance. When $R^2_{Y \cdot S}$ is relatively small,[2] the $\bar{Y}_p$ values vary about as much from row to row as they

---

[2] As a specific guide, when all correlations between pairs of columns are exactly zero, $R^2_{Y \cdot S} = (1 - R^2_{Y \cdot C})/c$. Thus, if as we assume later, $R^2_{Y \cdot C} = .15$, then zero correlation between conditions would produce an $R^2_{Y \cdot S} = (1 - .15)/4 = .2125$, while the one obtained was .60, indicating large S (or "blocking") effects. Unless $R^2_{Y \cdot S}$ substantially exceeds $(1 - R^2_{Y \cdot C})/c$, there is little if any practical advantage in this design. See Footnote 12.

would if they were based on c different randomly selected subjects, instead of all coming from the same subject (or matched set of subjects). In such circumstances, the correlations between *columns* would be about zero, instead of the substantial positive value on which this experimental strategy seeks to capitalize.

10.2.2 Within-Subjects Variance

Since the between-S variance and the within-S variance add up to the total Y variance, the within-S variance proportion is simply $1 - R^2_{Y \cdot S}$. The variation of each subject's Y values across the c levels of C go to make up this portion of the variance.

Our interest in the S by C design is in the systematic effects of C. If we now string out all $N = nc$ (= 200) observations, and code C by any appropriate method as $c - 1$ IVs, we can proceed in the usual way to find $R^2_{Y \cdot C}$ in an MRC analysis with $k_C = c - 1$ (= 3) IVs; let us assume $R^2_{Y \cdot C} = .15$. But $R^2_{Y \cdot C}$ is the proportion of the *total* Y variance accounted for by C, which is not what we want. The appropriate base for C variance is the *within-S* portion of the Y variance; hence,

$$(10.2.3) \qquad\qquad R^2_{(Y \cdot S) \cdot C} = \frac{R^2_{Y \cdot C}}{1 - R^2_{Y \cdot S}} .$$

These relationships may be clarified by the modified ballantine depicted in Figure 10.2.1. Taking the total Y variance as being of unit area, the vertical line demarcates the fundamentally different between-S and within-S portions. The Y variance for which C accounts is necessarily exclusively within-S variance; hence, its area a is taken as a proportion of the combined areas a + e. In our running

Between S + Within S = 1

$R^2_{Y \cdot S} + (1 - R^2_{Y \cdot S})$ =1

b + (a + e)   =1

$$R^2_{Y \cdot C} = \frac{a}{1} = a$$

$(10.2.3)\ R^2_{(Y \cdot S) \cdot C} = \dfrac{R^2_{Y \cdot C}}{1 - R^2_{Y \cdot S}} = \dfrac{a}{a + e}\quad (= R^2_{YC \cdot S});$

$(10.2.4)\quad R^2_{Y \cdot SC} = R^2_{Y \cdot S} + R^2_{Y \cdot C}$

$= b + a,$

$1 - R^2_{Y \cdot SC} = 1 - (b + a) = e;$

$(10.2.5)\ F = \dfrac{R^2_{Y \cdot C}}{1 - R^2_{Y \cdot SC}} \times (n - 1) = \dfrac{a}{e} \times (n - 1).$

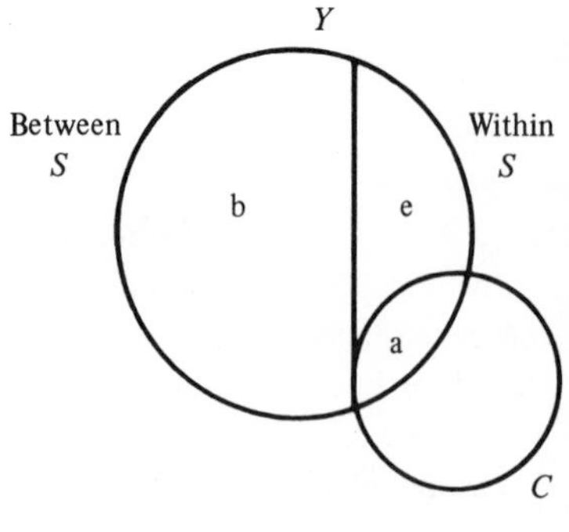

FIGURE 10.2.1 The partitioning of the Y variance for the subjects by conditions design.

example, the proportion of total variance due to C is $R^2_{Y \cdot C} = a/1 = .15$, but the more relevant proportion of within-S variance due to C is

$$R^2_{(Y \cdot S) \cdot C} = R^2_{Y \cdot C}/(1 - R^2_{Y \cdot S}) = a/(a + e) = .15/(1 - .60) = .3750.$$

The notation used in $R^2_{(Y \cdot S) \cdot C}$ indicates that it is a squared multiple semi-partial correlation, with $Y \cdot S$ the dependent variable and set C making up the IVs. $Y \cdot S$ is Y from which S has been partialled, the residual values which would result if from each subject's c scores his $\bar{Y}_p$ were subtracted. Since the structure assures that C and S are orthogonal, there is no difference between C and $C \cdot S$, hence no difference between $R^2_{(Y \cdot S) \cdot C}$ and $R^2_{YC \cdot S}$, the squared multiple *partial* correlation.

The orthogonality of C and S also means that the Y variance for which they account is additive, that is,

(10.2.4) $$R^2_{Y \cdot SC} = R^2_{Y \cdot S} + R^2_{Y \cdot C},$$

so that for our running example, $R^2_{Y \cdot SC} = .60 + .15 = .75$.

10.2.3 Significance Tests

To test the null hypothesis that C accounts for no variance in the population, the F ratio is adapted to

(10.2.5) $$F = \frac{R^2_{Y \cdot C}}{1 - R^2_{Y \cdot SC}} \times (n - 1),$$

with $df = c - 1$ and $(n - 1)(c - 1)$. As can be seen from Figure 10.2.1, the left-hand term is the ratio of the variance due to C, area a, to the error ("residual") variance, $1 - (b + a) = $ area e.

This is equivalently a test of the significance of $R^2_{(Y \cdot S) \cdot C}$ (or $R^2_{YC \cdot S}$), and therefore may also be written

(10.2.6) $$F = \frac{R^2_{(Y \cdot S) \cdot C}}{1 - R^2_{(Y \cdot S) \cdot C}} \times (n - 1),$$

with the same df. For our running example, these equations produce

$$F = \frac{.15}{1 - .75} \times (50 - 1) = .6(49) = 29.4,$$

and, equivalently,

$$F = \frac{.375}{1 - .375} \times (50 - 1) = .6(49) = 29.4,$$

which, with $df = (c - 1 =) 3$ and $[(n - 1)(c - 1) =] 147$ is highly significant (for $df = 3, 120, F = 3.95$ is the $\alpha = .01$ criterion value found in Appendix Table D.1).

Note that the F test of $R^2_{Y \cdot C}$ provided by a standard MRC computer program when used in this way will not give the correct result of Eqs. (10.2.5) and (10.2.6), but will apply the standard fixed-model test of Eq. (3.7.1), assume a sample size of $N = nc$, and use the wrong error term $(1 - R^2_{Y \cdot C}$ instead of $1 - R^2_{Y \cdot SC})$ and error df $[c(n - 1)$ instead of $(c - 1)(n - 1)]$. For these data, the incorrect F would therefore be found as

$$F' = \frac{.15}{1 - .15} \times \frac{200 - 3 - 1}{3} = .1765(65.33) = 11.529,$$

with $df = 3, 196$.

Similarly, the standard MRC computer program will produce incorrect t tests on the partial coefficients of the X_is used to code conditions (and, if they are given, incorrect pr_i and sr_i). Depending on the nature of C and our hypotheses about it, it will have been coded by one of the methods of nominal scale coding of Chapter 5, or as an orthogonal polynomial (Section 6.3). However coded, the B_is for the X_is in set C will correctly give the relevant contrast functions on the condition means, but the standard errors and hence the t tests of the B_is (and their df) will again be incorrect because of the incorrect error variance presumed by the computer program. We can write the correct standard error[3] for a B_i in the S by C design as

(10.2.7) $$SE_{B_i} = \frac{sd_Y}{sd_i} \sqrt{\frac{1 - R^2_{Y \cdot SC}}{(n - 1)(c - 1)}} \sqrt{\frac{1}{1 - R^2_i}}.$$

This differs from the standard fixed-model SE of Eq. (3.7.8) only in the middle term, which carries the error variance proportion and its df. Since $t_i = B_i/SE_{B_i}$, one can convert the incorrect t'_i given by the computer to the correct t for the S by C design by substituting the t'_i provided by the computer in

(10.2.8) $$t_i = t'_i \sqrt{\frac{1 - R^2_{Y \cdot C}}{1 - R^2_{Y \cdot SC}} \times \frac{c - 1}{c}},$$

[3] The correctness of this standard error and the t value for B_i rests on a further assumption. Each B_i represents some linear contrast function of the condition means, for example, $\bar{Y}_i - \bar{Y}_4$ if C was dummy coded. The residual error term may be partitioned into $c - 1$ interaction terms with subjects, that is, individual differences in these contrasts, for example, $(Y_1 - Y_4) \times S$, $(Y_2 - Y_4) \times S$, and $(Y_3 - Y_4) \times S$. The single error term we use is a pooling of these (and their df). We are in effect assuming that these $c - 1$ interaction variances are equal, otherwise our test of a given B_i should use its own error term, for example, the variance due to $(Y_i - Y_4) \times S$ with $df = n - 1$. We could generate such separate subject interactions by coding S as a nominal scale and explicitly representing as IVs each of the $X_i S$ product sets (where the X_is code C), and proceed as with any other interaction set to find the proportion of Y variance for which it accounts, with $df = n - 1$ for each. As noted above, unless n is small, the number of IVs which result will likely exceed the capacity of most computer programs. In most applications, however, such elaboration into separate error terms is not necessary, and one may proceed, as in the test, with a single pooled error term.

with $df = (n-1)(c-1)$. As an alternative to Eq. (10.2.7), SE_{B_i} can be found by dividing the B_i by its correct t_i.

10.2.4 sr and pr for a Single Independent Variable

Since all partial coefficients for X_i share the same t_i, we can use it together with some algebraic relationships we have already exploited to determine the semi-partial and partial correlation coefficients for X_i. It should be kept in mind that the dependent variable here is not Y, but $Y \cdot S$, so that the basic variance being accounted for by the X_i which code C is within-S (not total) Y variance. Using the t_i of Eq. (10.2.8), we can find

$$(10.2.9) \qquad sr_i^2 = \frac{t_i^2 \, (1 - R_{(Y \cdot S) \cdot C}^2)}{(n-1)(c-1)}.$$

This is the proportion of the *within-S* Y variance uniquely accounted for by X_i, that is, by X_i from which the other $c-2$ IVs have been partialled.

For pr_i^2, we again use the t_i of Eq. (10.2.8), substituting in

$$(10.2.10) \qquad pr_i^2 = \frac{t_i^2}{t_i^2 + (n-1)(c-1)}.$$

The pr_i^2 relate to the sr_i^2 in this context the same as always—the numerator is the same (variance due uniquely to X_i, hence some portion of area a in Figure 10.2.1), but the denominator is reduced to the within-S Y variance minus Y variance due to the other IVs. Both sr_i and pr_i take the same sign as B_i, as always.

The exact interpretations of B_i, sr_i, and pr_i depend upon the type of coding used to code the $c-1$ IVs of set C (as well, of course, as upon the substantive content of Y and of the levels of C). The reader is referred to the detailed discussions in Chapter 5 (and for orthogonal polynomial coding, Section 6.3), but must keep in mind that references to the Y (or total Y) variance there must be changed to within-S Y (or $Y \cdot S$) variance for application in the present context of the S by C design.

For numerical illustration, we continue with our running example of 50 ($= n$) subjects by 4 ($= c$) conditions, where we have found thus far that $R_{Y \cdot S}^2 = .60$ and $R_{Y \cdot C}^2 = .15$; therefore, $R_{Y \cdot SC}^2 = .75$ and $R_{(Y \cdot S) \cdot C}^2 = .3750$. Assume that C_4 is a control condition relative to the other three, and that therefore C had been coded by dummy variables (Section 5.3), with C_4 as the reference condition. The coding thus was exactly that shown in Table 5.3.1, with "condition" replacing "group." Assume further that in the computer run based on $N = nc = 200$ observations, it is found that $B_1 = 9.33$. This indicates (correctly) that $\bar{Y}_1 - \bar{Y}_4 = 9.33$, that is, the subjects' mean under C_1 is 9.33 points higher than it is under the control condition C_4. It is accompanied by an incorrect t_1' of 1.822 (or incorrect "F_1' to remove" = $1.822^2 = 3.320$), based on an incorrect error

$df = c(n - 1) = 196$. The correct t_1 is found from Eq. (10.2.8) as

$$t_1 = 1.822 \sqrt{\frac{1 - .15}{1 - .75} \times \frac{4 - 1}{4}} = 1.822(1.597) = 2.910,$$

which, for $df = (n - 1)(c - 1) = 49(3) = 147$, is significant at $P < .01$ (Appendix Table A).

Using Eq. (10.2.9), we determine that

$$sr_1^2 = \frac{2.910^2 (1 - .3750)}{(50 - 1)(4 - 1)} = .0360,$$

which indicates that 3.6% of the within-S variance is accounted for by the $C_1 - C_4$ distinction. Equivalently, we can report that $sr_1 = \sqrt{.0360} = .190$ is the correlation between Y and the C_1 versus C_4 dichotomy, holding subjects constant (since between-S variance has been removed).

Finally, using Eq. (10.2.10), we can find

$$pr_1^2 = \frac{2.910^2}{2.910^2 + (50 - 1)(4 - 1)} = .0545,$$

which gives the variance due to the $C_1 - C_4$ distinction as a proportion of the within-S variance not accounted for by the other ($C_2 - C_4$ and $C_3 - C_4$) distinctions. Its square root, $pr_1 = .233$, is the correlation between Y and the C_1 versus C_4 dichotomy, with subjects, C_2 versus C_4, and C_3 versus C_4 all held constant.

In closing this section, we reiterate that all the equations presented are quite general for S by C designs and hold for any coding method employed to represent C. The reader can interpret the B_i, sr_i, and pr_i that result by reference to the earlier discussions of the coding methods, substituting $Y \cdot S$ for Y and within-S variance for total variance. The possibilities afforded by different coding methods confer the same flexibility in S by C designs as was seen earlier in fixed-model designs. Thus, if the four conditions were structured as a 2 × 2 factorial, they could be contrast coded and interpreted accordingly (see Section 5.5.4). As another example, C might be four successive trial blocks in a learning experiment or four levels of a stimulus, and could be usefully coded by orthogonal polynomials, which would yield a partition of $R^2_{(Y \cdot S) \cdot C}$ into sr^2s for linear, quadratic, and cubic portions of the within-S variance (see Section 6.3).

10.2.5 Power Analysis

The methods of power analysis discussed in Sections 4.5 and 3.8 may be applied in S by C designs, with some adaptation of the formulas. The reader would likely find a review of those sections useful at this point.

When the focus is on the variance accounted for by set C, the relevant ES is given by Eq. (4.5.1), restated for S and C as

(10.2.11)
$$f^2 = \frac{R^2_{Y \cdot SC} - R^2_{Y \cdot S}}{1 - R^2_{Y \cdot SC}} = \frac{R^2_{Y \cdot C}}{1 - R^2_{Y \cdot SC}},$$

or equivalently,

(10.2.12)
$$f^2 = \frac{R^2_{(Y \cdot S) \cdot C}}{1 - R^2_{(Y \cdot S) \cdot C}},$$

where the R^2 are alternate-hypothetical or estimated *population* values. Equation (10.2.12) requires specifying only one quantity, $R^2_{(Y \cdot S) \cdot C}$, the proportion of within-S variance accounted for by C in the population. The value of L is found by reference to the appropriate Appendix Table (E.1 for $\alpha = .01$, E.2 for $\alpha = .05$), entering with desired power (column), and the numerator df, k_B. In S by C designs, $k_B = c - 1$. Then, to find n^*, the number of *rows* (that is, subjects or sets of matched subjects) required by the above specification of α, power, f^2 and k_B, substitute in

(10.2.13)
$$n^* = \frac{L}{f^2(c-1)} + 1.$$

For example, if in planning a research where we expect that the proportion of within S variance accounted for by 4 (= c) conditions in the population is .20 (= $R^2_{(Y \cdot S) \cdot C}$), then by Eq. (10.2.12), $f^2 = .20/(1 - .20) = .25$. (Alternatively, we could have reached this result by having specified that $R^2_{Y \cdot S} = .50$, $R^2_{Y \cdot C} = .10$, and therefore $R^2_{Y \cdot SC} = .60$, these values then being substituted in Eq. 10.2.11.) If we plan to use an $\alpha = .05$ criterion and set power at .80, we find from Appendix Table E.1 at column .80, row $k_B = c - 1 = 3$, that $L = 10.90$. Substituting in Eq. (10.2.13),

$$n^* = \frac{10.90}{.25(4-1)} + 1 = 15.53,$$

so we would require 16 rows, giving rise to a total of $nc = 16(4) = 64$ observations, to meet these specifications.

It is worth comparing this result with what would be obtained for a simple fixed design, where n (unmatched) subjects would be assigned randomly to the four conditions. Since we entertained the estimate $R^2_{Y \cdot C} = .10$, it follows in this case from Eq. (3.8.1) that $f^2 = .10/(1 - .10) = .1111$, and, employing the same $L = 10.90$, we find by Eq. (3.8.2) that $n^* = 103$, which is both the total number of subjects and also the number of observations in this design. If we compare this with the repeated measurements version of the S by C design where each row is a single subject, the disparity in subjects required for the same specifications, 103 versus 16, is spectacular. However, carryover and sequence effects of conditions often preclude the use of repeated measurement designs. When $n^* = 103$ is compared with the S by C case where each row is a set of c matched subjects, so that the total $n^*c = 64$ observations come from 64 different subjects, the saving in subjects is still quite impressive. However, it is often not possible to accomplish the matching of subjects on relevant extraneous variables so as to produce the substantial value for $R^2_{Y \cdot S}$ necessary to make the number of observations required in an S by C design ($O^*_1 = n^*c$) materially smaller than the number required in a fixed design ($O_2^* = n_2^*$). More specifically, for the same specifications of power and α, the ratio of numbers of observations required by the two designs is approximately (a generally slight *under*estimation)

(10.2.14)
$$\frac{O_1^*}{O_2^*} \approx \frac{c(1 - R_{Y \cdot SC}^2)}{(c-1)(1 - R_{Y \cdot C}^2)}.$$

For the illustrative data, this approximation gives $4(1 - .60)/3(1 - .10) = .59$; the exact value is $64/103 = .62$. Thus, for the same specifications, the S by C design requires only .6 as many observations as does the fixed design in this instance. The above ratio will be less than one when $(1 - R_{Y \cdot SC}^2)/(1 - R_{Y \cdot C}^2)$ is smaller than $(c - 1)/c$ (approximately).

To determine power as a function of n for the test of variance due to C in the S by C design, we proceed as described in Section 4.5.7, but find L from the equation

(10.2.15)
$$L^* = f^2(n - 1)(c - 1),$$

where f^2 is found from Eqs. (10.2.11) or (10.2.12), and n is again the number of rows (subjects or sets of matched subjects). If, in our illustrative example, positing $f^2 = .25$ and $c = 4$ (as before), we wish to estimate the power at $\alpha = .05$ for $n = 24$, we find from Eq. (10.2.15) that $L^* = .25(24 - 1)(4 - 1) = 17.25$. Entering Appendix Table E.2 at row $k_B = c - 1 = 3$, we find that $L = 17.25$ just exceeds the criterion value of 17.17 for power of .95.

For power analysis of tests of the unique contribution of a single X_i of set C in an S by C design, we use as the ES measure an adaptation of Eq. (3.8.3),

(10.2.16)
$$f^2 = \frac{sr_i^2}{1 - R_{(Y \cdot S) \cdot C}^2},$$

where sr_i^2 is defined as above, that is, as the proportion of within-S variance due uniquely to X_i (in the population). To find n^*, this f^2 is substituted in Eq. (10.2.13), as before, but note that the L called for is now for $k_B = 1$ (not $c - 1$). For power as a function of n for a single X_i, use Eq. (10.2.15) and look up L^* in the appropriate Appendix Table E for $k_B = 1$. For the reconciliation of varying n^* for different X_i, proceed in the spirit of Section 4.5.6.

10.3   SUBJECTS WITHIN GROUPS BY CONDITIONS

To the basic S by C design, we introduce a new element: the n subjects (or sets of matched subjects) are assigned to two or more groups, and a research factor G made up of g mutually exclusive and exhaustive groups is thus defined. All n subjects (or subject sets) are then observed under the c conditions, for a total of $N = nc$ observations, as before, with $n_1 c$ coming from G_1, $n_2 c$ for G_2, etc., where $n = n_1 + n_2 + \cdots + n_g$. In the AV-experimental design literature, these designs are variously called "split-plot," "randomized blocks factorial," and by yet other names, and the term "nested" is often used to describe the group-membership relationship of the rows to G, for example "subjects nested within groups repeatedly measured on a factor." We characterize this experimental structure as subjects within groups by conditions, and abbreviate it as $S(G)$ by C. To avoid annoying repetition, we will in the discussion again use the term "subjects" to include both the case where each row represents a set of c matched subjects as well as the case where each row represents a single repeatedly measured subject. It is also true for the $S(G)$ by C design that the statistical assumptions underlying the model are more likely to be satisfied in the former case.

The details of the MRC analysis and the formal meaning of the results do not depend upon the nature of G. It may be a preexisting ("organismic") property of the subjects (for example, diagnosis, age, sex, religion, or factorial combinations of such properties) or the consequence of differential experimental manipulations assigned randomly to subjects (for example, treatment, stimulus level, instructions, and factorial combinations of such manipulations). Of course, the nature of G greatly effects the substantive interpretation of the results. The discussion of this issue in Chapter 5 applies with equal force to G in the $S(G)$ by C design.

The consequences to the analytic structure of introducing G as a research factor distinguishing among subjects are twofold: first, a portion of the between-S variance can be identified as attributable to G, and, second, a portion of the within-S variance can be identified as attributable to the $C \times G$ interaction.

As in the S by C design, we approach the analysis of the $S(G)$ by C design by effecting the fundamental distinction between the between-S and within-S components of the total Y variance (based on $N = nc$ observations) and analyze them separately.

10.3.1   The Analysis of the Between-Subjects Variance

Whereas in the S by C design, our interest in the between-S variance was to determine its proportion of the total Y variance ($R_{Y \cdot S}^2$) so as to exclude it in order to attend solely to the within-S portion ($1 - R_{Y \cdot S}^2$), in the $S(G)$ by C design, the between-S portion now contains identifiable variance due to G. We proceed as before to find each subject's mean over the c conditions, $\overline{Y}_p$, as per Eq. (10.2.1). The analysis of the between-S variance is quite simply an MRC analysis of these n subject means: $\overline{Y}_p$ is the dependent variable and G is coded into $g - 1$ IVs ($X_j; j = 1, 2, \ldots, g - 1$) by whatever method suits the purposes of the investigator. In other words, one proceeds exactly as described throughout the book where categorical research factors are represented as IVs: Chapter 5 for nominal scales, Section 6.3 for orthogonal polynomially coded quantitative variables, Sections 8.7 and 8.8 for factorial structure with interactions, etc. One simply replaces Y in those discussions by $\overline{Y}_p$; both for G treated as a set (or as a set made up of subsets) and for its constituent single IVs, in the significance testing, power analysis, and interpretation of correlation and regression coefficients, the formulas and discussions hold. No corrections are required in the standard MRC computer output.

Again we note that $\bar{Y}_p$ carries no information about systematic differential effects of C (or of the $C \times G$ interaction), since the c levels of C contribute equally to each $\bar{Y}_p$. The variable $\bar{Y}_p$ is a measure of the general factor which underlies the subjects' (or matched sets of subjects') responses to the c levels, that is, of whatever it is which produces correlation between the columns (conditions) of the data matrix. The variance of $\bar{Y}_p$ (between S) represents "individual differences" (or matched set differences) in this general factor, and its analysis as a function of G is interpreted accordingly. In some applications of the $S(G)$ by C design, $\bar{Y}_p$ as such may have little meaning or interest, and its analysis as a function of G need not proceed. In others, it may well provide important information of substantive interest to supplement the analysis of the within-S variance.

In coding G (as in coding C), there is available the full flexibility and appositeness to the structure and hypotheses of the research afforded by the MRC system. Concretely, for $g = 6$ (hence, $k_G = g - 1 = 5$), it may be as simple as dummy coding the X_js for a single research factor made up of 5 different experimental and one control condition (Section 5.2), or G may be coded into X_js which represent a $D \times E$ factorial structure: an experimental-control dichotomy D ($k_D = 1$), three levels of a quantitative treatment E coded as linear and quadratic orthogonal polynomials ($k_E = 2$), and their interaction-bearing DE set product ($k_{D \times E} = 2$); thus, $k_G = 1 + 2 + 2 = 5$ (see Sections 6.3, 8.6, and 8.7).

In Figure 10.3.1 below, we show the basic partition of the total Y variance in the $S(G)$ by C design into between-S and within-S portions, with G overlapping the between-S portion as a single set circle. G may be conceptually expanded to indicate any subset structure which it may have (as in $D \times E$ above) into multiple nonoverlapping (for example, in orthogonal factorial design) or overlapping (for example, in nonorthogonal factorial design) circles. In the latter case, the investigator may elect to perform the MRC analysis of the $\bar{Y}_p$ hierarchically (as in Section 8.7).

10.3.2 The Analysis of the Within-Subjects Variance

As before, the total $N = nc$ observations of Y are strung out and analyzed by MRC. There are now, however, three sets of IVs: C, G, and the CG product set which contains the $C \times G$ interaction (that is, $C \times G = CG \cdot C,G$). C is coded into $c - 1$ ($= k_C$) IVs (X_i; $i = 1, 2, \ldots, c - 1$) by whatever scheme suits the analytic needs of the investigation, all n observations of a given level of C being identically coded. Whatever coding scheme was used to code G into its $g - 1$ ($= k_G$) IVs (X_j; $j = 1, 2, \ldots, g - 1$) is again used here; now, however, each subject is represented by c observations of Y all of which receive the same group-membership coding as that subject's $\bar{Y}_p$ had in the preceding analysis. Finally, the CG product IVs are generated by multiplying each of the $c - 1$ variates X_i which code C by each of the $g - 1$ variates X_j which code G, thus producing $(c - 1)$ $(g - 1)$ ($= k_{C \times G}$) variates which are literally of the form $X_i X_j$. Thus, we proceed to code the $(c - 1)(g - 1)$ interaction-bearing IVs *exactly* as for any two-way factorial: each Y observation falls in one of cg cells depending on the specific condition and group in which it occurs, and its coding values for the $(c-1)$ $(g - 1)$ product IVs are found by multiplying its X_i values (for that condition) by its X_j values (for that group).

To illustrate the coding, consider an $S(G)$ by C design in which there are $c = 4$ (effects-coded) conditions and $g = 3$ (dummy-coded) groups which contain a total of $n = 50$ subjects. For the analysis of the resulting $N = nc = 200$ Y observations, there will be $c - 1 = 3$ IVs (X_i) to represent C, $g - 1 = 2$ IVs (X_j) to represent G, and $(c - 1)(g-1) = 6$ IVs (X_iX_j) for the CG product set. In Chapter 8, Table 8.7.1 lays out the coding for C and G (called there respectively H and T) and for their joint coding so as to fully represent the $cg = 12$ condition–group cells. For the present analysis, the coding diagram for these twelve cells would be *exactly* as shown in Table 8.7.1c. There are a total of $k = 3 + 2 + 6 = 11$ (generally, $cg - 1$) IVs to represent the 12 cells. Each of the 200 observations is coded on X_1 through X_{11} according to the cell to which it belongs.

The analysis by MRC is then performed exactly as in any factorial design, as illustrated in Section 8: Proceed hierarchically, first using set C, then sets C and G, and finally sets C, G, and CG. Since the dependent variable here is Y, and our primary interest is in $Y \cdot S$, it will again be necessary to subject some of the results of standard MRC programs (R^2s, significance tests) to further manipulation. The nature of the partitioning of variance in the $S(G)$ by C design is shown in Figure 10.3.1, which will be the focus of the ensuing discussion.

Note first that the total variance of Y (taken as unity or 100%) is again partitioned into between-S (d + f) and within-S (a + i + e) portions. Note also that the Y variance accounted for by G (area d) is wholly in the between-S portion, and that the Y variance accounted for by C (area a) and by $C \times G$ (area i) is wholly in the within-S portion. Moreover, not only can the Y variance of G therefore not overlap with that of C and $C \times G$, but the latter two cannot overlap with each other. Since each group has equal numbers of Y observations in each condition, the difficulties occasioned by the common overlapping in Y among sets of IVs which we have encountered (see, in general, Section 4.3.4 and, in particular, Section 8.7) simply cannot occur. Specifically, the $S(G)$ by C design is an *orthogonal* $C \times G$ factorial design because of its balanced structure— all five lettered areas of Figure 10.3.1 are necessarily nonnegative.

From the hierarchical MRC of the nc observations of Y, we determine the total Y variance (sd_Y^2), and, with the between-S variance ($sd_{\bar{Y}_p}^2$) found in the analysis of $\bar{Y}_p$, find $R_{Y \cdot S}^2$ from Eq. (10.2.2): the proportion of sd_Y^2 which $sd_{\bar{Y}_p}^2$ constitutes. Although we have already fully analyzed the between-S variance, we need $R_{Y \cdot S}^2$, and more particularly $1 - R_{Y \cdot S}^2$, in the adjustment of the results of the MRC analysis of Y so that they are oriented to the within-S (i.e., $Y \cdot S$) variance.

The present analysis yields the cumulative $R_{Y \cdot C}^2$ (= a), $R_{Y \cdot C,G}^2$ (= a + d), and $R_{Y \cdot C,G,CG}^2$ (= a + d + i). From these we obtain, as increments, $R_{Y \cdot G}^2 = R_{Y \cdot C,G}^2 - R_{Y \cdot C}^2$ (= d), and $R_{Y \cdot C \times G}^2 = R_{Y \cdot C,G,CG}^2 - R_{Y \cdot C,G}^2$ (= i). Note that $R_{Y \cdot G}^2$

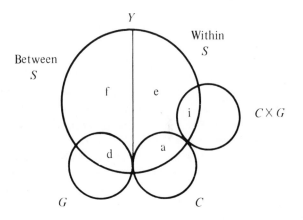

$$d + f + a + i + e = 1$$

Areas as functions of R^2 s R^2 s as functions of areas

Between S:

$$d = R^2_{Y \cdot G}$$ $$R^2_{Y \cdot S} = d + f,$$

$$f = R^2_{Y \cdot S} - R^2_{Y \cdot G}$$ $$R^2_{\bar{Y}_p \cdot G} = \frac{d}{d + f},$$

Within S:

$$a = R^2_{Y \cdot C}$$ $$1 - R^2_{Y \cdot S} = a + i + e;$$

$$i = R^2_{Y \cdot C \times G}$$ (10.2.3) $$R^2_{(Y \cdot S) \cdot C} = \frac{R^2_{Y \cdot C}}{1 - R^2_{Y \cdot S}} = \frac{a}{a + i + e}$$

$$e = 1 - R^2_{Y \cdot S} - R^2_{Y \cdot C} - R^2_{Y \cdot C \times G}$$ (10.3.1) $$R^2_{(Y \cdot S) \cdot C \times G} = \frac{R^2_{Y \cdot C \times G}}{1 - R^2_{Y \cdot S}} = \frac{i}{a + i + e}$$

Fs as Functions of Areas

(5.3.5) $$F = \frac{R^2_{\bar{Y}_p \cdot G}}{1 - R^2_{\bar{Y}_p \cdot G}} \times \frac{n - g}{g - 1} = \frac{d}{f} \times \frac{n - g}{g - 1},$$

(10.3.2) $$F = \frac{R^2_{Y \cdot C}}{1 - R^2_{Y \cdot S} - R^2_{Y \cdot C} - R^2_{Y \cdot C \times G}} \times (n - g) = \frac{a}{e} \times (n - g),$$

(10.3.4) $$F = \frac{R^2_{Y \cdot C \times G}}{1 - R^2_{Y \cdot S} - R^2_{Y \cdot C} - R^2_{Y \cdot C \times G}} \times \frac{n - g}{g - 1} = \frac{i}{e} \times \frac{n - g}{g - 1}.$$

FIGURE 10.3.1 The partitioning of the Y variance for the subjects within groups by conditions design.

(= d) is of no special interest to us; we have already fully analyzed the between-S variance, in which we found the more relevant $R^2_{\bar{Y}_p \cdot G}$, which is d/(d + f).[4] But we do need $R^2_{Y \cdot C}$ (= a) and $R^2_{Y \cdot C \times G}$ (= i).

For the within-S analysis, we need to relate C and $C \times G$ to $Y \cdot S$ instead of Y. We find $R^2_{(Y \cdot S) \cdot C}$ by Eq. (10.2.3) (that is, by simply dividing $R^2_{Y \cdot C}$ by $1 - R^2_{Y \cdot S}$), and interpret it as in the S by C design—the proportion of within-S variance accounted for by C. In terms of the areas of Figure 10.3.1, it is a/(a + i + e). To find the analogous relationship to $Y \cdot S$ of the $C \times G$ interaction, we proceed in exactly the same way, that is,

$$(10.3.1) \qquad R^2_{(Y \cdot S) \cdot C \times G} = \frac{R^2_{Y \cdot C \times G}}{1 - R^2_{Y \cdot S}},$$

which, in areas, is i/(a + i + e). (Note that $R^2_{Y \cdot C \times G} \neq R^2_{Y \cdot (CG)}$ in general; see Footnote 7 below). This is the proportion of within-S variance accounted for by the differential response of the g groups to the c conditions. Like any other two-factor interaction, it reflects the extent to which the effect of one factor depends on the status of the other factor.

10.3.3 Significance Tests

All the significance tests involving G and its constituent X_j were found in the preceding standard MRC analysis of $\bar{Y}_p$ (that is, of the between-S variance). Like all the results of that analysis, the computer output is correct as it stands.

On the other hand, neither the Fs nor the ts of the standard MRC analysis of Y are correct, since (as in the S by C design) they use the wrong error variance. To test the statistical significance of C and $C \times G$, one constructs the usual ratio of source variance to the correct error variance and multiples by a function of the df. For both C and $C \times G$, the appropriate error variance is the "residual" within S variance, area e in Fig. 10.3.1 (and not $1 - R^2_{Y \cdot C, G, CG}$ = e + f, as the standard MRC computer program assumes). For C, the test may be written in terms of R^2s as

$$(10.3.2) \qquad F = \frac{R^2_{Y \cdot C}}{1 - R^2_{Y \cdot S} - R^2_{Y \cdot C} - R^2_{Y \cdot C \times G}} \times (n - g),$$

with $df = c - 1$ for the numerator, and $(n - g)(c - 1)$ for the denominator. (Note that, as is conventional in AV, we use Model II error throughout this chapter.) From the construction of the left-hand term, it can be seen that it is the ratio of

---

[4] It may occur to the reader that since this analysis yields $R^2_{Y \cdot G}$, and since $R^2_{\bar{Y}_p \cdot G}$ = $R^2_{Y \cdot C}/R^2_{Y \cdot S}$, we could dispense with the MRC analysis of $\bar{Y}_p$ on G, simply finding $sd^2_{\bar{Y}_p}$ to obtain $R^2_{Y \cdot S}$. This is true, but this route would also necessitate additional calculation for the other results (F and for each X_j its t_j, sr_j, and pr_j) to correct them to $R^2_{Y \cdot S}$, just as we must correct the within-S results to $1 - R^2_{Y \cdot S}$. It seems far easier to perform the relevant MRC analysis of $\bar{Y}_p$, which may itself be complicated, and avoid these unnecessary additional complications. (Also see Footnote 7 below.)

areas a/e.[5] Alternatively, an equivalent F test of C can be found from

$$(10.3.3) \qquad F = \frac{R^2_{(Y \cdot S) \cdot C}}{1 - R^2_{(Y \cdot S) \cdot C} - R^2_{(Y \cdot S) \cdot C \times G}} \times (n - g),$$

with, of course, the same $df.$[6]

Similarly, the test of significance of $C \times G$ may be written in terms of the R^2s as

$$(10.3.4) \qquad F = \frac{R^2_{Y \cdot C \times G}}{1 - R^2_{Y \cdot S} - R^2_{Y \cdot C} - R^2_{Y \cdot C \times G}} \times \frac{n - g}{g - 1},$$

with $df = (g - 1)(c - 1)$ for the numerator, and $(n - g)(c - 1)$ for the denominator. The left-hand term is the ratio of areas i/e. Again, an equivalent F for $C \times G$ may be found from

$$(10.3.5) \qquad F = \frac{R^2_{(Y \cdot S) \cdot C \times G}}{1 - R^2_{(Y \cdot S) \cdot C} - R^2_{(Y \cdot S) \cdot C \times G}} \times \frac{n - g}{g - 1},$$

with the same $df.$

The results for individual IVs are given in the final MRC analysis in which Y is regressed simultaneously on the X_is (for C), the X_js (for G), and the $X_i X_j$ products (for $C \times G$). The raw score B coefficients for all $k = cg - 1$ of these IVs are correct, but not so their SE_B and t values, since they, too, are generated using incorrect error variance. Before proceeding to their correction, we first note that the (correct) B coefficients for the X_js coding G and their incorrect ts are of no interest to us, since both G and its constituent X_j were fully analyzed in the preceding between-S analysis.[7] We thus are left needing to correct the SE_Bs and the ts for the individual X_is that code C and the individual $X_i X_j$ products that code $C \times G$. Since both types are corrected in the same way, we will use the i subscript to represent both X_i and $X_i X_j$ IVs. The correct standard error[8] of these Bs is

$$(10.3.6) \qquad SE_{B_i} = \frac{sd_Y}{sd_i} \sqrt{\frac{1 - R^2_{Y \cdot S} - R^2_{Y \cdot C} - R^2_{Y \cdot C \times G}}{(n - g)(c - 1)}} \sqrt{\frac{1}{1 - R^2_i}}.$$

[5] Since the R^2s in the denominator are additive, it can be more compactly written as $1 - R^2_{Y \cdot S, C, C \times G}$, that is, area e. This is analogous to Eq. (10.2.4), which asserts the additivity of $R^2_{Y \cdot S}$ and $R^2_{Y \cdot C}$ in the S by C design.

[6] As in Footnote 5, we can write the denominator here as $1 - R^2_{(Y \cdot S) \cdot C, C \times G}$.

[7] The reason for the inclusion of G in the analysis of Y is so that it be partialled (together with C) from the CG product set in order to make of the latter a true interaction, a fundamental principle of Chapter 8. The inclusion of G as well as C here confers freedom in the coding of both C and G, since, however they may be coded, $C \times G = CG \cdot C, G$. Thus, the product set as a whole and also its constituent IVs in the final equation may be unambiguously interpreted in terms of interactions. Note that although however coded, C and G are orthogonal in Y (i.e., $R^2_{Y \cdot C, G} = R^2_{Y \cdot C} + R^2_{Y \cdot G}$), in general CG and C, and CG and G are not. Thus, $R^2_{Y \cdot C \times G}$ must, in general, be found hierarchically—it does not, in general, equal $R^2_{Y \cdot (CG)}$.

[8] The same assumption of homogeneity of Footnote 3 above is made here, replacing interactions with subjects by interactions with subjects within groups.

The correct SE again differs from the standard fixed model SE of Eq. (3.7.8) only in the error variance proportion and its df in the middle term. Again, since $t_i = B_i/SE_{B_i}$, the incorrect t_i' given in the computer output can be converted to the correct t_i for the IVs constituting the sets C and $C \times G$ by substituting t_i' in

(10.3.7) $$t_i = t_i' \sqrt{\frac{1 - R_{Y \cdot C,G,CG}^2}{1 - R_{Y \cdot S}^2 - R_{Y \cdot C}^2 - R_{Y \cdot C \times G}^2} \times \frac{c-1}{c}},$$

with $df = (n - g)(c - 1)$. As an alternative to Eq. (10.3.6), SE_{B_i} can be found by dividing the B_i by its t_i.

10.3.4 *sr* and *pr* for a Single Independent Variable

We again take advantage of the relationship of t_i to sr_i and pr_i to find the latter. Keep in mind that the dependent variable here is $Y \cdot S$, so that the variance being accounted for uniquely by an X_i or $X_i X_j$ is within-S, not total, Y variance.

(10.3.8) $$sr_i^2 = \frac{t_i^2 \ (1 - R_{(Y \cdot S) \cdot C}^2 - R_{(Y \cdot S) \cdot C \times G}^2)}{(n - g)(c - 1)},$$

with i again standing for either X_i or $X_i X_j$. This is the proportion of the within-S Y variance accounted for uniquely by the IV, that is, by the IV from which the remaining $g(c - 1) - 1$ IVs accounting for within-S variance have been partialled. In Figure 10.3.1, this is some part of area a or i expressed as a proportion of a + i + e.

For pr_i^2, we again use the t_i of Eq. (10.3.7), substituting in

(10.3.9) $$pr_i^2 = \frac{t_i^2}{t_i^2 + (n - g)(c - 1)}.$$

This expresses the variance due uniquely to X_i (or $X_i X_j$) as a proportion of the within-S variance not accounted for by the remaining $g(c - 1) - 1$ IVs. In Figure 10.3.1, this is some part of area a or i expressed as a proportion of the sum of area e and that part of a or i. pr_i and sr_i, as always, take the same sign as B_i.

No new principles are involved in the interpretation of the partial coefficients (B, sr, pr) of the single IVs constituting C and $C \times G$ (and for that matter G) in the $S(G)$ by C design. They are interpreted in the usual way in accordance with the coding methods used. Each IV of the product set is interpreted as a single interaction facet (see footnote 7), an "aspect of C by aspect of G" effect, the aspects being determined by the coding methods used. The discussion and illustration of these discrete interaction effects in Section 8.7 is fully applicable here, with the usual proviso that the dependent variable is $Y \cdot S$ and not Y.

10.3.5 Illustrative Example

Let us assume the $S(G)$ by C design already used to illustrate the coding in Section 10.3.2: There are a total of 50 ($= n$) subjects (or subject sets) belonging to 3 ($= g$) groups (G) and giving rise to observations (Y) under each of 4 ($= c$) conditions (C), thus yielding a total of 200 ($= nc = N$) observations. The 3 variates of set C $(X_i: X_1, X_2, X_3)$ are effects-coded, the 2 variates of set G $(X_j: X_4, X_5)$ are dummy-coded, and the 6 variates of the product set CG $(X_iX_j: X_6, X_7, \ldots, X_{11})$ are found in the usual manner.

The Between-Subjects Analysis

The $\bar{Y}_p$ for each subject (or set of matched subjects) is found and serves as the dependent variable in an MRC analysis in which it is regressed against the dummy-coded X_4 and X_5 of set G, with $n = 50$, $k_G = g - 1 = 2$. Assume we find $sd^2_{\bar{Y}_p} = 32$ and $R^2_{\bar{Y}_p \cdot G} = .15$ [= d/(d + f) in Figure 10.3.1]. The F value by the standard general Eq. (3.7.1) or its equivalent, Eq. (5.3.5), equals 4.147, which gives $P < .05$ for $df = 2, 47$. Thus, the three groups differ significantly on the general factor running over conditions reflected in $\bar{Y}_p$, accounting for 15% of the variance of individual (or matched set) differences in this factor. Given the dummy coding, X_4 when partialled carries the $G_1 - G_3$ distinction in regard to $\bar{Y}_p$, so that B_4 is simply the overall Y mean difference between G_1 and the control G_3. Assume $B_4 = 3.81$, with $t_4 = 2.084$, $df = 47$; hence, $P < .05$. Thus, mean value of the G_1 for $\bar{Y}_p$ is 3.81 units higher than that of G_3, and significantly so. This can also be expressed in proportion of variance terms as

$$sr^2_4 = \frac{2.084^2 (1 - .15)}{50 - 3} = .0785,$$

or as

$$pr^2_4 = \frac{2.084^2}{2.084^2 + (50 - 3)} = .0846,$$

using the simple Appendix 3 formulas, which give sr and pr as functions of t. These are interpreted as described in Section 5.3; indeed, all the discussion and equations in Section 5.3 obtain here, with $\bar{Y}_p$ substituted for Y. Note, for example, that the number of subjects per group can not be equal here, since 50 subjects are divided into 3 groups, but we have seen in Chapter 5 that this does not affect the analysis.

We stress again that the standard computer output is correct as it stands for the between-S analysis.

The Within-Subjects Analysis

The hierarchical MRC analysis of Y for the 200 observations provides the results which, when corrected to reflect $Y \cdot S$, constitutes the within-S analysis.

Assume that $sd^2_Y = 80$. Taken with the previously determined $sd^2_{\bar{Y}_p} = 32$, by Eq. (10.2.2), $R^2_{Y \cdot S} = 32/80 = .40$.[9] This is area d + f in Figure 10.3.1. Its

---

[9] Unfortunately, many computer programs give as the sd^2 of a variable what is actually the estimate of its population variance, the sum of squared deviations divided by the df, the number of observations *minus one*. Our definition of sd^2 is as a sample descriptor which requires division by the number of observations. Thus, $sd^2_{\bar{Y}_p}$ implies division by n (= 50), and sd^2_Y division by N (= 200). To correct a variance estimate to a variance, multiply by the ratio: number of observations minus one/number of observations.

complement, $1 - R^2_{Y \cdot S} = .60$, is the proportion of within-S variance, area a + i + e. Although not as large as in the S by C numerical example (where $R^2_{Y \cdot S}$ was .60), this still represents a considerable "blocking" effect, and hence efficiency of the design, as we shall see below.[10]

In the hierarchical setwise MRC analysis, we find

$$R^2_{Y \cdot C} = .18 \ (= \text{area a}),$$

$$R^2_{Y \cdot C,G} = .24 \ (= \text{area a + d}),$$

$$R^2_{Y \cdot C,G,CG} = .36 \ (= \text{area a + d + i}).$$

From these we find the increments

$$R^2_{Y \cdot G} = R^2_{Y \cdot C,G} - R^2_{Y \cdot C} = .24 - .18 = .06 \ (= \text{area d}),$$

$$R^2_{Y \cdot C \times G} = R^2_{Y \cdot C,G,CG} - R^2_{Y \cdot C,G} = .36 - .24 = .12 \ (= \text{area i}).$$

Since the effect of G on $\bar{Y}_p$ has been fully analyzed in the between-S analysis, we have no need for $R^2_{Y \cdot G}$, but note that $R^2_{\bar{Y}_p \cdot G} = R^2_{Y \cdot G}/R^2_{Y \cdot S}$, that is, .15 = .06/.40, as it should; in terms of areas of Figure 10.3.1, $R^2_{Y \cdot G} = d$, while $R^2_{\bar{Y}_p \cdot G} = d/(d + f)$. We do however, require $R^2_{Y \cdot C}$ and $R^2_{Y \cdot C \times G}$, which must now be corrected to $Y \cdot S$ as the dependent variable. Thus, from Eq. (10.2.3), we find

$$R^2_{(Y \cdot S) \cdot C} = \frac{.18}{1 - .40} = .30.$$

Thus, C accounts for 30% of the within-S variance, a/(a + i + e). Similarly, the $C \times G$ interaction accounts for .12 (= i) of the total Y variance, but, by Eq. (10.3.1), it accounts for

$$R^2_{(Y \cdot S) \cdot C \times G} = \frac{.12}{1 - .40} = .20$$

of the within-S variance, i/(a + i + e).

The F test value for C may be found by either of the equivalent Eqs. (10.3.2) or (10.3.3):

$$F = \frac{.18}{1 - .40 - .18 - .12} \times (50 - 3) = \frac{.18}{.30} \times 47 = 28.2,$$

$$F = \frac{.30}{1 - .30 - .20} \times (50 - 3) = \frac{.30}{.50} \times 47 = 28.2,$$

with numerator $df = c - 1 = 3$ and denominator $df = (n - g)(c - 1) = 47(3) = 141$, and is highly significant. Similarly, the F test value for $C \times G$ may be found

---

[10] When all correlations between pairs of columns (conditions) in all groups equal zero, $R^2_{Y \cdot S} = [1 + (c - 1)R^2_{Y \cdot G} - R^2_{Y \cdot C} - R^2_{Y \cdot C \times G}]/c$. For the values of the R^2s posited below, this gives $R^2_{Y \cdot S} = .22$, substantially less than the actual $R^2_{Y \cdot S} = .40$ found. See Footnote 12.

by either of the equivalent Eqs. (10.3.4) or (10.3.5):

$$F = \frac{.12}{1 - .40 - .18 - .12} \times \frac{50 - 3}{3 - 1} = \frac{.12}{.30} \times \frac{47}{2} = 9.4,$$

$$F = \frac{.20}{1 - .30 - .20} \times \frac{50 - 3}{3 - 1} = \frac{.20}{.50} \times \frac{47}{2} = 9.4,$$

with numerator $df = (c - 1)(g - 1) = 6$ and denominator $df = (n - g)(c - 1) = 141$, and is also highly significant.

Since both C and $C \times G$ were found to be significant, the assessment of the significance of the partial coefficients of their constituent IVs is warranted by our protected t procedure. Assume, for example, that $B_1 = 5.41$. Since X_1 is the effects-coded variable on which C_1 is coded $+ 1$, this means that the effect of C_1 is 5.41, specifically that the mean of the 50 observations on C_1 is 5.41 units higher than the mean of all four condition means which, since all conditions have the same n, is the grand mean of all 200 observations. The incorrect t'_1 for this B_1 is given by the computer as 3.281 (with incorrect $df = c[n - g] = 188$). The correct t_1 is found from Eq. (10.3.7) as

$$t_1 = 3.281 \sqrt{\frac{1 - .36}{1 - .40 - .18 - .12} \times \frac{4 - 1}{4}} = 4.150,$$

with $df = (n - g)(c - 1) = 141$, which gives $P < .01$.[11] Thus the effect of C_1 is significant.

With t_1 determined, we can find by Eq. (10.3.8)

$$sr_1^2 = \frac{4.150^2 \ (1 - .30 - .20)}{(50 - 3)(4 - 1)} = .0611.$$

Recalling that .30 ($= R^2_{(Y \cdot S) \cdot C}$) of the within-$S$ variance was accounted for by the set C, we note that .0611 out of that proportion is accounted for by the effect of C_1. (The sr_i^2s are, however, not additive; see Section 5.4.3.)

For pr_1^2, we substitute in Eq. (10.3.9), finding

$$pr_1^2 = \frac{4.150^2}{4.150^2 + (50 - 3)(4 - 1)} = .1089,$$

which is the proportion of the within-S variance, excluding the portion accounted for by the effects of the other conditions and of the $C \times G$ interaction, which is accounted for by the C_1 effect.

To exemplify the results for an interaction IV, we consider those for $X_6 = X_1 X_4$. From the coding of C (effects) and G (dummy) described in Section 10.3.2, X_6 when partialled carries the difference in the effect of C_1 (represented in X_1) between G_1 and the control G_3 (represented in X_4). Assume $B_6 = 3.69$,

---

[11] If the correct SE_{B_i} is desired, it may be found from Eq. (10.3.6), or more easily (once t_i has been determined) as B_i/t_i, so for X_1, $SE_{B_1} = 5.41/4.150 = 1.30$.

that is, the C_1 effect, which we found above averaged 5.41 $(= B_1)$ over all 50 cases, is 3.69 units higher on the average in G_1 than it is in G_3. Equivalently, we can say that the $G_1 - G_3$ difference, which in the between-S analysis we found to average 3.81 $(= B_4)$ over all conditions, is 3.69 units higher in C_1.

Assume that the incorrect computer-generated t_6' for B_6 is 1.633 (again for incorrect $df = 188$); by Eq. (10.3.7), the correct

$$t_6 = 1.633 \sqrt{\frac{1 - .36}{1 - .40 - .18 - .12} \times \frac{4 - 1}{4}} = 2.066,$$

which, for $df = 141$, is significant at $P < .05$. We can now, as before, use Eq. (10.3.8) to find

$$sr_6^2 = \frac{2.066^2 (1 - .30 - .20)}{(50 - 3)(4 - 1)} = .0151,$$

and Eq. (10.3.9) to find

$$pr_6^2 = \frac{2.066^2}{2.066^2 + (50 - 3)(4 - 1)} = .0294,$$

small but statistically significant proportions of variance that are interpreted in the usual way.

The final regression equation for Y contains all 11 $(= cg - 1)$ B coefficients and A. Although all the significance test results given by a standard MRC computer program are incorrect, the Bs and A are all correct, and may be interpreted and combined as in any two-way factorial, as illustrated in Section 8.7.

For our illustrative example, we exemplified both C and G as simply structured research factors. We remind the reader that this need not be the case. Either or both may be complex. All that is required is that the coding of the complex factor correctly reflect its structure. For example, if G is itself structured as a $U \times V$ factorial, it is coded accordingly. When the CG product IVs are found, they are $CG = C(UV) = CU, CV,$ and $CUV,$ which when partialled become the $C \times U, C \times V,$ and $C \times U \times V$ interaction sources, respectively. It then may be desirable to perform the between-S analysis hierarchically, and in the MRC analysis of Y to proceed hierarchically through the $CU, CV,$ and CUV subsets. A similar argument holds for a complex C. No new principles are involved, since complex structure (for example, G as a $U \times V$ factorial) differs from simple structure (for example, G as dummy variables) only in degree and not in kind. The coding of a research factor *is* its structure, and, together with causal-substantive considerations, determine the details of the analysis and its interpretation. Also, it is when the structure is complex that the MRC has clear computational advantages over the AV approach.

10.3.6 Power Analysis

The procedures for power analysis of a $S(G)$ by C design are as described in Sections 4.5 and 3.8, with some adaptation of the formulas for the within-S effects similar to that of the S by C design (Section 10.2.5).

For the analysis of Y_p as a function of G or its constituent X_j one simply follows the methods of Section 3.8 (and when G is made up of subsets, Section 4.5), keeping in mind only that the dependent variable is $\bar{Y}_p$ and not Y.

For the within-S analysis, as was the case for significance testing, we follow the AV convention and presume Model II error in the power analysis. For set C, the ES is given by an adaptation of Eq. (4.5.3), which, for this design, is

$$(10.3.10) \qquad f^2 = \frac{R^2_{Y \cdot C}}{1 - R^2_{Y \cdot S} - R^2_{Y \cdot C} - R^2_{Y \cdot C \times G}},$$

or, equivalently,

$$(10.3.11) \qquad f^2 = \frac{R^2_{(Y \cdot S) \cdot C}}{1 - R^2_{(Y \cdot S) \cdot C} - R^2_{(Y \cdot S) \cdot C \times G}},$$

where the R^2 are alternate-hypothetical or estimated *population* values. To determine n^* as a function of power, first find L as usual from Appendix Tables E by entering the table for the specified α with the power desired (column) and the numerator df, k_B (row); for C, $k_B = c - 1$. To find n^*, the number of *rows* (subjects or subject sets) required by the above specification of α, power, f^2, and k_B, substitute in

$$(10.3.12) \qquad n^* = \frac{L}{f^2 (c - 1)} + g.$$

The number of *observations* required therefore will be n^*c.

To determine power as a function of n for set C, the process is reversed: Substitute the f^2 for C in

$$(10.3.13) \qquad L^* = f^2 (n - g)(c - 1),$$

and refer L^* to the appropriate Appendix Table E in the row $k_B = c - 1$ and find the power values between which it falls, interpolating if desired.

To determine n^* for the test of $C \times G$, f^2 is found by replacing the numerator of Eq. (10.3.10) by $R^2_{Y \cdot C \times G}$ or, equivalently, the numerator of Eq. (10.3.11) by $R^2_{(Y \cdot S) \cdot C \times G}$. L is found in the usual way (although not necessarily for the same power and α specified for the test of C), with $k_B = (c - 1)(g - 1)$.

If the problem is inverted and one wishes to determine power as a function of n for a test of $C \times G$, one proceeds as above, substituting its f^2 in Eq. (10.3.13), and assessing the resulting L^* by reference to the relevant Appendix Table E at row $k_B = (c - 1)(g - 1)$.

When the power analysis for set C in the S by C design was discussed (Section 10.2.5) it was shown how the blocking effect of this design increased its efficiency relative to that of a simple fixed design when $R^2_{Y \cdot S}$ was substantial. This was expressed in Eq. (10.2.14) as the ratio of numbers of observations required by the S by C design, $O^*_1 = n^*_1 c$, to the number required by the simple fixed design, $O^*_2 = n^*_2$, when ES, α and power are the same. The same advantage accrues in the power analysis of both C and $C \times G$ to the $S(G)$ by C design compared with a fixed $C \times G$ factorial. The ratio here is approximately (again, a generally slight underestimate)

(10.3.14)
$$\frac{O_1^*}{O_2^*} \approx \frac{c(1 - R_{Y \cdot S}^2 - R_{Y \cdot C}^2 - R_{Y \cdot C \times G}^2)}{(c-1)(1 - R_{Y \cdot C, G, CG}^2)}.$$

This ratio will be less than one (and, hence, blocking is advantageous) when the ratio of error terms is smaller than $(c-1)/c$ (approximately). As $R_{Y \cdot S}^2$ increases, the ratio decreases, that is, the $S(G)$ by C design increases in relative power efficiency.[12]

For the power analysis of single IVs in sets C or $C \times G$ in the $S(G)$ by C design, the ES is

(10.3.15)
$$f^2 = \frac{sr_i^2}{1 - R_{(Y \cdot S) \cdot C}^2 - R_{(Y \cdot S) \cdot C \times G}^2},$$

where the terms are defined as above but, of course, as estimated for the population. To find n^* as a function of power, this f^2 is substituted in Eq. (10.3.12), with L looked up at $k_B = 1$. To find power as a function of n, L^* is found from Eq. (10.3.13), and looked up in the appropriate Appendix Table E for $k_B = 1$.

Although the formulas given in Sections 3.8 and 4.5 require adaptation for the $S(G)$ by C design, the general principle and procedures presented there are fully applicable. Indeed, the whole point of the $S(G)$ by C and S by C designs is their greater power efficiency relative to their fixed-model counterparts.

10.4 SUMMARY

This chapter considers the MRC analysis of "mixed-model" research designs in which each of n subjects or sets of matched subjects (S) gives rise to c observations, one under each condition (C), thus resulting in $N = nc$ observations in an "S by C" design ("repeated measurements" or "matched subjects"). When, in addition, the subjects are distinguished as members of g groups (G), the result is called a "$S(G)$ by C" design ("split plot"). Depending on the nature of the investigation, C and G may each be coded variously as simple nominal scales, orthogonal polynomials ("trend analysis"), or factorial combinations (Section 10.1).

---

[12] For both the S by C and $S(G)$ by C designs, the estimation of the population $R_{Y \cdot S}^2$ requires comment. As Footnotes 2 and 10 indicate, in both designs, $R_{Y \cdot S}^2$ depends upon c, and even when there is no blocking effect, $R_{Y \cdot S}^2$ will typically be sizable when c is small. In estimating $R_{Y \cdot S}^2$ by reference to past results, this dependence on c must be taken into account. In any given study, unless $R_{Y \cdot S}^2$ can be estimated as being materially larger than the values found from the formulas in Footnotes 2 (for S by C) or 10 [for $S(G)$ by C], or equivalently, unless the O_1^*/O_2^* ratios of Eqs. (10.2.14) or (10.3.14) are distinctly less than unity (say, no more than .8), it hardly pays to proceed with these mixed-model designs.

In the S by C design, a fundamental distinction is made between the between-S variance, $R^2_{Y \cdot S}$ (Section 10.2.1) and the within-S variance, $1 - R^2_{Y \cdot S}$, which are separately analyzed. The within-S variance is analyzed through an MRC analysis for the $N = nc$ observations with Y as the dependent variable and set C (made up of appropriately coded X_i) as the IVs. Since it is the partialled $Y \cdot S$ whose variance is the within-S variance and not Y, the results of the standard MRC analysis must be revised accordingly with regard to both the base for proportions of variance, for example, $R^2_{(Y \cdot S) \cdot C}$ rather than $R^2_{Y \cdot C}$ (Section 10.2.2), and significance tests (Section 10.2.3). Also, the sr and pr of the individual X_i which make up C are defined and interpreted with $Y \cdot S$ as the dependent variable (Section 10.2.4). Finally, the methods of power analysis for C and its constituent X_i are described, and the generally greater power efficiency of the S by C design relative to the simple fixed design is demonstrated (Section 10.2.5).

In the $S(G)$ by C design, the same between-S–within-S distinction is made, and separate MRC analyses are performed. In the between-S analysis, the mean of the c observations of each subject (or matched set of subjects), $\overline{Y}_p$, is the dependent variable in a standard fixed-model MRC analysis for n cases with set G (coded as X_j) as the IVs (Section 10.3.1). For the within-S analysis, using all $N = nc$ observations, a standard MRC analysis for a two-way factorial is first performed: with Y as the dependent variable, sets C, G, and CG are entered hierarchically. The structure of the design is such that the Y variance due to G is part of the between-S variance, the Y variance due to C and to $C \times G$ is part of the within-S variance, and all three sources are additive (nonoverlapping in Y). The relationship among these and error sources of variance are represented by areas in a modified ballantine, and formulas for various R^2s and F tests are given and expressed in terms of these areas (Figure 10.3.1). The revisions of the standard computer output of the analysis of Y as a function of C and $C \times G$ which are necessary in order to shift the base to within-S variance (that is, the variance of $Y \cdot S$) are given both with regard to the R^2s (Section 10.3.2) and significance tests (Section 10.3.3). The analogous corrections to $Y \cdot S$ as the dependent variable are given for the sr and pr of the individual IVs which make up C and $C \times G$ (Section 10.3.4). An illustrative numerical example is given to concretize the relationships (Section 10.3.5). The chapter closes with a section that describes power analysis in the $S(G)$ by C design and demonstrates its power efficiency relative to a fixed $C \times G$ factorial (Section 10.3.6).

11

Multiple Regression/Correlation and Other Multivariate Methods

11.1 INTRODUCTION

Our goal in this chapter is modest, of necessity. Having spent a book's worth in the exposition of MRC, obviously several books would be necessary for comparable exposition of the other multivariate (MV) methods. All we mean to do here is sketch out the relationship of MRC to these, provide some information about their purposes and characteristics, and show how MRC can accomplish some of these and, in some circumstances, more. In doing this, we show that the MRC method, as general as we have demonstrated it to be, is nevertheless a special case of the yet more general method of canonical correlation and regression analysis (CA), which (inverting the relationship) not only specializes to MRC, but also to the other major MV methods. The mathematical structure is beautiful, provocative of richly rewarding insights, but, inevitably, quite complicated—about an order of magnitude more so than MRC.

In the relatively few pages which we devote to MV methods, we will not penetrate this complexity, but seek to bound it and describe it. Here we will not consider "how," but rather "what." Our treatment will be completely verbal–intuitive, without computational methods or worked examples. We presume the existence of tests of statistical significance, without describing them, and ignore issues of distribution assumptions. The result should serve as a conceptual introduction to this field for our audience of behavioral scientists. For those whose appetites are whetted, we strongly recommend Cooley and Lohnes' *Multivariate Data Analysis* (1971), whose approach differs from ours, but shares with it the data-analytic spirit and the emphasis on worked and interpreted examples. Another good treatment, particularly for readers with a strong preparation in AV, is provided by Tatsuoko (1971). Bock (1975) is excellent for those with a good mathematical statistics background.

11.2   THE CANONICAL GENERALIZATION

MRC treats the relationship between a single dependent variable Y and a group of k independent variables, X_i. We have seen that by the use of various techniques, we can package in the Xs information of varying structure and analytic function as research factors and describe their whole and partial relationships to Y. This affords much generality—the MRC system developed in this book incorporates and extends AV and ACV, partial and multiple partial correlation analysis, and the analysis of curvilinear relationships. But all this generality is accomplished using the multiplicity of the X variables—MRC deals with only one Y variable.

Canonical correlation and regression analysis (CA) is that generalization of MRC which provides for multiplicity of Y variables as well: it provides a general method for relating a group of k different X variables (designated X_i) to a group of p different Y variables (designated Y_j). Note that this is not merely a matter of successively relating the X_i first to Y_1, then to Y_2, etc., up to Y_p, in a series of MRC analyses, no more so than MRC relates its Y first to X_1, then to X_2, etc., up to X_k in a series of bivariate analyses. Rather, in CA, the p Y variables are analyzed simultaneously, as the k X variables are.

That such analytic freedom is useful in behavioral science hardly needs demonstrating. We have seen throughout the book the utility, for both formal and structural purposes, of multiplicity. It makes possible representing such research factors as treatments, socioeconomic status, college major, and length of hospitalization on the X side of the equation. What's good for the Xs is good for the Ys.

Multiple regression/correlation analysis solves the problem of relating k X variables to a single Y by finding a set of regression coefficients or weights with which to fashion of the X variables a linear composite that is maximally correlated with Y, as expressed in the regression equation. Canonical analysis begins by solving the more difficult MV problem of finding *two* sets of weights, one set for the X variables and another for the Y variables such that, when separate linear composites are found, they are maximally correlated. We assume that all variables have been standardized[1]; thus, CA solves for the b_is and c_js of the *canonical variates x and y*, ·

(11.2.1) $$x = \sum b_i X_i \quad (i = 1 \text{ to } k),$$

(11.2.2) $$y = \sum c_j Y_j \quad (j = 1 \text{ to } p),$$

where x and y each have variance of unity. The b and c weights are quite analogous to the βs of MRC. Most particularly, they are *partial* regression

---

[1] The standardization is purely a matter of convenience in exposition and the avoidance of unduly complicated notation, and loses no generality. As we have seen in MRC, the scaling information carried by the means and *sd*s of the variables is readily reintroduced when and as necessary.

coefficients, each being the optimal *net* contribution of its variable in the context of the other variables in its group. Equations (11.2.1) and (11.2.2) are thus much like regression equations, except that instead of yielding least-squares estimates of a dependent variable, they yield the canonical variates x and y, which are least-squares estimates of, and hence correlate maximally with, each other. This (zero-order) correlation between the canonical variates x and y is the canonical correlation, designated R_c. Explicity,

$$(11.2.3) \qquad\qquad R_c = r_{xy},$$

and R_c^2 is the proportion of variance of y accounted for by x, and vice versa. Just as in MRC, R must be at least as large as the largest r_{Yi}, so in CA, R_c must be at least as large as the largest r between the X and Y variables, and also as the largest R of either set with a variable of the other.

Before considering the interpretation of the canonical variates, another feature of CA must be considered. Each canonical variate accounts for only a portion of the total variance of its own group of variables. Thus, the k standardized X variables each has $sd^2 = 1$; hence, the total variance of the X variables is k. Similarly, the total variance of the Y variables is p. It is possible to extract or partial out in each group its (first) canonical variate, and then repeat the entire process on the two groups of residual variables. New weights can be determined which produce canonical variates x_2 and y_2 from the two sets of residuals, such that *their* correlation is a maximum. This is R_{c2}, the second canonical correlation, which is necessarily smaller than the first canonical correlation (which we can now rechristen R_{c1}) between the first pair of canonical variates (now x_1 and y_1). The process is repeated, yielding progressively smaller R_cs. There can be no more pairs than the lesser of k and p, and although all are typically found, for interpretive purposes those with nonsignificant (or small) R_cs are discarded, and only q pairs are retained. The residualizing process assures a condition of orthogonality such that all correlations among the x variates and all correlations among the y variates are zero, as are correlations between unmatched x, y variates, for example, $r_{x_1 y_2} = 0$.

A recent development in CA provides another useful measure of effect size. Recall that each canonical variate accounts for a definable proportion of its group's total variance. Thus, we symbolize as P_{Yy_h} the proportion of the total variance of the Y variables extracted by the hth y variate. Now, R_{ch}^2 is the proportion of the variance of the hth y variate accounted for by its paired x variate. We can now define an index of redundancy,

$$(11.2.4) \qquad\qquad R_{dy_h}^2 = P_{Yy_h} R_{ch}^2,$$

the proportion of the total Y variance accounted for by x_h via y_h. Concretely, assume that y_1 extracts 40% of the total Y variance (i.e., $P_{Yy_1} = .40$), and x_1 accounts for half of the variance of y_1 (i.e., $R_{c1}^2 = .50$). Then Eq. (11.2.4) simply states that x_1 will account for 20% of the total Y variance, that is, $R_{dy_1}^2$

= .40(.50) = .20. Because of the orthogonality condition among nonpaired canonical variates, the redundancy indices for $x_1, x_2, \ldots, x_q$ are additive to the total redundancy, R^2_{dY}, the proportion of the total Y variance accounted for by all the X variables via their q canonical variates:

$$(11.2.5) \qquad R^2_{dY} = R^2_{dy_1} + R^2_{dy_2} + \cdots + R^2_{dy_q} .$$

R^2_{dY} provides a measure of the "total" relationship to the Y group analogous to MRC's $R^2_{Y \cdot 123 \ldots k}$.

All the above holds when one interchanges the xs and ys, and the X group and Y group. Note, however, that although the R^2_cs are symmetrical ($r^2_{xy} = r^2_{yx}$), this is *not*, in general, the case either for the per-variate redundancies ($R^2_{dy_h} \neq R^2_{dx_h}$), or for the total redundancies ($R^2_{dY} \neq R^2_{dX}$). This is simply because it is not true, in general, that (for example) x_1 accounts for the same proportion of the total X group variance as y_1 does of the total Y group variance. If, for example, in the illustration of the preceding paragraph, x_1 accounts for only .30 ($= P^2_{Xx_1}$) of its group's total variance, then $R^2_{dx_1} = .30(.50) = .15$, whereas $R^2_{dy_1} = .40(.50) = .20$. With multiple variables for both Y and X, we can choose to consider either group causally dependent on the other or make no causal assumptions, in accordance with the conceptual framework underlying the analysis.

This already constitutes a formidable amount of analytic apparatus, and we have not yet addressed ourselves to the substantive interpretation of the canonical variates. We have q "aspects" of the X group, packaged as xs, and their paired "aspects" of the Y group, packaged as ys. Each aspect is produced by a different set of partial weights (the b_is and c_js), which mathematically define it, as in Eqs. (11.2.1) and (11.2.2). Efforts to substantively interpret a given x in terms of the X_is by means of the relative magnitudes of its b_is (and analogously for the c_js of a given y for the Y_js) run into the problems we have seen in the interpretation of partial coefficients in MRC—they are subject to the same effects of redundancy and suppression among the variables in the group. They can, as in MRC, be rescaled as Bs, prs, or srs, which, however useful they might be for certain interpretive purposes, do not solve the problem of canonical variate interpretation, since they are all *partial* coefficients.

Instead, interpretation of a given canonical variate is best undertaken by means of the "structure" coefficients, which are simply the (zero-order) correlations of that variate with its constituent variables (as was $r_{\hat{y}i}$ in MRC—see Section 3.7.6). Consider, as a concrete example, that the Y variables are being interpreted as dependent on the X variables, and we find that y_1 correlates .60 ($= R_{c1}$) with x_1. To understand the nature of the y_1 aspect of the Y dependent variables, one can find the simple correlation of each Y_j variable with y_1, that is, r_{y_1j} is the structure coefficient of Y_j for y_1. By observing which Y_j are more and which less correlated with y_1, one has a basis for a substantive interpretation of y_1 couched in terms of the Y variables. This tells us "what" it is that x_1 relates to with a correlation of .60.[2] Similarly, the "whatness" of x_1 can be interpreted by reference to the structure coefficients of the X_i variables, that is, the string of $r_{x_1 i}$s. Another useful relationship lies in the fact that the proportion of the total variance of the group of variables which is accounted for by one of its canonical variates is the mean of the squared structure coefficients for that variate, that is,

$$(11.2.6) \qquad P_{Yy_h} = \frac{\sum r^2_{yhj}}{p},$$

$$(11.2.7) \qquad P_{Xx_h} = \frac{\sum r^2_{xhi}}{k}.$$

The squared structure coefficients for any variable Y_j or X_i may also be summed over its q canonical variates to determine the extent to which its variance is accounted for by them, and hence the degree to which it is contributing to its canonical variates which in turn carry the relationship between the groups of variables.

To aid in the interpretation of the correlation between a given pair of canonical variates, for example, of R_{c1}, one can determine the simple correlations of each X_i with y_1, and of each Y_j with x_1, that is, the $r_{y_1 i}$s and $r_{x_1 j}$s. Thus, the relationship between paired canonical variates is analyzed using the same kind of ingredients as those employed for each canonical variate. An interesting further relationship that helps in interpretation lies in the fact that the mean of a set of these squared cross correlations with a canonical variate x_h or y_h is the redundancy index of that variate

$$(11.2.8) \qquad R^2_{dy_h} = \frac{\sum r^2_{yhi}}{k},$$

$$(11.2.9) \qquad R^2_{dx_h} = \frac{\sum r^2_{xhj}}{p}.$$

These cross r^2s may also be summed for a given variable across the q canonical variates; this sum is the proportion of its variance accounted for by its relationship, via the canonical model, to the other group of variables, and hence directly indexes its overall role in the relationship between the two groups of variables.

Finally, an intermediate result that facilitates interpretation in CA gives the multiple regression equation for each of the p Y variables as dependent on the group of X variables, and for each of the k X variables as dependent on the group of Y variables.

---

[2] This is exactly analogous to how one proceeds to interpret a factor in factor analysis. Canonical variates are, indeed, factors of a certain kind. For those readers with a background in factor analysis, we note that canonical variates are *unrotated* factors.

The above provides a complex array of statistics which exhaust the data (and quite possibly the reader). For our limited purposes of introducing CA, we can broadly summarize the kinds of information potentially available from CA in a three-tier array. We will do so from a perspective in which the Y variables are treated as dependent on the X variables, so the focus will be on "explaining" the Y variables by means of the information in the X variables.

1. At the highest level, we take as our base the total Y-group variance. The total redundancy index, R_{dY}^2, is the proportion of this total which is accounted for by the group of X variables via the q x variates of the canonical model. This may be partitioned into portions due to the first, second, etc. x variate by means of the R_{dy}^2s, as given in Eq. (11.2.5).

2. At the middle level, we focus on the y variates considered as units. Each accounts for a portion of the total Y variance, for example, for y_1 it is P_{Yy_1}. The portion of the variance of each of these y variates which is accounted for by its paired x variate (hence by the X variables) is the squared canonical correlation, for example, for y_1 it is R_{c1}^2.

3. Finally, the substantive interpretation of the y (and x) canonical variates proceeds primarily by means of the structure coefficients, their simple correlations with their constituent Y (and X) variables, for example, $r_{y_1 j}$ (and $r_{x_1 i}$). The interpretative analysis of the canonical correlation proceeds similarly with cross correlations of y variate with X variable, for example, $r_{y_1 i}$ (and conversely). An additivity property of both of these coefficients (when squared) makes it possible to determine the makeup of a given y variate in terms of the Y variables on the one hand (Eq. 11.2.6) and the contributions to its R_{dy}^2 by each X_i on the other (Eq. 11.2.8).

4. The substantive interpretation of a canonical variate by reference to the weights of Eqs. (11.2.1) and (11.2.2), rather than to the structure coefficients, may be seriously misleading because these are *partial* regression coefficients, analogous to the βs in MRC, and thus subject to redundancy and suppression effects. (We shall see below that partial coefficients of all kinds do have important interpretive uses for other purposes in certain CA applications.)

The computer program for CA provided by Cooley and Lohnes (1971) provides rather complete and well-labeled output along the above lines. Their full mathematical description is accompanied by several real illustrative examples which greatly clarify this complex method.

11.3   SPECIALIZING AND EXPANDING CANONICAL ANALYSIS

Just as it was found that AV and ACV are special cases of the MRC system, the case will be sketched out for the proposition that the major MV methods are special cases of CA. The ingredients of the argument will be essentially the same—the exploitation of the methods of nominal scale coding (Chapter 5), curvilinear representation (Chapter 6), and interactions (Chapter 8), together with the use of sets of variables as units of analysis (Chapter 4) and the logic of the hierarchical model, make for an expansion of CA like that of MRC, which includes (but is not exhausted by) the MV methods of discriminant analysis (DA) and multivariate AV and ACV (MANOVA/MANACOVA).

11.3.1 Multiple Regression/Correlation

First, the obvious specialization to MRC, which can be viewed as CA with 1 (= p) Y variable and k X variables. With only one Y variable, $R_{c1}^2 = R_{dy_1}^2 = R_{dY}^2$, and simply becomes $R_{Y\cdot12\ldots k}^2$. The b_is are proportional to the MRC's β_is, and the sole x of Eq. (11.2.1) is proportional to and hence perfectly correlated with $\hat{z}_Y$. The structure coefficients of the X_i become the $r_{\hat{Y}i}$ of Eq. (3.7.12) for MRC, and the cross correlations reduce to the simple r_{Yi} "validity" coefficients. The fact that there is only one Y variable results in there being only one canonical variate "pair" (of which $y_1 = 1.00$ $Y = z_Y$), and thus all the other coefficients discussed in Section 11.2 vanish. The utility of stating this specialization of CA to MRC lies in the insight it affords to certain aspects of CA; for example, proportion of total Y variance has about the same interpretation whether Y is a single variable or a group of variables. Given a good understanding of MRC, at least half the battle to understand CA is won, and once the latter is grasped, it is fairly easy to understand its special cases, for example, DA and MANOVA. In fact, the understanding is likely to be deeper than when each MV method is approached as an ad hoc gimmick.

We have seen that CA is much more than a series of multiple regression analyses, one for each of the p Y variables, with the X variable as IVs, or, conversely, k regression analyses of the X variables with the Y variables as IVs. However, these $p + k$ analyses may be recovered as a by-product of a complete CA, thus indicating in another way the superordinate–subordinate relationship of CA to MRC.

11.3.2 Discriminant Analysis with g Groups

The circumstances in which DA is employed are usually described in somewhat the following manner: For subjects in g groups (for example, experimental condition or diagnostic group), there is available information on k X variables. How do these groups relate to each other with regard to these variables? The interest lies particularly in the g population means for the X variables. Doing a series of k F tests among the g sample means is unsatisfactory because the inevitable correlations among the variables render such tests nonindependent and create difficulties in statistical inference. The DA solution is to treat the k X variables *jointly* and *analytically* by solving for a set of k weights, which produce a linear composite that maximally discriminates between the groups, that is, by a linear *discriminant function* of the variables. Applying these weights to the scores on the variables produces a "discriminant score" for each subject, and these scores are such as to maximize (among other things) the F ratio of between-/within-group mean squares in a simple AV. Interest usually attaches to the nature of the function, and substantive interpretations are usually made by reference to the discriminant weights. When this (first) discriminant function is partialled from the scores, a second set of k weights then may be determined that, when applied to the X residuals, yields a second discriminant score which maximally discriminates among the groups. And so on. These discriminant functions are decreasingly discriminating, they are mutually uncorrelated, and there can be no more of them than the lesser of $g - 1$ and k. The latter is a mathematical maximum; in practice, only the q statistically significant discriminant functions are included in the model.

Interest in the results may center on the interpretation of the discriminant functions for theoretical purposes and/or on using them for classifying new subjects by generating their discriminant scores, and assigning each on the basis of his proximity to the discriminant score means of the groups. Results are frequently displayed in a figure in which the group means are plotted on coordinate axes representing the first two discriminant functions, which, in most applications, exhaust the significant discrimination.

It is instructive to view all this from a different perspective. In MRC, we relate a group of k X variables to a single *quantitative* variable, Y. In the above data structure, we have k X variables which we mean to relate to group membership, that is, to a *nominal* scale variable, G, made up of g groups. So, we would like to do MRC with a nominal scale as the "dependent" variable. But we have seen that it takes $g - 1$ variables to represent all the information in G. Thus, we need, in general, *multiple* "dependent" variables, and the problem may be attacked by the generalization of MRC to CA, with $p = g - 1$ Y variables to be related to the k X variables.

To conceive of DA as that special case of CA wherein one of the two groups of variables constitutes the coding of a nominal scale is not merely a nice theoretical idea, but one that carries important practical applications in data analysis. By judiciously selecting one of the nominal scale coding methods of Chapter 5 (dummy, effects, or contrast), and proceeding to do a CA, one has the yield of a DA, and much more. There is available the analytic richness of CA combined with that which we have seen derives from nominal scale coding.

A concrete example will no doubt help here. Imagine a psychiatric investigation in which, for 200 (= n) patients, the relationship between G, group membership in 4 (= g) diagnostic categories, and a battery of 8 (= k) psychological tests of cognition and personality is to be studied. Let the latter be designated the X variables (X_i: $i = 1, 2, \ldots, 8$), and let G be coded into 3 (= $g - 1 = p$) Y variables (Y_j: $j = 1, 2, 3$).[3] Since these diagnostic categories are to be

[3] Our assignment of X and Y designations to the two groups of variables is quite arbitrary. In particular, using the Y variables to represent G does not necessarily mean that we believe that group membership is "dependent" on the X scores in a causal sense. This is purely a

treated as being on an equal footing, we choose the method of effects coding of Section 5.4; Table 5.4.1 shows exactly how membership in the 4 diagnostic categories is rendered in effects coding (except that they are Y_1, Y_2, and Y_3 here, not X_1, X_2 and X_3).[4] A full CA is then performed on 3 $(= p)$ Y variables and 8 $(= k)$ X variables.

Assume that the first and second R_cs are statistically significant; the third (and only remaining, since $3 = p < k = 8$) is both nonsignificant and trivial. Our canonical model then will explain the relationships between the X and Y variables using 2 $(= q)$ pairs of canonical variates, or, equivalently, the 4 diagnostic groups are significantly discriminated along 2 $(= q)$ orthogonal dimensions (discriminant functions) which are generated from the 8 psychological tests.

Recall from Eqs. (11.2.1) and (11.2.2) that the b_is are the canonical weights applied to the standardized X_is to produce an x canonical variate, and, similarly, the c_js produce y when applied to the standardized Y_js. Consider the first canonical variate of the X scores, x_1. Since it is guaranteed by its construction to be the linear composite most highly correlated with y_1, its b_i weights are necessarily a set which produces discriminant scores from the tests which maximally distinguish among the groups, that is, they are (standardized) discriminant function weights. R_{c1}^2 is the proportion of the variance of x_1 accounted for by diagnostic category membership, that is, what MRC would give as the R^2 for x_1 as the dependent variable and Y_1, Y_2, and Y_3 the independent variables. The b_i for x_2 and R_{c2}^2 are similarly interpreted. P_{Xx_1} and P_{Xx_2} are the proportions of the total variance of the 8 tests which x_1 and x_2 account for; their sum is therefore the proportion of this total implicated by the $(q =)$ 2-factor canonical model. Interpretation of x_1 and x_2 proceeds best by means of their 8 structure coefficients, for example, $r_{x_1 3}$ is the zero-order correlation of X_3, an introversion score, with the first canonical variate/discriminant function. The practice of using the b_is for this purpose is prevalent, but likely to be misleading.

Consider now the question "What proportion of the total variance of the 8 (standardized) scores on cognition and personality tests is accounted for by diagnosis (G)?" Via the first canonical relationship, this is $R_{dx_1}^2$ $(= P_{Xx_1} R_{c1}^2)$, and via the second, $R_{dx_2}^2$. Note that the base of these proportions is the total X variance $(= k)$. Each y, which may be understood as a facet of G, accounts for an independent portion of this total, and these sum to R_{dX}^2, the overall "redundancy of the X group given the Y group," or, more analogously with MRC

---

matter of interpretation and assumption, and is not preempted by the assignment to X and Y. As a matter of fact, in this example, we chose to consider the X-variables dependent. Recall that CA yields effect size measures which are quite symmetrical in meaning (e.g., R_{dY}^2 and R_{dX}^2), although not generally equal in value (e.g., $R_{dY}^2 \neq R_{dX}^2$). In interpreting CA results, we may choose to view either group of variables as causally dependent on the other, make other causal assumptions, or none at all.

[4] As described in Section 11.2, these Y_j would then be standardized, as called for in Eq. (11.2.2).

language, the proportion of total X variance accounted for by G via the canonical model.

To appraise the fine grain of the total relationship, and see the advantage of the CA approach with strategically coded G, consider the makeup of the y variates. Their structure coefficients (for example, the r of y_1 with Y_3) are not helpful, since they only reflect the group's sample sizes and the arbitrary assignment of diagnostic categories in the effects coding. We found in Section 5.4 that the raw-score partial regression coefficients of effects-coded variables (B_i there) gave contrasts on the dependent variable of a group's mean minus the unweighted mean of all the g means. To achieve the same analytic result here, we must rescale the c_js of the standardized Y_js to their raw-score analogues by multiplying them by R_c/sd_j, that is, $C_j = c_j R_c/sd_j$. These are now interpretable exactly as the B_i of Section 5.4, for example, C_2 for x_1 is the "effect" of G_2 on x_1, literally, the x_1 mean of the G_2 cases minus the unweighted mean of the means of x_1 of all four diagnostic groups. (For G_4, it is $-\Sigma\ C_j$.) The same relationships hold, of course, for x_2.

We are not constrained to the C_js for our interpretation of partial relationships. One could determine, as in MRC, the $3\ (= p)$ sr_js and pr_js for each x variate. They may be found from

(11.3.1)
$$sr_j^2 = c_j^2 R_c^2\ (1 - R_j^2),$$

where R_j^2 is the R^2 for Y_j as dependent, and the other $(p - 1 = 2)$ Y variables as independent, and

(11.3.2)
$$pr_j^2 = \frac{sr_j^2}{1 - R_c^2 + sr_j^2}.$$

The interpretation would be exactly as in Section 5.4, with x_1 or x_2 replacing Y as the dependent variable. For example, for x_1, sr_3^2 expresses the effect of G_3 as a proportion of x_1 variance, and pr_3^2 expresses the effect of G_3 as a proportion of that part of the x_1 variance not accounted for by the effects of the other diagnostic groups.

We thus have, from CA, a rather complete picture of the MV relationship between the tests and the diagnostic categories: global measures of total test variance accounted for, means of interpreting the discriminant functions (the x canonical variates), and, given the effects coding of G, explicit measures of each diagnostic group's "effect" or "eccentricity" on these functions by means of its C_j, sr_j, and pr_j. Were G of a different nature, we would have used dummy (Section 5.3) or contrast (Section 5.5) coding, which would affect only our interpretation of the resulting partial coefficients (C_j, sr_j, pr_j); their interpretation would be exactly as described in those sections, replacing by x_1 and x_2 the Y dependent variable there. For example, with dummy coding (as in Table 5.3.1), C_3 for x_1 would be the mean of G_3 on x_1 minus that of G_4, the reference group; sr_3^2 would express the G_3–G_4 distinction as the proportion of x_1 variance for which it accounts.

11.3.3 Multivariate AV (MANOVA)

Multivariate AV, or MANOVA, is presented in the literature as a direct extension of univariate AV, the word "multivariate" referring to the simultaneous attention to a group of "dependent"[5] variables. It is possible to generalize the concept of variance for a single variable to that of MV "dispersion" for a group of k variables, which may for some purposes be treated as a matrix (for example, of correlations for standardized variables, or of variances and covariances) and for others as a single value (the determinant—see Appendix 1) derived from such a matrix, called the "generalized variance." (Note that this is not the same as the total variance, k, discussed above.) The details aside, it suffices for our purposes to note that we can meaningfully conceive of the dispersion of a group of variables much as we have of the variance of a single variable, including the possibility of partitioning it, and assigning portions of it to various sources of research interest.

In a simple one-way univariate AV, we assess the proportion of variance of the single dependent variable accounted for by group membership. We can analogously view a simple one-way MANOVA as an appraisal of the proportion of the dispersion of a group of dependent variables which is accounted for by group membership. We saw (throughout Chapter 5 and elsewhere) that univariate AV is readily accomplished by MRC with nominal scale coding. In turn, MANOVA is readily accomplished by CA.

Just as the MRC system yields more information than AV, so does CA yield more information than MANOVA. The one-way MANOVA addresses the dependent variables globally, appraising their *overall* relationship to groups. In CA terms, it is devoted to the *total* redundancy of the dependent variables (R_{dX}^2 if the X variables are dependent), or equivalently, the issue of differences among groups in their sets of X-variable means. In fact, MANOVA uses the same significance test for this purpose as CA (or DA) uses for the first R_c (or discriminant function). But CA goes on, as we have seen, to a rather full analysis of the dimensions underlying the overall relationship and the role therein of the variables and groups. It is difficult to conceive of circumstances in which an investigator's interest would not extend beyond the global relationships to the details of its analytic structure.

The parallel extends beyond one-way to factorial and other complex designs. MANOVA apportions MV dispersion to main effect and interaction (or other) sources, but appraises each globally. CA may be extended to deal with factorial design MANOVA in much the same way as MRC handles factorial design AV. In direct parallel, one may proceed hierarchically, and find the increments in R_{dX}^2, for example, as the Y variables include first a nominally coded set G, then G and H, and finally, G, H, and the GH product sets. Also, the variance (dispersion) accounted for by each source can be analyzed into its canonical variate (discrimi-

---

[5] We use the term "dependent" here only for convenience, noting again that it carries no necessary causal implications in the ensuing discussion.

nant function) components for a detailed view of the hows and wherefores of the discrimination afforded by each main effect and interaction. (In Section 11.4.3, we describe how this may also be accomplished by MRC analyses under certain conditions.)

11.3.4 Multivariate ACV (MANACOVA)

The ACV, as Chapter 9 liberally demonstrated, invokes the idea of partialling from a single dependent variable and a set B of independent variables, the variance which can be accounted for by a set (A) of covariates. MANACOVA (multivariate ACV) works with the generalized variance of a group of dependent variables, and like conventional univariate ACV, a set B which carries group-membership information. The previously described parallels hold—MANACOVA globally analyzes the covariate-partialled (or residualized) dispersion of a group of dependent variables, assigning portions to the covariate-partialled sources (for example, main effects, interactions) represented in set B.

One may employ a CA approach in such design circumstances which is a direct parallel to the hierarchical MRC–ACV: perform a hierarchical series of CAs, tracking the increment in proportion of total variance (R_{dX}^2) accounted for by the covariates and set B combined over what is accounted for by the covariates alone.[6] If set B is itself made up of functional subsets, proceed hierarchically through them. Interpretations below the level of increments in total redundancy is made difficult by the fact that in each successive CA, the makeup of the canonical variates (discriminant functions) on the dependent variables change, and one is thus shooting at a moving target.

Another CA approach, which makes up in interpretability what it lacks in elegance, is to work directly with residuals. First, do a CA of the dependent variables with the covariates, and compute the actual residuals for each subject on the dependent variables after all canonical variates have been removed. Then do a CA of the coded set B variables with the covariates, again computing residuals per subject after extracting the canonical variates. (They will be identical for all subjects in the same group.) With data on n subjects on the two covariate-adjusted groups of variables, perform a third and final CA. This is interpreted as the usual CA (or DA), but, of course, for covariance-adjusted relationships. Using either CA approach, one can construct multivariate analogues to the multiple partial and semipartial correlations of MRC; with multiple dependent variables, these might be dubbed (one hesitates) *multiple* multiple partial and semipartial correlations, with their squares as proportions of (multivariate) dispersion. (In Section 11.4.3, we offer an MRC method that accomplishes the equivalent of a CA with covariates under certain conditions.)

---

[6] The homogeneity of the MV regression of covariates on Y variables between groups is assessed as is its univariate counterpart in ACV: form an AB product set to carry the $A \times B$ interaction and check the increment in R_{dX}^2 it produces over that provided by sets A plus B.

11.3.5 Expanding Canonical Analysis

In Chapters 4 through 9, it was seen that when, to the basic structure of MRC, one added (*a*) methods of representing information as sets of IVs, (*b*) the use of these sets as units of analysis, (*c*) the concepts of partialling and of the hierarchical model, and (*d*) the generalization of interactions as conditional relationships, there emerged a general system which included novel forms and methods of data analysis. CA may be expanded in a very similar way for multiple dependent variables and with a similar result. For our present limited purposes, we will simply point out some of the more obvious possibilities. For simplicity of exposition, we will treat Y as having become multiple in going from MRC to CA; the X variables remain multiple, and cover the same range of possibilities as in MRC.

We have shown that the multiple groups (G) of DA, MANOVA, and MANACOVA require only a rendering into coded variates by any of the methods of Chapter 5. The several methods of Chapter 6 may be used (in various ways) to represent curvilinear aspects on the Y side of the equation, thus permitting, for example, discriminant functions of standardized v (= Y_1), v^2 (= Y_2), and v^3 (= Y_3), with the X_is representing group membership. The structure coefficients of the Y_j and X_i would facilitate the interpretation, or one might proceed hierarchically. With quantitative variables on both sides of the equation, curvilinear aspects could be represented in either the Y_js or the X_is, or both. The other options for curvilinear representation by multiple variables of a single research factor (orthogonal polynomials, nominalization) are also available for either the Y variables or the X variables.

Similarly, the missing-data methods of Chapter 7 may be applied. We noted there that missing data on Y, the dependent variable, could not be handled by MRC, and discussed the difficulties in inference that result when subjects with missing Y data are dropped (Section 7.1.1). But CA offers the possibility of representing missing data for a dependent variable in the same way as MRC represents it for an IV: let Y_1 be the missing data dichotomy and Y_2 the constant-plugged quantitative variable. When the mean is used as the plugging constant, Y_1 and Y_2 correlate zero, and a special case occurs which can be analyzed by MRC (see Section 11.4.3). One may similarly employ the missing-data methods for nominal scales as the Y variables; for example, in survey research, when responses are coded in qualitative categories, "no response" can be incorporated as part of a dependent variable.

Not only single sets but multiple sets of variables may constitute the Y group. One might proceed hierarchically, or with canonical residuals as in MANACOVA, or simultaneously, relying primarily on the structure coefficients of the Y_js for interpreting the y canonical variates. The Y variables may also be employed to carry interactions of single variables or sets, representing them as described in Chapter 8.

MANACOVA was described above as the MV analogue of univariate ACV. Just as the latter was generalized, via MRC, to APV (Section 9.4), so may MANA-COVA be generalized to a MV–APV, and largely by the same means.

Finally, it is possible to usefully expand CA into data-analytic territory now occupied by "contingency table analysis," with its global chi-square test of independence and equally global measures of relationship, for example, the contingency coefficient or Cramèr phi (Hays, 1973, Chapter 17). A two-way table of frequencies provides the basic data portraying the relationship between two nominal scales. By coding each using the methods of Chapter 5, they may then be related as two groups of variables via CA. Here again, with judicious selection of coding methods, the information produced by a CA should provide the basis for a far more detailed examination of the relationship than do the standard methods.

We should make clear what is implied by the shift to a more tentative tone in this section. There exists no fully worked-out systematic expansion of CA along the lines of MRC, and our suggestions along these lines must be taken as provisional. They must be subjected to trial by application to appropriate data and generally firmed up theoretically. They do, however, constitute exciting possibilities for more scientifically relevant data analysis in the future.

11.4 SOME MULTIPLE REGRESSION/CORRELATION ALTERNATIVES TO CANONICAL ANALYSIS

Having stretched the reader's imagination to CA and its possibilities, we may now appropriately conclude this section with a consideration of how the MRC system may be used to accomplish some of the purposes for which CA methods are sometimes employed. Apart from the practical data-analytic procedures we can briefly describe under this rubric, this material should also serve to further the reader's understanding of the relationship between MRC and other MV methods.

11.4.1 Multiple Multiple Regression/Correlation

In CA, the p Y variables are regressed jointly on the k X variables (and vice versa). An obvious alternative analytic strategy is to perform p MRCs, one at a time for each Y variable, treating the X variables consistently through all analyses (that is, same sets, same hierarchical, intermediate, or simultaneous model).

The procedure of multiple MRC (MMRC) analyses has some obvious disadvantages in that it fails to take into account, as does CA, the relationships among the Y variables. This failure may result in the analyst being misled in his interpretations. For example, substantive conclusions about Y_j and Y_f may be quite incorrect, if, for example, Y_f is merely an effect (or epiphenomenon) of Y_j. Relationships of the X variables to Y_f are then "spurious" in the sense that they would not hold for the partialled variable $Y_f \cdot Y_j$. When CA is applied in such a situation, it automatically mutually partials the Y variables, and it could be found that Y_f's (partial) canonical weight (but not its structure coefficient) is trivial. The features of the pattern of relationships among the Y variables which bear on their relationship to the IVs—not only complete redundancy, but also partial redundancy and suppression—are lost to the view of the analyst in MMRC.

On the other hand, there is much to be said for the greater simplicity and familiarity of MRC, and circumstances exist (or can be created) in which MMRC analyses may be used in place of, or even in preference to, CA. The failure of MMRC to take into account the structure of the Y variables is no impediment when there is little or no correlation among them. In the extreme case where their correlations are literally zero, CA in fact reduces to nothing more than a series of p MRC analyses and when their correlations are small, MMRC will produce results closely approximating those of CA, and hence hardly likely to be misleading. We will see (Section 11.4.3) how zero correlation among the Y variables may be created in order to make possible a special form of MMRC analyses, and instances in which the latter is superior to CA as a data-analytic procedure.

Under certain conditions, a case can be made for MMRC even though correlation exists among the Y variables. An investigator may be interested in each Y variable in its own right, and proceed to MMRC. He may be able to cope with their intercorrelations by taking them into account when he interprets the p MRC analyses. It may be possible to make some credible causal assumptions and interpret the results accordingly. For example, if he assumes that no Y variable causes any other, but that they are rather all effects of the research factors embodied in the X variables, then the kind of spuriousness described earlier does not obtain (subject, of course, to the validity of these assumptions).

A special case occurs when the dependent variable is a nominal scale G (with g categories) of the kind one encounters in nonexperimental research. Using all n cases, one can create g Y variables, each a dichotomy distinguishing one category (coded 1) from all the remaining categories (coded 0), and perform all g MRC analyses. Note that all categories, not only the $g - 1$ necessary for full MV representation, are thus made into dichotomous variables. Since each category in the set is a meaningful entity (for example, Protestant versus all other religious affiliations, or paranoid schizophrenia vs. all other diagnoses), each of the g MRC analyses is readily interpretable in its own right (including the gth, despite its mathematical redundancy with the others). A comparison across the MRC analyses of the g Y variable's relationships with the X variables then ties the picture together. MMRC works well here.

Thus, it can be seen that the presence of multiple correlated Y variables does not necessarily commit the data analyst to CA—the data may well yield up their meaning to MMRC analyses.

11.4.2   Discriminant Analysis with Two Groups

A few moments of reflection will make it apparent that for the special case where two groups are to be discriminated using k X variables, the problem is not multivariate on the Y side at all. When $g = 2$, $p = g - 1 = 1$, and the analysis reduces to a single MRC for a single dichotomous Y (which can be coded 1–0, or with any other pair of different values). The MRC analysis is mathematically and statistically identical with a CA when $p = 1$; hence, it is identical with a DA for 2 groups. $R^2_{Y \cdot 12 \ldots k}$ equals the (sole) R^2_c ($= R^2_{dY}$) and the multiple regression equation is proportional to the discriminant function and hence perfectly correlated with it.

11.4.3   Uncorrelated Y Variables: MMRCYO

Since when the Y variables are uncorrelated, CA (and hence its DA special case) is equivalent to a series of MRC analyses, it is worth considering how one might create this state of affairs, and the application of the resulting MMRC with Y variables correlated zero (hence, MMRCYO). We shall see that there are several circumstances wherein MMRCYO is a more relevant data-analytic method than ordinary CA.

Nominal Scales

Consider first the case where an experiment has been performed on g groups (or "cells") of equal (or proportional) numbers of cases. Following the contrast coding methods of Section 5.5, it is possible to express all the information of G as a set of ($g - 1 = p$) Y variables using orthogonal contrast coding coefficients. Moreover, equality (proportionality) of group sizes assures that the coded Y *variables* will be uncorrelated, so that each represents a wholly distinct contrast between sets of groups. Thus, we have accomplished the goal of uncorrelated Y variables, and have exhausted the group-membership information into p independent parcels, each an issue or hypothesis about G which is of research interest. Each of these can now serve as a dependent variable for an MRC on the k X variables, the latter organized as is appropriate to the analysis (by single IVs or sets, with or without covariates, simultaneously or hierarchically). Here CA offers no advantage over MMRCYO analyses, and conventional DA provides much less relevant information, since it does not reflect the structure of G as do the contrast Y variables.[7]

---

[7] Note the difference between this procedure and the treatment of G as g dichotomous Y variables described above. When G is produced by experimental manipulation ("treatments"), contrast coding is more likely to capture the research issues and sample sizes are likely to be equal, thus satisfying the requirement of uncorrelated Y variables for MMRC-YO. When G is a natural preexisting attribute which the subjects bring to the investigation, as in surveys or observational studies, each of the g categories is likely to be meaningfully compared with all others, and the use of the g dichotomously coded (and necessarily correlated) Y variables in a MMRC analysis will accomplish the data-analytic goal.

The above bare statement does not do justice to the generality of MMRCYO. It is applicable to a considerable variety of research designs and design features. Consider, for example, a $G \times H$ (concretely, $g = 2$, $h = 3$) factorial design with equal or proportional cell frequencies, such as would be appropriate for MANO-VA. Assume that the research issues underlying G and H make it useful to contrast code each, thus producing Y_1 for G, and Y_2 and Y_3 for H. Now, the set products are formed in the usual manner, yielding $(g - 1)(h - 1) = 2$ Y variables (Y_4 and Y_5), each an interpretable interaction contrast. We thus have a total of $gh - 1 = 5$ Y variables that correlate zero with each other, and can perform a meaningful MRC for each against the appropriately organized X variables. Similarly, repeated measurement (Chapter 10), other nested designs (Section 5.5.1), Latin square designs, and others can be treated by MMRCYO. It is, of course, not necessary to take all the contrasts of a set (number of cells minus one) as Y variables. Those not of central interest may be excluded in the interest of minimizing spuriously "significant" results (see Section 4.6.2).

Groups along a Quantitative Continuum

Yet another circumstance which permits the useful application of MMRCYO analysis is the presence of a research factor made up of groups of equal size along an equally spaced quantitative variable W, a frequent occurrence in psychological experiments. As was seen in Section 6.3, such research factors may be rendered as linear, quadratic, cubic, etc. orthogonal polynomials by appropriate coding (Table 6.3.1). Recall that these are, in fact, a form of contrast coding, and with groups of equal size, the variables they produce (Y_j) are zero correlated. Therefore, if such a quantitative W is to be related to a group of X variables, aspects of its shape may be represented as separate Y variables, each in a separate MRC. One is not likely to be interested in more than the first two (or, conceivably three) orthogonal polynomials in the resulting MMRCYO analysis. Note that W, so rendered, could be combined with other research factors in a factorial or other multiple research factor design.

Multiple Quantitative Variables

As was noted in the early discussion of functional sets of IVs (Section 4.6) and elsewhere, the use of a large number of variables as a set of IVs for the purpose of covering a theoretical construct (for example, "anxiety," "prejudice") is poor research practice. Here, too, what holds for the Xs also holds for the Ys. The casual use of CA invites this practice, since it allows for the inevitable correlation among such a group of variables. One can make a virtue of MMRCYO's necessity that the Y variables be uncorrelated by reducing such a large group factor analytically.[8] A principle components analysis of these variables followed by

---

[8] We apologize to readers not familiar with factor analysis, who are asked to simply accept the fact that it can be used to achieve the desired goal by means of widely available computer programs. See, for example, Cooley and Lohnes (1971, Chapter 5).

varimax (or other orthogonal simple-structure type) rotation and the computation of factor scores would typically result in a very few, interpretable, and zero-correlated variables to use as the Y variables for an MMRCYO. The "very few" is not infrequently one, and MMRCYO reduces to MRC.

It is instructive to compare this approach with CA. Canonical analysis also dimensionalizes the original Y variables, but its canonical variate (y) "factors" are designed to be maximally correlated with the X variables. It is often the case that the canonical variates of the Y variables are difficult to interpret, or lead to awkward and uncompelling constructs. (See Footnote 2 above.) They will, in any case, change (in general) as one goes from one group of X variables to another. In contrast, the simple structure factor scores, which are also linear combinations of the original Y variables, are almost certain to be readily interpretable aspects of the domain which they represent, and in no way dependent for their meaning on other variables. It would seem that under the conditions described, MMRCYO is a generally superior form of analysis to CA, and not merely a device for evading it.

Hierarchical Y Structure

Another circumstance in which MMRCYO is more suitable than CA occurs when the Y variables have a hierarchical structure, that is, when the substantive issues make desirable the analysis of u, $v \cdot u$, $w \cdot uv$, etc. After analyzing u (= Y_1) in an MRC analysis, one computes $v \cdot u$ (= $v - \hat{v}_u$), the residuals of v from which u has been partialled, which then serve as Y_2 in a second MRC analysis. $w \cdot uv$ (= $w - \hat{w}_{uv}$) is then determined and MRC analyzed. And so on. These Y variables are necessarily zero correlated, and each MRC analysis is independently and unambiguously interpretable, given, of course, the validity of the causal model implicit in the hierarchy.

Yet another application of the above hierarchical MMRCYO procedure is in nonlinear representation of a dependent variable V by power polynomials (Section 6.2). Let $Y_1 = v$, $Y_2 = v^2 \cdot v$, etc., and proceed as above.

Unique Parts of Dependent Variables

When an analyst's interest extends only to the unique aspects of each of a group of p dependent variables, and therefore not to any variance they may share in common, an MMRCYO analysis may be performed. The Y variables are the residuals of each variable from the remaining $p - 1$ variables, thus for $p = 3$, $Y_1 = u \cdot vw$, $Y_2 = v \cdot uw$, and $Y_3 = w \cdot uv$. For these special conditions, too, CA is not suitable, since it analyzes *all* the variance of the dependent variable, not only the variance unique or specific to each.

Instead of literally creating the residual values for each of the n subjects on each dependent variable that are called for in the preceding three paragraphs, it is possible through matrix-algebraic manipulation to transform the original matrix of simple correlations among all variables into one which contains the appropriate correlations (semipartial, partial, and simple) among the variables as and where residualized. As a practical matter, it is probably easier to proceed as previously described, particularly since most statistical program packages facilitate the creation of residuals and their further manipulation.

11.5 SUMMARY

A great deal of material has been covered in a relatively few pages. We have sketched out the operation of CA, which is a powerful analytic device for the study of the relationships between two groups of variables (Section 11.2), seen how it can be specialized to cover the major MV methods (Sections 11.3.1– 11.3.4), and how it may be expanded along the same general lines as MRC has been (Section 11.3.5). Finally, a closer examination has revealed how special applications of MRC (MMRC and MMRCYO) are possible in various circumstances which might ordinarily be thought of as requiring CA or one of its special cases. In some of these, MMRC analysis provides an approximation to CA, in others its equivalent. In yet others, we have shown how MMRCYO is uniquely appropriate, while ordinary CA is not (Section 11.4).

For the reader who is a relative neophyte in research, this chapter has no doubt been tough going, indeed. It would defeat our purpose if he responded to its complexity with discouragement. Rather, we would wish to leave him with the conviction that CA and MRC provide the data-analytic means whereby many substantive research issues can be conceptually illuminated and constructively attacked. This material should serve as an introduction and possibly a future reference.

For the experienced researcher, we would hope that this brief exposition may have provided further insights into the nature of MV data structures, and both general guidelines and specific means for coping with problems that may be encountered in the analysis of complex data.

APPENDICES

APPENDIX 1

The Mathematical Basis
for Multivariate Regression/Correlation
and Identification of the
Inverse Matrix Elements

The multiple regression model requires the determination of a set of weights for the k independent variables, which, when used in the linear regression equation, minimizes the average squared deviation of the estimated $\hat{Y}$ scores from the observed Y scores. This solution is somewhat simplified by standardizing all variables. Thus the problem is to find a set of β_i weights such that

$$\sum (z_Y - \beta_1 z_1 - \beta_2 z_2 - \beta_3 z_3 - \cdots - \beta_k z_k)^2$$

is a minimum, as we have seen. By means of the differential calculus, the partial derivative of the function with respect to each unknown β_i is found and set to zero. The β_i weights are then expressed as a system of k "normal" equations in k unknowns of the form

(A1.1)

$$\beta_1 = r_{Y1} - r_{12}\beta_2 - r_{13}\beta_3 - \cdots - r_{1k}\beta_k$$

$$\beta_2 = r_{Y2} - r_{12}\beta_1 - r_{23}\beta_3 - \cdots - r_{2k}\beta_k$$

$$\cdot$$
$$\cdot$$
$$\cdot$$

$$\beta_k = r_{Yk} - r_{1k}\beta_1 - r_{2k}\beta_2 - \cdots - r_{(k-1)k}\beta_{k-1}.$$

A computer program may be used directly to solve this set of simultaneous equations. For those familiar with matrix algebra the problem may be stated usefully in matrix notation. The set of normal equations may be rearranged as follows:

$$r_{Y1} = \beta_1 + r_{12}\beta_2 + r_{13}\beta_3 + \cdots + r_{1k}\beta_k,$$

$$r_{Y2} = \beta_2 + r_{12}\beta_1 + r_{23}\beta_3 + \cdots + r_{2k}\beta_k,$$

and so on.

The right hand side of this set of equations may be recognized as the product of the square, symmetric matrix of correlation coefficients between independent variables $(\mathbf{R}_{ij})$ and the vector of $\boldsymbol{\beta}_{Yi}$. Thus, the equation set may be restated in matrix form as a single equation,

$$\mathbf{R}_{ij}\boldsymbol{\beta}_{Yi} = \mathbf{r}_{Yi} \,.$$

The solution may then be seen to lie in the premultiplication of the vector of correlations of IVs with Y, r_{Yi}, by the inverse of the matrix of correlations among IVs, $\mathbf{R}_{ij}^{-1}$:

$$\boldsymbol{\beta}_{Yi} = \mathbf{R}_{ij}^{-1}\mathbf{r}_{Yi} \,.$$

The problem, therefore, is to invert the correlation matrix, for which many computer programs have been written. Calculation by desk calculator of the inverse matrix is practical when there are no more than five or six independent variables. A method for doing so is presented in Appendix 2. Note that once the inverse matrix is determined, it is a relatively easy matter to apply MRC analysis to a new *dependent* variable W (for the same IVs), since all one needs to do is substitute in this equation for the new dependent variable its r_{Wi} vector.

Identification of the elements of the inverse matrix will reveal further identities in the multiple regression/correlation system. Also, since some computer programs provide the inverse matrix as standard or optional output, coefficients not provided by a given program may be readily determined as simple functions of these elements. Concrete illustration of the determination of these coefficients is provided in Appendix 2. The diagonal elements of $\mathbf{R}_{ij}^{-1}$, the inverse of the correlation matrix among IVs, are

$$(A1.2) \qquad\qquad r^{ii} = \frac{1}{1 - R_i^2},$$

where R_i^2 is the squared multiple correlation of the $k - 1$ remaining independent variables with X_i. Thus,

$$(A1.3) \qquad\qquad R_i^2 = 1 - \frac{1}{r^{ii}}.$$

This quantity is, of course, very useful itself in understanding the system of relationships with Y, especially in cases of high redundancy among some or all IVs. In addition, r^{ii} may be used to determine sr_i, which is frequently not provided in program output, by

$$(A1.4) \qquad\qquad sr_i = \frac{\beta_{Yi}}{\sqrt{r^{ii}}}.$$

Finally, we have seen in Eqs. (3.7.8) and (3.7.9) that this quantity is a necessary part of the standard errors of partial regression and standardized partial regression coefficients, which then may be written

(A1.5)
$$SE_{B_i} = \frac{sd_Y}{sd_i} \sqrt{\frac{1-R_Y^2}{n-k-1}} \sqrt{r^{ii}},$$

and

(A1.6)
$$SE_{\beta_i} = \sqrt{\frac{1-R_Y^2}{n-k-1}} \sqrt{r^{ii}},$$

respectively.

The off-diagonal elements of $\mathbf{R}_{ij}^{-1}$ may be identified as

(A1.7)
$$r^{ij} = \frac{-\beta_{ij}}{1-R_i^2} = \frac{-\beta_{ji}}{1-R_j^2} = r^{ji},$$

where β_{ij} is the standardized partial regression coefficient of X_i on X_j, other IVs having been partialled, and β_{ji} is the corresponding coefficient of X_j on X_i. Note that $r^{ij} = r^{ji}$ and thus, like the correlation matrix itself, its inverse is symmetrical about the diagonal. In determining the standard error of a single $\hat{Y}_o$ predicted from a new observed set of values $X_{1o}, X_{2o}, X_{3o}, \ldots, X_{ko}$, as given in Eq. (3.7.11), both diagonal and off-diagonal elements of the inverted matrix are needed. This equation may be restated as

(A1.8)
$$\widetilde{sd}_{Y_o-\hat{Y}_o} = \frac{\widetilde{sd}_{Y-\hat{Y}}}{\sqrt{n}} \sqrt{n+1+\sum r^{ii}z_{io}^2 + 2\sum r^{ij}z_{io}z_{jo}},$$

where the first summation is over the k diagonal elements and the second is over the $k(k-1)/2$ off-diagonal elements above (or below, since they are symmetrical) the diagonal.

The inverted matrix is used to determine the β_{Yi} by postmultiplying it by the vector of validity coefficients. Thus

(A1.9)
$$\beta_{Y1} = r^{11}r_{Y1} + r^{12}r_{Y2} + r^{13}r_{Y3} + \cdots + r^{1k}r_{Yk},$$
$$\beta_{Y2} = r^{21}r_{Y1} + r^{22}r_{Y2} + r^{23}r_{Y3} + \cdots + r^{2k}r_{Yk},$$

and so on. Restating the first of these in terms of the equivalents of the inverted matrix elements,

$$\beta_{Y1} = \frac{r_{Y1}}{1-R_1^2} - \frac{r_{Y2}\beta_{12}}{1-R_1^2} - \frac{r_{Y3}\beta_{13}}{1-R_1^2} - \cdots - \frac{r_{Yk}\beta_{1k}}{1-R_1^2}$$
$$= (r_{Y1} - r_{Y2}\beta_{12} - r_{Y3}\beta_{13} - \cdots - r_{Yk}\beta_{1k})/(1-R_1^2).$$

A mathematically equivalent version of the off-diagonal elements is given by

(A1.10)
$$r^{ij} = r^{ji} = \frac{-pr_{ij}}{\sqrt{(1-R_i^2)(1-R_j^2)}},$$

where pr_{ij} is the correlation between X_i and X_j, all other independent variables having been partialled from each. pr_{ij} may be obtained directly by

(A1.11)
$$pr_{ij} = \frac{-r^{ij}}{\sqrt{r^{ii}r^{jj}}},$$

and this matrix of partial correlations among IVs is provided by some MRC programs.

In most MRC studies in which an original set of validity coefficients and a final set of partial relationships with Y are provided in the computer output, a means of sorting out the effects of the independent variables on each other's relationship with Y is needed. It is particularly difficult to surmise these effects from the zero-order correlation matrix when there are more than two or three IVs, especially when their intercorrelations are not trivial. It is often useful to know which of the remaining variables are the source of the redundancy in the relationship between a given X_i and Y. It is also important to be able to detect and to identify sources of suppression among the IVs. A means of determining these effects which uses the inverse matrix elements is presented in Appendix 2, together with an illustration of its use in a concrete example.

ALTERNATIVE MATRIX METHODS

Some programs or hand calculation methods choose to invert matrices which express the relationships among the independent variables in raw score form. One such matrix is the variance–covariance matrix $\mathbf{V}_{ij}$, which starts with the sd_i^2 in the diagonal and the cov_{ij} ($= r_{ij}\,sd_i\,sd_j$) in the off-diagonal positions. The elements of the inverted matrix $\mathbf{V}_{ij}^{-1}$ are consequently

(A1.12)
$$v^{ii} = \frac{1}{sd_i^2(1 - R_i^2)} = \frac{r^{ii}}{sd_i^2},$$

and

(A1.13)
$$v^{ij} = \frac{-\beta_{ij}}{sd_i\,sd_j(1 - R_i^2)} = \frac{r^{ij}}{sd_i\,sd_j}.$$

When $\mathbf{V}_{ij}$ rather than $\mathbf{R}_{ij}$ is used, the entire matrix equation is

(A1.14)
$$\mathbf{B}_{Yi} = \mathbf{V}_{ij}^{-1}\mathbf{v}_{Yi},$$

where $\mathbf{v}_{Yi}$ is the vector of covariances of the independent variables with Y, and $\mathbf{B}_{Yi}$ is the vector of raw score regression coefficients.

Yet another matrix frequently employed for this purpose, $\mathbf{C}_{ij}$, is made up of the summed (over n) deviation (from mean) squares and cross products, $\Sigma\,x_i^2$ and $\Sigma\,x_i x_j$. This matrix differs from $\mathbf{V}_{ij}$ only in that its elements have not been divided by n. When its inverse is postmultiplied by the vector of summed deviation cross products with Y, c_{Yi}, it also yields the vector of raw score regression coefficients, $\mathbf{B}_{Yi}$. Similarly, the elements of $\mathbf{C}_{ij}^{-1}$ when divided by n equal the corresponding elements of $\mathbf{V}_{ij}^{-1}$.

One final matrix inversion method may be mentioned. The entire correlation matrix including Y may be inverted and R_Y^2, and the pr_{Yi} and β_{Yi} are directly determined from the row (or column) corresponding to Y. Except for small problems, this method has the disadvantage that the already lengthy computations involved in inverting a $k \times k$ matrix are increased in a $k + 1 \times k + 1$ matrix. More important, the potential analytic uses of the remaining elements of the inverted matrix are lost, since they include the effects of Y on the independent variables.

DETERMINANTS

Although an understanding of calculus or matrix algebra is not necessary for the intelligent use of MRC, it is useful to be aware of the characteristics of one numerical value resulting from this system. Every square matrix can be characterized by a unique number, its determinant, which is a complicated function of products of its elements. For the correlation matrix $\mathbf{R}_{ij}$ the determinant $|\mathbf{R}_{ij}|$ may take on any value from zero to one. The size of $|\mathbf{R}_{ij}|$ is a function of the R_i^2 values. When it is zero, at least one variable is a perfect linear function of the others, and the matrix is said to be *singular*. The inverse of a singular matrix does not exist and the multiple regression problem can not be solved until the (one or more) offending variables are removed. When $|\mathbf{R}_{ij}| = 1$, the variables all have intercorrelations of zero. As $|\mathbf{R}_{ij}|$ approaches zero, the researcher should be wary for two reasons. The situation is clearly one of high multicollinearity among two or more variables, with all the attendent problems in sorting out the meaning of the results and of the expected decrease in the sampling stability of coefficients. Furthermore, the researcher may have cause to worry about the accuracy of his MRC program under such conditions, especially if it operates in single precision (see Appendix 3).

Since the determinant is often provided in MRC output, one additional equation may prove useful:

(A1.15)
$$R_p^2 = 1 - \frac{|\mathbf{R}_{12\ldots p}|}{|\mathbf{R}_{12\ldots (p-1)}|},$$

that is, the squared multiple correlation of $X_1, X_2, X_3, \ldots, X_{p-1}$ with X_p is equal to one minus the ratio of the determinant of the matrix including X_p to the determinant of the matrix excluding X_p. If we let $X_p = Y$, this gives the standard R_Y^2. If we let X_p be the last entered variable in a hierarchical model, we may determine its R_p^2 with previously entered variables whenever determinants are provided.

APPENDIX 2

Desk Calculator Solution of the Multiple Regression/Correlation Problem: Determination of the Inverse Matrix and Applications Thereof

Although most practitioners of MRC analysis are likely to turn over the necessary computational work to a computer program, it is quite possible to perform these operations with a desk calculator. The major problems in doing so are the many opportunities for error, and the amount of time required, both of which increase rapidly as a function of n and k. Several methods for determining the MRC solution are available, of which the best known is the Doolittle solution. We prefer the method which is presented here because it provides the inverse of the correlation matrix among IVs, whose elements provide results which are of analytic interest. Using the correlations among IVs given in Table 3.5.1, we will illustrate the computation of this inverse matrix, and of the various multiple and partial coefficients which its elements yield, and then give the substantive interpretation of the entire set of results.

The method presented here may be generalized to more than, or fewer than, the illustrated four IVs. A mathematical understanding of the purpose of the various operations is not necessary for correct determination of the solution. It is, however, useful to have ways of controlling the error which may creep in. One of these is provided by carrying a sufficient number of decimal places for each of the original correlation coefficients and the subsequent operations. In this example we have used 5; with more variables more decimal places would be appropriate. A check against human error is provided by the last column which, at the points indicated by check marks, should equal the sum of the other numbers in the row, within at most 3 units in the last decimal.

Table A2.1 illustrates the entire procedure for calculation of the inverse. The first column contains the instructions. The method begins by copying the full correlation matrix among IVs ($\mathbf{R}_{ij}$) and, to its right, a parallel matrix in which the diagonal elements are all 1 and off-diagonal elements are 0 (the identity matrix, $\mathbf{I}$), as shown in the table. Each operation is carried out for every column, with the exception that once a left hand column contains 1, no further computations are carried out in that column. Thus, on line 6 each of the line 5 numbers is multiplied by the line 5, column r_2 value, .68318. On line 7 these line 6 values have been subtracted from the line 2 numbers. Note that the check sum number .92046 equals the sum of the row values, as well as 2.98950 − 2.06904. On line 8 the line 7 numbers are divided by the left hand value in line 7, and so on through the table. With a little study, the pattern of computation becomes clear and readily generalized for k IVs. Had there been only three independent variables column r_4 would be missing, the column headings r^{14}, r^{24}, r^{34}, and r^{44} would be replaced by r^{13}, r^{23} and r^{33}, and the lines would end with line 12. Five independent variables would require an additional row and column in the $\mathbf{R}_{ij}$ matrix as well as in the $\mathbf{I}$ matrix, and six more lines of calculation. Each entry in the next to last line would equal the bottom row value of the correlation or identity matrix minus the four values in the lines immediately above.

Line 17 provides the inverse matrix elements for X_4, that is, the r^{i4} values (see Table A2.2). From these numbers and working backward, the elements for X_3 are determined next by subtracting from the right hand figures in line 12 the product of the left hand figure and the corresponding r^{i4}. Since $r^{34} = r^{43}$ this element is already provided in row 17. Next, the figures in line 8 are combined with the previously determined r^{i4} and r^{i3} values to determine r^{22} and r^{21}. Finally, line 5 and the r^{i1} are used to produce r^{11}.

Some insight into the manner in which the method proceeds to systematically remove the proportion of the variance accounted for by each of the variables from the others may be gained by noting some by-products of the method. Looking first at the numbers below the correlation matrix, we see that the r_2 column includes r_{21}, r_{21}^2, and $1 - r_{21}^2$ in lines 5, 6, and 7, respectively. In column r_3 we find r_{31}^2, $sr_{3(2 \cdot 1)}^2$, and $1 - R_{3 \cdot 12}^2$ in lines 9, 10, and 11, respectively. Similarly, in the r_4 column we find r_{41}^2, $sr_{4(2 \cdot 1)}^2$, $sr_{4(3 \cdot 12)}^2$, and $1 - R_{4 \cdot 123}^2$ in lines 13 to 16, respectively. From the numbers under the identity matrix it may be determined that $1.87522 = 1/(1 - r_{12}^2)$, $1.07302 = 1/(1 - R_{3 \cdot 12}^2)$, and $1.27823 = 1/(1 - R_{4 \cdot 123}^2)$. The interested reader may determine more such identities by substitution of the appropriate r_{ij} into the numerical equivalents.

Table A2.2 reproduces the entire inverse matrix, $\mathbf{R}_{ij}^{-1}$, as well as the validity coefficients, r_{Yi}. The β_is are determined by summing the products of the row elements with the corresponding validity coefficients, according to Eq. (A1.9), for example,

$$\beta_1 = r^{11} r_{Y1} + r^{12} r_{Y2} + r^{13} r_{Y3} + r^{14} r_{Y4}.$$

etc. These products and their sums are given at the bottom of the table.

Finally, it is always useful to have a check against specific accumulated error, or even a computational blunder. The normal equations (A1.1) provide us with such a check, since, for example,

$$\beta_1 = r_{Y1} - \beta_2 r_{12} - \beta_3 r_{13} - \beta_4 r_{14},$$

etc.

TABLE A2.1

Computing the Inverse of the IV Correlation Matrix for the Academic Ranks Example

Instruction	Line	R_{ij}				I				Check Sum
		r_1	r_2	r_3	r_4	r^{14}	r^{24}	r^{34}	r^{44}	
	1.	1.	.68318	.04859	.29677	1	0	0	0	3.02854
	2.	.68318	1.	-.15396	.46028	0	1	0	0	2.98950
	3.	.04859	-.15396	1.	-.00616	0	0	1	0	1.88847
	4.	.29677	.46028	-.00616	1.	0	0	0	1	2.75089
Copy 1.	5.	1	.68318	.04859	.29677	1	0	0	0	3.02854
5. × .68318	6.		.46673	.03320	.20275	.68318	0	0	0	2.06904
2.-6.	7.		.53327	-.18716	.25753	-.68318	1	0	0	.92046 √
7. ÷ .53327	8.		1	-.35097	.48293	-1.28111	1.87522	0	0	1.72607 √
5. × .04859	9.			.00236	.01442	.04859	0	0	0	.14716
7. × -.35097	10.			.06569	-.09039	.23978	-.35097	0	0	-.32305
3.-9.-10.	11.			.93195	.06981	-.28837	.35097	1	0	2.06436 √
11. ÷ .93195	12.			1	.07491	-.30943	.37660	1.07302	0	2.21510 √
5. × .29677	13.				.08807	.29677	0	0	0	.89878
7. × .48293	14.				.12437	-.32993	.48293	0	0	.44452
11. × .07491	15.				.00523	-.02160	.02629	.07491	0	.15464
4.-13.-14.-15.	16.				.78233	.05476	-.50922	-.07491	1	1.25292 √
16. ÷ .78233	17.				1	.07000	-.65090	-.09575	1.27823	1.60156 √

From Line 12: $r^{33} = 1.07302 -(.07491)(-.09575) = 1.08019$,

$\qquad\qquad\ r^{32} = .37660 -(.07491)(-.65090) = .42536$,

$\qquad\qquad\ r^{31} = -.30943 -(.07491)(.07000) = -.31467$.

From Line 8: $r^{22} = 1.87522 -(.48293)(-.65090)-(-.35097)(.42536) = 2.33885$,

$\qquad\qquad\ r^{21} = -1.28111 -(.48293)(.07000)-(-.35097)(-.31467) = -1.42535$.

From Line 5: $r^{11} = 1 - ((.68318)(-1.42535) - (.04859)(-.31467) - (.29677)(.07000) = 1.96829$.

Subtracting β_i from both sides of the equation should yield zero. Checking on the obtained β_i, we find

$.46304 - .35243(.68318) + .19300(.04859) - .27904(.29677) - \beta_1 = -.00002,$

$.61226 - .14885(.68318) + .19300(-.15396) - .27904(.46028) - \beta_2 = .00001,$

etc. The checks for β_3 and β_4 are, respectively, .00000 and .00001. All the βs check well within acceptable limits.

As we saw in Eq. (A1.3) the proportion of variance in each IV which is shared with the other IVs may be determined by

$$R_i^2 = 1 - \frac{1}{r^{ii}}.$$

Determining this value for each of our variables we find that $R_{1 \cdot 234}^2 = 1 - (1/1.96829) = .49194$, $R_{2 \cdot 134}^2 = 1 - (1/2.33885) = .57244$, $R_{3 \cdot 124}^2 = .07424$, and $R_{4 \cdot 123}^2 = .21767$. In this rather modest example these R_i^2s may easily be seen to be consistent with the correlation matrix. Thus, the number of years since Ph.D. is fairly substantially associated with two of the other IVs, especially with the number of publications which in turn is modestly correlated with number of citations. Sex is relatively independent of the other IVs.

The diagonal elements of the inverse matrix may also be used, together with the β_i, to determine the sr_i, since by Eqs. (A1.4) and (3.5.6),

$$sr_i = \frac{\beta_i}{\sqrt{r^{ii}}} = \beta_i \sqrt{1 - R_i^2}.$$

TABLE A2.2
Inverted Correlation Matrix, Validity Coefficients, and β_i

	$\mathbf{R}_{ij}^{-1}$				
	Number Publ. X_1	Yrs Ph.D. X_2	Sex X_3	Citations X_4	Acad. rank r_{Yi}
X_1	1.96829	−1.42535	−.31467	.07000	.46304
X_2	−1.42535	2.33885	.42536	−.65090	.61226
X_3	−.31467	.42536	1.08019	−.09575	−.24175
X_4	.07000	−.65090	−.09575	1.27823	.48661

$\beta_1 = .91140 - .87268 + .07607 + .03406 = .14885$

$\beta_2 = -.65999 + 1.43198 - .10283 - .31673 = .35243$

$\beta_3 = -.14570 + .26043 - .26114 - .04659 = -.19300$

$\beta_4 = .03241 - .39852 + .02315 + .62200 = .27904$

For example,

$$sr_1 = \frac{.14885}{\sqrt{1.96829}} = .14885\sqrt{1 - .49194} = .10610,$$

and similarly, $sr_2 = .23045$, $sr_3 = -.18570$, and $sr_4 = .24681$, all of which values agree with those provided in Table 3.5.2. By including the information on the ms, sds, and n, all of the other coefficients and their significance tests may be determined by the equations presented in Chapter 3.

If the partial correlations *between IVs* are desired, they may be determined from the off-diagonal elements of the inverse matrix. The partial correlations, as given by Eq. (A1.11), are

$$pr_{12 \cdot 34} = \frac{-r^{12}}{\sqrt{r^{11}r^{22}}} = \frac{1.42535}{\sqrt{(1.96829)(2.33885)}} = .66432,$$

$$pr_{23 \cdot 14} = \frac{-r^{23}}{\sqrt{r^{22}r^{33}}} = \frac{-.42536}{\sqrt{(2.33885)(1.08019)}} = -.26761,$$

etc.

How do these partial relationships among the independent variables affect their relationships with Y? Equivalently, how may we identify redundancy, independence, or suppression among the IVs' relationships with Y? We note that each β_i is smaller than and of the same sign as its corresponding r_{Yi}, and thus conclude that on the whole the variables may be characterized as showing some mutual redundancy in accounting for Y in this example. However, this overall conclusion does not tell us how each variable affects each other *when it is a part of the entire system* used to estimate Y. For the answer to these questions we may return to the inverted matrix, and more particularly to Eq. (A1.9), illustrated in Table A2.2, where the β_is are expressed as a sum of the $r^{ii}r_{Yi}$ products. This application is also illustrated by means of the academic rank example. While the sample size was deliberately made small in order to allow the reader to follow each step in the computations of this example as presented in Chapter 3, in the following discussion each value is treated as if it were determined for the population. A discussion of the problem of assessing the statistical significance of the redundancy and suppression effects as determined on a sample will follow the interpretation of these effects.

The term $r^{ii}r_{Yi} = r_{Yi}/(1 - R_i^2)$, which appears for each β_i, may be seen to be the β_i coefficient which would have been obtained if all the linear relationship between X_i and the (other) X_j (i.e., $i \neq j$) were independent of X_i's relationship with Y (equivalently, if at least one of every r_{ij}, r_{Yj} pair were zero). The case where $\beta_i = r^{ii}r_{Yi}$ is equivalent to that of classical suppression (see Section 3.4.3), in which all other variables which correlate with X_i have zero correlations with Y. Obviously this number represents no real population; however, it provides a reference point for other variables in the equation. In the example, for X_1, the number of publications, Table A2.2 shows the $r^{11}r_{Y2}$ value to be .91140. The second value in the equation for β_1, that is, $-.87268$ ($= r^{22}r_{Y2}$), indicates the effect of X_2, years since Ph.D., on the relationship between publications and academic rank, *in the context of* sex of faculty member and number of citations. The fact that this value very substantially reduces β_1 reflects the fact that number of publications is redundant to a substantial degree with the number of years since Ph.D. Those faculty members who had published a lot and were Associate or full Professor also tended to have been in the field longer.

The third and fourth values in the β_1 equations are positive; thus, both the sex of the faculty member and the number of citations have partial correlations with the number of publications which are of opposite sign to their validity coefficients. Equivalently, when the effect of years since Ph.D. is partialled, these two variables tend to suppress variance in the number of publications which is irrelevant to academic rank. Another equivalent and necessary sign of suppression is that if either of these IVs were added to the regression equation which included the other three IVs, the β_1 value would increase. Looking ahead, we may see that, as always, such suppression is mutual—the inclusion of X_1 also increases the (absolute) value of β_3 and β_4. Finally, with a little computation the reader may satisfy himself that although X_1 and X_3 suppress each other even when no other variables are in the regression equation which estimates Y, X_1 and X_4 do not. It is the pattern of (linear) association among the variables, especially their correlations with X_2, which causes the emergence of the suppressing relationship between X_1 and X_4 with regard to Y.

The remaining β_i equations may be similarly interpreted. Looking at β_2 we see that $r^{22}r_{Y2} = r_{Y2}/(1 - R_2^2) = 1.43198$. It is not surprising that this value exceeds 1; as we have seen, when suppression exists β values may exceed unity. The effects associated with X_1, X_3, and X_4 in this equation are all negative, indicating that each of their associations with Y is in part redundant with the association with Y of years since Ph.D.

Turning to the structure of β_3, we note that this variable had the smallest zero-order relationship with Y, and given the contribution to B_3 associated with X_2, $+.26043$, it would appear that most of this relationship was due to the tendency of women faculty members to be younger. However, the negative contributions to β_3 from X_1 and X_4 indicate that *controlling for years since Ph.D.*, women tended to publish a little more and to be cited a little more frequently than men. Since these two variables relate positively to academic rank, their partial correlations with sex show shared variance that is independent of their correlations with Y. The net result of the redundancy with X_2 and the suppression of X_1 and X_4 is a relatively small reduction in β_3 relative to r_{Y3}.

Finally X_4, the number of citations, has the second largest zero-order relationship with academic rank. Although fairly substantial redundancy with years since Ph.D. greatly reduces the net relationship of X_4, the suppressing partial correlations of number of publications and sex with number of citations serve to slightly increase the value of β_4.

In summary, the identification and relative magnitude of redundancy and suppression effects of other IVs in the equation on the regression of Y on X_i may be conveniently made by means of the $r^{ij}r_{Yj}$ terms as follows:

1. Independence between X_i and X_j with respect to Y exists whenever $r^{ij}r_{Yj} = 0$. (Note that two variables which correlate zero with each other may have a nonzero partial r and therefore $r^{ij} \neq 0$.)

2. Redundancy between X_i and X_j with respect to Y exists whenever $r^{ij}r_{Yj}$ is of the *opposite* sign and smaller than $r^{ii}r_{Yi}$.

3. Suppression between X_i and X_j with respect to Y exists whenever $r^{ij}r_{Yj}$ is of the *same* sign as $r^{ii}r_{Yi}$, *or* when of opposite sign, $r^{ij}r_{Yj}$ is of larger magnitude than $r^{ii}r_{Yi}$.

Thus far, we have approached this analysis of redundancy and suppression as though the values had been obtained on the entire population. The actual sample size of the original example was 15, a number so small that $r < .514$ is not significant at (two-tailed) $\alpha = .05$, and only the validity coefficient of years since Ph.D. among the r_{Yi} exceeds this value. Suppose, however, that the sample had been much larger, and we wished to assess the significance of the various redundancy and suppression effects on β_i carried by the $r^{ij}r_{Yj}$ product terms. In this case, although the exact test would be very complicated, an appropriately conservative test of the significance of $r^{ij}r_{Yj}$ may be made by requiring that each of the components of the product be significant at the desired α criterion. r_{Yj} is the zero-order validity coefficient and thus may be tested by the conventional t test of Eq. (2.8.1). As we saw in Eq. (A1.10), r^{ij} may be expressed as $-pr_{ij}/\sqrt{(1 - R_i^2)(1 - R_j^2)}$, and may be tested for significance either by computing pr_{ij} from Eq. (A1.11) and applying the standard t test, or directly from the elements of the inverted matrix as

(A2.1)
$$t = \frac{-r^{ij}\sqrt{n-k}}{\sqrt{r^{ii}r^{jj} - r^{ij\,2}}},$$

with $df = n - k$.

By way of illustration, suppose that the results of the running example came from a sample of $n = 100$ faculty members from a large liberal arts college, and we wish to assess the significance of the effect of X_2 on X_3's relationship to Y, that is, of the product $r^{32}r_{Y2} = (.42536)(.61226) = .26043$ on β_3. We first test $r^{32} = .42536$ by Eq. (A2.1):

$$t = \frac{-.42536\sqrt{100-4}}{\sqrt{(2.33885)(1.08019) - .42536^2}} = -2.721.$$

Checking Appendix Table A for $\alpha = .05, df = 96$, we find that 2.721 exceeds the criterion $t = 1.98$, and conclude that the r^{32} component of the product differs significantly from zero. We then test the $r_{Y2} = .61226$ component by the standard t test for an r of Eq. (2.8.1):

$$t = \frac{.61226\sqrt{100-2}}{\sqrt{1 - .61226^2}} = 7.666,$$

which with df = 98 also meets the significance criterion. Since both components of the product .26043 were found significant, we can conclude on the basis of this conservative test that the product is significant, that is, that years since Ph.D. (X_2) is significantly redundant with sex (X_3) in the latter's relationship to academic rank (Y), given the presence of X_1 and X_4.

Note that although $r^{ij} = r^{ji}$, $r^{ij}r_{Yj} \neq r^{ji}r_{Yi}$ and thus the effect on β_i of X_j is not in general equal to the effect on β_j of X_i. We can thus inquire, as a separate matter, into $r^{23}r_{Y3} = (.42536)(-.24175) = -.10283$ as an element in β_2. This describes the effect of sex (X_3) on β_2, the partial regression of academic rank on years since Ph.D. (X_2). We have already found that r^{32} $(= r^{23})$ is significant, and the test of $r_{Y3} = -.24175$ gives $t = -2.466$, which is also significant $(df = 98, P <$ .05). Thus, we accept the product $r^{23}r_{Y3} = -.10283$ as significant, and conclude that sex affects (here, in the direction of redundancy) the relationship of years since Ph.D. with academic rank.

APPENDIX 3

Computer Analysis
of Multiple Regression/Correlation

There are probably more than a hundred existing programs for MRC analysis at various computer installations around the country. All of the problems described in this book can be done on any of these programs. However, programs may vary considerably with regard to their accuracy, flexibility, demands on the user, and completeness of output. The purpose of this appendix is first, to give the reader some general guidelines for making a choice among those programs available, second, to suggest ways of accomplishing analytical goals not explicitly provided for by programs, and third, to describe in a general way some of the most widely available programs.

CONSIDERATIONS RELEVANT TO MULTIPLE REGRESSION/ CORRELATION PROGRAM SELECTION

1. *The accuracy of the output.* There are really several distinguishable ways in which computer output may be faulty. First, and most serious, the program itself may have errors, "bugs," that reflect errors in writing the program or in the cards or tape which transmit the program to the computer, or possibly in an incompatibility between the program and the computer installation. Most often, but not always, such errors will result in a "bomb"—a failure of the program to carry out the analysis called for. Reliance on well established programs such as those described later in this appendix will probably eliminate most of these problems. The surest way to avoid them is to use programs which have been established, tested, and repeatedly used on the computer to which you have access.

Second is the issue of the accuracy of the computations reflected in the number of digits carried along in the mathematical operations. Many older and small programs written for IBM computers are set up for "single precision"—no more than *eight* digits are carried at each computational step. The resulting rounding errors are not likely to be serious when both k and the correlations (and multiple correlations) among independent variables are small. When the MRC calls for power polynomials or products of IVs (especially of order higher than two), or extensive nominal scale coding, it is wise to make sure that the selected program when used by your computer will provide more than eight digit accuracy. Some programs have "double precision" options available to the user; the precision of other programs depends upon the particular computer installation.

A third issue of accuracy arises by program limits on the number of digits allowed for each input variable. As discussed in Section 6.2.4, when the use of higher order polynomial powers results in variables with more digits than the program allows, it will probably not hurt to simply drop right-hand digits as necessary.

No matter what program is used, the user himself is often a source of output inaccuracy. Computers are very reliable and fast, but stupid. The computer will faithfully try to carry out whatever instructions it receives, whether they make sense or not. With any program, the user had better check his instructions, as well as the output, to make sure that the problem was correctly specified.

There is a final question of accuracy, the accuracy level which may be instrinsic to the algorithm or computational recipe that the particular program uses for achieving its goals. Since it is neither clear what differences may exist among various algorithms, nor is there often a way for the user to determine what algorithm has been used, there is little a user can do about this question at present. A study now under way by a committee of the American Statistical Association may eventually provide some answers to this question for the major package programs.

2. *The capacity of the program.* Virtually all programs have some limit to the number of variables, the number of cases, or some combination of the two. These limits are not likely to be serious for most social science problems which do not even approach the relatively modest limits of 25 variables and 9,999 cases. (Some programs include both the original data variables and any other variables created as functions of these in the count.) Since these limits are often determined not only by the way the program was written but more seriously by the available capacity of the particular computer installation, it pays to check.

3. *The analytic strategy of the program.* Multiple regression/correlation programs are set up to determine simultaneous regression equations, to do stepwise regression, or to do both. When the program does simultaneous regression analysis, hierarchical analyses may be accomplished by calling for the analysis of the relationship of a dependent variable with each of several overlapping cumulative sets of IVs. Stepwise regression programs typically enter variables in a sequence dependent on their contribution to R^2 (Section 3.6.2). Using stepwise programs, hierarchical analysis may be accomplished as with simultaneous regression, ensuring the entry of all variables (in each of several overlapping cumulative

sets) by requiring very small criterion stopping values (such as $F = .01$). Many stepwise regression programs also allow the user to specify the sequence of variable entry. These "forced" stepwise options permit the direct performance of hierarchical MRC analysis.

4. *Recode options.* It is very often the case that the form of the data as recorded on computer cards or tape is not the same as that required for analysis, and that some "recoding," transformation, or additional variable construction is needed. Many of the MRC programs which come as a part of a larger statistical package will allow for recoding variables at the beginning of each analysis. These programs almost always include, on option, the creation of variables as power functions of other variables (needed in power polynomial analyses), of variables which are products of other variables (needed for representing interactions), and of nonlinear transformations, such as trigonometric or logarithmic functions. The suitability of the recode options available for creating dummy-, contrast-, or effects-coded nominal scale values varies. Only rarely do these recode options include transformations of variables which require information not provided by the user, such as subtracting means from all scores, or standardizing. The user must also beware of the program's recognition of missing data codes. For example, some programs will treat blanks as zero values, many will not accept cards some of whose columns contain codes representing letters of the alphabet, or with columns containing any multiple punches. It is prudent to find out about such constraints before having your data punched.

5. *Treatment of missing data and subsample selection.* It is frequently an unavoidable fact that some cases have no information on some independent variables. Chapter 7 provides a presentation of ways of coping with this problem without losing cases (and the consequent representativeness of the sample) by coding explicitly for the "missingness" of data. Some programs allow for dropping subjects with incomplete data from the analysis. Yet another solution offered by some programs is to determine each correlation using only those cases having information on both variables and proceeding with MRC on the resulting matrix. This last procedure is quite risky when missing data are extensive and/or nonrandom. See Chapter 7 for a full discussion of the issues involved.

Many of the larger programs also provide for subsample selection, so that analyses may be done on cases representing some portion of the total sample, chosen by the researcher on the basis of some distinctive characteristic(s). This option is undoubtedly useful; however, the researcher should recognize that tests of the statistical significance of differences of regression coefficients between subsamples are usually also desired. These tests are not provided when each group is subjected to a separate analysis, whereas they are performed automatically when such characteristics are included in interaction terms (see Chapter 8).

6. *Output.* As the reader has seen, the analysis of the relationship between a dependent variable and a set of IVs involves many different coefficients that are useful at various points in the analysis. In addition, many of these have standard

errors and associated significance tests. It is very often useful to have raw recoded data, estimated $\hat{Y}$ scores, or $Y - \hat{Y}$ residuals printed out or punched on cards. One may also wish to be able to call for bivariate plots of Y, $\hat{Y}$, or $Y - \hat{Y}$ with various IVs. On the one hand, no program provides all of these. On the other hand, at least with regard to the results of the analysis itself, all programs give enough information to permit more or less easy hand calculation of information not provided as output. The following lists provide the information which is (or is not) relatively standard in the MRC output of programs which are a part of the larger popular statistical packages. In addition, coefficients which may not be provided in some output are presented as a function of other information that is nearly always available.

Practically all these programs give

Variables identified by name

Zero-order r matrix, means, and sds

B_i and SE_{B_i} for each X_i

$$\beta_i \quad \left[= B_i \frac{sd_i}{sd_Y} \right]$$

A (sometimes identified as B_o), the Y intercept

t_i or F_i $(= t_i^2)$ on partial coefficients for each X_i

R_Y and/or R_Y^2 and its F

$$\widetilde{sd}_{Y-\hat{Y}} \quad \left[= \sqrt{ sd_Y^2(1 - R_Y^2) \frac{n}{n-k-1} } \right]$$

$$pr_i \quad \left[= \frac{t_{B_i}}{\sqrt{t_{B_i}^2 + n - k - 1}} \right]$$

$Y - \hat{Y}$, residuals (errors) from regression.

Some of these programs give:

Raw data, or $\hat{Y}$

$$sr_i \text{ or } sr_i^2 = \text{unique contribution} \quad \left[sr_i = t_{B_i} \sqrt{\frac{1 - R_Y^2}{n-k-1}} \right]$$

Probabilities for significance tests (or look up in a table)

$$\text{Shrunken } \hat{R}_Y \text{ or } \hat{R}_Y^2 \quad \left[\hat{R}_Y^2 = 1 - (1 - R_Y^2)\left(\frac{n-1}{n-k-1} \right) \right]$$

The determinant of the $\mathbf{R}_{ij}$ matrix (see Appendix 1)

The inverted r matrix, $\mathbf{R}_{ij}^{-1}$ (useful for obtaining multiple or partial relationships among IVs, see Appendices 1 and 2)

Multiple and/or partial correlations among IVs

Bivariate plots of $Y - \hat{Y}$ and IVs

Bivariate plots of other variables (e.g., Y with X_i)

CHARACTERISTICS OF SOME
MULTIPLE REGRESSION/CORRELATION PROGRAMS

The following is a description of some of the most widely available MRC programs that come as elements of larger statistical packages or systems. The list is not comprehensive, nor can we vouch for the accuracy of detail—these programs tend to be frequently "updated," and such changes may or may not be a part of any given computer installation's version of the program. Second, we have not had personal experience with some of these programs and have relied on descriptions in the published manuals, which are, alas, often inadequate. Third, although we have attempted to be mostly objective in our descriptions, subjective evaluations of the ease with which programs may be used are also helpful, although subject to dispute. The Committee on Evaluation of Program Packages of the American Statistical Association is carrying out a major review and evaluation of the statistical program packages most widely used in the United States. This study, scheduled for completion in 1976, should provide an objective and comprehensive evaluation.

The programs are herewith listed, but not in any special order of priority or preference:

Statistical Package for the Social Sciences (SPSS)
(Nie, Bent, & Hull, 1970; Nie & Hull, 1973)

This program will take raw data or output from other programs in the package. It has a very flexible recoding and sample selection system. It will treat missing data by dropping cases, by recode, or by working from a missing data r matrix. (Beware of the latter, since it may not "bomb" even when the resulting matrix is impossible—see Section 7.1.2.) This program is easy to use without substantial computer or mathematical sophistication and is moderately efficient and flexible.

A final version is projected for 1975.

Data Text
(Armer & Couch, 1972)

The revised version of this long established system was not complete or reliably "debugged" as of this writing. It will take raw data or output from other programs in the system. It has a very flexible recoding system. It will treat missing data by dropping cases, by recode, or by using a missing data r matrix.

This system is a relatively easy one to use for unsophisticated users. It is particularly easy to move from one program to another. Some problems may require many user-supplied program cards. The MRC output is especially complete, including unique variance (sr_i^2), significance tests with associated exact probability values, and $\hat{Y}$ scores. It requires a computer system with a large core memory and, since it is highly user oriented, is relatively inefficient in computer time.

BMD 02R and 03R
(Dixon, 1968–1970)

This package is an old standby, somewhat updated, with many of the faults of older programs but with the virtues of wide familiarity. Each program stands by itself and requires separate runs. 03R determines simultaneous regression equations, although it also provides the successive contributions to R^2 of IVs determined by their sequence in the data set. It handles a maximum of 80 variables, and provides the inverted r matrix which is useful for finding R^2 and pr^2 among IVs. 02R does stepwise regression in computer-selected sequence or in a sequence determined ("forced") by the user who may assign to each variable one of up to seven levels of priority, and thus perform most hierarchical MRC analyses. It will handle a maximum of 50 variables. Neither program provides for a very flexible recode system, and they are especially poor for nominal scale recoding (but, of course, nominal scale coding may be done by hand). A separate program in this package performs polynomial regression. These are relatively difficult programs to use, although they do not require mathematical or computer sophistication. Machine efficiency is poor.

OSIRIS III
(Institute for Social Research, University of Michigan, 1973)

In this program, stepwise regression variables are program selected and dropped from the equation when they no longer contribute as specified by a criterion. Alternatively, simultaneous regression equations are produced. It has a relatively full recode system and handles missing data. Output options include the inverted r matrix and the matrix of prs among IVs. It handles up to 75 variables and is relatively machine efficient.

PSTAT
(Princeton, Buhler)

This MRC program accepts the r matrix from another program in the system, including missing data matrices. A relatively full recode system is available. It computes simultaneous equations. The number of variables is limited to 50 in total, 24 IVs in a single equation. Single precision only. No manual exists, but instructions are a part of the program package which may be printed out by the computer.

OMNITAB II
(Hogben, Peavy, & Varner, 1971)

This system is matrix oriented and thus difficult to use by those not mathematically sophisticated. For the sophisticated user, this system as a whole provides almost anything, and with a high degree of accuracy (by repute). The user specifies the sequence of IVs. The output from the regression program as such is relatively limited, but the extended system provides much supplementary information. On option it will test automatically for curvilinearity.

Regression Analysis Program for Economists (RAPE)
(Raduchel, 1972, 1973)

This program determines simultaneous regression equations for up to 29 variables per problem. It includes an adequate, but not easy, system for recoding. It provides full output including R^2_Y and pr among IVs, $\hat{Y}$ and $Y - \hat{Y}$ residual values, and also bivariate plots of any variable pairs, including residuals.

Scientific Subroutine Package
(IBM Corp.)

These programs are suitable only for the user who is mathematically sophisticated and familiar with computers. In the IBM 360 version, stepwise regressions are determined for up to 35 variables in a program-determined or user-specified sequence.

Appendix Tables

TABLE A
t Values for α = .01, .05 (Two Tailed)[a]

df	α .01	α .05	df	α .01	α .05
6	3.707	2.447	36	2.720	2.028
7	3.499	2.365	37	2.715	2.026
8	3.355	2.306	38	2.712	2.024
9	3.250	2.262	39	2.708	2.023
10	3.169	2.228	40	2.704	2.021
11	3.106	2.201	42	2.698	2.018
12	3.055	2.179	44	2.692	2.015
13	3.012	2.160	46	2.687	2.013
14	2.977	2.145	48	2.682	2.011
15	2.947	2.131	50	2.678	2.009
16	2.921	2.120	52	2.674	2.007
17	2.898	2.110	54	2.670	2.005
18	2.878	2.101	56	2.666	2.003
19	2.861	2.093	58	2.663	2.002
20	2.845	2.086	60	2.660	2.000
21	2.831	2.080	64	2.655	1.998
22	2.819	2.074	68	2.650	1.996
23	2.807	2.069	72	2.646	1.994
24	2.797	2.064	76	2.642	1.992
25	2.787	2.060	80	2.639	1.990
26	2.779	2.056	90	2.632	1.987
27	2.771	2.052	100	2.626	1.984
28	2.763	2.048	120	2.617	1.980
29	2.756	2.045	150	2.609	1.976
30	2.750	2.042	200	2.601	1.972
31	2.744	2.040	300	2.592	1.968
32	2.738	2.037	400	2.588	1.966
33	2.733	2.034	600	2.584	1.964
34	2.728	2.032	1000	2.581	1.962
35	2.724	2.030	∞	2.576	1.960

[a]This table is abridged from Table 2.1 in Owen (1962). Reproduced with the permission of the publishers. (Courtesy of the U.S. Atomic Energy Commission.)

TABLE B
z' Transformation of r

r	z'	r	z'	r	z'	r	z'
.00	.000	.30	.310	.60	.693	.850	1.256
.01	.010	.31	.321	.61	.709	.855	1.274
.02	.020	.32	.332	.62	.725	.860	1.293
.03	.030	.33	.343	.63	.741	.865	1.313
.04	.040	.34	.354	.64	.758	.870	1.333
.05	.050	.35	.365	.65	.775	.875	1.354
.06	.060	.36	.377	.66	.793	.880	1.376
.07	.070	.37	.388	.67	.811	.885	1.398
.08	.080	.38	.400	.68	.829	.890	1.422
.09	.090	.39	.412	.69	.848	.895	1.447
.10	.100	.40	.424	.70	.867	.900	1.472
.11	.110	.41	.436	.71	.887	.905	1.499
.12	.121	.42	.448	.72	.908	.910	1.528
.13	.131	.43	.460	.73	.929	.915	1.557
.14	.141	.44	.472	.74	.950	.920	1.589
.15	.151	.45	.485	.75	.973	.925	1.623
.16	.161	.46	.497	.76	.996	.930	1.658
.17	.172	.47	.510	.77	1.020	.935	1.697
.18	.182	.48	.523	.78	1.045	.940	1.738
.19	.192	.49	.536	.79	1.071	.945	1.783
.20	.203	.50	.549	.800	1.099	.950	1.832
.21	.213	.51	.563	.805	1.113	.955	1.886
.22	.224	.52	.576	.810	1.127	.960	1.946
.23	.234	.53	.590	.815	1.142	.965	2.014
.24	.245	.54	.604	.820	1.157	.970	2.092
.25	.255	.55	.618	.825	1.172	.975	2.185
.26	.266	.56	.633	.830	1.188	.980	2.298
.27	.277	.57	.648	.835	1.204	.985	2.443
.28	.288	.58	.662	.840	1.221	.990	2.647
.29	.299	.59	.678	.845	1.238	.995	2.994

TABLE C
Normal Distribution[a]

z	p	h	z	p	h	z	p	h
.00	.500	.399	1.25	.106	.183	2.50	.006	.018
.05	.480	.398	1.30	.097	.171	2.55	.005	.015
.10	.460	.397	1.35	.089	.160	2.60	.005	.014
.15	.440	.394	1.40	.081	.150	2.65	.004	.012
.20	.421	.391	1.45	.074	.139	2.70	.003	.010
.25	.401	.387	1.50	.067	.130	2.75	.003	.009
.30	.382	.381	1.55	.061	.120	2.80	.003	.008
.35	.363	.375	1.60	.055	.111	2.85	.002	.007
.40	.345	.368	1.65	.049	.102	2.90	.002	.006
.45	.326	.361	1.70	.045	.094	2.95	.002	.005
.50	.309	.352	1.75	.040	.086	3.00	.0014	.0044
.55	.291	.343	1.80	.036	.079	3.50	.0002	.0009
.60	.274	.333	1.85	.032	.072	4.00	.0000	.0001
.65	.258	.323	1.90	.029	.066	4.50	.0000	.0000
.70	.242	.312	1.95	.026	.060		Fractiles	
.75	.227	.301	2.00	.023	.054	.253	.40	.386
.80	.212	.290	2.05	.020	.049	.431	.333	.364
.85	.198	.278	2.10	.018	.044	.524	.30	.348
.90	.184	.266	2.15	.016	.040	.674	.25	.318
.95	.171	.254	2.20	.014	.035	.842	.20	.280
1.00	.159	.242	2.25	.012	.032	1.282	.10	.176
1.05	.147	.230	2.30	.011	.028	1.645	.05	.103
1.10	.136	.218	2.35	.009	.025	1.960	.025	.058
1.15	.125	.206	2.40	.008	.022	2.326	.01	.027
1.20	.115	.194	2.45	.007	.020	2.576	.005	.014

[a]This table is abridged from Tables 1.1, 1.2, and 1.3 in Owen (1962). Reproduced with the permission of the publishers. (Courtesy of the U.S. Atomic Energy Commission.)

$df_{den.}$	\ $df_{num.}$ 1	2	3	4	5	6	7	8	9	10	11	12	13
15	8.68	6.36	5.42	4.89	4.56	4.32	4.14	4.00	3.89	3.80	3.73	3.67	3.61
16	8.53	6.23	5.29	4.77	4.44	4.20	4.03	3.89	3.78	3.69	3.62	3.55	3.50
17	8.40	6.11	5.18	4.67	4.34	4.10	3.93	3.79	3.68	3.59	3.52	3.46	3.40
18	8.29	6.01	5.09	4.58	4.25	4.01	3.84	3.71	3.60	3.51	3.43	3.37	3.32
19	8.18	5.93	5.01	4.50	4.17	3.94	3.77	3.63	3.52	3.43	3.36	3.30	3.24
20	8.10	5.85	4.94	4.43	4.10	3.87	3.70	3.56	3.46	3.37	3.29	3.23	3.18
21	8.02	5.78	4.87	4.37	4.04	3.81	3.64	3.51	3.40	3.31	3.24	3.17	3.12
22	7.95	5.72	4.82	4.31	3.99	3.76	3.59	3.45	3.35	3.26	3.18	3.12	3.07
23	7.88	5.66	4.76	4.26	3.94	3.71	3.54	3.41	3.30	3.21	3.14	3.07	3.02
24	7.82	5.61	4.72	4.22	3.90	3.67	3.50	3.36	3.26	3.17	3.09	3.03	2.98
25	7.77	5.57	4.68	4.18	3.86	3.63	3.46	3.32	3.22	3.13	3.06	2.99	2.94
26	7.72	5.53	4.64	4.14	3.82	3.59	3.42	3.29	3.18	3.09	3.02	2.96	2.90
27	7.68	5.49	4.60	4.11	3.78	3.56	3.39	3.26	3.15	3.06	2.99	2.93	2.87
28	7.64	5.45	4.57	4.07	3.75	3.53	3.36	3.23	3.12	3.03	2.96	2.90	2.84
29	7.60	5.42	4.54	4.04	3.73	3.50	3.33	3.20	3.09	3.00	2.93	2.87	2.81
30	7.56	5.39	4.51	4.02	3.70	3.47	3.30	3.17	3.07	2.98	2.90	2.84	2.79
32	7.50	5.34	4.46	3.97	3.65	3.43	3.26	3.13	3.02	2.93	2.86	2.80	2.74
34	7.45	5.29	4.42	3.93	3.61	3.39	3.22	3.09	2.98	2.90	2.82	2.76	2.70
36	7.40	5.25	4.38	3.89	3.58	3.35	3.18	3.05	2.95	2.86	2.79	2.72	2.67
38	7.35	5.21	4.34	3.86	3.54	3.32	3.15	3.02	2.92	2.83	2.75	2.69	2.64
40	7.31	5.18	4.31	3.83	3.51	3.29	3.12	2.99	2.89	2.80	2.73	2.66	2.61
42	7.28	5.15	4.29	3.80	3.49	3.27	3.10	2.97	2.86	2.78	2.70	2.64	2.59
44	7.25	5.12	4.26	3.78	3.47	3.24	3.08	2.95	2.84	2.75	2.68	2.62	2.56
46	7.22	5.10	4.24	3.76	3.45	3.22	3.06	2.93	2.82	2.73	2.66	2.60	2.54
48	7.20	5.08	4.22	3.74	3.43	3.20	3.04	2.91	2.80	2.72	2.64	2.58	2.53
50	7.17	5.06	4.20	3.72	3.41	3.19	3.02	2.89	2.79	2.70	2.63	2.56	2.51
55	7.12	5.01	4.16	3.68	3.37	3.15	2.98	2.85	2.75	2.66	2.59	2.53	2.47
60	7.08	4.98	4.13	3.65	3.34	3.12	2.95	2.82	2.72	2.63	2.56	2.50	2.44
65	7.04	4.95	4.10	3.62	3.31	3.09	2.93	2.80	2.69	2.61	2.53	2.47	2.42
70	6.98	4.92	4.08	3.60	3.29	3.07	2.91	2.78	2.67	2.59	2.51	2.45	2.40
80	6.96	4.88	4.04	3.56	3.26	3.04	2.87	2.74	2.64	2.55	2.48	2.42	2.36
90	6.93	4.85	4.01	3.54	3.23	3.01	2.85	2.72	2.61	2.53	2.45	2.39	2.33
100	6.90	4.82	3.98	3.51	3.21	2.99	2.82	2.69	2.59	2.50	2.43	2.37	2.31
120	6.85	4.79	3.95	3.48	3.17	2.96	2.79	2.66	2.56	2.47	2.40	2.34	2.28
150	6.81	4.75	3.92	3.45	3.14	2.93	2.76	2.63	2.53	2.44	2.37	2.31	2.25
200	6.76	4.71	3.88	3.42	3.11	2.89	2.73	2.60	2.50	2.41	2.34	2.28	2.22
300	6.72	4.68	3.85	3.38	3.08	2.86	2.70	2.57	2.47	2.38	2.31	2.25	2.19
400	6.70	4.66	3.83	3.37	3.06	2.85	2.69	2.56	2.45	2.37	2.29	2.23	2.17
1000	6.66	4.63	3.80	3.34	3.04	2.82	2.66	2.53	2.43	2.34	2.26	2.20	2.15
∞	6.63	4.61	3.78	3.32	3.02	2.80	2.64	2.51	2.41	2.32	2.25	2.18	2.1

					$df_{num.}$							
14	15	16	18	20	24	30	40	50	60	80	120	∞
3.56	3.52	3.48	3.42	3.37	3.29	3.21	3.13	3.08	3.05	3.00	2.96	2.87
3.45	3.41	3.37	3.31	3.26	3.18	3.10	3.02	2.97	2.93	2.84	2.84	2.75
3.35	3.31	3.27	3.21	3.16	3.08	3.00	2.92	2.87	2.83	2.79	2.75	2.65
3.27	3.23	3.19	3.13	3.08	3.00	2.92	2.84	2.78	2.75	2.70	2.66	2.57
3.19	3.15	3.12	3.05	3.00	2.92	2.84	2.76	2.71	2.67	2.63	2.58	2.49
3.13	3.09	3.05	2.99	2.94	2.86	2.78	2.69	2.64	2.61	2.56	2.52	2.42
3.07	3.03	2.99	2.93	2.88	2.80	2.72	2.64	2.58	2.55	2.50	2.46	2.36
3.02	2.98	2.94	2.88	2.83	2.75	2.67	2.58	2.53	2.50	2.45	2.40	2.31
2.97	2.93	2.89	2.83	2.78	2.70	2.62	2.54	2.48	2.45	2.40	2.35	2.26
2.93	2.89	2.85	2.79	2.74	2.66	2.58	2.49	2.44	2.40	2.36	2.31	2.21
2.89	2.85	2.81	2.75	2.70	2.62	2.54	2.45	2.40	2.36	2.32	2.27	2.17
2.86	2.82	2.78	2.71	2.66	2.58	2.50	2.42	2.36	2.33	2.28	2.23	2.13
2.82	2.78	2.74	2.68	2.63	2.55	2.47	2.38	2.33	2.29	2.25	2.20	2.10
2.79	2.75	2.72	2.65	2.60	2.52	2.44	2.35	2.30	2.26	2.22	2.17	2.06
2.77	2.73	2.69	2.62	2.57	2.49	2.41	2.33	2.27	2.23	2.19	2.14	2.03
2.74	2.70	2.66	2.60	2.55	2.47	2.39	2.30	2.24	2.21	2.16	2.11	2.01
2.70	2.66	2.62	2.55	2.50	2.42	2.34	2.25	2.20	2.16	2.11	2.06	1.96
2.66	2.62	2.58	2.51	2.46	2.38	2.30	2.21	2.16	2.12	2.07	2.02	1.91
2.62	2.58	2.54	2.48	2.43	2.35	2.26	2.18	2.12	2.08	2.03	1.98	1.87
2.59	2.55	2.51	2.45	2.40	2.32	2.23	2.14	2.09	2.05	2.00	1.95	1.84
2.56	2.52	2.48	2.42	2.37	2.29	2.20	2.11	2.06	2.02	1.97	1.92	1.80
2.54	2.50	2.46	2.40	2.34	2.26	2.18	2.09	2.03	1.99	1.94	1.89	1.78
2.52	2.48	2.44	2.37	2.32	2.24	2.16	2.07	2.01	1.97	1.92	1.87	1.75
2.50	2.46	2.42	2.35	2.30	2.22	2.14	2.05	1.99	1.95	1.90	1.84	1.72
2.48	2.44	2.40	2.33	2.28	2.20	2.12	2.03	1.97	1.93	1.87	1.82	1.70
2.46	2.42	2.38	2.32	2.27	2.18	2.10	2.01	1.95	1.91	1.86	1.80	1.68
2.42	2.38	2.34	2.28	2.23	2.15	2.06	1.97	1.91	1.87	1.82	1.76	1.64
2.39	2.35	2.31	2.25	2.20	2.12	2.03	1.94	1.88	1.84	1.78	1.73	1.60
2.37	2.33	2.29	2.22	2.17	2.09	2.00	1.91	1.85	1.81	1.75	1.70	1.57
2.35	2.31	2.27	2.20	2.15	2.07	1.98	1.89	1.83	1.78	1.73	1.67	1.54
2.31	2.27	2.23	2.17	2.12	2.03	1.94	1.85	1.79	1.75	1.69	1.63	1.49
2.29	2.25	2.21	2.14	2.09	2.01	1.92	1.82	1.76	1.72	1.66	1.60	1.45
2.26	2.22	2.18	2.12	2.07	1.98	1.89	1.80	1.73	1.69	1.63	1.57	1.43
2.23	2.19	2.15	2.09	2.03	1.95	1.86	1.76	1.70	1.66	1.59	1.53	1.37
2.20	2.16	2.12	2.06	2.00	1.92	1.83	1.73	1.66	1.62	1.56	1.49	1.33
2.17	2.13	2.09	2.02	1.97	1.89	1.79	1.69	1.63	1.58	1.52	1.45	1.28
2.14	2.10	2.06	1.99	1.94	1.85	1.76	1.66	1.59	1.55	1.48	1.41	1.22
2.13	2.08	2.04	1.98	1.93	1.84	1.75	1.64	1.57	1.53	1.46	1.39	1.19
2.10	2.06	2.02	1.95	1.90	1.81	1.72	1.61	1.54	1.49	1.42	1.35	1.11
2.08	2.04	2.00	1.93	1.88	1.79	1.70	1.59	1.52	1.47	1.40	1.32	1.00

^{a}This table is partly abridged from Table 4.1 in Owen (1962), and partly computed by linear interpolations in reciprocals of df. Reproduced with the permission of the publishers. (Courtesy of the U.S. Atomic Energy Commission.)

$df_{den.}$	\multicolumn{13}{c}{$df_{num.}$}												
	1	2	3	4	5	6	7	8	9	10	11	12	13
15	4.54	3.68	3.29	3.06	2.90	2.79	2.71	2.64	2.59	2.54	2.51	2.48	2.45
16	4.49	3.63	3.24	3.01	2.85	2.74	2.66	2.59	2.54	2.49	2.46	2.42	2.40
17	4.45	3.59	3.20	2.96	2.81	2.70	2.61	2.55	2.49	2.45	2.41	2.38	2.35
18	4.41	3.55	3.16	2.93	2.77	2.66	2.58	2.51	2.46	2.41	2.37	2.34	2.31
19	4.38	3.52	3.13	2.90	2.74	2.63	2.54	2.48	2.42	2.38	2.34	2.31	2.28
20	4.35	3.49	3.10	2.87	2.71	2.60	2.51	2.45	2.39	2.35	2.31	2.28	2.25
21	4.32	3.47	3.07	2.84	2.68	2.57	2.49	2.42	2.37	2.32	2.28	2.25	2.22
22	4.30	3.44	3.05	2.82	2.66	2.55	2.46	2.40	2.34	2.30	2.26	2.23	2.20
23	4.28	3.42	3.03	2.80	2.64	2.53	2.44	2.38	2.32	2.27	2.24	2.20	2.17
24	4.26	3.40	3.01	2.78	2.62	2.51	2.42	2.36	2.30	2.25	2.22	2.18	2.15
25	4.24	3.39	2.99	2.76	2.60	2.49	2.40	2.34	2.28	2.24	2.20	2.16	2.14
26	4.23	3.37	2.98	2.74	2.59	2.47	2.39	2.32	2.27	2.22	2.18	2.15	2.12
27	4.21	3.35	2.96	2.73	2.57	2.46	2.37	2.31	2.25	2.20	2.16	2.13	2.10
28	4.20	3.34	2.95	2.71	2.56	2.45	2.36	2.29	2.24	2.19	2.15	2.12	2.09
29	4.18	3.33	2.93	2.70	2.55	2.43	2.35	2.28	2.22	2.18	2.14	2.10	2.07
30	4.17	3.32	2.92	2.69	2.53	2.42	2.33	2.27	2.21	2.16	2.13	2.09	2.06
32	4.15	3.30	2.90	2.67	2.51	2.40	2.31	2.24	2.19	2.14	2.10	2.07	2.04
34	4.13	3.28	2.88	2.65	2.49	2.38	2.29	2.23	2.17	2.12	2.08	2.05	2.02
36	4.11	3.26	2.87	2.63	2.48	2.36	2.28	2.21	2.15	2.11	2.07	2.03	2.00
38	4.10	3.24	2.85	2.62	2.46	2.35	2.26	2.19	2.14	2.09	2.05	2.02	1.99
40	4.08	3.23	2.84	2.61	2.45	2.34	2.25	2.18	2.12	2.08	2.04	2.00	1.97
42	4.07	3.22	2.83	2.59	2.44	2.32	2.24	2.17	2.11	2.07	2.02	1.99	1.96
44	4.06	3.21	2.82	2.58	2.43	2.31	2.23	2.16	2.10	2.05	2.01	1.98	1.95
46	4.05	3.20	2.81	2.57	2.42	2.30	2.22	2.15	2.09	2.04	2.00	1.97	1.94
48	4.04	3.19	2.80	2.57	2.41	2.30	2.21	2.14	2.08	2.03	1.99	1.96	1.9
50	4.03	3.18	2.79	2.56	2.40	2.29	2.20	2.13	2.07	2.03	1.99	1.95	1.9
55	4.02	3.17	2.77	2.54	2.38	2.27	2.18	2.11	2.06	2.01	1.97	1.93	1.90
60	4.00	3.15	2.76	2.53	2.37	2.25	2.17	2.10	2.04	1.99	1.95	1.92	1.8
65	3.99	3.14	2.75	2.51	2.36	2.24	2.15	2.08	2.03	1.98	1.94	1.90	1.8
70	3.98	3.13	2.74	2.50	2.35	2.23	2.14	2.07	2.02	1.97	1.93	1.89	1.86
80	3.96	3.11	2.72	2.49	2.33	2.21	2.13	2.06	2.00	1.95	1.91	1.88	1.8
90	3.95	3.10	2.71	2.47	2.32	2.20	2.11	2.04	1.99	1.94	1.90	1.86	1.8
100	3.94	3.09	2.70	2.46	2.31	2.19	2.10	2.03	1.98	1.93	1.89	1.85	1.8
120	3.92	3.07	2.68	2.45	2.29	2.18	2.09	2.02	1.96	1.91	1.87	1.83	1.8
150	3.90	3.06	2.67	2.43	2.27	2.16	2.07	2.00	1.94	1.89	1.85	1.82	1.7
200	3.89	3.04	2.65	2.42	2.26	2.14	2.06	1.99	1.93	1.88	1.84	1.80	1.7
300	3.87	3.03	2.64	2.40	2.24	2.13	2.04	1.97	1.91	1.86	1.82	1.79	1.7
400	3.86	3.02	2.63	2.39	2.23	2.12	2.03	1.96	1.90	1.85	1.81	1.78	1.7
1000	3.85	3.00	2.61	2.38	2.22	2.11	2.02	1.95	1.89	1.84	1.80	1.76	1.7
∞	3.84	3.00	2.60	2.37	2.21	2.10	2.01	1.94	1.88	1.83	1.79	1.75	1.7

					$df_{num.}$							
14	15	16	18	20	24	30	40	50	60	80	120	∞
2.42	2.40	2.38	2.35	2.33	2.29	2.25	2.20	2.18	2.16	2.14	2.11	2.07
2.37	2.35	2.33	2.30	2.28	2.24	2.19	2.15	2.12	2.11	2.08	2.06	2.01
2.33	2.31	2.29	2.26	2.23	2.19	2.15	2.10	2.08	2.06	2.03	2.01	1.96
2.29	2.27	2.25	2.22	2.19	2.15	2.11	2.06	2.04	2.02	1.99	1.97	1.92
2.26	2.23	2.21	2.18	2.16	2.11	2.07	2.03	2.00	1.98	1.95	1.93	1.88
2.23	2.20	2.18	2.15	2.12	2.08	2.04	1.99	1.97	1.95	1.92	1.90	1.84
2.20	2.18	2.16	2.12	2.10	2.05	2.01	1.96	1.94	1.92	1.89	1.87	1.81
2.17	2.15	2.13	2.10	2.07	2.03	1.98	1.94	1.91	1.89	1.86	1.84	1.78
2.15	2.13	2.11	2.07	2.05	2.00	1.96	1.91	1.88	1.86	1.84	1.81	1.76
2.13	2.11	2.09	2.05	2.03	1.98	1.94	1.89	1.86	1.84	1.82	1.79	1.73
2.11	2.09	2.07	2.03	2.01	1.96	1.92	1.87	1.84	1.82	1.80	1.77	1.71
2.09	2.07	2.05	2.02	1.99	1.95	1.90	1.85	1.82	1.80	1.78	1.75	1.69
2.08	2.06	2.04	2.00	1.97	1.93	1.88	1.84	1.81	1.79	1.76	1.73	1.67
2.06	2.04	2.02	1.99	1.96	1.91	1.87	1.82	1.79	1.77	1.74	1.71	1.65
2.05	2.03	2.01	1.97	1.94	1.90	1.85	1.81	1.77	1.75	1.73	1.70	1.64
2.04	2.01	1.99	1.96	1.93	1.89	1.84	1.79	1.76	1.74	1.71	1.68	1.62
2.02	1.99	1.97	1.94	1.91	1.86	1.82	1.77	1.74	1.71	1.69	1.66	1.59
2.00	1.97	1.95	1.92	1.89	1.84	1.80	1.75	1.71	1.69	1.66	1.63	1.57
1.98	1.95	1.93	1.90	1.87	1.82	1.78	1.73	1.69	1.67	1.64	1.61	1.55
1.96	1.94	1.92	1.88	1.85	1.81	1.76	1.71	1.68	1.65	1.62	1.59	1.53
1.95	1.92	1.90	1.87	1.84	1.79	1.74	1.69	1.66	1.64	1.61	1.58	1.51
1.93	1.91	1.89	1.85	1.83	1.78	1.73	1.68	1.65	1.62	1.59	1.56	1.49
1.92	1.90	1.88	1.84	1.81	1.77	1.72	1.67	1.63	1.61	1.58	1.55	1.48
1.91	1.89	1.87	1.83	1.80	1.76	1.71	1.65	1.62	1.60	1.57	1.53	1.46
1.90	1.88	1.86	1.82	1.79	1.75	1.70	1.64	1.61	1.59	1.55	1.52	1.45
1.89	1.87	1.85	1.81	1.78	1.74	1.69	1.63	1.60	1.58	1.54	1.51	1.44
1.88	1.85	1.83	1.79	1.76	1.72	1.67	1.61	1.58	1.55	1.52	1.49	1.41
1.86	1.84	1.81	1.78	1.75	1.70	1.65	1.59	1.56	1.53	1.50	1.47	1.39
1.85	1.82	1.80	1.76	1.73	1.69	1.63	1.58	1.54	1.52	1.48	1.45	1.37
1.84	1.81	1.79	1.75	1.72	1.67	1.62	1.56	1.53	1.50	1.47	1.44	1.35
1.82	1.79	1.77	1.73	1.70	1.65	1.60	1.54	1.51	1.48	1.45	1.41	1.32
1.80	1.78	1.76	1.72	1.69	1.64	1.59	1.53	1.49	1.46	1.43	1.39	1.30
1.79	1.77	1.74	1.71	1.68	1.63	1.57	1.52	1.48	1.45	1.41	1.37	1.28
1.77	1.75	1.73	1.69	1.66	1.61	1.55	1.50	1.46	1.43	1.39	1.35	1.25
1.76	1.73	1.71	1.67	1.64	1.59	1.54	1.47	1.43	1.41	1.37	1.33	1.22
1.74	1.72	1.69	1.65	1.62	1.57	1.52	1.45	1.41	1.38	1.34	1.30	1.19
1.73	1.70	1.68	1.64	1.61	1.55	1.50	1.43	1.39	1.36	1.32	1.27	1.15
1.72	1.69	1.67	1.63	1.60	1.54	1.49	1.42	1.38	1.35	1.31	1.26	1.13
1.70	1.68	1.65	1.61	1.58	1.53	1.47	1.41	1.36	1.33	1.28	1.24	1.08
1.69	1.67	1.64	1.60	1.57	1.52	1.46	1.39	1.35	1.32	1.27	1.22	1.00

[a]This table is partly abridged from Table 4.1 in Owen (1962), and partly computed by linear interpolation in reciprocals of df. Reproduced with the permission of the publishers. (Courtesy of the U.S. Atomic Energy Commission.)

L Values for $\alpha = .01$

k_B						Power					
	.10	.30	.50	.60	.70	.75	.80	.85	.90	.95	.99
1	1.67	4.21	6.64	8.00	9.61	10.57	11.68	13.05	14.88	17.81	24.03
2	2.30	5.37	8.19	9.75	11.57	12.64	13.88	15.40	17.43	20.65	27.42
3	2.76	6.22	9.31	11.01	12.97	14.12	15.46	17.09	19.25	22.67	29.83
4	3.15	6.92	10.23	12.04	14.12	15.34	16.75	18.47	20.74	24.33	31.80
5	3.49	7.52	11.03	12.94	15.12	16.40	17.87	19.66	22.03	25.76	33.50
6	3.79	8.07	11.75	13.74	16.01	17.34	18.87	20.73	23.18	27.04	35.02
7	4.08	8.57	12.41	14.47	16.83	18.20	19.79	21.71	24.24	28.21	36.41
8	4.34	9.03	13.02	15.15	17.59	19.00	20.64	22.61	25.21	29.29	37.69
9	4.58	9.47	13.59	15.79	18.30	19.75	21.43	23.46	26.12	30.31	38.89
10	4.82	9.88	14.13	16.39	18.97	20.46	22.18	24.25	26.98	31.26	40.02
11	5.04	10.27	14.64	16.96	19.60	21.13	22.89	25.01	27.80	32.16	41.09
12	5.25	10.64	15.13	17.51	20.21	21.77	23.56	25.73	28.58	33.02	42.11
13	5.45	11.00	15.59	18.03	20.78	22.38	24.21	26.42	29.32	33.85	43.09
14	5.65	11.35	16.04	18.53	21.34	22.97	24.83	27.09	30.03	34.64	44.03
15	5.84	11.67	16.48	19.01	21.88	23.53	25.43	27.72	30.72	35.40	44.93
16	6.02	12.00	16.90	19.48	22.40	24.08	26.01	28.34	31.39	36.14	45.80
18	6.37	12.61	17.70	20.37	23.39	25.12	27.12	29.52	32.66	37.54	47.46
20	6.70	13.19	18.45	21.21	24.32	26.11	28.16	30.63	33.85	38.87	49.03
22	7.02	13.74	19.17	22.01	25.21	27.05	29.15	31.69	34.99	40.12	50.51
24	7.32	14.27	19.86	22.78	26.06	27.94	30.10	32.69	36.07	41.32	51.93
28	7.89	15.26	21.15	24.21	27.65	29.62	31.88	34.59	38.11	43.58	54.60
32	8.42	16.19	22.35	25.55	29.13	31.19	33.53	36.35	40.01	45.67	57.07
36	8.92	17.06	23.48	26.80	30.52	32.65	35.09	38.00	41.78	47.63	59.39
40	9.39	17.88	24.54	27.99	31.84	34.04	36.55	39.56	43.46	49.49	61.57
50	10.48	19.77	27.00	30.72	34.86	37.23	39.92	43.14	47.31	53.74	66.59
60	11.46	21.48	29.21	33.18	37.59	40.10	42.96	46.38	50.79	57.58	71.12
70	12.37	23.05	31.25	35.45	40.10	42.75	45.76	49.35	53.99	61.11	75.27
80	13.22	24.51	33.15	37.55	42.43	45.21	48.36	52.11	56.96	64.39	79.13
90	14.01	25.89	34.93	39.53	44.62	47.52	50.80	54.71	59.75	67.47	82.76
100	14.76	27.19	36.62	41.41	46.70	49.70	53.11	57.16	62.38	70.37	86.18

L Values for $\alpha = .05$

k_B	Power										
	.10	.30	.50	.60	.70	.75	.80	.85	.90	.95	.99
1	.43	2.06	3.84	4.90	6.17	6.94	7.85	8.98	10.51	13.00	18.37
2	.62	2.78	4.96	6.21	7.70	8.59	9.64	10.92	12.65	15.44	21.40
3	.78	3.30	5.76	7.15	8.79	9.77	10.90	12.30	14.17	17.17	23.52
4	.91	3.74	6.42	7.92	9.68	10.72	11.94	13.42	15.41	18.57	25.24
5	1.03	4.12	6.99	8.59	10.45	11.55	12.83	14.39	16.47	19.78	26.73
6	1.13	4.46	7.50	9.19	11.14	12.29	13.62	15.26	17.42	20.86	28.05
7	1.23	4.77	7.97	9.73	11.77	12.96	14.35	16.04	18.28	21.84	29.25
8	1.32	5.06	8.41	10.24	12.35	13.59	15.02	16.77	19.08	22.74	30.36
9	1.40	5.33	8.81	10.71	12.89	14.17	15.65	17.45	19.83	23.59	31.39
10	1.49	5.59	9.19	11.15	13.40	14.72	16.24	18.09	20.53	24.39	32.37
11	1.56	5.83	9.56	11.58	13.89	15.24	16.80	18.70	21.20	25.14	33.29
12	1.64	6.06	9.90	11.98	14.35	15.74	17.34	19.28	21.83	25.86	34.16
13	1.71	6.29	10.24	12.36	14.80	16.21	17.85	19.83	22.44	26.55	35.00
14	1.78	6.50	10.55	12.73	15.22	16.67	18.34	20.36	23.02	27.20	35.81
15	1.84	6.71	10.86	13.09	15.63	17.11	18.81	20.87	23.58	27.84	36.58
16	1.90	6.91	11.16	13.43	16.03	17.53	19.27	21.37	24.13	28.45	37.33
18	2.03	7.29	11.73	14.09	16.78	18.34	20.14	22.31	25.16	29.62	38.76
20	2.14	7.65	12.26	14.71	17.50	19.11	20.96	23.20	26.13	30.72	40.10
22	2.25	8.00	12.77	15.30	18.17	19.83	21.74	24.04	27.06	31.77	41.37
24	2.36	8.33	13.02	15.87	18.82	20.53	22.49	24.85	27.94	32.76	42.59
28	2.56	8.94	14.17	16.93	20.04	21.83	23.89	26.36	29.60	34.64	44.87
32	2.74	9.52	15.02	17.91	21.17	23.04	25.19	27.77	31.14	36.37	46.98
36	2.91	10.06	15.82	18.84	22.23	24.18	26.41	29.09	32.58	38.00	48.96
40	3.08	10.57	16.58	19.71	23.23	25.25	27.56	30.33	33.94	39.54	50.83
50	3.46	11.75	18.31	21.72	25.53	27.71	30.20	33.19	37.07	43.07	55.12
60	3.80	12.81	19.88	23.53	27.61	29.94	32.59	35.77	39.89	46.25	58.98
70	4.12	13.79	21.32	25.20	29.52	31.98	34.79	38.14	42.48	49.17	62.53
80	4.41	14.70	22.67	26.75	31.29	33.88	36.83	40.35	44.89	51.89	65.83
90	4.69	15.56	23.93	28.21	32.96	35.67	38.75	42.14	47.16	54.44	68.92
100	4.95	16.37	25.12	29.59	34.54	37.36	40.56	44.37	49.29	56.85	71.84

TABLE F.1
Power of Significance Test of r at $\alpha = .01$ (Two Tailed)[a]

n	.10	.20	.30	.40	.50	.60	.70	.80	.90
15	01	03	06	13	25	44	68	90	*
16	01	03	07	14	28	48	73	93	
17	01	03	08	16	30	52	77	95	
18	01	04	08	17	33	56	80	96	
19	02	04	09	19	36	59	83	97	
20	02	04	09	20	38	62	85	98	
21	02	04	10	21	41	66	88	98	
22	02	04	11	23	43	68	90	99	
23	02	04	12	25	46	71	91	99	
24	02	05	12	26	49	74	93	99	
25	02	05	13	28	51	76	94	*	
26	02	05	14	30	53	78	95		
27	02	06	14	31	55	80	96		
28	02	06	15	33	57	82	96		
29	02	06	16	34	60	84	97		
30	02	06	17	36	62	85	98		
31	02	07	17	37	64	87	98		
32	02	07	18	39	66	88	98		
33	02	07	19	40	67	89	99		
34	02	07	20	42	69	90	99		
35	02	08	20	43	71	91	99		
36	02	08	21	45	72	92	99		
37	02	08	22	47	74	93	99		
38	02	08	23	48	76	94	*		
39	02	09	24	49	77	95			
40	02	09	25	50	78	95			
42	03	09	26	53	81	96			
44	03	10	28	56	83	97			
46	03	11	29	58	85	98			
48	03	11	31	61	87	98			

n	.10	.20	.30	.40	.50	.60	.70	.80	.90
50	03	12	33	63	89	99	*	*	*
52	03	12	34	66	90	99			
54	03	13	36	68	91	99			
56	03	14	38	70	93	99			
58	03	14	39	72	94	*			
60	03	15	41	74	94				
64	04	16	44	77	96				
68	04	17	47	80	97				
72	04	19	50	83	98				
76	04	20	53	85	98				
80	04	21	56	87	99				
84	05	23	59	89	99				
88	05	24	61	91	99				
92	05	25	64	92	*				
96	05	27	66	94					
100	06	29	69	95					
120	07	35	78	98					
140	08	42	85	99					
160	09	49	90	*					
180	11	55	94						
200	12	61	96						
250	16	73	99						
300	20	82	*						
350	24	89							
400	28	93							
500	37	97							
600	45	99							
700	53	*							
800	60								
1000	72								

Note: Decimal points omitted in power values.

*Power values at and below this point exceed .995.

[a]Slightly abridged from Table 3.3.4 in Cohen (1969). Reproduced with the permission of the publisher.

TABLE F.2
Power of Significance Test of r at $\alpha = .05$ (Two Tailed) [a]

n	.10	.20	.30	.40	.50	.60	.70	.80	.90
5	06	11	19	32	50	70	88	98	*
6	07	11	21	35	53	73	90	98	
7	07	12	22	37	56	76	92	99	
8	07	12	23	39	59	79	94	99	
9	07	13	24	41	62	81	95	99	
10	07	14	25	43	64	83	96	*	
11	07	14	27	45	66	85	96		
12	07	15	28	47	69	87	97		
13	07	15	29	49	71	89	98		
14	07	16	30	51	73	90	98		
15	08	16	31	53	75	91	99		
16	08	17	33	54	76	92	99		
17	08	17	34	56	78	93	99		
18	08	18	35	58	80	94	99		
19	08	18	36	59	81	95	99		
20	08	19	37	61	83	95	*		
21	08	19	38	62	84	96			
22	08	20	39	64	85	97			
23	09	20	40	65	86	97			
24	09	21	42	67	87	97			
25	09	21	43	68	88	98			
26	09	22	44	69	89	98			
27	09	22	45	70	90	98			
28	09	23	46	72	91	99			
29	09	23	47	73	91	99			
30	09	24	48	74	92	99			
32	10	25	50	76	93	99			
34	10	26	52	78	94	99			
36	10	27	54	80	95	*			
38	10	28	55	82	96				

n	.10	.20	.30	.40	.50	.60	.70	.80	.90
50	11	29	57	83	97	*	*	*	*
52	11	30	59	85	97				
54	11	31	61	86	98				
56	11	32	62	87	98				
58	12	33	64	89	98				
60	12	34	65	90	99				
64	12	36	68	91	99				
68	13	38	71	93	99				
72	13	39	73	94	*				
76	14	41	76	95					
80	14	43	78	96					
84	15	45	80	97					
88	15	47	82	98					
92	16	48	83	98					
96	16	50	85	98					
100	17	52	86	99					
120	19	59	92	*					
140	22	66	95						
160	24	72	97						
180	27	77	98						
200	29	81	99						
250	35	89	*						
300	41	94							
350	46	97							
400	52	98							
500	61	99							
600	69	*							
700	76								
800	81								
1000	89								

Note: Decimal points omitted in power values.
*Power values at and below this point exceed .995.
[a]Slightly abridged from Table 3.3.5 in Cohen (1969). Reproduced with the permission of the publisher.

TABLE G.1

n^* to Detect r by t Test at $\alpha = .01$ (Two Tailed)[a]

Desired power	Population r								
	.10	.20	.30	.40	.50	.60	.70	.80	.90
.25	362	90	40	23	15	11	8	6	5
.50	662	164	71	39	24	16	12	8	6
.60	797	197	86	47	29	19	13	9	7
2/3	901	222	96	53	32	21	15	10	7
.70	957	236	102	56	34	23	15	11	7
.75	1052	259	112	61	37	25	17	11	8
.80	1163	286	124	67	41	27	18	12	8
.85	1299	320	138	75	45	30	20	13	9
.90	1480	364	157	85	51	34	22	15	9
.95	1790	440	190	102	62	40	26	17	11
.99	2390	587	253	136	82	52	34	23	13

[a]Reproduced from Table 3.4.1, in Cohen (1969) with permission of the publisher.

TABLE G.2

n^* to Detect r by t Test at $\alpha = .05$ (Two Tailed)[a]

Desired power	Population r								
	.10	.20	.30	.40	.50	.60	.70	.80	.90
.25	166	42	20	12	8	6	5	4	3
.50	384	95	42	24	15	10	7	6	4
.60	489	121	53	29	18	12	9	6	5
2/3	570	141	62	34	21	14	10	7	5
.70	616	152	66	37	23	15	10	7	5
.75	692	171	74	41	25	17	11	8	6
.80	783	193	84	46	28	18	12	9	6
.85	895	221	96	52	32	21	14	10	6
.90	1046	258	112	61	37	24	16	11	7
.95	1308	322	139	75	46	30	19	13	8
.99	1828	449	194	104	63	40	27	18	11

[a]Reproduced from Table 3.4.1 in Cohen (1969) with permission of the publisher.

References

Acton, F. S. *Analysis of straight-line data.* New York: Wiley, 1959.

Althauser, R. P. Multicollinearity and non-additive regression models. In H. M. Blalock, Jr. (Ed.), *Causal models in the social sciences.* Chicago: Aldine Atherton, 1971.

Anderson, R. L., & Houseman, E. E. *Tables of orthogonal polynomial values extended to N = 104.* Research Bulletin No. 297. Ames, Iowa: Iowa Agricultural Experiment Station, 1942.

Armor, D. J., & Couch, A. S. *Data-text primer: An introduction to computerized social data analysis.* New York: Free Press, 1972.

Baker, B., Hardyck, C. D., & Petrinovich, L. F. Weak measurements vs. strong statistics: An empirical critique of S.S. Stevens' proscriptions on statistics. *Educational and Psychological Measurement,* 1966, **26**, 291–309.

Bartlett, M. S. Square-root transformation in analysis of variance. *Journal of the Royal Statistical Society Supplement,* 1936, **3**, 68–78.

Binder, A. M. Considerations of the place of assumptions in correlational analysis. *American Psychologist,* 1959, **14**, 504–510.

Blalock, H. M., Jr. *Causal inferences in nonexperimental research.* Chapel Hill: University of North Carolina Press, 1964.

Blalock, H. M., Jr. (Ed.) *Causal models in the social sciences.* Chicago: Aldine Atherton, 1971.

Blalock, H. M., Jr., & Blalock, A. B. (Eds.) *Methodology in social research.* New York: McGraw-Hill, 1968.

Bock, R. D. *Multivariate statistical methods in behavioral research.* New York: McGraw-Hill, 1975.

Boneau, C. A. The effects of violations of assumptions underlying the *t* test. *Psychological Bulletin,* 1960, **57**, 49–64.

Campbell, D. T., & Erlebacher, A. How regression artifacts in quasi-experimental evaluations can mistakenly make compensatory education look harmful. In J. Hellmuth (Ed.), *Disadvantaged child, Vol. 3: Compensatory education: A national debate.* New York: Bruner/Mazel, 1970.

Carmer, S. G., & Swanson, M. R. An evaluation of ten pairwise multiple comparison procedures by Monte Carlo methods. *Journal of the American Statistical Association,* 1973, **68**, 66–74.

Chesire, L., Safir, M., & Thurstone, L. L. *Computing diagrams for the tetrachoric correlation coefficient.* Chicago: University of Chicago Bookstore, 1933.

Cochran, W. G. Some consequences when the assumptions for the analysis of variance are not satisfied. *Biometrics*, 1947, 3, 22–38.

Cohen, J. The statistical power of abnormal-social psychological research: A review. *Journal of Abnormal and Social Psychology*, 1962, 65, 145–153.

Cohen, J. Some statistical issues in psychological research. In B. B. Woleman (Ed.), *Handbook of clinical psychology*. New York: McGraw-Hill, 1965. Pp. 95–121.

Cohen, J. Multiple regression as a general data-analytic system. *Psychological Bulletin*, 1968, 70, 426–443.

Cohen, J. *Statistical power analysis for the behavioral sciences*. New York: Academic Press, 1969.

Cohen, J. Causation ⟶ correlation. *Contemporary Psychology*, 1973, 18, 453–455. (a)

Cohen, J. Statistical power analysis and research results. *American Educational Research Journal*, 1973, 10, 225–229. (b)

Coleman, J. S. *Introduction to mathematical sociology*. Glencoe, Illinois: Free Press, 1964.

Conger, A. J. A revised definition for suppressor variables: A guide to their identification and interpretation. *Educational and Psychological Measurement*, 1974, 34, 35–46.

Cooley, W. W., & Lohnes, P. R. *Multivariate data analysis*. New York: Wiley, 1971.

Cronbach, L. J. *Essentials of psychological testing*. (3rd ed.) New York: Harper & Row, 1970.

Dawes, R. M., & Corrigan, B. Linear models in decision making. *Psychological Bulletin*, 1974, 81, 95–106.

Dixon, W. J. (Ed.) *Biomedical computer programs*. Berkeley and Los Angeles: University of California Press, 1968–1970.

Donaldson, T. S. Robustness of the *F* test to errors of both kinds and the correlation between the numerator and denominator of the *F* ratio. *Journal of the American Statistical Association*, 1968, 63, 660–676.

Draper, N. R., & Smith, H. *Applied regression analysis*. New York: Wiley, 1966.

Edwards, A. E. *Experimental design in psychological research*. (4th ed.) New York: Holt, Rinehart & Winston, 1972.

Ezekiel, M., & Fox, K. A. *Methods of correlation and regression analysis*. (3rd ed.) New York: Wiley, 1959.

Fisher, R. A., & Yates, F. *Statistical tables for biological, agricultural and medical research*. (6th ed.) New York: Hafner, 1963.

Freeman, M. F., & Tukey, J. W. Transformations related to the angular and the square root. *Annals of Mathematical Statistics*, 1950, 21, 607–611.

Games, P. A. Multiple comparisons of means. *American Educational Research Journal*, 1971, 8, 531–565.

Harris, C. W. (Ed.) *Problems in measuring change*. Madison: University of Wisconsin Press, 1963.

Hays, W. L. *Statistics for the social sciences*. (2nd ed.) New York: Holt, Rinehart & Winston, 1973.

Hogben, D., Peavy, S. T., & Varner, R. N. *Omnitab II. User's reference manual*. NBS Technical Note 552. Washington, D.C.: National Bureau of Standards, 1971.

Hotelling, H. The selection of variates for use in prediction, with some comments on the general problem of nuisance parameters. *Annals of Mathematical Statistics*, 1940, 11, 271–283.

Institute for Social Research. *OSIRIS III*. Ann Arbor: University of Michigan Press, 1973.

Johnston, J. *Econometric methods*. New York: McGraw-Hill, 1963.

Jones, M. B. Moderated regression and equal opportunity. *Educational and Psychological Measurement*, 1973, 33, 591–602.

Lewis, D. *Quantitative methods in psychology*. New York: McGraw-Hill, 1960.

Lord, F. M. Statistical adjustment when comparing preexisting groups. *Psychological Bulletin*, 1969, 72, 336–337.

Lord, F. M., & Novick, M. R. *Statistical theories of mental test scores.* Reading, Mass.: Addison-Wesley, 1968.

Luce, R. D., Bush, R. R., & Galanter, E. (Eds.) *Handbook of mathematical psychology.* New York: Wiley, 1963.

Miller, R. G., Jr. *Simultaneous statistical inference.* New York: McGraw-Hill, 1966.

Nie, N. H., Bent, D. H., & Hull, C. H. *Statistical package for the social sciences.* New York: McGraw-Hill, 1970.

Nie, N. H., & Hull, C. H. *Statistical package for the social sciences. Update manual.* Chicago: National Opinion Research Center, University of Chicago, 1973.

Nunnally, J. C. *Psychometric theory.* New York: McGraw-Hill, 1967.

Owen, D. B. *Handbook of statistical tables.* Reading, Massachusetts: Addison-Wesley, 1962.

Pearson, E. S., & Hartley, H. O. (Eds.) *Biometrika tables for statisticians.* Vol. 1. Cambridge: Cambridge University Press, 1954.

Raduchel, W. J. *The regression analysis program for economists. Reference guide.* Cambridge, Massachusetts: Harvard Institute of Economic Research, Harvard University Press, 1972. (Revised 1973)

Ryan, T. A. Multiple comparisons in psychological research. *Psychological Bulletin,* 1959, 56, 26–47.

Stephenson, W. *The study of behavior.* Chicago: University of Chicago Press, 1953.

Stevens, S. S. Mathematics, measurement, and psychophysics. In S. S. Stevens (Ed.), *Handbook of experimental psychology.* New York: Wiley, 1951. Pp. 1–49.

Stevens, S. S. The psychophysics of sensory function. In W. A. Rosenblith (Ed.), *Sensory communication.* New York: Wiley, 1961. Pp. 1–33.

Tatsuoka, M. M. *Multivariate analysis: Techniques for educational and psychological research.* New York: Wiley, 1971.

Timm, N. H. The estimation of variance-covariance and correlation matrices from incomplete data. *Psychometrika,* 1970, 35, 417–438.

Tucker, L. R., Damarin, F., & Messick, S. A base-free measure of change. *Psychometrika,* 1966, 31, 457–473.

Tukey, J. The future of data analysis. *Annals of Mathematical Statistics,* 1962, 33, 1–67.

Van de Geer, J. P. *Introduction to multivariate analysis for the social sciences.* San Francisco: Freeman, 1971.

Walker, H. M., & Lev, J. *Statistical inference.* New York: Holt, 1953.

Wampler, R. H. A report on the accuracy of some widely used least squares computer programs. *Journal of the American Statistical Association,* 1970, 65, 549–565.

Wilson, W. A note on the inconsistency inherent in the necessity to perform multiple comparisons. *Psychological Bulletin,* 1962, 59, 296–300.

Winer, B. J. *Statistical principles in experimental design.* (2nd ed.) New York: McGraw-Hill, 1971.

Author Index

485

Subject Index

Matched subjects, 403–404, *see also* Repeated measures
Matrix algebra, 13, 449–453
Maximum of fitted curves, 222–223
Means, *see also* Significance
 adjusted, 329–330, 348
 of cases missing data, 272, 273, 277–278
 multiple comparison of, 155–165
 prediction to, 180–186, 194–195, 199–200
Minimum of fitted curves, 222–223
Missing data, 125, 265–290, 357–361
Model
 analytic, 97–104, 127–129
 error, *see* Error; Error term
 regression, 3–7, 49, 176–179, 345–348, 368–378
 theoretical, 98–102, 242–261, 265–266, 345–348
Moderator variable, 314, *see also* Interaction
Multicollinearity, 100–102, 115–117, 228, 286
Multiple comparisons, *see* Means
Multivariate ACV (MANACOVA), 438
Multivariate analysis, 427–445
Multivariate AV (MANOVA), 437–438

N

Nested design, 197, 413–425
Nominal scales, 10–11, 124, 171–211, 241–242, *see also* Coding; Means
 as covariates, 361–363
Nonexperimental research, 395–399
Nonlinearity, 27, 70, 125
 ACV and, 354–357
 interactions and, 320
 missing data and, 281–282
 tests for, 212–264
 transformation and, 242–261
Nonsense coding, 207–210
Normal distribution, table for, 471
Normal equations, 449
Normalization, 259–261, 262–263
Notation, 28–32, 73, *see also* Subscripts

O

Orthogonal coding, 197
Orthogonal comparisons, 157–158
Orthogonal polynomials, 231–241, 281
Orthogonal variables, 7–8, 442–443

P

Parallel slopes, *see* Interaction
Part correlation, *see* Correlation, semipartial
Partial correlation, *see* Correlation, partial
Partialling, 76, 393–400
Partial variance, *see* Analysis of partial variance

Partitioning variance, 80–82, 95, 96–97, 98
Part–whole correlation, 66–67
Path analysis, 368
Pearson product moment r, *see* Correlation
Phi coefficient (r_ϕ), 37
Point biserial r, 35–37
Population estimates and df, 30, 47, *see also* Shrunken $\tilde{R}^2$
Power analysis, 54–55, 144, 334–337
 and inference, 155–165
 for partial coefficients, 119–120
 for R, 117–118, 150–151
 for r and B, 55–56
 for sets, 144–155
 tables for bivariate, 478–480
 tables for MRC, 476–477
Power polynomial, 213–231, 241, 279–281, 319, 321
Predicted score, 4, 39, *see also* Standard error of predicted score
Prediction
 equations for, *see* Regression
 use of MRC for, 113–115
Probit transformation, 257–258
Product moment correlation, *see* Correlation
Product terms, *see* Interaction
Proportions, transformation of, 254–259
Protected t test, 157, 162–165, 185, 351

Q

Quadratic fit, 214–219
Quantitative variable, 10, 60, 212–213

R

Randomized groups in ACV, 399–400
Range, restriction of, 64–66
Rank order, transformation of, 259–261
Rank order correlation, 37–39
Ratios as variables, 67–70, *see also* Scale level
Reciprocal transformation, 253, 261–262
Redundancy, 8, 85–87, 458, 460–461
 in CA, 429–430
Regression
 comparisons of, 52–53
 curvilinear, graph of, 214–222
 linear, graph of, 39–43
 toward mean, 43
Regression coefficient
 and interaction, 303–308, 315–319
 multiple IVs, 73–75, 92–94
 one IV, 42, 48
Regression weight, *see* Regression coefficient
Reliability, effects of, 62–64, 369, 371–374, 376